Natural Additives in Food

Natural Additives in Food

Editors

Lillian Barros
Isabel C.F.R. Ferreira

MDPI • Basel • Beijing • Wuhan • Barcelona • Belgrade • Manchester • Tokyo • Cluj • Tianjin

Editors

Lillian Barros
Centro de Investigação
de Montanha (CIMO)
Instituto Politécnico
de Bragança
Bragança
Portugal

Isabel C.F.R. Ferreira
Centro de Investigação
de Montanha
Instituto Politécnico
de Bragança
Bragança
Portugal

Editorial Office
MDPI
St. Alban-Anlage 66
4052 Basel, Switzerland

This is a reprint of articles from the Special Issue published online in the open access journal *Molecules* (ISSN 1420-3049) (available at: www.mdpi.com/journal/molecules/special_issues/natural_additives).

For citation purposes, cite each article independently as indicated on the article page online and as indicated below:

LastName, A.A.; LastName, B.B.; LastName, C.C. Article Title. *Journal Name* **Year**, *Volume Number*, Page Range.

ISBN 978-3-0365-4106-8 (Hbk)
ISBN 978-3-0365-4105-1 (PDF)

Contents

About the Editors

Lillian Barros

Lillian Barros is an assistant researcher at the Centro de Investigação de Montanha (CIMO), Instituto Politécnico de Bragança and vice-coordinator of CIMO. She obtained her Ph.D. in Pharmacy (Nutrition and Bromatology) at the University of Salamanca (2008). She has published more than 515 indexed papers in the Food Science and Technology area (61 h-index), edited 4 books, written several book chapters, registered 8 national patents, is principal investigator of several research projects and has supervision several post-doc, Ph.D. and master students. She has reached top positions in the world rankings, such as Highly Cited Researcher Clarivate, since 2016. Her research targets are mainly in the identification, separation and recovery of functional molecules from different natural products.

Isabel C.F.R. Ferreira

Isabel Ferreira is currently the Secretary of State for Inland Enhancement of Portugal. She was a Vice-President at the Polytechnic Institute of Bragança, the Director of the Mountain Research Centre and Mentor of the Collaborative Laboratory Mountains of Research. She is a Full Professor at the IPB, with a Ph.D. in Biochemistry and one of the world's most cited scientists (top 1%) since 2015 by Clarivate Analytics in the Essencial Science Indicators index. She has published more than 700 papers in the Food Science and Technology area (61h-index), edited 4 books, written 60 book chapters, registered several national and international patents, most focusing on the transfer of technology to the industry.

Preface to "Natural Additives in Food"

The controversy and ambiguity related with chemical additives, allied to sporadic scares, have paved the way for natural additives to gain interest and funding. Today, most consumers prefer foods added with natural additives, rather than artificial ones, which is seen by the food industry as an opportunity to find new and more efficient natural-based solutions, while fighting to reduce the overall use of additives, producing minimally processed goods. The benefits of natural additives are endless, their synergy and effectiveness are a great leap over artificial additives that carry out, in most cases, only one effect over the food.

The plant and fungi kingdoms are great sources of bioactive compounds, that can be used to develop natural food ingredients. These natural compounds can be added as extracts, taking advantage of the synergistic effects between compounds, or as individual molecules, after purification, thus adding the most bioactive ones to the foodstuff. Although quite promising, natural additives still face some drawbacks and limitations, availability of natural resources, exhaustive and not very efficient extraction methods, and several intrinsic and extrinsic factors that can affect the stability, availability, and bioactivity of natural compounds. Therefore, an important research topic is the discovery of new alternative sources of natural additives fulfilling the different classes: preservatives (antimicrobials, antioxidants, and anti-brownings), nutritional additives, coloring agents, flavoring agents, texturizing agents, and miscellaneous agents.

In order to address the advantages and challenges the use of natural additives in foods, an enormous amount of work, divided into three review and 15 original articles, involving diverse expert teams of from different parts of the world, embodied this book "Natural Additives in Foods", where current issues on natural additives are discussed and explored. Briefly, topics on alternatives for sustainable obtaining of bioactive molecules from agro-food byproducts, an important cheap source of added value compounds, which can contribute to total exploration of natural source, and for reduction of environmental impact, are presented and discussed in some studies. Likewise, optimized extraction techniques, aiming at greater extraction efficiency with less use of natural resources, are proposed by for some authors. Several potential natural colorants to replace artificial ones are covered in topics that discuss their main sources and bioactivities; and the sustainable production methods and the chemical stability of these compounds for later commercial use. In addition, interesting studies approach the valorization of wild species to obtain molecules with biological proprieties, both for human health and for addition in foods. And, finally, the production of foodstuff healthier than the traditional, is proposed through the use of natural ingredients rich in phytochemicals and through processes that aim to minimize the absorption of compounds harmful to health.

Lillian Barros and Isabel C.F.R. Ferreira
Editors

molecules

Review

Agro-Food Byproducts as a New Source of Natural Food Additives

Margarida Faustino, Mariana Veiga, Pedro Sousa, Eduardo M. Costa, Sara Silva and Manuela Pintado *

CBQF–Centro de Biotecnologia e Química Fina–Laboratório Associado, Escola Superior de Biotecnologia, Universidade Católica Portuguesa, Rua Arquiteto Lobão Vital 172, 4200-374 Porto, Portugal; afaustino@porto.ucp.pt (M.F.); mveiga@porto.ucp.pt (M.V.); pedro.miguel.c.s@hotmail.com (P.S.); emcosta@porto.ucp.pt (E.M.C.); snsilva@porto.cup.pt (S.S.)

* Correspondence: mpintado@porto.ucp.pt; Tel.: +351-225-580-097

Academic Editors: Lillian Barros and Isabel C.F.R. Ferreira

Received: 4 February 2019; Accepted: 12 March 2019; Published: 18 March 2019

Abstract: Nowadays, the agro-food industry generates high amounts of byproducts that may possess added value compounds with high functionality and/or bioactivity. Additionally, consumers' demand for healthier foodstuffs has increased over the last years, and thus the food industry has strived to answer this challenge. Byproducts are generally secondary products derived from primary agro-food production processes and represent an interesting and cheaper source of potentially functional ingredients, such as peptides, carotenoids, and phenolic compounds, thus promoting a circular economy concept. The existing body of work has shown that byproducts and their extracts may be successfully incorporated into foodstuffs, for instance, phenolic compounds from eggplant can be potentially used as a mulfitunctional food additive with antimicrobial, antioxidant, and food colorant properties. As such, the aim of this review is to provide insights into byproducts and their potential as new sources of foodstuffs additives.

Keywords: byproducts; food additives; antimicrobial; antioxidant; colorants; texturizing agents; foaming capacity and emulsifiers

1. Introduction

Food functionalization is an ever-increasing market that requires new bioactive ingredients that can be used by the food industry for the development of innovative functional products with scientifically sustained claims. In this regard, much attention has been paid in recent years to natural compounds and their associated bioactivities. However, natural sources are finite, and new alternatives have to be sought to sustain the ever-growing needs for ingredients and additives of the food industry [1,2].

The European Union (EU) action plan for the circular economy to reduce food waste includes a strategic approach based on the reduction, reuse, recovery, and recycling of materials and energy, enhancing the value and consequently the useful life of products, materials, and resources in the economy. The reuse of agro-industrial byproducts can represent a renewable source for some already in use food additives or even originate new added-value ingredients with functional compounds and properties, which will benefit the entire food system [3]. For instance, byproducts contain polysaccharides, organic acids, proteins, and other compounds, which, at no additional production cost and at a reduced industrial cost, make them a rich source of natural compounds that can potentially be applied in the food industry as food additives sources (summarized in Table 1) [4].

Furthermore, these natural compounds may also be regarded as nutraceutical ingredients or complements, allowing for the development of products with enhanced nutritional value, potential health benefits, longer shelf-life, as well as a good sensory profile [5–7].

Table 1. Potential applications of agro-food byproducts compounds.

Byproduct	Origin	Compounds	Bioactivity	Types of Extract	Dose of Byproduct Extract	Reference
Buffalo horn	Animal	Peptides	Antioxidant	Aqueous		[8]
Bovine Achilles tendon collagen	Animal	Peptides	Antihypertensive	Aqueous		[9]
Bovine haemoglobin hydrolysate	Animal	Peptides	Antibacterial Antihypertensive	Ethanolic	187.1 and 35.2 μM 42.55 to 1095 μM	[10]
Salmon fish	Animal	Hydrolysates	Emulsifying capacity Emulsion stability Fat adsorption capacity	Aqueous	0.02–5% mg/mL	[11]
Herring fish	Animal	Protein	Emulsifying capacity Emulsion stability Fat absorption capacity	Aqueous	5–50 mg/mL	[12]
		Hydrolysates	Emulsifying capacity Emulsion stability Fat adsorption capacity Antioxidant activity	Aqueous and ethanolic	0.01–50 mg/mL	[13]
Poultry	Animal	Hydrolysates	Water holding capacity Oil absorption capacity Emulsifying capacity Foaming capacity	Aqueous	7.5–100 mg/mL	[14]
Rainbow trout viscera	Animal	Hydrolysates	Water holding capacity Oil absorption capacity Emulsifying capacity Emulsion stability Foaming capacity	Aqueous	7.5–100 mg/ml	[14]
Grape peels	Fruit	Phenolic compounds Anthocyanins	Antioxidant	Ethanolic	1.652 mg/mL	[15]
Pomegranate seed	Fruit	Phenolic compounds	Antioxidant	Ethanolic	25 mg/L	[16]
Pineapple peels powder	Fruit	Dietary fibres	Texturizing agent	Aqueous	1%	[17]
Banana peels	Fruit	Flavonoids. Tannins B-carotene	Antioxidant	Aqueous	5 mg of banana peel extracts/1 mL	[18]
Grape seed	Fruit	Phenolic compounds Flavonoids	Antioxidant Antimicrobial	Ethanolic	100 mL/25 g	[19]
Wine pomace, skin and seed extracts	Fruit	Total Phenolic Flavonoid Compounds	Source of phenols	Ethanolic	0.1, 0.2 and 0.3 wt/vol	[20]
Orange byproducts	Fruit	Dietary fibres	Texturizing agent	Aqueous	0.2 to 1 g/mL	[21]
Wine pomace extract and flour	Fruit	Dietary fibres Phenolic compounds	Antioxidant Colorant	Ethanolic	1% to 3% 1% to 2%	[22]
Wine pomace flour	Fruit	Polyphenols Prebiotics Phenolic compounds	Source of phenols, Probiotic protection	Aqueous	10, 20 and 50 g/L	[23]
Grape seed	Fruit	Phenolic acids Flavonoids Aromatic compound	Antioxidant Antimicrobial	Ethanolic	10 mL/25 g	[19]
Banana inflorescence bracts	Fruit	Anthocyanin cyanidin-3-rutinoside	Antioxidant Colorant	Ethanolic	14–32 mg/100 g	[24]
Pear stones	Fruit	Dietary fibres	Texturizing agent	Aqueous	3–5%	[25]
Male flower (*Musa paradisiaca*)	Vegetable	Epigallocatechin and derivatives	Antioxidant Antimicrobial	Ethanolic Aqueous	12.93–2.34%	[26]
Soy milk	Vegetable	Protein	Water and fat binding capacity Foaming capacity Emulsifying capacity Emulsion stability	Aqueous		[27]

Table 1. *Cont.*

Byproduct	Origin	Compounds	Bioactivity	Types of Extract	Dose of Byproduct Extract	Reference
Tiger nut milk	Vegetable	Dietary fibres	Water and oil holding capacity Water and oil absorption capacity Emulsifying capacity Emulsion stability	Aqueous	1.5% mg/mL	[28]
Tomato byproducts	Vegetable	Dietary fibres, Proteins, Carotenoids, Tocopherols, Polyphenols Lycopene	Anti-inflammatory, Antiallergenic, Antimicrobial, Vasodilatory, Antithrombotic, Cardioprotective Antioxidant Colorant	Ethanolic	55.70 to 28.64 mg/100 g	[29,30]
Overly ripe berries	Vegetable	Phenolic compounds Dietary fibres	Antioxidant Anti-inflammatory Antimicrobial Antidiabetic Neuroprotective	Ethanolic		[31]
Pomegranate peel	Vegetable	Phenolic compounds Proanthocyanidins Tannins Flavonoids Oligomeric ellagitannins	Antimicrobial Antioxidant	Aqueous-methanolic	5-g portions in 80% methanol Antimicrobial activity tested at 0.1% (v/v)	[32,33]
Tomato processing byproduct	Vegetable	Lycopene Phenolic compounds	Antioxidant	Ethanolic	400 and 800 mg/kg	[34]
Rice bran	Cereal	γ-oryzanol Anthocyanins Phenolic compounds	Colorant	Ethanolic	0.2–0.6%	[35]
Whey protein	Dairy products	Caseinates	Texturizing agent Source of protein	Aqueous	2%	[36]

Taking this into account, this review aims to provide a broader look into the potential use of byproducts as new sources of food additives (already in use or potential new ones) to be used by the industry.

2. Consumer Perspectives

In the 1960s, the E number system was introduced to assure consumers that the additives included into their foodstuffs were safe for consumption. However, the use of this code made some consumers reticent in regard to these compounds with false allegations (on their lack of safety) being made in some publications [37–39]. Moreover, with the increase in life expectancy, concerns grew in regard to overall life quality. This, coupled with the widespread link between diet and health, made consumers particularly aware of the foodstuffs they ingested and increased the demand for healthier solutions. One trend associated with this shift in perception is "clean labels", i.e., products that are perceived as "natural", such as "free-range", "less processed", "organic", or "biological" foods [39,40]. Overall, this means that not adding additives has become a differentiating factor for food products, and consequently that the industry has become more interested in new solutions that, while exerting the same technological effect as traditional additives, have no negative perception. Agro-food byproducts present an interesting source of bioactive and technologically relevant compounds that, given their low commercial value, pose as a relevant potential source of new natural additives [41–44].

3. Applicable Legislation

Food additives have an essential role in the current industry and consumption habits, as they not only make food products more appealing, but they increase their stability and inherent safety. Overall, food additives may be defined as compounds/extracts that are added to a food product in order to accomplish a specific technological goal but are not ingested as a food product themselves. According to the European Food Safety Agency (EFSA), an additive must not pose a safety concern for the consumers health (when ingested) while fulfilling a specific technological need that cannot be satisfied through other reasonable means. Examples of these needs are the enhancement of the sensory quality, the fulfillment of specific dietary needs, or the ease of production, packaging, transport, and/or storage of food products [45]. Overall in the EU, the use of additives (non-enzymatic) is regulated by European Commission (EC) No 1333/2008 with the additives, the list of allowed additives, and subsequent limitations always dependent on the appearance of new evidence regarding their safety. In this legislation, the different groups are defined (Table 2) along with rules on how an additive must be referred to in a product (e.g., the information must be present in the label with the compounds referred to either by their name or their E-number and by the function they play in the final product). Moreover, food additives must follow specific purity criteria that are described in three different directives: Directive 2008/60/EC for sweeteners; Directive 2008/128/EC for colors; and Directive 2008/84/EC for other additives [46–48]. After the inclusion of the list of approved additives and food carriers (and the conditions associated with their use) into Regulation (EC) No 1333/2008, a revision of the purity criteria of food additives was undertaken, resulting in a new regulation, Regulation (EU) No 231/2012, that repealed the previous directives for sweeteners, colors, and other additives [45,48–50].

Table 2. Types of food additives and their functions as described by European Food Safety Agency (EFSA) in Regulation (EC) No 1333/2008.

	Function
Sweeteners	Increase the sweetness (can be added or table-top)
Colors	Add or restore color
Preservatives	Prolong shelf-life by inhibiting microbial deterioration or the growth of pathogens
Antioxidants	Prolong shelf-life by inhibiting oxidative deterioration (e.g., color changes or rancidity)
Carriers	Physically modify a compound to ease its application/handling, without compromising the activity of the added compounds and having no technological effect by themselves
Acids	Increase the acidity and/or impart a sour taste
Acidity regulators	Alter/control the pH of a foodstuff
Anti-caking agents	Reduce particle agglomeration
Anti-foaming agents	Prevent/reduce foam formation
Bulking agents	Increase the volume of a foodstuff without significantly increasing its energetic value
Emulsifiers	Ease the formation/maintenance of an homogenous mixture of two immiscible phases
Emulsifying salts	Convert cheese proteins into a dispersed form contributing to an homogenous distribution of other components (e.g., fat)
Firming agents	Either keep fruit and vegetables firm/crisp or produce/strengthen gels
Flavor enhancers	Enhance taste/odor
Foaming agents	Ease the dispersion of a gaseous phase in a liquid/solid
Gelling agents	Form a gel and improve texture
Glazing agents (including lubricants)	Give a shiny appearance or provide protective coating
Humectants	Prevent drying or promote the dissolution of powders in an aqueous media
Modified starches	Chemically treated edible starches
Packaging gases	Gases (not air) introduced into containers before placing the foodstuff in them
Propellants	Gases (not air) that expel a foodstuff from a container
Raising agents	Release gas therefore increasing the volume or a dough or batter
Sequestrants	Complex metallic ions
Stabilizers	Maintain the physico-chemical state of a foodstuff
Thickeners	Increase the viscosity
Flour treatment agents	Improve the baking quality of flours/doughs (non-emulsifiers)

Some of the additives currently allowed under the scope of Regulation (EC) No 133/2008 may be found in agro-food byproducts. Namely, anthocyanins (E163) may be found in grape/winemaking byproducts, chlorophylls (E140) may be found in almost all green leaf vegetable byproducts or mango peels along with all green leaf wastes that result from pruning during agricultural production, and lycopene (E160d) can be found in tomato wastes [51–57]. There is a consensus that as long as the additive compound/molecule is already part of the list of authorized compounds, it can be used [58]. However, if the production process is varied significantly (by using a significantly different raw matter or using new production procedures), the "new" additive must be evaluated again by EFSA. This means that the focus given to the development of new and more efficient green technologies to attain additives from agro-food byproducts may result in potential new additives that must be subjected to a new evaluation in order to ensure their safety. This path starts with a thorough safety (short and long term) evaluation of any potential metabolic, genotoxic, reproductive, and chronic or carcinogenic side effects [59]. Following this, it is possible to define a no observable effect level (NOEL) and then an allowable/acceptable daily intake (ADI). Once all the relevant information is gathered, EFSA or other similar organizations (like the Food and Drug Administration—FDA) can be petitioned to validate the introduction of the additive through an amendment of the legislation in order to add the substance to the list of authorized food additives. If this authorization is granted, the additive will be eligible to be used on the market under direct supervision of the agency that granted the permission [49,59,60]. In the EU, the submission of a potential new additive for validation must start with an application submitted to the EC, who will verify it. If valid, EFSA must then give an opinion within a timeframe of nine months, a period that may be extended if further information is required from whoever submitted the application (for risk management purposes, EC may also require further elucidation). If EFSA gives a positive opinion, EC has nine months to submit a regulatory draft aimed at the inclusion of the substance in the allowed additives list, whose approval is dependent on the votes of member states. If approved, as with all decisions of the EU, it must then pass a three month long period of scrutiny. Overall, this process is very long and, in the new era of circular economy where food byproducts valorization is of the upmost importance, legislation approaches should be reanalyzed to facilitate and speed up the process of new additives approval while still guaranteeing the safety of the final additive [49,61].

4. Preservatives

Microbiological processes can adversely affect the quality of food, leading to its spoilage. For this to occur, conditions that favor the growth and development of spoilage microorganisms must be met, such as bioavailable nutrients, favorable water activity, adequate pH value, presence/absence of oxygen, and redox potential [62]. The term "food spoilage" is only applied if the changes in the foodstuffs due to the microorganisms' potentially harmful metabolic products become recognizable, thus making the product unsafe for consumption and augmenting the risk of foodborne illness [62–65]. However, not all microbiological change in food is considered harmful (for example, fermentation of grape juice in order to produce wine) [66].

Taking this into account, preservatives are widely used in the food industry in order to prevent microbial contaminations, demonstrating a significant impact upon a product's shelf-life as well as food safety [58,66–68]. There are different antimicrobial compounds that can potentially be used as preservatives ranging from enzymes, bacteriocins, fungicides, and salts to essential oils and other components, some of which may be found agro-food byproducts [7,69–76]. The use of natural compounds to replace traditional additives is an emerging trend that has been driven by the consumer's preferences for "clean labels", with the scientific community striving to provide natural alternatives, some of which may be attained from agro-food byproducts (e.g., phenolic compounds) [7,44,77]. Nitrates (E240-E259) and nitrites (E249-E250) are the most commonly used preservatives in foodstuffs. Both have been associated with the formation of nitrosamine (a carcinogenic compound responsible for the development of gastric and other types of cancer). Therefore, actions have been taken to reduce their intake [the current daily intake for nitrates is 3.7 milligrams per kilogram of body weight (mg/kg bw/day), while for nitrites it was re-established to 0.07 mg/kg bw/day] [78–80]. However, EFSA determined that there was insufficient evidence to ban the use of nitrates and nitrites as food additives due to health concerns, particularly with them being the only additives capable of exerting antimicrobial activity against *Clostridium botulinum* and preventing botulinic toxin production/accumulation [81].

Agro-food byproducts, particularly fruit peels and seeds, have been regarded as a potential source of preservatives with several reports reporting on the potential antimicrobial activity of different fruit and vegetable byproduct extracts, which could potentially be translated into an industrial application if the appropriate regulatory body gives a positive opinion [7,41,67,82]. For instance, Gul and Bakht [83] reported how an ethanolic turmeric extract possessed antibacterial activity against *Escherichia coli* and *Staphylococcus aureus*, an effect that has been attributed to its phenolic content [84–86]. Additionally, turmeric oil, a byproduct from curcumin manufacture, has also been described as possessing antibacterial and antifungal activity [86–88]. Berries are fruits with high phenolic content, particularly anthocyanins. While by themselves they possess an interesting commercial value, if the fruits fall from the bushes (overly ripe berries), they will not be commercialized [89–91]. However, they remain a phenolic rich fruit that can be used as a source of potential antimicrobial additives. For instance, blueberry and cranberry anthocyanin-rich extracts have been reported as possessing vast antimicrobial activity and could potentially be exploited as new natural food additives [92–97]. Olive leaves are also a good source of phenolic compounds and have been reported as possessing some antimicrobial activity against *Bacillus cereus*, *E. coli*, *S. aureus*, and some fungi such as *Candida albicans* and *Cryptococcus neoformans* [98–101]. Wang, et al. [102] reported how the addition of green tea polyphenols (mainly constituted by catechins) and tocopherol to dry-cured bacon resulted in significantly lower *Enterobacteriaceae* content. Green tea and black tea wastes have been studied for their potential nutritive, antimicrobial, and antioxidant values due to their high tannin and catechin content [103,104].

Wine pomace, a well-known byproduct, also showcases some potential as a new source of antimicrobial food additives, as its activity has also been associated with its high phenolic content and anthocyanins in particular [22,105]. Pomegranate peel extracts, reported to be natural inhibitors of food-borne pathogens such as *Listeria monocytogenes*, *E. coli*, and *Yersinia enterocolitica*, have been added to poultry products with the results showing good antimicrobial activity against *S. aureus* and *B. cereus* and permitting the increase of shelf-life by two weeks [32,33,106–109]. Avocado, a tropical fruit, has also been described as possessing a relevant antimicrobial activity, with several reports focusing on the biological activity of its peel and seed [110,111]. For instance, Calderón-Oliver et al. [112] reported how a nisin (an antimicrobial peptide) avocado peel mixture resulted in an enhancement of nisin's antimicrobial activity against food-borne pathogens such as *Listeria* sp. Overall, the reported results favor the use of natural byproduct extracts as potential new preservatives at an industrial level, helping to reduce costs and environmental impact, although the leap to an industrial setting is limited by a lack of regulatory framework for their use.

Currently, the only animal derived antimicrobial additive used in the EU and United States (US) is lysozyme (E1105). Lysozyme originates from eggs, and while it is mainly used in cheese conservation, studies concerning eggs, milk, and beef have been carried out. However, it does not exert any action against yeasts or fungi [113–115].

5. Antioxidant Additives

Oxidation is a not a process exclusive to the human body. It occurs in every living organism and biological system, such as food products. Food oxidation may result in altered flavor, color, nutritional value, and texture, as well as create toxic compounds [82,116,117]. As such, antioxidant compounds are one of the most important conservation technologies used by the food industry with their main function being the prevention of oxidative induced degradation of foods, therefore allowing for extended shelf times [82,117,118]. These additives help stabilize lipids (avoiding lipidic peroxidation) as well as other compounds and can neutralize free radicals, avoiding a cascade of oxidative reactions. [117,119].

As previously mentioned, due to a shift in consumer preferences, in recent years there has been an increase in the demand for more natural (i.e., with less additives) food products [120]. As such, there have been studies comparing synthetic and natural antioxidants with results showing that natural phenolic antioxidants are capable of inhibiting oxidation and toxin formation, meaning that they present an interesting natural alternative to the traditionally used antioxidant additives [117,121]. Butylated hydroxy anisole (BHA), butylated hydroxytoluene, ethoxyquin, tert-butylhydroquinone, and propyl gallate are the most common synthetic antioxidants used in foods. Reports on their potential health impact are divided [121–124].

Since plants are one of the main sources of antioxidants compounds, agricultural byproducts are among the most relevant potential sources of natural antioxidants that could be exploited for product quality preservation. Phenolic compounds, besides being associated with antimicrobial activity, are known for their high antioxidant capacity. They are ubiquitous to plants and therefore present one interesting class of antioxidant compounds to be exploited, although other compounds with a strong antioxidant capacity can also be found, such as some vitamins (vitamin C, E, and A), bioactive peptides, polysaccharides, some minerals, and enzymes. Any byproduct with a high content of any of these compounds may be regarded as a possible source of new antioxidant food additives, e.g., overly ripe berries, or citric and exotic fruits, peels, and seeds [77,116,121,125]. Meat byproducts (including blood, bones, meat trimmings, and viscera) can result in protein hydrolysates with a relevant bioactivity, namely antioxidant bioactive peptides [126,127]. Onion byproducts (namely onion peels and stems) have been regarded as potential food additives due to their antioxidant and anti-browning properties [128]. Larrosa et al. [129] reported that adding an artichoke byproduct extract (namely artichoke blanching waters) to a tomato juice resulted in higher antioxidant activity (measured by the $DPPH^{\bullet}$ and $ABTS^{\bullet+}$ methods) and consequently a longer shelf life for this product. Similarly, eggplant aqueous acetone extracts have also been studied, with reports describing a high antioxidant potential of its peels (evaluated by FRAP and TEAC) likely due to its rich anthocyanin content [130]. Mango byproducts are an example of a vastly studied tropical fruit with a high antioxidant capacity and a wide range of potential applications [54,131]. An example is the inclusion of mango peel powder in macaroni and bakery products (such as biscuits) to provide some added functional value as well as function as a natural antioxidant (as the supplemented products exhibit a higher capacity to quench $DPPH^{\bullet}$) [54,132,133].

The potential for the use of natural alternatives to antioxidant additives has been supported by the work of Caleja, Barros, Antonio, Oliveira, and Ferreira [121], who reported no significant differences between the use of natural extracts (chamomile and fennel) and a synthetic (BHA) antioxidant additive in biscuits, with no significant changes in color or nutritional value observed after 60 days of storage. Similarly, there have been reports on the successful addition of natural antioxidant extracts to bakery, dairy, and meat products, which also confer some added functionality to the foodstuffs [79,121,134–137]. Overall, byproducts of industrial fruit processing consist mainly of seeds, peels, and unused flesh. Some of these residues have been reported as possessing a higher concentration of bioactive compounds than the used fruit flesh [108,111,132,138–140]. Furthermore, the antioxidant compounds of natural origin, when attained using adequate solvents, are considered as generally recognized as safe (GRAS). Moreover, some of the antioxidant compounds naturally found in these byproducts are already approved for use as antioxidant additives and possess an E number, namely ascorbic acid (E300), tocopherol (E306), and β-carotene (E160a) [68,102,113].

6. Food Colorants

Although the flavor and nutritional value tend to be the most studied and appreciated components of a food product, its appearance is also an important sensory aspect [141,142]. Colorants are food additives used to impart color to foodstuffs to make them look more appetizing and/or help compensate color loss due to exposure to natural elements (light, air, temperature, etc.) [143,144]. Color plays an important role in the consumer's emotional reaction and acceptance of food. Color is appreciated both for its aesthetic and quality indicator role, as an adequate color is frequently used for quality assessment due to its association with flavor, nutritional value, and food safety [145]. Color provides visual suggestions to flavor identification and taste thresholds, influencing food preference, food acceptability, and ultimately, food choice [146]. Current market trends include the substitution of synthetic colorants for natural compounds found in certain foodstuffs (such as fruits) or in food byproducts, a trend that is reinforced by studies regarding possible detrimental effects of synthetic colorant usage in foods [142,144]. Most commercial colorants are produced synthetically, including erythrosine (red), cantaxanthin (orange), amaranth (azoic red), tartrazine (azoic yellow), and annatto bixine (yellow orange) [67]. Nonetheless, a few colorants like carotenoids (β-carotene, astaxanthin, canthaxanthin, and zeaxanthin) are obtained from natural sources, such as tomato, paprika, and algae [147]. However, synthetic colorants are still used due to their stability and low cost [44]. As agro-food byproducts are usually discarded, their use as a new source of these coloring agents could be a means to shift to more natural colors while still maintaining a low production cost (Table 3).

Table 3. Food byproducts sources of potential colorant food additives. Adapted from Iriondo-DeHond, Miguel, and del Castillo [142].

Food Industry	Byproducts	Function	Color	Chromophore	Reference
Winery	Wine pomace extract and flour	Antioxidant and colorant agent	Red to purple to blue	Anthocyanins	[22]
Cereal	Rice bran	Colorant agent	Yellow to light brown	Carotenoids	[35]
Vegetable	Lycopene from tomato byproducts	Colorant agent, antioxidant, and antimicrobial	Red	Carotenoids	[148]
	Carotenoids from tomato peels	Antioxidant and colorant agent	Red	Carotenoids	[149]
	Phenolic compounds from eggplant	Colorant agent, antioxidant, and antimicrobial	Purple-blue	Anthocyanins	[130]
	Phenolic compounds from potato peels	Colorant agent, antioxidant, and antimicrobial	White	Anthoxanthins	[150]
	Leafy green vegetables	Colorant agent and antioxidant	Green	Chlorophyll	[151]
Fruit	Anthocyanins from berries' peels Phenolic compounds from tropical and citrus fruits peels, seeds, and unused flesh	Colorant agent, antioxidant, and antimicrobial	Purple-blue Yellow-orange	Anthocyanins Carotenoids	[33,131, 133,152–154]

As previously mentioned, consumers have been demanding the replacement of synthetic colorants by natural alternatives. Authors like Siegrist and Sütterlin [155] reported that symbolic information such as the E-numbers on the foodstuff's label influences a consumer's perception of different foodstuff and its origin, with consumers being hesitant to accept the addition of synthetic food additives. Additionally, there have been several reports pertaining to synthetic colorants side effects, including hypersensitive and allergic reactions as well as potential toxicity and carcinogenicity claims [144,156,157]. Natural additives have been associated with health promoting benefits, as they are a part of the bioactive compounds present in fruit and vegetable byproducts. However, the use of these natural pigments can be limited by their lower stability and weaker color strength (when compared to their synthetic counterpart). Additionally, natural additives may confer an undesirable flavor or odor to the food products, which will negatively impact the consumer's acceptance [142,145,158]. Nonetheless, fruit and vegetable byproducts have become an important potential source of natural pigments, as they are colored by green chlorophylls, yellow-orange-red carotenoids, red-blue-purple anthocyanins, and red betanins [158]. Overall, the main groups of coloring substances found in nature are carotenoids, anthocyanins, porphyrins, and chlorophylls [145,158–160].

Anthocyanins are a good example of natural color additives. These compounds are a group of natural pigments responsible for the blue, red, purple, violet, and magenta coloration of several species in the plant kingdom. They can also be found in extracts of their byproducts. Some examples are winery byproducts, radishes, red potatoes, red cabbage, black carrots, purple sweet potatoes, coffee husks, and berries, among others [106,161].

Carotenoids stand as the major group of compounds used as color additives. These natural pigments are responsible for many of the colors seen in edible fruits, vegetables, mushrooms, flowers, and even lobster and trout hues from the animal kingdom [158]. Much like anthocyanins, carotenoids are produced synthetically (β-carotene, astaxanthin, canthaxanthin, and zeaxanthin), although some are obtained from natural sources, namely annatto, marigold, tomato, algae, and microbial fermentation [157]. In addition, these compounds function as sources of provitamin A and are capable of absorbing solar light, oxygen transporters, powerful quenchers of singlet oxygen, as well as other functions not yet studied [160].

The natural pigments were defined in the Regulation (EC) No 1333/2008 of the European Parliament and of the Council of 16 December 2008 and are listed in the Annexes of said regulation [161]. This document includes detailed information on the application of individual pigments in defined food products, their doses, and limitations of use. Presently, 16 natural pigments are permitted: betalains–betanin, quinones–cochineal, flavonoids–anthocyanins, isoprenoids–carotene, annatto (bixin, norbixin), paprika extract, lutein, canthaxanthin, porphyrins–chlorophylls and chlorophyllins, and copper complexes of these compounds, among others, like caramels, curcumin, or plant coal. According to the Regulation (EC) No 1129/2011 [162] of the European Parliament, in the EU, there are 40 color additives legislated for food use.

New technologies such as pulsed-light, high pressure, pulsed-electric, magnetic fields high pressure processing, ionizing radiation, and ultraviolet radiation are being studied and could allow for the use of byproducts as natural source pigments, which could then be exploited as potential new food colorants in the food industry with the advantage of imparting potential health benefits to the consumer as well as contributing to an economical valorization of residues and avoiding waste [163]. For instance, there have been studies regarding the addition of banana peels to biscuits, which resulted in a product with low calories and high dietary fiber content without any significant differences in color, aroma, and taste observed. The banana peel was incorporated at a 10% and 20% concentration into the biscuits [164–166]. The peel is the main byproduct of the banana, rich in phytochemical compounds with high antioxidant capacity, such as phenolic compounds, anthocyanins (delphinidin and cyanidin), carotenoids (β-carotenoids, α-carotenoid, and xanthophylls), catecholamines, sterols, and triterpenes, which, as previously mentioned, could provide different functions as potential food additives besides coloring, namely as antioxidant and antimicrobial components [18,167]. Mango is another example of a fruit with biologically interesting compounds (including phenolic compounds, carotenoids, and dietary fiber) that could be used in the food industry. Most mango byproducts result from the epicarp and endocarp, but it is the mango seed kernel residue with the highest amount of carotenoids in its composition, which is likely due to the amount of fruit pulp left around the kernel by the chopping machine. The carotenoid content was found to be four to eight times higher in ripe mango peels compared to raw fruit peels [133,165]. The high levels of bioactive compounds in the mango peel makes this byproduct a potentially valuable raw material for the formulation of additives and supplements for the food industry [168]. Likewise, tomato peel is the main byproduct resulting from the tomato processing industry [169]. The carotenoid pigment lycopene present is the compound responsible for its red color. Tomato lycopene extract and tomato lycopene concentrate from tomato peels have been approved for use as colorants exempt from certification [170]. Oleoresins, powders, and water-dispersible preparations that can impart colors from yellow to orange to red are commercially available. An example of the utilization of lycopene from tomato byproducts includes dairy foods, where this compound is applied in the coloring of butter and ice-cream, maintaining a stable reddish color for up to four months [146,148,149].

Even though the use of the aforementioned compounds as food colorants could present an ecological solution to current production issues in addition to possessing an added advantage of potential health benefits, their use has been limited. Regulated colorants from natural sources include anthocyanins (E163), betanin (E162), and carotenoids (E160), including β-carotene (E160a), lycopene (E160d) (its obtention from tomato processing byproducts has been optimized), lutein (E161b), canthaxanthin (E161g), chlorophyll and chlorophyllin (E140 and E141), and curcumin (E100) [according to Regulation (EC) No 1129/2011] [163]. However, the list of anthocyanin colorants in the Codex Alimentarius includes only grape skin extract (E163), and in the FDA, "grape color extract" and "grape skin extract" (enocyanin) [146,148,149].

Regardless, to include a new pigment as a food colorant additive according to Regulation (EC) No 1333/2008 of the European Parliament and of the Council of 16 December 2008 [162], these new pigments need to be capable of restoring the original appearance of food whose color has been affected by processing, storage, packaging, and distribution, leading to impaired visual acceptability. Thus, colorants need to make food more visually appealing as well as give color to food that is otherwise colorless.

7. Texturizing Agents

Texturizing agents, such as emulsifiers, stabilizers, thickeners, and bulking agents, are used in the food industry to modify the overall texture and mouth feel of foodstuffs [171]. Thickeners, when added to the food mixture, increase the viscosity without modifying other food properties, while bulking agents increase the bulk of a food without affecting its nutritional value. Emulsifiers, on the other hand, allow water and oil to remain mixed together in an emulsion. These agents are used to add or modify the texture of food products by modifying the creaminess, thickness, viscosity, or by stabilizing foodstuffs structure [67]. These agents are used in frozen desserts, dairy products, cakes, puddings, gelatin mixes, dressings, jams, jellies, and sauces [172]. An example of the use of these food additives is their incorporation into hydrocolloids like fermented milks, dairy desserts, cream, and ice-cream to stabilize and thicken them. Another example of texture additives are phosphates and coagulation agents that are used in the curdling of milk in cheese production [173]. Most of the hydrocolloids used in the dairy industry result from the isolation of seaweeds and plant cells and are obtained and/or extracted from byproducts such as plant food wastes. The natural agar obtained from algae is the most researched texture agent used as a food additive in bakery products, confectionery, ice cream, peanut butter, and beverages [171]. Table 4 discloses some examples of byproducts used as source of texture additives.

Table 4. Food byproducts sources of potential texturizing agents' additives. Adapted from Iriondo-DeHond, Miguel and del Castillo [142].

Food Industry	Byproducts	Function	Reference
Fruit	Olive pomace	Texturizing agent	[142,174,175]
	Passion fruit peels		[142,174,175]
	Grape pomace		[142,174,175]
	Pineapple peel powder		[17]
	Pineapple fruit peels		[176]
	Orange byproducts		[21,177]
	Citrus peels		[142]
	Apple pomace		[142]
	Pear stones		[25]
Dairy	Whey protein		[36,178]
	Whey protein and Buttermilk		[179]
Vegetable	Onion hulls		[142]
	Spinach		[25]

Citrus fruits and their byproducts (such as peels and seed powders) have been studied as possible sources of texturizing agents due to their natural high pectin content and dietary fiber [142,176,180]. Currently, there are some examples of citrus byproducts being used in the industry. Oranges are being used as a texturizing agents in yogurts and/or ice creams [21,179,181], and lemon byproducts are being used as thickening, texturizing, gelling, and stabilizing agents [182]. Furthermore, citrus byproducts have the added advantage of being rich in bioactive compounds, which possess nutritional and functional benefits including reducing the risk of certain pathologies such as obesity, cardiovascular disease, and colon cancer, as well as preventing neurodegenerative diseases and osteoporosis [139,182–185]. Additionally, their high dietary fiber content is an added bonus, as it can be used as fat replacers and thus functions as a food additive to impart texture to the final product [186–188]. In fact, Crizel et al. [189] showed that incorporation of fiber from orange byproducts into yogurts allowed for the manufacture of low-fat yogurts, and Dervisoglu and Yazici [190] reported that while citrus fiber as a single stabilizer did not improve the viscosity, overrun, and sensory properties of ice cream, it improved the melting resistance of these foodstuffs. Similarly, the industrial processing of tomato leads to high amounts of unused matter (mostly peels and seeds), which are byproducts rich in lycopene and dietary fiber. These byproducts have been incorporated in tomato sauce as a food texturizer, with sensorial tasting panels deeming it as acceptable [149,191]. Another example is the ß-glucans resulting from cereals such as oat and barley, which have also been used as fat replacers in a variety of foodstuff ranging from baked goods and pasta to beverages and soups with promising results [192–194]. Additionally, the presence of β-glucans in foods has also been shown to lead to an increase in fiber intake, which in turn prevents constipation, reduces intestinal transit time, reduces the risk of colorectal cancer, and promotes the growth of beneficial intestinal bacteria [195].

Other potential sources of dietary fiber and pectin are cocoa (Theobroma cacao L.) pod husk (an abundant industrial waste with potential application in the food industry), and oat bran. Studies have shown that cocoa byproducts can be used as a texturizing agents after drying and grinding, while juice resulting from the pods can be used to prepare hydrocolloids [196]. On the other hand, oat bran extract has also been studied as a natural emulsifier, with results showing stability through a different range of pH values, heat treatments, and storage life up to 40 days [197].

On a different note, fat plays an important role in the structural integrity and mouth feeling of foodstuffs (ice-cream and yogurt in particular) due to its interaction with casein micelles [198]. Many different types of fat replacers have been explored in bovine and goat milk yogurts, including the addition of inulin, β-glucan, high milk protein powder, and whey protein concentrates [199–204]. Whey proteins, obtained as a byproduct of the dairy industry, have many functional properties including gelation, thickening, and water-holding capacity [205]. In the study by Wang et al. [206], whey protein isolate was used to produce a goat's milk yogurt of acceptable quality. Milk fortification with whey protein improved the textural and microstructural characteristics and some sensory characteristics of yogurts. In addition, whey protein concentrates caused some interactions between globular proteins and caseins, which led to an improved texture of goat's milk yogurt and higher water retention capacity [36,206–208]. Whey proteins are also present in high amounts in a byproduct of butter-making—buttermilk. This product is now considered valuable because of its high content in fragments of milk fat globule membrane in addition to phospholipids [209,210]. Studies indicated that the moisture content of cheese supplemented with buttermilk remained high due largely to phospholipids improving its texture [211,212].

8. Foaming Capacity

Foam is a colloidal dispersion in which a gaseous phase is dispersed in a liquid or solid phase. Food foams are dependent on the surface activity and film-forming properties of specific protein compounds [213,214]. Proteins have to be either very hydrophobic or hydrophilic to possess good foaming properties, and therefore their chemical or enzymatic modification can make them more active on the surface. As such, most foaming agents commonly used in the food industry are mainly natural modified food proteins such as soy, casein, egg white, whey, serum proteins isolated from lactoglobulins, and lysozyme [214–216]. Globulins are excellent foaming agents, but their foaming is significantly affected by interactions of the proteins with ovomucine, lysozyme, and, to a lesser extent, ovomucoid, ovotransferrin, and ovalbumin [217]. A novel source of possible foaming agents is the fishery industry, as fish processing leads to high amounts of byproducts rich in collagen and gelatin. This gelatin foaming capacity has been studied, with reports showing that gelatin from shark cartilage possessed foaming properties similar to those of porcine skin [218]. According to Muzaifa et al. [219], fish byproducts (dark muscle, cut offs, viscera, skin, scales, small bones, and fins) could potentially be used to obtain protein hydrolysates through an enzymatic hydrolysis using Alcalase® 2.4 L and Flavourzyme® 500 L, leading to compounds with foaming capacity. Protein hydrolysates obtained from poultry byproducts (head and leg) and rainbow trout (*Onchorhynchus mykiss*) viscera after an enzymatic hydrolysis using Alcalase® 2.4 L also demonstrated foaming capacity [14]. On another work, Kotlar et al. [220] reported on the use of brewer's spent grain (BSG) hydrolysates (attained using a *B. cereus* extracellular peptidase) to improve the foaming expansion in brewery products. Okara, a byproduct obtained from the soy milk production, was also analyzed for its functional properties, with the authors observing that the isolated proteins from this byproduct could potentially be used as a foaming agent [27]. When it comes to slaughter byproducts, there are several residues, including skin, bones, hooves, muscles, and blood. Blood represents up to 4% of the animal live weight. However, the direct use of blood in foods is not useful due to the dark color given to the food. In practice, blood is separated by centrifugation into cellular and plasma fractions. Plasma proteins have relevant and interesting properties for food processing [221], e.g., they contribute to cross-link proteins and gelling [222], proteins enrichment [223], as well as emulsifying and foaming agents [224].

9. Emulsifiers

Emulsifiers, molecules such as polysaccharides (e.g., gum arabic) or phospholipids (e.g., lecithins) with a surface activity capable of mixing and stabilizing two immiscible phases like water and oil, are largely used in food technology [225,226].

Emulsifier additives can be obtained from a variety of food products (e.g., milk protein isolates) [227] and byproducts (e.g., okara) [27]. Gbogouri, Linder, Fanni, and Parmentier [11] suggested that salmon (Salmo salar) head hydrolysates treated with the commercial enzyme Alcalase® 2.4 L could potentially be a new source of compounds with great emulsifying capacity and stability. Using the same enzyme mix, Sathivel et al. [13] analyzed the potential of herring (Clupea harengus) byproducts hydrolysates. Although the emulsifying capacity was lower than that of egg albumin and soy protein, the hydrolysates still demonstrated some emulsifying capacity and stability, an effect that was also observed for the protein extracts before hydrolysis [12]. A potential emulsifier additive could be obtained from okara, a byproduct obtained from soymilk production. Even though okara protein isolates had poor solubility, they exhibited other functional properties (emulsification, foaming, and binding properties) that were comparable to those of a commercial soy isolate, further demonstrating the potential use of these isolates as a food ingredient [27]. The Horchata production, a vegetable milk obtained from tiger nuts (Cyperus esculentus), also originates a solid waste byproduct rich in dietary fiber that could potentially be used as a new ingredient for its emulsifying capacity and high emulsifying stability [28]. Emulsifiers can also be found in meat industry byproducts. For example, bovine blood derivate products (plasma and globulin) may be used as a potential new emulsifier agents additive in meat products and others [28,228,229]. As such, compounds obtained directly or indirectly from byproducts could potentially be used as new emulsifying agents in the food industry.

10. Conclusions

Given the consumer's demand for "clean label" products and the environmental constraints that reinforce the need to change the traditional industrial raw matters with renewable sources, agro-food byproducts have appeared as one of the most relevant potential solutions. In fact, some of the additives used nowadays (like anthocyanins and carotenes) can be found in these materials, which makes their extracts interesting from a consumer's perspective (some would prefer a tomato extract instead of traditional lycopene), particularly when considering the possibilities opened up by green, safe, new extraction methodologies like high pressure extraction, ohmic extraction, pulsed electric field, or supercritical extraction. However, their direct inclusion into commercial products may depend on the limitations posed by the legislation itself because, even if the additive itself is already approved for use, should its production process or raw material differ significantly from the one currently used, its future as an additive will be dependent on a new safety evaluation.

Overall, it is possible to see the potential of byproducts derived food additives and potential new additives for application in the food industry. They are an integrated solution with low cost and reduced environmental impact capable of providing alternatives for an industry that relies heavily upon the chemical synthesis compounds. Thus, the use of byproducts as a source of food additives stands out as an economically and environmentally conscious choice and will promote the new era of circular economy.

Author Contributions: M.V., M.F. and P.S. carried out the research and wrote the document, E.M.C. and S.S. were responsible for defining the paper's scope, structure and contributed to its writing and, along with M.P., were responsible for the overall definition and reviewing of the document. All authors reviewed and agreed on the final version.

Funding: This work was supported by National Funds from FCT (grant number UID/Multi/50016/2013, POCI-01-0145-FEDER-016403) from FEDER (grant number PDR2020-101-030775) and from QREN-ANI (grant number 17819).

Conflicts of Interest: The authors declare no conflict of interest.

References

1. Herrero, M.; Sánchez-Camargo, A.d.P.; Cifuentes, A.; Ibáñez, E. Plants, seaweeds, microalgae and food by-products as natural sources of functional ingredients obtained using pressurized liquid extraction and supercritical fluid extraction. *TrAC Trends Anal. Chem.* **2015**, *71*, 26–38. [CrossRef]
2. Fleuri, L.F. Orange and mango by-products: Agro-industrial waste as source of bioactive compounds and botanical versus commercial description—A review AU-Okino Delgado, Clarissa Hamaio. *Food Rev. Int.* **2016**, *32*, 1–14. [CrossRef]
3. European Commission. *Closing the Loop—An EU Action Plan for the Circular Economy*; European Commission: Brussels, Belgium, 2015.
4. Szabo, K.; Cătoi, A.-F.; Vodnar, D.C. Bioactive Compounds Extracted from Tomato Processing by-Products as a Source of Valuable Nutrients. *Plant Foods Hum. Nutr.* **2018**, *73*, 268–277. [CrossRef] [PubMed]
5. Abuajah, C.I.; Ogbonna, A.C.; Osuji, C.M. Functional components and medicinal properties of food: A review. *J. Food Sci. Technol.* **2015**, *52*, 2522–2529. [CrossRef] [PubMed]
6. Mohamed, S. Functional foods against metabolic syndrome (obesity, diabetes, hypertension and dyslipidemia) and cardiovasular disease. *Trends Food Sci. Technol.* **2014**, *35*, 114–128. [CrossRef]
7. Carocho, M.; Morales, P.; Ferreira, I.C. Natural food additives: Quo vadis? *Trends Food Sci. Technol.* **2015**, *45*, 284–295. [CrossRef]
8. Liu, R.; Wang, M.; Duan, J.-A. Antipyretic and antioxidant activities of the aqueous extract of Cornu bubali (water buffalo horn). *Am. J. Chin. Med.* **2010**, *38*, 293–306. [CrossRef]
9. Banerjee, P.; Shanthi, C. Isolation of novel bioactive regions from bovine Achilles tendon collagen having angiotensin I-converting enzyme-inhibitory properties. *Process Biochem.* **2012**, *47*, 2335–2346. [CrossRef]
10. Adje, E.Y.; Balti, R.; Kouach, M.; Dhulster, P.; Guillochon, D.; Nedjar-Arroume, N. Obtaining antimicrobial peptides by controlled peptic hydrolysis of bovine hemoglobin. *Int. J. Biol. Macromol.* **2011**, *49*, 143–153. [CrossRef]
11. Gbogouri, G.; Linder, M.; Fanni, J.; Parmentier, M. Influence of hydrolysis degree on the functional properties of salmon byproducts hydrolysates. *J. Food Sci.* **2004**, *69*, C615–C622. [CrossRef]

12. Sathivel, S.; Bechtel, P.J.; Babbitt, J.; Prinyawiwatkul, W.; Negulescu, I.I.; Reppond, K.D. Properties of Protein Powders from Arrowtooth Flounder (*Atheresthes stomias*) and Herring (*Clupea harengus*) Byproducts. *J. Agric. Food Chem.* **2004**, *52*, 5040–5046. [CrossRef]
13. Sathivel, S.; Bechtel, P.J.; Babbitt, J.; Smiley, S.; Crapo, C.; Reppond, K.D.; Prinyawiwatkul, W. Biochemical and Functional Properties of Herring (*Clupea harengus*) Byproduct Hydrolysates. *J. Food Sci.* **2003**, *68*, 2196–2200. [CrossRef]
14. Taheri, A.; Anvar, S.; Ahari, H.; Fogliano, V. Comparison the functional properties of protein hydrolysates from poultry by-products and rainbow trout (*Onchorhynchus mykiss*) viscera. *Iran. J. Fish. Sci.* **2013**, *12*, 154–169.
15. Ghafoor, K.; Park, J.; Choi, Y.-H. Optimization of supercritical fluid extraction of bioactive compounds from grape (*Vitis labrusca* B.) peel by using response surface methodology. *Innov. Food Sci. Emerg. Technol.* **2010**, *11*, 485–490. [CrossRef]
16. Ersouml, E.; Yerlikaya, O.; Accedil, M. Effect of phenolic compounds on characteristics of strained yoghurts produced from sheep milk. *Afr. J. Agric. Res.* **2011**, *6*, 5351–5359.
17. Sah, B.; Vasiljevic, T.; McKechnie, S.; Donkor, O. Physicochemical, textural and rheological properties of probiotic yogurt fortified with fibre-rich pineapple peel powder during refrigerated storage. *LWT-Food Sci. Technol.* **2016**, *65*, 978–986. [CrossRef]
18. Someya, S.; Yoshiki, Y.; Okubo, K. Antioxidant compounds from bananas (*Musa cavendish*). *Food Chem.* **2002**, *79*, 351–354. [CrossRef]
19. Shan, B.; Cai, Y.-Z.; Brooks, J.D.; Corke, H. Potential application of spice and herb extracts as natural preservatives in cheese. *J. Med. Food* **2011**, *14*, 284–290. [CrossRef]
20. da Silva, D.F.; Matumoto-Pintro, P.T.; Bazinet, L.; Couillard, C.; Britten, M. Effect of commercial grape extracts on the cheese-making properties of milk. *J. Dairy Sci.* **2015**, *98*, 1552–1562. [CrossRef]
21. Sendra, E.; Kuri, V.; Fernandez-Lopez, J.; Sayas-Barbera, E.; Navarro, C.; Perez-Alvarez, J. Viscoelastic properties of orange fiber enriched yogurt as a function of fiber dose, size and thermal treatment. *LWT-Food Sci. Technol.* **2010**, *43*, 708–714. [CrossRef]
22. Tseng, A.; Zhao, Y. Wine grape pomace as antioxidant dietary fibre for enhancing nutritional value and improving storability of yogurt and salad dressing. *Food Chem.* **2013**, *138*, 356–365. [CrossRef] [PubMed]
23. Frumento, D.; do Espirito Santo, A.P.; Aliakbarian, B.; Casazza, A.A.; Gallo, M.; Converti, A.; Perego, P. Development of milk fermented with Lactobacillus acidophilus fortified with *Vitis vinifera* marc flour. *Food Technol. Biotechnol.* **2013**, *51*, 370.
24. Pazmiño, A.; Giusti, M.; Wrolstad, R.; Gloria, M.B.A. Anthocyanins from banana bracts (*Musa X Paradisiaca*) as potential food colorants. *Food Chem.* **2001**, *73*, 327–332.
25. Saraç, M.G.; Dogan, M. Incorporation of dietary fiber concentrates from fruit and vegetable wastes in butter: Effects on physicochemical, textural, and sensory properties. *Eur. Food Res. Technol.* **2016**, *242*, 1331–1342. [CrossRef]
26. Padam, B.; Tin, H.; Chye, F.; Abdullah, M. Antibacterial and Antioxidative Activities of the Various Solvent Extracts of Banana (*Musa paradisiaca* cv. Mysore) Inflorescences. *J. Biol. Sci.* **2012**, *12*, 62–73.
27. Ma, C.-Y.; Liu, W.-S.; Kwok, K.C.; Kwok, F. Isolation and characterization of proteins from soymilk residue (okara). *Food Res. Int.* **1996**, *29*, 799–805. [CrossRef]
28. Sanchez-Zapata, E.; Fuentes-Zaragoza, E.L.; Fernandez-Lopez, J.; Sendra, E.; Sayas, E.; Navarro, C.; Pérez-Álvarez, J.A. Preparation of dietary fiber powder from tiger nut (*Cyperus esculentus*) milk (*"Horchata"*) byproducts and its physicochemical properties. *J. Agric. Food Chem.* **2009**, *57*, 7719–7725. [CrossRef]
29. Ciurlia, L.; Bleve, M.; Rescio, L. Supercritical carbon dioxide co-extraction of tomatoes (*Lycopersicum esculentum* L.) and hazelnuts (*Corylus avellana* L.): A new procedure in obtaining a source of natural lycopene. *J. Supercrit. Fluids* **2009**, *49*, 338–344. [CrossRef]
30. Viuda-Martos, M.; Sanchez-Zapata, E.; Sayas-Barberá, E.; Sendra, E.; Pérez-Álvarez, J.; Fernández-López, J. Tomato and tomato byproducts. Human health benefits of lycopene and its application to meat products: A review. *Crit. Rev. Food Sci. Nutr.* **2014**, *54*, 1032–1049. [CrossRef]
31. Jurikova, T.; Skrovankova, S.; Mlcek, J.; Balla, S.; Snopek, L. Bioactive Compounds, Antioxidant Activity, and Biological Effects of European Cranberry (*Vaccinium oxycoccos*). *Molecules* **2018**, *24*, 24. [CrossRef]
32. Al-Zoreky, N. Antimicrobial activity of pomegranate (*Punica granatum* L.) fruit peels. *Int. J. Food Microbiol.* **2009**, *134*, 244–248. [CrossRef]

33. Akhtar, S.; Ismail, T.; Fraternale, D.; Sestili, P. Pomegranate peel and peel extracts: Chemistry and food features. *Food Chem.* **2015**, *174*, 417–425. [CrossRef]
34. Abid, Y.; Azabou, S.; Jridi, M.; Khemakhem, I.; Bouaziz, M.; Attia, H. Storage stability of traditional Tunisian butter enriched with antioxidant extract from tomato processing by-products. *Food Chem.* **2017**, *233*, 476–482. [CrossRef]
35. Nontasan, S.; Moongngarm, A.; Deeseenthum, S. Application of Functional Colorant Prepared from Black Rice Bran in Yogurt. *APCBEE Procedia* **2012**, *2*, 62–67. [CrossRef]
36. Akalın, A.; Unal, G.; Dinkci, N.; Hayaloglu, A. Microstructural, textural, and sensory characteristics of probiotic yogurts fortified with sodium calcium caseinate or whey protein concentrate. *J. Dairy Sci.* **2012**, *95*, 3617–3628. [CrossRef]
37. Peters, H. M. Hanssen und J. Marschen: E for Additives. 223 Seiten. Thorsons Publishers Limited, Wellingborough, Northamptonshire. 1984. Preis: 2.95 £. *Food Nahr.* **1986**, *30*, 204. [CrossRef]
38. Saltmarsh, M. Recent trends in the use of food additives in the United Kingdom. *J. Sci. Food Agric.* **2015**, *95*, 649–652. [CrossRef]
39. Aschemann-Witzel, J.; Varela, P.; Peschel, A.O. Consumers' categorization of food ingredients: Do consumers perceive them as 'clean label' producers expect? An exploration with projective mapping. *Food Qual. Prefer.* **2019**, *71*, 117–128. [CrossRef]
40. Asioli, D.; Aschemann-Witzel, J.; Caputo, V.; Vecchio, R.; Annunziata, A.; Næs, T.; Varela, P. Making sense of the "clean label" trends: A review of consumer food choice behavior and discussion of industry implications. *Food Res. Int.* **2017**, *99*, 58–71. [CrossRef]
41. Ayala-Zavala, J.; Vega-Vega, V.; Rosas-Domínguez, C.; Palafox-Carlos, H.; Villa-Rodriguez, J.; Siddiqui, M.W.; Dávila-Aviña, J.; González-Aguilar, G. Agro-industrial potential of exotic fruit byproducts as a source of food additives. *Food Res. Int.* **2011**, *44*, 1866–1874. [CrossRef]
42. Ofori, J.A.; Hsieh, Y.-H.P. The use of blood and derived products as food additives. In *Food Additive*; IntechOpen: Rijeka, Croatia, 2012.
43. Schieber, A.; Stintzing, F.; Carle, R. By-products of plant food processing as a source of functional compounds—Recent developments. *Trends Food Sci. Technol.* **2001**, *12*, 401–413. [CrossRef]
44. Gowe, C. Review on potential use of fruit and vegetables by-products as a valuable source of natural food additives. *Food Sci. Qual. Manag.* **2015**, *45*, 47–61.
45. EU. Regulation (EC) No 1333/2008 of the European Parliament and of the Council of 16 December 2008 on food additives (1). *Off. J. Eur. Union* **2009**, *51*, 16–33.
46. EU. Commission Directive 2008/60/EC of 17 June 2008 laying down specific purity criteria concerning sweeteners for use in foodstuffs (Codified version). *Off. J. Eur. Union* **2008**, *158*, 17–40.
47. EU. Commission Directive 2008/128/EC of 22 December 2008 laying down specific purity criteria concerning colours for use in foodstuffs (Codified version). *Off. J. Eur. Union* **2009**, *52*, 20–63.
48. EU. Commission Directive 2008/84/EC of 27 August 2008 laying down specific purity criteria on food additives other than colours and sweeteners (Codified version). *Off. J. Eur. Union* **2009**, *51*, 1–175.
49. Saltmarsh, M. *Essential Guide to Food Additives*; Royal Society of Chemistry: London, UK, 2013.
50. EU. Commission Regulation (EU) No 231/2012 of 9 March 2012 laying down specifications for food additives listed in Annexes II and III to Regulation (EC) No 1333/2008. *Off. J. Eur. Union* **2012**, *55*, 1–295. [CrossRef]
51. Rajha, H.N.; El Darra, N.; Hobaika, Z.; Boussetta, N.; Vorobiev, E.; Maroun, R.G.; Louka, N. Extraction of total phenolic compounds, flavonoids, anthocyanins and tannins from grape byproducts by response surface methodology. Influence of solid-liquid ratio, particle size, time, temperature and solvent mixtures on the optimization process. *Food Nutr. Sci.* **2014**, *5*, 397. [CrossRef]
52. Corrales, M.; Toepfl, S.; Butz, P.; Knorr, D.; Tauscher, B. Extraction of anthocyanins from grape by-products assisted by ultrasonics, high hydrostatic pressure or pulsed electric fields: A comparison. *Innov. Food Sci. Emerg. Technol.* **2008**, *9*, 85–91. [CrossRef]
53. Kalogeropoulos, N.; Chiou, A.; Pyriochou, V.; Peristeraki, A.; Karathanos, V.T. Bioactive phytochemicals in industrial tomatoes and their processing byproducts. *LWT-Food Sci. Technol.* **2012**, *49*, 213–216. [CrossRef]
54. Ajila, C.; Naidu, K.; Bhat, S.; Rao, U.P. Bioactive compounds and antioxidant potential of mango peel extract. *Food Chem.* **2007**, *105*, 982–988. [CrossRef]
55. Dabbou, S.; Dabbou, S.; Flamini, G.; Pandino, G.; Gasco, L.; Helal, A.N. Phytochemical compounds from the crop byproducts of Tunisian globe artichoke cultivars. *Chem. Biodivers.* **2016**, *13*, 1475–1483. [CrossRef]

56. Gapor, A.; Kato, A.; Ong, A. Tocopherol content in oil palm leaflet. *J. Am. Oil Chem. Soc.* **1986**, *63*, 330–331. [CrossRef]
57. Baysal, T.; Ersus, S.; Starmans, D. Supercritical CO_2 extraction of β-carotene and lycopene from tomato paste waste. *J. Agric. Food Chem.* **2000**, *48*, 5507–5511. [CrossRef]
58. Gherezgihier, B.A.; Mahmud, A.; Admassu, H.; Shui, X.W.; Fang, Y.; Tsighe, N.; Mohammed, J.K. Food additives: Functions, effects, regulations, approval and safety evaluation. *J. Acad. Ind. Res.* **2017**, *6*, 62–68.
59. Millstone, E. A. Larry Branen, P. Michael Davidson, Seppe Salminen and John H. ThorngateIII (editors). *Food Additives*, 2nd ed. New York and Basel: Marcel Dekker Inc.. 2001. US$225. pp. XV + 938, ISBN 0 8247 9343 9. *Br. J. Nutr.* **2007**, *89*, 157–158. [CrossRef]
60. Food and Drug Administration Center for Food Safety and Applied Nutrition. *Guidance for Industry Considerations Regarding Substances Added to Foods, Including Beverages and Dietary Supplements*; Food and Drug Administration Center for Food Safety and Applied Nutrition, U.S. Department of Health and Human Services: College Park, MD, USA, 2014.
61. EFSA Panel on Food Additives and Nutrient Sources added to Food (ANS). Guidance for submission for food additive evaluations. *EFSA J.* **2012**, *10*, 2760. [CrossRef]
62. Gram, L.; Ravn, L.; Rasch, M.; Bruhn, J.B.; Christensen, A.B.; Givskov, M. Food spoilage—Interactions between food spoilage bacteria. *Int. J. Food Microbiol.* **2002**, *78*, 79–97. [CrossRef]
63. Perumalla, A.; Hettiarachchy, N.S. Green tea and grape seed extracts—Potential applications in food safety and quality. *Food Res. Int.* **2011**, *44*, 827–839. [CrossRef]
64. McCabe-Sellers, B.J.; Beattie, S.E. Food safety: Emerging trends in foodborne illness surveillance and prevention. *J. Am. Diet. Assoc.* **2004**, *104*, 1708–1717. [CrossRef]
65. Pitt, J.I.; Hocking, A.D. The ecology of fungal food spoilage. In *Fungi and Food Spoilage*; Springer: New York, NY, USA, 2009; pp. 3–9.
66. Lück, E.; Jager, M. *Antimicrobial Food Additives: Characteristics, Uses, Effects*; Springer Science & Business Media: New York, NY, USA, 1997; Volume 2.
67. Abdulmumeen, H.A.; Risikat, A.N.; Sururah, A.R. Food: Its preservatives, additives and applications. *Int. J. Pharm. Chem. Biol. Sci.* **2012**, *1*, 36–47.
68. Smith, J.; Hong-Shum, L. *Food Additives Data Book*; John Wiley & Sons: New York, NY, USA, 2011.
69. O'sullivan, L.; Ross, R.; Hill, C. Potential of bacteriocin-producing lactic acid bacteria for improvements in food safety and quality. *Biochimie* **2002**, *84*, 593–604. [CrossRef]
70. Burt, S. Essential oils: Their antibacterial properties and potential applications in foods—A review. *Int. J. Food Microbiol.* **2004**, *94*, 223–253. [CrossRef]
71. Ohtsu, N.; Kohari, Y.; Gotoh, M.; Yamada, R.; Nagata, Y.; Murata, M. Utilization of the Japanese Peppermint Herbal Water Byproduct of Steam Distillation as an Antimicrobial Agent. *J. Oleo Sci.* **2018**, *67*, 1227–1233. [CrossRef]
72. Sanz-Puig, M.; Pina-Pérez, M.C.; Criado, M.N.; Rodrigo, D.; Martínez-López, A. Antimicrobial potential of cauliflower, broccoli, and okara byproducts against foodborne bacteria. *Foodborne Pathog. Dis.* **2015**, *12*, 39–46. [CrossRef]
73. Li, C.; Li, X.; Bai, C.; Zhang, Y.; Wang, Z. A chitinase with antifungal activity from naked oat (*Avena chinensis*) seeds. *J. Food Biochem.* **2018**, e12713. [CrossRef]
74. Kirbaşlar, F.G.; Tavman, A.; Dülger, B.; Türker, G. Antimicrobial activity of Turkish citrus peel oils. *Pak. J. Bot.* **2009**, *41*, 3207–3212.
75. Smith-Palmer, A.; Stewart, J.; Fyfe, L. Antimicrobial properties of plant essential oils and essences against five important food-borne pathogens. *Lett. Appl. Microbiol.* **1998**, *26*, 118–122. [CrossRef]
76. Fisher, K.; Phillips, C. Potential antimicrobial uses of essential oils in food: Is citrus the answer? *Trends Food Sci. Technol.* **2008**, *19*, 156–164. [CrossRef]
77. Ayala-Zavala, J.F.; González-Aguilar, G.A. Use of additives to preserve the quality of fresh-cut fruits and vegetables. In *Advances in Fresh-Cut Fruits and Vegetables Processing*; CRC Press: Boca Raton, FL, USA, 2010; pp. 233–256.
78. Correa, P. Diet and gastric cancer. In *Diet, Nutrition and Cancer: A Critical Evaluation*; CRC Press: Boca Raton, FL, USA, 2018; pp. 1–10.
79. Gassara, F.; Kouassi, A.P.; Brar, S.K.; Belkacemi, K. Green alternatives to nitrates and nitrites in meat-based products—A review. *Crit. Rev. Food Sci. Nutr.* **2016**, *56*, 2133–2148. [CrossRef]

80. Keeton, J.T. History of nitrite and nitrate in food. In *Nitrite and Nitrate in Human Health and Disease*; Springer: New York, NY, USA, 2017; pp. 85–97.
81. Adam, A.H.B.; Mustafa, N.E.M.; Rietjens, I.M. Nitrite in processed meat products in Khartoum, Sudan and dietary intake. *Food Addit. Contam. Part B* **2017**, *10*, 79–84. [CrossRef] [PubMed]
82. Jiang, J.; Xiong, Y.L. Natural antioxidants as food and feed additives to promote health benefits and quality of meat products: A review. *Meat Sci.* **2016**, *120*, 107–117. [CrossRef]
83. Gul, P.; Bakht, J. Antimicrobial activity of turmeric extract and its potential use in food industry. *J. Food Sci. Technol.* **2015**, *52*, 2272–2279. [CrossRef]
84. Veiga, R.; De Mendonça, S.; Mendes, P.; Paulino, N.; Mimica, M.; Lagareiro Netto, A.; Lira, I.; López, B.C.; Negrão, V.; Marcucci, M. Artepillin C and phenolic compounds responsible for antimicrobial and antioxidant activity of green propolis and Baccharis dracunculifolia DC. *J. Appl. Microbiol.* **2017**, *122*, 911–920. [CrossRef] [PubMed]
85. Cueva, C.; Moreno-Arribas, M.V.; Martín-Álvarez, P.J.; Bills, G.; Vicente, M.F.; Basilio, A.; Rivas, C.L.; Requena, T.; Rodríguez, J.M.; Bartolomé, B. Antimicrobial activity of phenolic acids against commensal, probiotic and pathogenic bacteria. *Res. Microbiol.* **2010**, *161*, 372–382. [CrossRef] [PubMed]
86. Maizura, M.; Aminah, A.; Wan Aida, W. Total phenolic content and antioxidant activity of kesum (*Polygonum minus*), ginger (*Zingiber officinale*) and turmeric (*Curcuma longa*) extract. *Int. Food Res. J.* **2011**, *18*.
87. Negi, P.; Jayaprakasha, G.; Jagan Mohan Rao, L.; Sakariah, K. Antibacterial activity of turmeric oil: A byproduct from curcumin manufacture. *J. Agric. Food Chem.* **1999**, *47*, 4297–4300. [CrossRef] [PubMed]
88. Hu, Y.; Zhang, J.; Kong, W.; Zhao, G.; Yang, M. Mechanisms of antifungal and anti-aflatoxigenic properties of essential oil derived from turmeric (*Curcuma longa* L.) on *Aspergillus flavus*. *Food Chem.* **2017**, *220*, 1–8. [CrossRef]
89. Nile, S.H.; Park, S.W. Edible berries: Bioactive components and their effect on human health. *Nutrition* **2014**, *30*, 134–144. [CrossRef]
90. Paredes-López, O.; Cervantes-Ceja, M.L.; Vigna-Pérez, M.; Hernández-Pérez, T. Berries: Improving human health and healthy aging, and promoting quality life—A review. *Plant Foods Hum. Nutr.* **2010**, *65*, 299–308. [CrossRef]
91. Nicoue, E.E.; Savard, S.; Belkacemi, K. Anthocyanins in wild blueberries of Quebec: Extraction and identification. *J. Agric. Food Chem.* **2007**, *55*, 5626–5635. [CrossRef] [PubMed]
92. Veiga, M.; Costa, E.M.; Silva, S.; Pintado, M. Impact of plant extracts upon human health: A review. *Crit. Rev. Food Sci. Nutr.* **2018**, 1–14. [CrossRef] [PubMed]
93. Silva, S.; Costa, E.M.; Mendes, M.; Morais, R.; Calhau, C.; Pintado, M. Antimicrobial, antiadhesive and antibiofilm activity of an ethanolic, anthocyanin-rich blueberry extract purified by solid phase extraction. *J. Appl. Microbiol.* **2016**, *121*, 693–703. [CrossRef] [PubMed]
94. Silva, S.; Costa, E.M.; Costa, M.R.; Pereira, M.F.; Pereira, J.O.; Soares, J.C.; Pintado, M.M. Aqueous extracts of Vaccinium corymbosum as inhibitors of Staphylococcus aureus. *Food Control* **2015**, *51*, 314–320. [CrossRef]
95. Puupponen-Pimiä, R.; Nohynek, L.; Meier, C.; Kähkönen, M.; Heinonen, M.; Hopia, A.; Oksman-Caldentey, K.M. Antimicrobial properties of phenolic compounds from berries. *J. Appl. Microbiol.* **2001**, *90*, 494–507. [CrossRef] [PubMed]
96. Ross, K.; Ehret, D.; Godfrey, D.; Fukumoto, L.; Diarra, M. Characterization of Pilot Scale Processed Canadian Organic Cranberry (*Vaccinium macrocarpon*) and Blueberry (*Vaccinium angustifolium*) Juice Pressing Residues and Phenolic-Enriched Extractives. *Int. J. Fruit Sci.* **2017**, *17*, 202–232. [CrossRef]
97. Neto, C.C. Cranberry and blueberry: Evidence for protective effects against cancer and vascular diseases. *Mol. Nutr. Food Res.* **2007**, *51*, 652–664. [CrossRef] [PubMed]
98. Talhaoui, N.; Taamalli, A.; Gómez-Caravaca, A.M.; Fernández-Gutiérrez, A.; Segura-Carretero, A. Phenolic compounds in olive leaves: Analytical determination, biotic and abiotic influence, and health benefits. *Food Res. Int.* **2015**, *77*, 92–108. [CrossRef]
99. Şahin, S.; Samli, R.; Tan, A.S.B.; Barba, F.J.; Chemat, F.; Cravotto, G.; Lorenzo, J.M. Solvent-free microwave-assisted extraction of polyphenols from olive tree leaves: Antioxidant and antimicrobial properties. *Molecules* **2017**, *22*, 1056. [CrossRef]
100. Paiva-Martins, F.; Correia, R.; Félix, S.; Ferreira, P.; Gordon, M.H. Effects of enrichment of refined olive oil with phenolic compounds from olive leaves. *J. Agric. Food Chem.* **2007**, *55*, 4139–4143. [CrossRef]

101. Pereira, A.P.; Ferreira, I.C.; Marcelino, F.; Valentão, P.; Andrade, P.B.; Seabra, R.; Estevinho, L.; Bento, A.; Pereira, J.A. Phenolic compounds and antimicrobial activity of olive (*Olea europaea* L. Cv. Cobrançosa) leaves. *Molecules* **2007**, *12*, 1153–1162. [CrossRef] [PubMed]
102. Wang, Y.; Li, F.; Zhuang, H.; Chen, X.; Li, L.; Qiao, W.; Zhang, J. Effects of plant polyphenols and α-tocopherol on lipid oxidation, residual nitrites, biogenic amines, and N-nitrosamines formation during ripening and storage of dry-cured bacon. *LWT-Food Sci. Technol.* **2015**, *60*, 199–206. [CrossRef]
103. Kondo, M.; Kita, K.; Yokota, H.-O. Feeding value to goats of whole-crop oat ensiled with green tea waste. *Anim. Feed Sci. Technol.* **2004**, *113*, 71–81. [CrossRef]
104. Farhoosh, R.; Golmovahhed, G.A.; Khodaparast, M.H. Antioxidant activity of various extracts of old tea leaves and black tea wastes (*Camellia sinensis* L.). *Food Chem.* **2007**, *100*, 231–236. [CrossRef]
105. García-Lomillo, J.; Gonzalez-SanJose, M.L.; Del Pino-García, R.; Rivero-Perez, M.D.; Muniz-Rodriguez, P. Antioxidant and antimicrobial properties of wine byproducts and their potential uses in the food industry. *J. Agric. Food Chem.* **2014**, *62*, 12595–12602. [CrossRef]
106. Teixeira, A.; Baenas, N.; Dominguez-Perles, R.; Barros, A.; Rosa, E.; Moreno, A.D.; Garcia-Viguera, C. Natural Bioactive Compounds from Winery By-Products as Health Promoters: A Review. *Int. J. Mol. Sci.* **2014**, *15*, 15638–15678. [CrossRef]
107. Çam, M.; İçyer, N.C.; Erdoğan, F. Pomegranate peel phenolics: Microencapsulation, storage stability and potential ingredient for functional food development. *LWT-Food Sci. Technol.* **2014**, *55*, 117–123. [CrossRef]
108. Li, Y.; Guo, C.; Yang, J.; Wei, J.; Xu, J.; Cheng, S. Evaluation of antioxidant properties of pomegranate peel extract in comparison with pomegranate pulp extract. *Food Chem.* **2006**, *96*, 254–260. [CrossRef]
109. Kanatt, S.R.; Chander, R.; Sharma, A. Antioxidant and antimicrobial activity of pomegranate peel extract improves the shelf life of chicken products. *Int. J. Food Sci. Technol.* **2010**, *45*, 216–222. [CrossRef]
110. Rodríguez-Carpena, J.-G.; Morcuende, D.; Andrade, M.-J.; Kylli, P.; Estévez, M. Avocado (*Persea americana* Mill.) phenolics, in vitro antioxidant and antimicrobial activities, and inhibition of lipid and protein oxidation in porcine patties. *J. Agric. Food Chem.* **2011**, *59*, 5625–5635. [CrossRef]
111. Raymond Chia, T.W.; Dykes, G.A. Antimicrobial activity of crude epicarp and seed extracts from mature avocado fruit (*Persea americana*) of three cultivars. *Pharm. Biol.* **2010**, *48*, 753–756. [CrossRef]
112. Calderón-Oliver, M.; Escalona-Buendía, H.B.; Medina-Campos, O.N.; Pedraza-Chaverri, J.; Pedroza-Islas, R.; Ponce-Alquicira, E. Optimization of the antioxidant and antimicrobial response of the combined effect of nisin and avocado byproducts. *LWT-Food Sci. Technol.* **2016**, *65*, 46–52. [CrossRef]
113. Baines, D.; Seal, R. *Natural Food Additives, Ingredients and Flavourings*; Elsevier: Amsterdam, The Netherlands, 2012.
114. Sung, K.; Khan, S.A.; Nawaz, M.S.; Cerniglia, C.E.; Tamplin, M.L.; Phillips, R.W.; Kelley, L.C. Lysozyme as a barrier to growth of Bacillus anthracis strain Sterne in liquid egg white, milk and beef. *Food Microbiol.* **2011**, *28*, 1231–1234. [CrossRef]
115. Barbiroli, A.; Bonomi, F.; Capretti, G.; Iametti, S.; Manzoni, M.; Piergiovanni, L.; Rollini, M. Antimicrobial activity of lysozyme and lactoferrin incorporated in cellulose-based food packaging. *Food Control* **2012**, *26*, 387–392. [CrossRef]
116. Lorenzo, J.M.; Munekata, P.E.; Gómez, B.; Barba, F.J.; Mora, L.; Pérez-Santaescolástica, C.; Toldrá, F. Bioactive peptides as natural antioxidants in food products—A review. *Trends Food Sci. Technol.* **2018**, *79*, 136–147. [CrossRef]
117. Carocho, M.; Morales, P.; Ferreira, I.C. Antioxidants: Reviewing the chemistry, food applications, legislation and role as preservatives. *Trends Food Sci. Technol.* **2018**, *71*, 107–120. [CrossRef]
118. Roca-Saavedra, P.; Mendez-Vilabrille, V.; Miranda, J.M.; Nebot, C.; Cardelle-Cobas, A.; Franco, C.M.; Cepeda, A. Food additives, contaminants and other minor components: Effects on human gut microbiota—A review. *J. Physiol. Biochem.* **2017**, 1–15. [CrossRef]
119. Pisoschi, A.M.; Pop, A. The role of antioxidants in the chemistry of oxidative stress: A review. *Eur. J. Med. Chem.* **2015**, *97*, 55–74. [CrossRef]
120. Di Pasquale, J.; Adinolfi, F.; Capitanio, F. Analysis of consumer attitudes and consumers' willingness to pay for functional foods. *Int. J. Food Syst. Dyn.* **2011**, *2*, 181–193.
121. Caleja, C.; Barros, L.; Antonio, A.L.; Oliveira, M.B.P.; Ferreira, I.C. A comparative study between natural and synthetic antioxidants: Evaluation of their performance after incorporation into biscuits. *Food Chem.* **2017**, *216*, 342–346. [CrossRef]

122. Williams, G.; Iatropoulos, M.; Whysner, J. Safety assessment of butylated hydroxyanisole and butylated hydroxytoluene as antioxidant food additives. *Food Chem. Toxicol.* **1999**, *37*, 1027–1038. [CrossRef]
123. Wu, D.; Yan, J.; Tang, P.; Li, S.; Xu, K.; Li, H. Binding properties and structure–affinity relationships of food antioxidant butylated hydroxyanisole and its metabolites with lysozyme. *Food Chem.* **2015**, *188*, 370–376. [CrossRef]
124. Taghvaei, M.; Jafari, S.M. Application and stability of natural antioxidants in edible oils in order to substitute synthetic additives. *J. Food Sci. Technol.* **2015**, *52*, 1272–1282. [CrossRef]
125. McCusker, M.M.; Durrani, K.; Payette, M.J.; Suchecki, J. An eye on nutrition: The role of vitamins, essential fatty acids, and antioxidants in age-related macular degeneration, dry eye syndrome, and cataract. *Clin. Dermatol.* **2016**, *34*, 276–285. [CrossRef]
126. Toldrá, F.; Mora, L.; Reig, M. New insights into meat by-product utilization. *Meat Sci.* **2016**, *120*, 54–59. [CrossRef]
127. Ryder, K.; Ha, M.; Bekhit, A.E.-D.; Carne, A. Characterisation of novel fungal and bacterial protease preparations and evaluation of their ability to hydrolyse meat myofibrillar and connective tissue proteins. *Food Chem.* **2015**, *172*, 197–206. [CrossRef]
128. Roldán, E.; Sánchez-Moreno, C.; de Ancos, B.; Cano, M.P. Characterisation of onion (*Allium cepa* L.) by-products as food ingredients with antioxidant and antibrowning properties. *Food Chem.* **2008**, *108*, 907–916. [CrossRef]
129. Larrosa, M.; Llorach, R.; Espín, J.C.; Tomás-Barberán, F.A. Increase of antioxidant activity of tomato juice upon functionalisation with vegetable byproduct extracts. *LWT-Food Sci. Technol.* **2002**, *35*, 532–542. [CrossRef]
130. Sadilova, E.; Stintzing, F.C.; Carle, R. Anthocyanins, colour and antioxidant properties of eggplant (*Solanum melongena* L.) and violet pepper (*Capsicum annuum* L.) peel extracts. *Z. Nat. C* **2006**, *61*, 527–535. [CrossRef]
131. Kim, H.; Moon, J.Y.; Kim, H.; Lee, D.-S.; Cho, M.; Choi, H.-K.; Kim, Y.S.; Mosaddik, A.; Cho, S.K. Antioxidant and antiproliferative activities of mango (*Mangifera indica* L.) flesh and peel. *Food Chem.* **2010**, *121*, 429–436. [CrossRef]
132. Vega-Vega, V.; Silva-Espinoza, B.A.; Cruz-Valenzuela, M.R.; Bernal-Mercado, A.T.; Gonzalez-Aguilar, G.A.; Ruiz-Cruz, S.; Moctezuma, E.; Siddiqui, M.W.; Ayala-Zavala, J.F. Antimicrobial and antioxidant properties of byproduct extracts of mango fruit. *J. Appl. Bot. Food Qual.* **2013**, *86*. [CrossRef]
133. Ajila, C.M.; Aalami, M.; Leelavathi, K.; Rao, U.J.S.P. Mango peel powder: A potential source of antioxidant and dietary fiber in macaroni preparations. *Innov. Food Sci. Emerg. Technol.* **2010**, *11*, 219–224. [CrossRef]
134. Caleja, C.; Barros, L.; Antonio, A.L.; Ciric, A.; Barreira, J.C.; Sokovic, M.; Oliveira, M.B.P.; Santos-Buelga, C.; Ferreira, I.C. Development of a functional dairy food: Exploring bioactive and preservation effects of chamomile (*Matricaria recutita* L.). *J. Funct. Foods* **2015**, *16*, 114–124. [CrossRef]
135. Caleja, C.; Barros, L.; Antonio, A.L.; Ciric, A.; Soković, M.; Oliveira, M.B.P.; Santos-Buelga, C.; Ferreira, I.C. Foeniculum vulgare Mill. as natural conservation enhancer and health promoter by incorporation in cottage cheese. *J. Funct. Foods* **2015**, *12*, 428–438. [CrossRef]
136. Shah, M.A.; Bosco, S.J.D.; Mir, S.A. Plant extracts as natural antioxidants in meat and meat products. *Meat Sci.* **2014**, *98*, 21–33. [CrossRef]
137. Bajaj, S.; Urooj, A.; Prabhasankar, P. Effect of incorporation of mint on texture, colour and sensory parameters of biscuits. *Int. J. Food Prop.* **2006**, *9*, 691–700. [CrossRef]
138. He, B.; Zhang, L.-L.; Yue, X.-Y.; Liang, J.; Jiang, J.; Gao, X.-L.; Yue, P.-X. Optimization of ultrasound-assisted extraction of phenolic compounds and anthocyanins from blueberry (*Vaccinium ashei*) wine pomace. *Food Chem.* **2016**, *204*, 70–76. [CrossRef]
139. Guimarães, R.; Barros, L.; Barreira, J.C.; Sousa, M.J.; Carvalho, A.M.; Ferreira, I.C. Targeting excessive free radicals with peels and juices of citrus fruits: Grapefruit, lemon, lime and orange. *Food Chem. Toxicol.* **2010**, *48*, 99–106. [CrossRef]
140. Prakash, C.V.S.; Prakash, I. Bioactive chemical constituents from pomegranate (*Punica granatum*) juice, seed and peel—A review. *IJRCE* **2011**, *1*, 1–18.
141. Verbeke, W. Functional foods: Consumer willingness to compromise on taste for health? *Food Qual. Prefer.* **2006**, *17*, 126–131. [CrossRef]
142. Iriondo-DeHond, M.; Miguel, E.; del Castillo, M. Food Byproducts as Sustainable Ingredients for Innovative and Healthy Dairy Foods. *Nutrients* **2018**, *10*, 1358. [CrossRef]

143. *Codex Alimentarius General Standard for Food Additives*; Food and Agriculture Organization of the United Nations: Rome, Italy, 1995; Volume 192.
144. Amchova, P.; Kotolova, H.; Ruda-Kucerova, J. Health safety issues of synthetic food colorants. *Regul. Toxicol. Pharmacol.* **2015**, *73*, 914–922. [CrossRef] [PubMed]
145. Sigurdson, G.T.; Tang, P.; Giusti, M.M. Natural colorants: Food colorants from natural sources. *Annu. Rev. Food Sci. Technol.* **2017**, *8*, 261–280. [CrossRef] [PubMed]
146. Bridle, P.; Timberlake, C.F. Anthocyanins as natural food colours—Selected aspects. *Food Chem.* **1997**, *58*, 103–109. [CrossRef]
147. Rodriguez-Amaya, D.B. Natural food pigments and colorants. *Curr. Opin. Food Sci* **2016**, *7*, 20–26. [CrossRef]
148. Kaur, D.; Wani, A.A.; Singh, D.P.; Sogi, D.S. Shelf Life Enhancement of Butter, Ice-Cream, and Mayonnaise by Addition of Lycopene. *Int. J. Food Prop.* **2011**, *14*, 1217–1231. [CrossRef]
149. Rizk, E.M.; El-Kady, A.T.; El-Bialy, A.R. Characterization of carotenoids (lyco-red) extracted from tomato peels and its uses as natural colorants and antioxidants of ice cream. *Ann. Agric. Sci.* **2014**, *59*, 53–61. [CrossRef]
150. De Ancos, B.; Colina-Coca, C.; González-Peña, D.; Sánchez-Moreno, C. Bioactive compounds from vegetable and fruit by-products. *Biotechnol. Bioact. Compd. Sources Appl.* **2015**, 1–36. [CrossRef]
151. Chaitanya Lakshmi, G. Food coloring: The natural way. *Res. J. Chem. Sci.* **2014**, *2231*, 606X.
152. Ajila, C.; Leelavathi, K.; Rao, U.P. Improvement of dietary fiber content and antioxidant properties in soft dough biscuits with the incorporation of mango peel powder. *J. Cereal Sci.* **2008**, *48*, 319–326. [CrossRef]
153. Tung, Y.-C.; Chang, W.-T.; Li, S.; Wu, J.-C.; Badmeav, V.; Ho, C.-T.; Pan, M.-H. Citrus peel extracts attenuated obesity and modulated gut microbiota in mice with high-fat diet-induced obesity. *Food Funct.* **2018**. [CrossRef]
154. Ghasemi, K.; Ghasemi, Y.; Ebrahimzadeh, M.A. Antioxidant activity, phenol and flavonoid contents of 13 citrus species peels and tissues. *Pak. J. Pharm. Sci.* **2009**, *22*, 277–281.
155. Siegrist, M.; Sütterlin, B. Importance of perceived naturalness for acceptance of food additives and cultured meat. *Appetite* **2017**, *113*, 320–326. [CrossRef]
156. Vojdani, A.; Vojdani, C. Immune reactivity to food coloring. *Altern. Ther.* **2015**, *21*, 1–100.
157. Bézu, C.; Coutant, C.; Salengro, A.; Daraï, E.; Rouzier, R.; Uzan, S. Anaphylactic response to blue dye during sentinel lymph node biopsy. *Surg. Oncol.* **2011**, *20*, e55–e59. [CrossRef]
158. Rodriguez-Amaya, D.B. Natural Food Pigments and Colorants. In *Bioactive Molecules in Food*; Mérillon, J.-M., Ramawat, K.G., Eds.; Springer International Publishing: Cham, Switzerland, 2017; pp. 1–35.
159. Gordon, H.T.; Bauernfeind, J.C.; Furia, T.E. Carotenoids as food colorants. *Crit. Rev. Food Sci. Nutr.* **1983**, *18*, 59–97. [CrossRef]
160. Wissgott, U.; Bortlik, K. Prospects for new natural food colorants. *Trends Food Sci. Technol.* **1996**, *7*, 298–302. [CrossRef]
161. Prata, E.R.B.A.; Oliveira, L.S. Fresh coffee husks as potential sources of anthocyanins. *LWT-Food Sci. Technol.* **2007**, *40*, 1555–1560. [CrossRef]
162. European Commission. *Commission Regulation (EU) No 1129/2011 of 11 November 2011 amending Annex II to Regulation (EC) No 1333/2008 of the European Parliament and of the Council by Establishing a Union List of Food Additives Text with EEA Relevance*; European Commission: Brussels, Belgium, 2011.
163. Gesundheitsamt, S.E.; Commission du Manuel Suisse des Denrées Alimentaires. *Manuel Suisse des Denrées Alimentaires: Méthodes d'Analyse et d'Appréciation des Denrées Alimentaires et des Objets Usuels*; Office Central Fédéral des Imprimés et du Matériel: Berne, Switzerland, 1969.
164. Carocho, M.; Barreiro, M.F.; Morales, P.; Ferreira, I.C. Adding molecules to food, pros and cons: A review on synthetic and natural food additives. *Compr. Rev. Food Sci. Food Saf.* **2014**, *13*, 377–399. [CrossRef]
165. Kodagoda, K.; Marapana, R. Utilization of fruit processing by-products for industrial applications: A review. *Int. J. Food Sci. Nutr.* **2017**, *2*, 24–30.
166. Joshi, R.V. Low calorie biscuits from banana peel pulp. *J. Solid Waste Technol. Manag.* **2007**, *33*, 142–147.
167. Kanazawa, K.; Sakakibara, H. High Content of Dopamine, a Strong Antioxidant, in Cavendish Banana. *J. Agric. Food Chem.* **2000**, *48*, 844–848. [CrossRef]
168. Serna-Cock, L.; García-Gonzales, E.; Torres-León, C. Agro-industrial potential of the mango peel based on its nutritional and functional properties. *Food Rev. Int.* **2016**, *32*, 364–376. [CrossRef]

169. Nobre, B.P.; Palavra, A.F.; Pessoa, F.L.; Mendes, R.L. Supercritical CO_2 extraction of trans-lycopene from Portuguese tomato industrial waste. *Food Chem.* **2009**, *116*, 680–685. [CrossRef]
170. Wrolstad, R.E.; Culver, C.A. Alternatives to Those Artificial FD&C Food Colorants. *Annu. Rev. Food Sci. Technol.* **2012**, *3*, 59–77. [CrossRef]
171. Saha, D.; Bhattacharya, S. Hydrocolloids as thickening and gelling agents in food: A critical review. *J. Food Sci. Technol.* **2010**, *47*, 587–597. [CrossRef]
172. Pressman, P.; Clemens, R.; Hayes, W.; Reddy, C. Food additive safety: A review of toxicologic and regulatory issues. *Toxicol. Res.* **2017**, *1*, 1–22. [CrossRef]
173. Mudgil, D.; Barak, S. Composition, properties and health benefits of indigestible carbohydrate polymers as dietary fiber: A review. *Int. J. Biol. Macromol.* **2013**, *61*, 1–6. [CrossRef]
174. Müller-Maatsch, J.; Bencivenni, M.; Caligiani, A.; Tedeschi, T.; Bruggeman, G.; Bosch, M.; Petrusan, J.; Van Droogenbroeck, B.; Elst, K.; Sforza, S. Pectin content and composition from different food waste streams. *Food Chem.* **2016**, *201*, 37–45. [CrossRef]
175. Jeong, H.-S.; Kim, H.-Y.; Ahn, S.H.; Oh, S.C.; Yang, I.; Choi, I.-G. Optimization of enzymatic hydrolysis conditions for extraction of pectin from rapeseed cake (*Brassica napus* L.) using commercial enzymes. *Food Chem.* **2014**, *157*, 332–338. [CrossRef]
176. Espírito-Santo, A.P.; Lagazzo, A.; Sousa, A.L.O.P.; Perego, P.; Converti, A.; Oliveira, M.N. Rheology, spontaneous whey separation, microstructure and sensorial characteristics of probiotic yoghurts enriched with passion fruit fiber. *Food Res. Int.* **2013**, *50*, 224–231. [CrossRef]
177. Yi, T.; Huang, X.; Pan, S.; Wang, L. Physicochemical and functional properties of micronized jincheng orange by-products (*Citrus sinensis* Osbeck) dietary fiber and its application as a fat replacer in yogurt. *Int. J. Food Sci. Nutr.* **2014**, *65*, 565–572. [CrossRef]
178. Zhang, T.; McCarthy, J.; Wang, G.; Liu, Y.; Guo, M. Physiochemical Properties, Microstructure, and Probiotic Survivability of Nonfat Goats' Milk Yogurt Using Heat-Treated Whey Protein Concentrate as Fat Replacer. *J. Food Sci.* **2015**, *80*, M788–M794. [CrossRef]
179. Saffon, M.; Richard, V.; Jiménez-Flores, R.; Gauthier, F.S.; Britten, M.; Pouliot, Y. Behavior of Heat-Denatured Whey: Buttermilk Protein Aggregates during the Yogurt-Making Process and Their Influence on Set-Type Yogurt Properties. *Foods* **2013**, *2*, 444–459. [CrossRef]
180. Sundar Raj, A.; Rubila, S.; Jayabalan, R.; Ranganathan, T. A review on pectin: Chemistry due to general properties of pectin and its pharmaceutical uses. *Sci. Rep.* **2012**, *1*, 550–551.
181. Crizel, T.D.M.; Araujo, R.R.D.; Rios, A.D.O.; Rech, R.; Flôres, S.H. Orange fiber as a novel fat replacer in lemon ice cream. *J. Food Sci. Technol.* **2014**, *34*, 332–340. [CrossRef]
182. M'hiri, N.; Ghali, R.; Nasr, I.B.; Boudhrioua, N. Effect of different drying processes on functional properties of industrial lemon byproduct. *Process Saf. Environ. Prot.* **2018**, *116*, 450–460. [CrossRef]
183. Beecher, G.R. Phytonutrients' Role in Metabolism: Effects on Resistance to Degenerative Processes. *Nutr. Rev.* **1999**, *57*, 3–6. [CrossRef]
184. Desmedt, A.; Jacobs, H. Soluble fiber. In *Guide to Functional Foods*; Young, J.E., Ed.; Leatherhead Publishing: Leatherhead, UK, 2001; pp. 112–132.
185. Santos, B.A.; Campagnol, P.C.B.; Pacheco, M.T.B.; Pollonio, M.A.R. Fructooligosaccharides as a fat replacer in fermented cooked sausages. *Int. J. Food Sci. Technol.* **2012**, *47*, 1183–1192. [CrossRef]
186. Brennan, C.S.; Tudorica, C.M. Carbohydrate-based fat replacers in the modification of the rheological, textural and sensory quality of yoghurt: Comparative study of the utilisation of barley beta-glucan, guar gum and inulin. *Int. J. Food Sci. Technol.* **2008**, *43*, 824–833. [CrossRef]
187. Radi, M.; Niakousari, M.; Amiri, S. Physicochemical, Textural and Sensory Properties of Low-Fat Yogurt Produced by Using Modified Wheat Starch as a Fat Replacer. *J. Appl. Sci. Res.* **2009**, *9*, 2194–2197.
188. Raftani, A.Z.; Mahmudi, M.; Alimi, M. Effect of maltodextrin as a fat replacer on the quality of non-fat yogurt. *J. Food Res.* **2013**, *23*, 133–142.
189. Crizel, T.D.M.; Jablonski, A.; Rios, A.D.O.; Rech, R.; Flôres, S.H. Dietary fiber from orange byproducts as a potential fat replacer. *LWT-Food Sci. Technol.* **2013**, *53*, 9–14. [CrossRef]
190. Dervisoglu, M.; Yazici, F. Note. The Effect of Citrus Fibre on the Physical, Chemical and Sensory Properties of Ice Cream. *Int. J. Food Sci. Technol.* **2006**, *12*, 159–164. [CrossRef]

191. Guerrero García Ortega, M.; Riballo Ruiz-Roso, M.; Crespo Bermejo, A. Obtaining enriched fried tomato sauce using tomato byproducts. In *Proceedings of the XIV International Symposium on Processing Tomato 1159*; ISHS: Leuven, Belgium, 2017; pp. 167–174.
192. Nussinovitch, A.; Hirashima, M. *More Cooking Innovations: Novel Hydrocolloids for Special Dishes*; CRC Press: Boca Raton, FL, USA, 2018.
193. Lee, S.; Inglett, G.E.; Palmquist, D.; Warner, K. Flavor and texture attributes of foods containing β-glucan-rich hydrocolloids from oats. *LWT-Food Sci. Technol.* **2009**, *42*, 350–357. [CrossRef]
194. Domagała, J.; Sady, M.; Grega, T.; Bonczar, G. Rheological properties and texture of yoghurts when oat-maltodextrin is used as a fat substitute. *Int. J. Food Prop.* **2006**, *9*, 1–11. [CrossRef]
195. Ahmad, A.; Munir, B.; Abrar, M.; Bashir, S.; Adnan, M.; Tabassum, T. Perspective of β-glucan as functional ingredient for food industry. *J. Nutr. Food Sci.* **2012**, *2*, 133. [CrossRef]
196. Campos-Vega, R.; Nieto-Figueroa, K.H.; Oomah, B.D. Cocoa (*Theobroma cacao* L.) pod husk: Renewable source of bioactive compounds. *Trends Food Sci. Technol.* **2018**, *81*, 172–184. [CrossRef]
197. Ralla, T.; Salminen, H.; Edelmann, M.; Dawid, C.; Hofmann, T.; Weiss, J. Oat bran extract (*Avena sativa* L.) from food by-product streams as new natural emulsifier. *Food Hydrocoll.* **2018**, *81*, 253–262. [CrossRef]
198. Everett, D.W.; McLeod, R.E. Interactions of polysaccharide stabilisers with casein aggregates in stirred skim-milk yoghurt. *Int. Dairy J.* **2005**, *15*, 1175–1183. [CrossRef]
199. Aryana, K.; Plauche, S.; Rao, R.; McGrew, P.; Shah, N. Fat-free plain yogurt manufactured with inulins of various chain lengths and Lactobacillus acidophilus. *J. Food Sci.* **2007**, *72*, M79–M84. [CrossRef]
200. Sahan, N.; Yasar, K.; Hayaloglu, A.A. Physical, chemical and flavour quality of non-fat yogurt as affected by a β-glucan hydrocolloidal composite during storage. *Food Hydrocoll.* **2008**, *22*, 1291–1297. [CrossRef]
201. Mistry, V.V.; Hassan, H.N. Manufacture of Nonfat Yogurt from a High Milk Protein Powder. *J. Dairy Sci.* **1992**, *75*, 947–957. [CrossRef]
202. Guzmán-González, M.; Morais, F.; Ramos, M.; Amigo, L. Influence of skimmed milk concentrate replacement by dry dairy products in a low fat set-type yoghurt model system. I: Use of whey protein concentrates, milk protein concentrates and skimmed milk powder. *J. Sci. Food Agric.* **1999**, *79*, 1117–1122. [CrossRef]
203. Lobato-Calleros, C.; Martínez-Torrijos, O.; Sandoval-Castilla, O.; Pérez-Orozco, J.P.; Vernon-Carter, E.J. Flow and creep compliance properties of reduced-fat yoghurts containing protein-based fat replacers. *Int. Dairy J.* **2004**, *14*, 777–782. [CrossRef]
204. Sandoval-Castilla, O.; Lobato-Calleros, C.; Aguirre-Mandujano, E.; Vernon-Carter, E.J. Microstructure and texture of yogurt as influenced by fat replacers. *Int. Dairy J.* **2004**, *14*, 151–159. [CrossRef]
205. Bryant, C.M.; McClements, D.J. Molecular basis of protein functionality with special consideration of cold-set gels derived from heat-denatured whey. *Trends Food Sci. Technol.* **1998**, *9*, 143–151. [CrossRef]
206. Wang, W.; Bao, Y.; Hendricks, G.M.; Guo, M. Consistency, microstructure and probiotic survivability of goats' milk yoghurt using polymerized whey protein as a co-thickening agent. *Int. Dairy J.* **2012**, *24*, 113–119. [CrossRef]
207. Lin, Y.; Kelly, A.L.; O'Mahony, J.A.; Guinee, T.P. Fortification of milk protein content with different dairy protein powders alters its compositional, rennet gelation, heat stability and ethanol stability characteristics. *Int. Dairy J.* **2016**, *61*, 220–227. [CrossRef]
208. Bruzantin, F.; Daniel, J.; da Silva, P.; Spoto, M. Physicochemical and sensory characteristics of fat-free goat milk yogurt with added stabilizers and skim milk powder fortification. *J. Dairy Sci.* **2016**, *99*, 3316–3324. [CrossRef]
209. Jiménez-Flores, R.; Brisson, G. The milk fat globule membrane as an ingredient: Why, how, when? *Dairy Sci. Technol.* **2008**, *88*, 5–18. [CrossRef]
210. Singh, H. The milk fat globule membrane—A biophysical system for food applications. *Curr. Opin. Colloid Interface Sci.* **2006**, *11*, 154–163. [CrossRef]
211. Mistry, V.V.; Metzger, L.E.; Maubois, J.L. Use of Ultrafiltered Sweet Buttermilk in the Manufacture of Reduced Fat Cheddar Cheese [1]. *J. Dairy Sci.* **1996**, *79*, 1137–1145. [CrossRef]
212. Turcot, S.; St-Gelais, D.; Turgeon, S.L. Affinage de fromages allégés de type Cheddar fabriqués à partir de laits enrichis en phospholipides. *Le Lait* **2002**, *82*, 209–223. [CrossRef]
213. Kinsella, J. Relationships between structure and functional properties of food proteins. *Food Proteins* **1982**, *1*, 51–103.

214. Yavuz, H.Z.; Ozcan, M.M.; Lemiasheuski, V.K. The Effect of Some Chemical Additives on the Foaming Performance of the Pasteurized Liquid Egg White. *J. Nutr. Metab.* **2018**, *5*, 1054.
215. Mita, T.; Ishida, E.; Matsumoto, H. Physicochemical studies on wheat protein foams. II. Relationship between bubble size and stability of foams prepared with gluten and gluten components. *J. Colloid Interface Sci.* **1978**, *64*, 143–153. [CrossRef]
216. Kralova, I.; Sjöblom, J. Surfactants Used in Food Industry: A Review. *J. Dispers. Sci. Technol.* **2009**, *30*, 1363–1383. [CrossRef]
217. Johnson, T.M.; Zabik, M.E. Ultrastructural examination of egg albumen protein foams. *J. Food Sci.* **1981**, *46*, 1237–1240. [CrossRef]
218. Karayannakidis, P.D.; Zotos, A. Fish processing by-products as a potential source of gelatin: A review. *J. Aquat. Food Prod. Technol.* **2016**, *25*, 65–92. [CrossRef]
219. Muzaifa, M.; Safriani, N.; Zakaria, F. Production of protein hydrolysates from fish by-product prepared by enzymatic hydrolysis. *Aquac. Aquar. Conserv. Legis.* **2012**, *5*, 36–39.
220. Kotlar, C.E.; Ponce, A.G.; Roura, S.I. Improvement of functional and antimicrobial properties of brewery byproduct hydrolysed enzymatically. *LWT-Food Sci. Technol.* **2013**, *50*, 378–385. [CrossRef]
221. Hsieh, Y.; Ofori, J.A. Food-grade proteins from animal by-products. Their usage and detection methods. In *Handbook of Analysis of Edible Animal by-Products*; CRC Press: Boca Raton, FL, USA, 2011; pp. 3–11.
222. Kang, I.S.; Lanier, T.C. Bovine Plasma Protein Functions in Surimi Gelation Compared with Cysteine Protease Inhibitors. *J. Food Sci.* **1999**, *64*, 842–846. [CrossRef]
223. Yousif, A.M.; Cranston, P.; Deeth, H.C. Incorporation of bovine dry blood plasma into biscuit flour for the production of pasta. *LWT-Food Sci. Technol.* **2003**, *36*, 295–302. [CrossRef]
224. Del Hoyo, P.; Rendueles, M.; Díaz, M. Effect of processing on functional properties of animal blood plasma. *Meat Sci.* **2008**, *78*, 522–528. [CrossRef]
225. Whitehurst, R.J. *Emulsifiers in Food Technology*; John Wiley & Sons: New York, NY, USA, 2008.
226. Ozturk, B.; McClements, D.J. Progress in natural emulsifiers for utilization in food emulsions. *Curr. Opin. Food Sci.* **2016**, *7*, 1–6. [CrossRef]
227. Imafidon, G.I.; Farkye, N.Y.; Spanier, A.M. Isolation, purification, and alteration of some functional groups of major milk proteins: A review. *Crit. Rev. Food Sci. Nutr.* **1997**, *37*, 663–689. [CrossRef]
228. Toldrà, M.; Lynch, S.A.; Couture, R.; Álvarez, C. Blood Proteins as Functional Ingredients. In *Sustainable Meat Production and Processing*; Elsevier: Amsterdam, The Netherlands, 2019; pp. 85–101.
229. Leoci, R. *Animal by-Products (ABPs): Origins, Uses, and European Regulations*; Universitas Studiorum S.r.l.: Mantova, Italy, 2014.

molecules MDPI

Review

Green Natural Colorants

Isabel Viera, Antonio Pérez-Gálvez and María Roca *

Food Phytochemistry Department, Instituto de la Grasa (CSIC), University Campus, Building 46, Carretera de Utrera km. 1, 41013 Sevilla, Spain; iviera@ig.csic.es (I.V.); aperez@ig.csic.es (A.P.-G.)
* Correspondence: mroca@ig.csic.es; Tel.: +34-954-611-550

Academic Editors: Lillian Barros and Isabel C. F. R. Ferreira
Received: 6 November 2018; Accepted: 30 November 2018; Published: 2 January 2019

Abstract: Although there is no legal and clear definition of the term "natural food colorant", the market trends, and consequently industrial and commercial interest, have turned to foods with added natural pigments. This progressive substitution of artificial colorants has faced chemical complications with some colors, with a lack of stable green hues being one of them. Several strategies have been applied for green color stabilization in processed foods, from the formation of metallochlorophylls to the microencapsulation of green pigments. However, at present, the utilization of green coloring foodstuffs, which are considered an ingredient in the EU, seems to be the more successful solution for the market. Besides those topics, the present review aims to clarify the current confusion between the different chlorophyll compounds that form part of the authorized green food colorants. In this sense, legislations from different countries are compared. Finally, and in line with current concerns, the knowledge gathered so far in relation to the absorption, distribution, metabolism and excretion of all green natural food colorants is reviewed.

Keywords: ADME; absorption; chlorophylls; chlorophyllin; green colorant; zinc-chlorophylls; copper-chlorophyll; coloring foodstuff; natural colorants; food colors

---

1. Introduction

The first moment of truth is a term coined by Procter & Gamble [1] to describe the 3–7-s gap that makes a customer select a product over the rest of its competitors. It has been established that color is responsible for 62–90% of the consumer's assessment [2]. This fact makes expertise in food colorants a very profitable activity. Specifically, some market research companies have prognosticated a global food colors market size of USD 2.97 billion by 2025, with an estimated USD 1.79 billion in 2016 [3]. In this sense, consumer claims are responsible for the progressive strengthening of the food industry. We want food products to be more delightful, nutritive, attractive and healthy, with "fun-foods" as the maximum expression of this appeal [4]. The response to this worldwide trend is the unavoidable application of food colorants.

Before going further in this review, several concepts should be defined. Colorants can be classified following different parameters [5]. For example, in the USA, food colorants are categorized into whether they require or not the batch certification carried out by the US Food and Drug Administration (FDA). The list with the approved colorants that require certification is published in Title 21 CFR (Code of Federal Regulations) 74, where a Foods, Drugs and Cosmetics (FD & C) number is assigned. Hence, the certified colors commonly defined as synthetic could be classified as dyes (food-grade water soluble colorant [6]) or lakes (oil-dispersible; generally, dye extended on alumina). On the other hand, the list of the colorants exempted from certification is published in the Title 21 CFR 73 document [7]. Colorants within this category are mainly natural, although a few of them are produced by synthesis, but considered "nature-identical". On the other hand, authorized food colorants in the

EU are mainly legislated by the Regulation (EU) No. 1333/2008 [8], which is amended by Regulation (EC) No. 1129/2011. EU legislation includes indistinctly natural or artificial colorants.

Food colorants can be allocated following their origin (vegetal, animal, bacterial, fungal, etc.), their hue (red, yellow, purple, blue, green, etc.) or their chemical structure: flavonoid derivatives (anthocyanins), isoprenoid derivatives (carotenoids), nitrogen–heterocyclic derivatives (betalains), and the subject of this review, pyrrole derivatives or chlorophylls, which are responsible for blue and green hues.

At this point, it is relevant to distinguish between natural colorants and artificial ones. Although we could have an intuitive and clear conception about what each term means, the truth is that there is no official or definitive definition of what a natural or an artificial colorant is [9]. Only natural flavorings, another class of food additives, have a specific definition given by the FDA and EU [10]. Although there have been suggestions to extend the distinction of natural flavorings to food colorants, it seems difficult to reach a consensus between the different interested parties. In fact, neither in the US nor in the EU is there a recognized legal advertisement for "natural color" [11]. Indeed, the limits could be even more ambiguous if we consider, for example, the food colorant copper chlorophyllin, which is considered as natural. As will be noted below, this colorant is extracted from a natural source (generally from edible green leaves), but its manufacture requires additional chemical batch processing. Consequently, the question is as follows: where do we set the limit between natural and synthetic colorants? However, even realizing the hurdle of setting a frontier between natural and artificial colorants, the consumer increasingly demands more information, transparency, naturalness, clarity and trustworthiness in food label specifications. Thus, it is evident that there is an increasing global trend towards the natural side of foods.

Although previous studies linked the consumption of artificial food colorants with behavioral disorders, the inflection point was probably the well-known "Southampton study" [12]. This randomized, double-blinded trial test was performed with 137 3-year-old and 130 8/9-year-old children and concluded that the intake of dietary artificial food colorants increased the hyperactivity of children. Since then, many other research works and meta-analysis studies have confirmed and even amplified the disorders which probably originated from artificial food colors, although the precise physiological mechanism is unknown to date. In any case, the food industry and food legislative authorities have implemented some actions. The consumers' concern about the safety of artificial food colorants, reinforced by the possible health benefits of natural pigments, have induced the food industry to withdraw artificial colorants. In fact, artificial food colorants represent only 16% in the EU and 29% in the North America of the food colorants portfolio [13].

The progressive substitution of synthetic colorants by natural ones faces the challenge of the low stability of some colorants. Probably, green is one of the most complicated colors to both naturally set-up and counteract. The aim of the present review is to specifically address the current status of natural green food colorants, including the legal issues, the chemistry of chlorophylls and the existing alternatives for the stabilization of green color in foods, as well as the acclaimed "coloring foodstuff". For a general study of other food colorants, high-quality reviews have recently been published [4,10,11,13–17].

2. Legal Aspects on Green Natural Colorants

As has been noted elsewhere [16], one of the main problems affecting food additives is the lack of a globally harmonized legislation. This fact creates problems for wholesalers in this globalized food market. Even the concepts "natural" and "artificial additives" are country-dependent. Consequently, although this review only deals with the natural green colorants, whose main legal considerations are firstly described, a brief outline also shows the authorized synthetic green food colorants in different countries at the end of this section. Generally, the regulations of food additives include the list of authorized standards, their specifications, as well as the conditions of use (limitations on specific foods and maximum amounts). In addition, other requirements that should be considered are the schemes through which authorization is obtained and the conditions for packaging and labelling, which are different in each country.

European current legislation (Regulation (EC) No 1333/2008 and its amendments) [8] has allowed the use of two natural green colorants (Table 1), E140 and E141, which are structurally related with the chlorophylls, since the first European food colorant legislation was published in 1962 (Council Directive 62/2645/EEC). E140 comprises direct chlorophyll derivative extraction with organic solvent from natural sources: grass, alfalfa, nettles, spinach and other edible plant materials (Figure 1). The end-product could contain other lipids, pigments and waxes, so that the final aspect of the colorant is waxy, and it is marketed according to its solubility. Thus, the E140i or chlorophyll derivative is lipid-soluble, while the alternative for water-soluble foods (marketed as a powder) is E140ii (or chlorophyllins), obtained by the saponification of the solvent-extracted yield from edible plant material. Saponification breaks the ester–phytol bond, as will be detailed below, increasing the polarity of the derived products. However, both colorants are rather unstable, and the color is prone to experience a drastic change from green to brown. Therefore, the food industry favors the use of the colorant E141, which results from the addition of copper to the corresponding lipid or water-soluble chlorophyll solutions. The insertion of copper into the chlorophyll molecule stabilizes the structure, and the green coloration does not change, independent from the processing conditions or the storage time of the colored food. E141 is mainly composed of copper complexes of chlorophyll derivatives, with E141i (or copper chlorophylls) being the lipid-soluble option and E141ii (copper chlorophyllins) the water-soluble alternative [18].

Table 1. Classification of authorized natural green colorants according to different regulations.

Country	Chlorophyll	Chlorophyllin	Cu-Chlorophyll	Cu-Chlorophyllin	Na-Fe-Chlorophyllin
EU	E140i	E140ii	E141i	E141ii	
USA				73.125	
Japan	177	116	266	265	257
India	6				
China			08.153	08.009	
CA [1]	INS 140	INS 140	INS 141i	INS 141ii	

[1] CA: Codex Alimentarius.

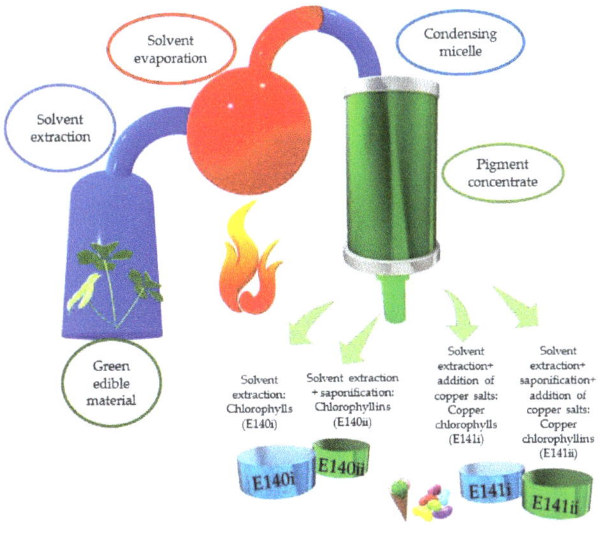

Figure 1. Scheme of the natural green colorant manufacture process.

In the USA, the regulation of natural food colorants is published in Title 21 CFR 73 [7]. Only copper chlorophyllin (Table 1) is authorized as a natural green colorant (CFR Section 73.125) and only for coloring citrus-based dry beverage mixes, the amount being limited to 0.2% in the dry mix. Current legislation in China [19] recognizes two green colorants: copper chlorophylls (CNS 08.153) and copper–sodium chlorophyllin (CNS 0.009), which is only allowed for use in determined food categories and with fixed maximum levels. At any rate, the China National Center for Food Safety Risk Assessment (CFSA) has published at the beginning of 2018 the first draft for a National Food Safety Standard for the Use of Food Additives, which revises and updates the current version issued in 2014. In Japan, the list of food additives from natural origin was compiled and published by the Ministry of Health and Welfare on 16 April 1996 [20]; in 2018, this accounts for 365 items, with chlorophyll (number 117 in the original Japanese list) and chlorophyllin (116) as the only authorized natural green colorants. In addition, Japanese legislation includes 455 designated food additives to date (obtained by chemical reactions), which include copper chlorophyll (266), sodium–copper chlorophyllin (265) and sodium–iron chlorophyllin (257). In both regulations, the target foods, the maximum limits and other requirements are detailed for each colorant. In India, the regulations are published in the Food Safety and Standards (Food Products Standards and Food Additives) Regulations (2011), which depend on the Food Safety and Standards Authority of India [21]. They only allow the use of chlorophyll as a coloring food matter which has the specification of having both chlorophyll *a* and chlorophyll *b*.

Finally, the Joint Food and Agriculture Organization of the United Nations/World Health Organization (FAO/WHO) Expert Committee on Food Additives (JECFA), as the international body responsible for evaluating the safety of food additives, has published the last version of the Codex Alimentarius general standard of food additives (41st meeting on 2018) [22]. In accordance with European legislation, chlorophylls (INS 140) and copper chlorophylls (INS 141) are registered. It is worthwhile to visit the corresponding websites to have access not only to the specification of each food additive, but also to the complete monographs and evaluations developed for each one.

Along with the colorants generally considered as natural, the legislations also authorize artificial green food colorants (Table 2), green S and fast green FCF, with both compounds being triarylmethane derivatives (Figure 2). In the EU, only green S is legal, under the code E142, while its use is banned in countries such as the USA, Japan, India, and China. It is also recognized by the Codex Alimentarius with the reference number INS 142. This colorant presents the molecular formula $C_{27}H_{25}N_2NaO_7S_2$ and a molecular weight equal to 576.63 Da (CAS number 3087-16-9). On the other hand, in the USA (along with other countries, Table 2), the authorized artificial green colorant is the so-called fast green FCF, a compound with the molecular formula $C_{37}H_{34}N_2O_{10}S_3Na_2$ and a molecular weight equal to 808.85 Da (CAS number 2353-45-9; PubChem CID 73557432), while its use is banned in the EU.

Table 2. Classification of authorized artificial green colorants by different legislations.

Denominations	Green S	Fast Green FCF
CI [1]	44090	42053
EU	E142	-
USA	-	FD&C green No. 3
Japan	-	Food green No. 3
India	-	Fast green FCF
China	-	-
CA	INS 142	INS 143

[1] The Color Index™ is a classification system for dyes and pigments globally used by manufacturers, researchers and users of dyes and pigments (https://colour-index.com/).

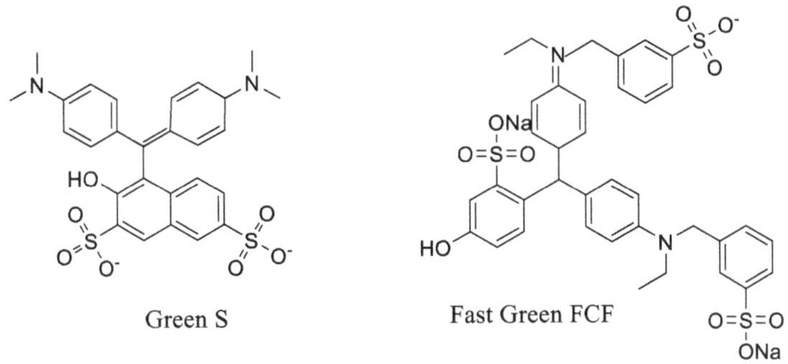

Figure 2. Structures of green S and fast green FCF.

3. Chlorophyll Derivatives in the Authorized Natural Green Colorants

Several chemical (conventional) terms have been commonly confused in the commercial and the industrial environments to tag chlorophyll colorants. We will explicitly detail the chlorophyll compounds, when described, that make up each natural green colorant, along with the conventional terms. To avoid misunderstanding in following the statements, Figure 3 depicts the skeleton of the chlorophyll structure showing the different substituent functions and the numbering system. The difference between chlorophyll *a* and chlorophyll *b* is located at C7, where chlorophyll *a* presents a methyl group while chlorophyll *b* includes a formyl group (Table 3). The central coordinated atom is magnesium in the chlorophylls, while two hydrogen atoms are present in pheophytins and pheophorbides. Otherwise, the phytyl chain ($C_{20}H_{40}$) is ester-bonded to the $C17^3$, which confers the lipophilic properties when present in non-polar derivatives: chlorophylls and pheophytins (Table 3). On the opposite set, the phytyl chain is absent in polar derivatives: chlorophyllides and pheophorbides. Lastly, the isocyclic ring (ring V) presents a carboxymethoxy arrangement at $C13^2$. The lack of this group yields the pyroderivatives. Additional unprecedented chlorophyll structures could be present in the colorants [23].

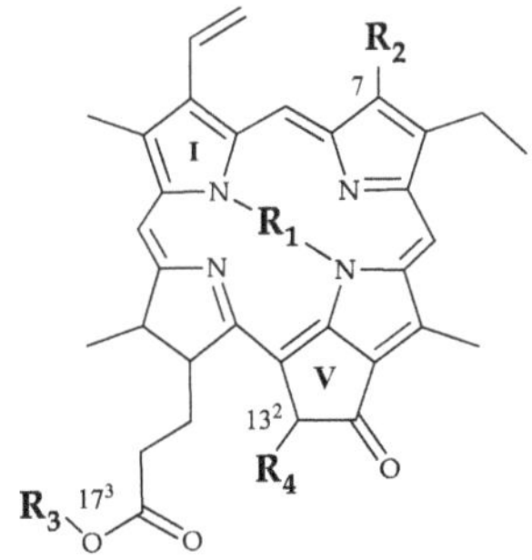

Figure 3. Chlorophyll skeleton with the different substituents.

The analysis of the chlorophyll derivatives contained in the green natural colorants started with the development of several chromatographic methods able to separate the different compounds in the mixtures [24,25]. The details of those methods have been recently compiled [23]. Due to the chromatic properties of chlorophylls, DAD is the main detector used, although Raman spectroscopy was also proposed [26]. However, the similarity of the UV-Vis spectrum between several chlorophyll derivatives, made essential the utilization of mass spectrometry to properly identify the different substituents of the chlorophyll compounds [27,28]. Details of the mass spectrometry characteristics of the chlorophyll compounds is not very abundant, although several studies have been developed [23,27].

The colorant named "chlorophyll" or "E140i", as it is marketed after direct solvent extraction from edible green plant materials, is mainly composed of chlorophyll *a* and *b*, along with their corresponding pheophytin derivatives, which are originated during the extraction at low pH, which catalyzes the substitution of the central magnesium by hydrogens (Figure 4). The relative proportion of chlorophyll to pheophytin present in the colorant depends on the conditions of the manufacturing practices, and it is highly variable between the different suppliers. In addition, the chlorophylls present in the colorant product are very labile, and the chemical conditions during the food processing accelerate the degradative process. The transformation causes a change in color from green to brown; a collateral effect which is not desired by the food industry.

Table 3. Main chlorophyll derivatives in function of their substituents.

Chlorophyll Compound	R_1	R_2	R_3	R_4
Chlorophyll *a*	Mg	CH_3	phytyl	$COOCH_3$
Chlorophyll *b*	Mg	CHO	phytyl	$COOCH_3$
Chlorophyllide *a*	Mg	CH_3	H	$COOCH_3$
Pheophytin *a*	H	CH_3	phytyl	$COOCH_3$
Pheophytin *b*	H	CHO	H	$COOCH_3$
Pheophorbide *a*	H	CH_3	H	$COOCH_3$
Pyropheophytin *a*	H	CH_3	phytyl	H

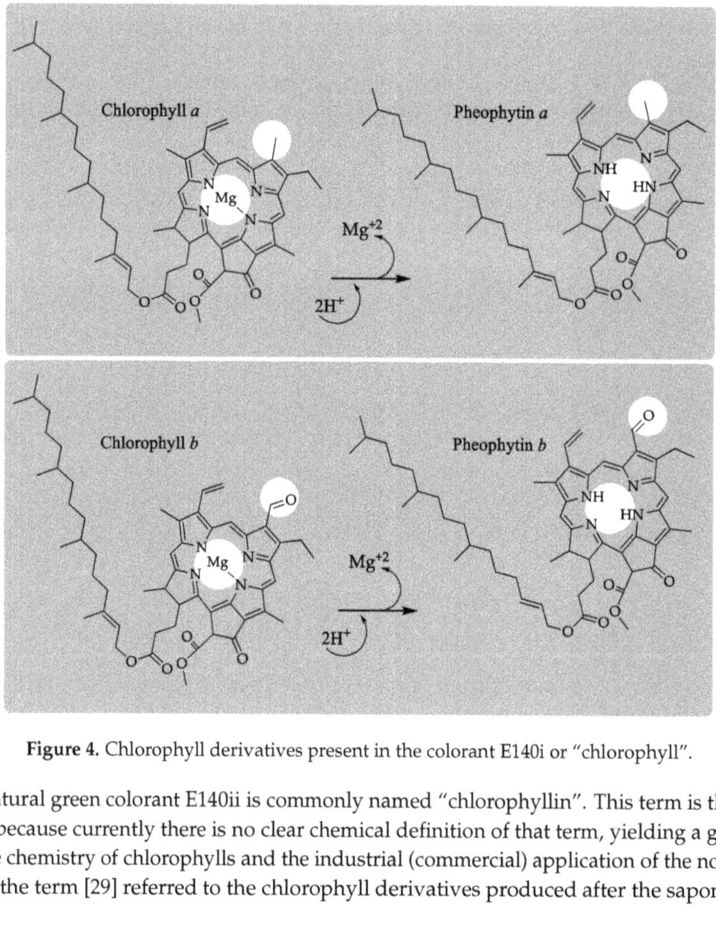

Figure 4. Chlorophyll derivatives present in the colorant E140i or "chlorophyll".

The natural green colorant E140ii is commonly named "chlorophyllin". This term is the source of confusion, because currently there is no clear chemical definition of that term, yielding a gap between the genuine chemistry of chlorophylls and the industrial (commercial) application of the nomenclature. Originally, the term [29] referred to the chlorophyll derivatives produced after the saponification of chlorophyll without a change in color. At that stage, when the definitive structure of chlorophyll was not described yet, chlorophyllin exclusively comprised those chlorophyll derivatives arising after the phytol ester breakdown. This means that chlorophyllins keep the central magnesium ion intact. Later, this term increased its application to other chlorophyll derivatives produced during the saponification; i.e., those arising from the cleavage of the isocyclic ring. However, the exact chemical definition of chlorophyllins covers those chlorophyll derivatives with an intact central magnesium ion. Therefore, the present commercial (industrial and sometimes legislative) use of the term "chlorophyllins" is incorrect, because the colorant products include chlorophyll structures without magnesium [30].

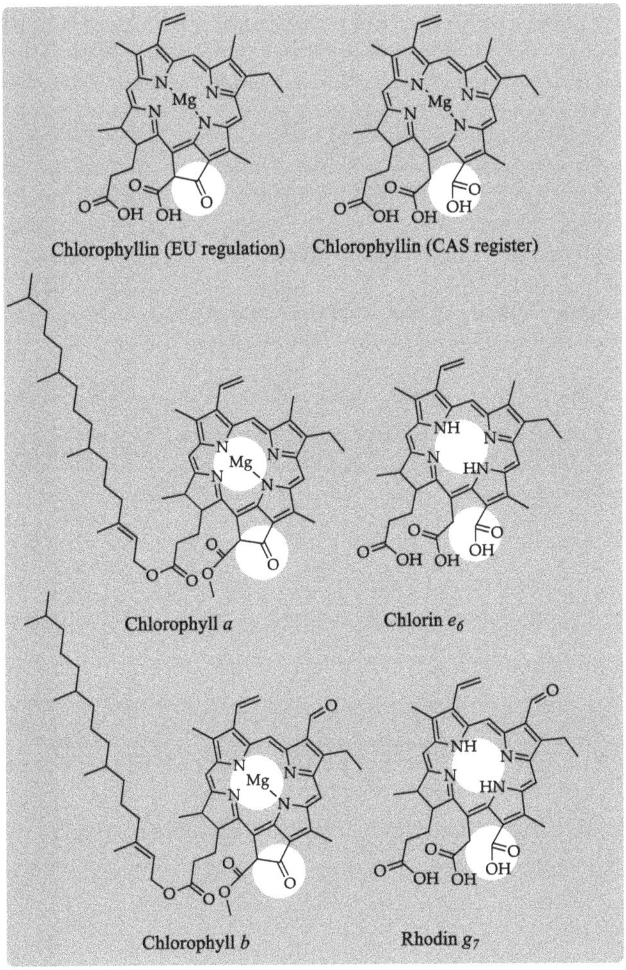

Figure 5. Chlorophyll derivatives present in the colorant E140ii or "chlorophyllins".

Figure 5 describes some of the different chlorophyll structures that have been assigned to the "chlorophyllin *a*" denomination. For example, EU regulation [8] describes a structure with elemental composition $C_{34}H_{32}MgN_4O_5$ and molecular weight 600.9467 Da that chemically corresponds to demethylated chlorophyllide *a*. This compound originates from chlorophyll after the breakdown of the phytyl ester bond at $C17^3$ (chlorophyllide *a*) and the fragmentation of the methoxy group at $C13^1$. Other examples are the compound described in the Chemical Abstract Service (CAS 15611-43-5) or in

PubChem (123798), which chemically corresponds with magnesium chlorin e_6. This structure with an elemental composition $C_{34}H_{34}MgN_4O_6$ and molecular weight 618.962 Da is formed from chlorophyll *a* after the fragmentation of the phytyl ester bond, loss of the carboxymethoxy group and opening of the ring V. However, during chlorophyll saponification, the reaction required for E140ii manufacturing, the central magnesium ion is replaced by two hydrogens. Consequently, the main chlorophyll compounds that form part of the colorant E140ii are the chlorin e_6 which originates from chlorophyll *a* and the rhodin g_7 formed from chlorophyll *b*, besides other chlorophyll derivatives such as pheophorbides.

The colorant E141i (Figure 6) is known as "copper chlorophyll", although it would be more accurate to label it as pheophytin with copper. The former nomenclature aims to denote the oily consistency of the colorant, which in chemical terms means that it is constituted by phytylated chlorophyll derivatives and, therefore, is lipid-soluble. This premise, in conjunction with the required copper treatment for its production, means that the main chlorophyll derivatives present in the colorant E141i are copper–pheophytin derivatives. In fact, an HPLC method was developed for the detection and quantification of copper chlorophylls *a* and *b* (copper pheophytins *a* and *b*) [25]. Later, several methods have been settled with the same aim [24,31]. At any rate, the complete characterization of several commercial E141i samples showed not only differences in the chlorophyll profile between them, but also that the common main compound present in all samples is copper–pyropheophytin *a* [32]. The MS characterization of this chlorophyll derivative has allowed us to propose a specific product ion as a probe for tracking the presence of the E141i food where its use is not authorized, particularly in olive oil [28]. Other chlorophyll derivatives that could be present in this colorant are copper–pheophytin *a*, copper–pheophytin *b*, copper–pyropheophytin *b*, and some others in trace amounts [32].

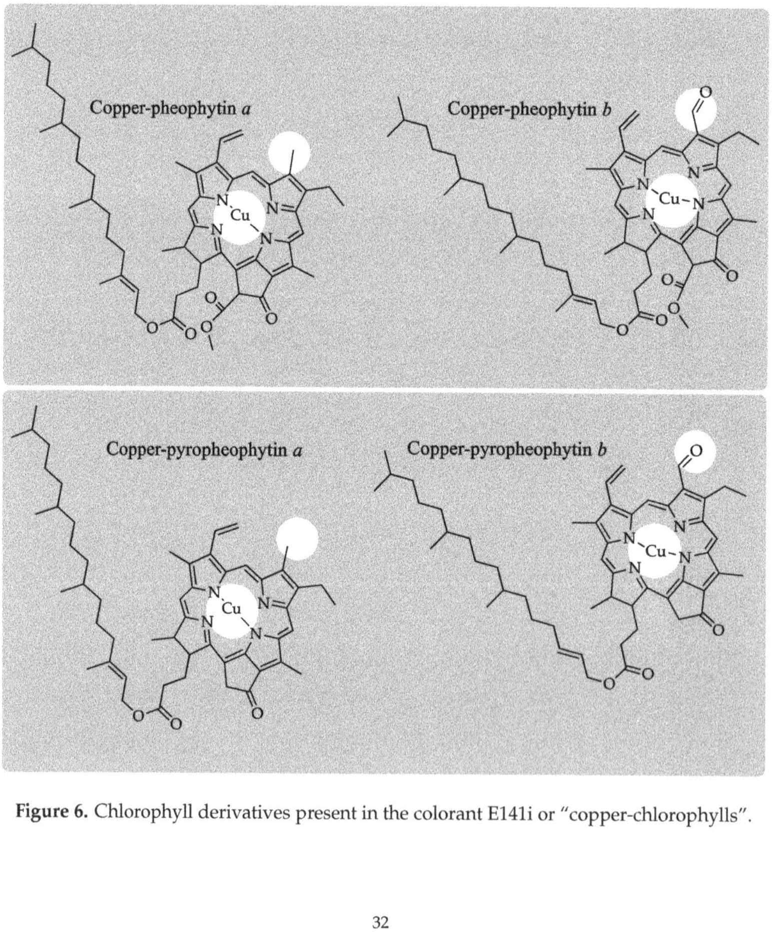

Figure 6. Chlorophyll derivatives present in the colorant E141i or "copper-chlorophylls".

The water-soluble green food colorant E141ii (Figure 7) has been the most studied as it is one of the most popular within the food industry. Its manufacture requires, besides solvent extraction, saponification and treatment with copper salts, which means a higher degree of transformation from chlorophylls. The first chlorophyll derivative identified in this colorant was the copper chlorin e_4 [33].

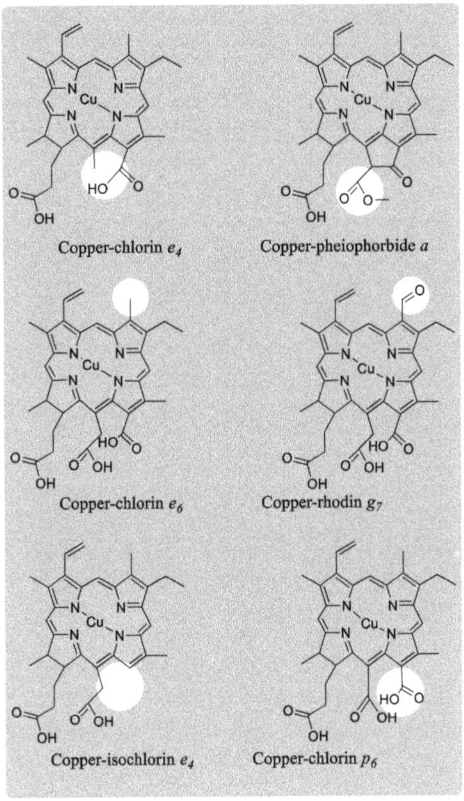

Figure 7. Chlorophyll derivatives present in the colorant E141ii or "copper-chlorophyllins".

Following this, three additional new chlorophyll derivatives (copper pheophorbide *a*, copper chlorin e_6 and copper rhodin g_7) were identified as the main components in addition to copper chlorin e_4, in the commercial "sodium copper chlorophyllin" (SCC), by comparison with authentic standards [34]. Authors determined not only absorption maxima but also molar extinction coefficients. Later, in five industrial samples of copper chlorophyllins, copper isochlorin e_4 was identified as the major chlorophyll derivative in most of the colorants [35], describing a high variability between commercial preparations. Different manuscripts have dealt with the presence of this colorant in processed foods [18,24,36–39]. At any rate, a further advance was the single HPLC-MS analysis published to date [27] of the chlorophyll derivatives present in several commercial colorants. The authors identified several new chlorophyll structures, such as Cu-chlorin p_6, although additional assignments were tentatively made. The lack of authentic standards and the fact that most of the substituents in the chlorophyll tetrapyrrole do not modify the features of the UV-Vis spectrum (chlorin/isochlorin, methyl/ethyl, ester or not, etc.) makes the correct identification of chlorophyll derivatives exclusively by their chromatographic properties difficult, or even leads to the misidentification of the chlorophyll compounds being possible.

4. In Vivo and In Vitro Adsorption, Distribution, Metabolism and Excretion (ADME)

In spite of the widespread presence of natural green colorants in foodstuffs, there is a general lack of information regarding their ADME. Although few studies have been developed either in vitro or in vivo with the aim of stablishing the ADME of E140i (chlorophylls) and E141ii ("copper chlorophyllin"), there are no data regarding the bioavailability of E140ii (chlorophyllins) and E141i (copper chlorophylls). Consequently, there are still several open questions about the in vivo ADME of natural green colorants.

4.1. ADME of E140i

At the in vitro level, the first study dealing with the digestion of green vegetables and subsequent absorption of the micellarized pigments by human intestinal Caco-2 cells was reported in 2001 [40]. The tested foods included fresh spinach puree, heat and acid-treated spinach puree, and spinach puree treated with $ZnCl_2$. The original chlorophylls were transformed into pheophytins during digestion as a consequence of the gastric pH. However, pheophytins with zinc were relatively stable during the simulated digestion. The authors demonstrated that around 5–10% of the micellarized pheophytins were absorbed by the cells and that micellarization and uptake changed significantly between the different kinds of chlorophyll derivatives. This was the first hint at determining that micellarization and absorption are greatly influenced by the lipophilicity of the compounds. Specifically, the more lipophilic molecules showed the highest accumulation in the intestinal cells.

Subsequently, these results were confirmed by applying the same in vitro protocol to pea preparations [41], pure standards of chlorophyll derivatives isolated from natural sources [42], and seaweeds [43]. Specifically, it has been shown in vitro that pheophorbides (dephytylated chlorophylls) are preferentially absorbed over pheophytins (phytylated chlorophylls) [42,43]. At this point, and by means of comparative absorption experiments with the Caco-2 cellular model at different concentrations of pheophytins, it was hypothesized that the absorption process could correlate with a passive facilitated mechanism [42]. On the contrary, to decipher the kind of cellular transport process for pheophorbides, Viera et al. [44] described the production of micelles rich in pheophorbide *a* to reach physiological micellar concentrations (7 μm). Pre-incubations of cell monolayers with different amounts of one specific inhibitor of the Scavenger Receptor class B type I (SR-BI) transporter (BLT1), significantly inhibited the uptake of pheophorbide *a*, which strongly suggests that SR-BI is involved in the transport of pheophorbide *a*. Consequently, the protein-mediated absorption of pheophorbide explains the preferential absorption of this chlorophyll derivative.

At the in vivo level, the pioneering studies on chlorophyll assimilation were based on the quantification of chlorophyll derivatives (E140i) in the feces of animals and by applying a balance matter (ingested against excreted) approach. Therefore, it has been historically assumed that chlorophylls were not absorbed by the organism, since almost all ingested chlorophylls were excreted. In this line, Brugsch and Sheard [45] estimated the quantitative decomposition of chlorophylls in the human body. Researchers supplied encapsulated crystalline chlorophyll to humans (100 mg per day for 4 days), showing that the highest percentage of degraded chlorophylls matched with fecal pheophytin [46]. Additional experiments [47] found scarce evidence of the subsequent hydrolysis of pheophytin to the water-soluble pheophorbide. The release of phytol by the colonic microflora in humans fed with pheophytin or spinach labelled with ^{14}C led to the conclusion that the main final digestive products of the ingested chlorophylls were the pheophytins. The significance of this finding lies in the fact that chlorophylls are the single source of phytanic acid, which accumulates in the Refsum disease [47].

After a great period of time, the in vivo absorption of chlorophyll derivatives was investigated in dogs fed with a diet containing 73 mg of chlorophyll (spinach)/kg of diet for 10 days [48], obtaining an apparent absorption of 3.4% of chlorophylls (determination of ingested against excreted). A second experiment was also carried out increasing the dose of chlorophyll; that is, the dogs consumed a diet containing 10% of dried spinach for 10 days. Under these conditions, no chlorophyll derivatives could be detected in the peripheral blood until 150 min after consumption, which suggests to the authors that chlorophylls were hardly absorbed and/or readily metabolized and excreted through the bile. In addition, chlorophylls *a* and *b* were transformed into their corresponding pheophytin derivatives in the gastrointestinal tract, and the authors concluded that beyond pheophytins, no other degradation products were produced.

However, opposite to the classical hypothesis, the in vivo ADME experiments of E140i developed with 30 mice fed with a diet supplemented with spirulina for four weeks [42] showed for the first time the existence of a first-pass chlorophyll metabolism in animals. The analyzed livers accumulated a diverse profile of chlorophyll derivatives, which were identified by HPLC-MS^2. The study highlighted that chlorophyll derivatives that retain the phytyl chain in their structure (apolar derivatives) are available for absorption from a dietary source and accumulate in the liver. Nevertheless, the explicit enrichment of the liver with pheophorbide *a* is particularly significant. Two possible mechanisms are proposed: that phytylated chlorophylls can be further metabolized in the liver to pheophorbide, or the existence of an intestinal transporter for this metabolite. If the pheophytin is de-esterified in the liver to yield pheophorbide, the authors reveal the enigmatic origin of phytol in the liver [49]. Indeed, as was mentioned above, the authors also presented data on the implication of the intestinal brush border transporter SR-BI in the absorption of pheophorbide. Therefore, independent of the exact mechanism, the present chlorophylls in the colorant E140i are at least absorbed, metabolized, accumulated and excreted in mammals.

4.2. ADME of E140ii and E141i

There are no scientific data regarding the in vitro or in vivo absorption, distribution, metabolism, excretion and toxicity of chlorophyllins (E140ii) or copper chlorophylls (E141i). This lack of relevant data was also revealed through different Scientific Opinions [50–52] by the European Food Safety Agency (EFSA). According to the conclusions of the panel of experts of the EFSA and in the sight of the great interest of business operators, the European Commission launched on October 2017 a call for scientific and technical data requested by EFSA to complete the risk assessment. The collected data within a three–four-year timeframe will be assessed by the EFSA, and later the European Commission will take the final decision on the status of the revised colorants [53].

4.3. ADME of E141ii

Following an experimental design similar to the analysis of the ADME of E140i, the Schwartz group [54] showed that part of the chlorophylls present in the E141ii colorant was absorbed through the human intestine. In detail, four chlorophyll derivatives—copper rhodin g_7, copper chlorin e_6, copper chlorin e_4 and copper pheophorbide *a*—from a commercial-grade sodium copper chlorophyllin were assayed. The copper chlorin e_4 (the main component of E141ii) was highly stable to the simulated in vitro digestion process while most of the copper chlorin e_6 was lost. However, the integration of the colorant in applesauce significantly improved the recovery of chlorin e_6, demonstrating the protective role of the food matrix. Moreover, copper chlorophyll derivatives were efficiently absorbed by the intestinal epithelium cells, probably through an active transport. Part of these chlorophyll compounds were even detectable in the basolateral compartment, which means that they are ready to be transported to peripheral tissues.

The first in vivo studies of E141ii were carried out by Henderson and Long in 1941 [55], who orally administered natural chlorophyll and SCC to rats, discovering the existence of uncharacterized derivatives dispersed throughout the liver, lymph nodes and the spleen. Reber and Willigan (in 1954) [56] obtained with their studies the first in vivo indications about the absorption of E141ii. Rats fed with 1% of copper chlorophyllins in their diet for 15 weeks showed, after euthanasia, a greenish tone throughout the skeletal muscle of the body, indicating a systemic distribution of the chlorophyll derivatives. In the same year [57], a study with different doses of copper chlorophyllin in the diet of rats resulted in the transport through the gastrointestinal membrane and accumulation in plasma of copper chlorophyllins in the μg range. The authors did not detect any copper derivative in the organs, and they assumed that copper chlorophyllins are excreted in the feces. Later, the chemopreventive activity against tumorigenesis of copper chlorophyllin was tested in female mice [58]. In this study, the sodium salt of copper chlorophyllin administered orally was rapidly distributed to the heart, liver, skin, kidneys, and lungs. A subsequent study estimated the accumulation of dietary E141ii (10 or 30 mg/kg) in different organs of 30 Wistar rats [59]. The results showed that while copper chlorin e_4 was effectively absorbed and accumulated in serum, liver and kidneys, copper chlorin e_6 was not detected by HPLC analysis in sera or tissues, according to the data published by Ferruzzi et al. [54].

Indirect evidence in humans suggests that any type of absorption could take place following a copper chlorophyllin diet. For example, urine discoloration has been described with incontinent patients subjected to an oral copper chlorophyllin intake (100–200 mg/day) [60,61]. However, the single evidence of the in vivo absorption and accumulation of E141ii in humans was provided by the studies of Egner et al. [62]. Thus, 182 volunteers ingested 100 mg of copper chlorophyllins for 16 weeks and three times per day (copper chlorin e_4, copper chlorin e_6 and copper chlorin e_4 ethyl ester). The study described for the first time the presence of copper chlorin e_4 ethyl ester as well as copper chlorin e_4, but not copper chlorin e_6, in the sera of all the participants. Therefore, certain components are able to be absorbed through the gastrointestinal membrane. Probably, the instability of copper chlorin e_6 to digestion could be responsible for the lack of appearance of this compound in human serum [63].

5. "Coloring Foodstuff"

The global trend of the replacement of synthetic colors with natural ones has created a new category in the market known as "coloring foodstuffs". This class includes food ingredients such as fruits or vegetables whose secondary effect is coloring. This new conception is bringing regulatory problems as these substances are not covered by the current regulation on food additives in the EU. As the situation is rather unclear and there is an ongoing debate on the distinction between color additives and coloring foodstuffs [16], the European Commission (Standing Committee on the Food Chain and Animal Health) endorsed the Guidance notes providing a tool for classification for when a substance should be considered a color additive or not [64]. The distinction in this guidance is based on the extraction methodology: if the method is a non-selective extraction procedure, then the obtained product is a food ingredient, not a food additive. On the contrary, if the extraction is selective for obtaining a pigment, the compound is considered to be a color additive and consequently covered by the regulation on food additives. Previous to this guidance, nettles and spinach were the preferred coloring foodstuff to provide green hues to foods. However, now these color solutions no longer satisfy the criteria of EU guidelines. Otherwise, blends with spirulina are the alternative to create brilliant shades of green for confectionary products and ice creams; for example, with safflower. Specifically, in the USA, spirulina derived from *Arthrospira platensis* has been recently added to the list of approved color additives exempt from certification in response to a Mars Inc. petition [16]. In relation to spirulina extract, the Joint FAO/WHO Expert Committee has requested, in the meeting held in June 2018, information on the products on the market by December 2019 in order to remove the tentative designation from the specifications. Specifically, it is necessary to provide data on the full compositional characterization of commercial products and regarding the validated analytical methods applied for the identification of the substance and for the determination of the purity of the substance.

In any case, there is a new growing sector in the market, with several different extracts sold by different suppliers. Table 4 list the coloring foodstuffs authorized in the EU and USA, describing the maximum amounts and the foods allowed. As they are considered "natural", to date, no amount limitations have been defined, and in the EU, as they are classified as ingredients, there are no restrictions for their use in any food category.

As a consequence of the natural trend in the coloring food market, recent research has developed looking for new strategies to develop new "natural" green colorants. For example, the utilization of an extract from the leaves of *Centella asiatica* L. after steaming and metal complexations has been proposed [15]. The authors assessed the stability and cytotoxicity in different beverages and food models as an alternative to synthetic colorants. Following the same aim, the properties of spray-dried microalgae have been analyzed (color, sensory and textural qualities) as a natural coloring agent in chewing gum [65]. Although positive results have been obtained with *Isochrysis galbana* and *Nannochloropsis oculata*, the spray-dryer technique requires an optimization in the base of the characteristics of each microalgae specie.

Table 4. Classification of "coloring foodstuff" in EU and USA.

	Vegetable Juice	Fruit Juice	Spirulina Extract
		European Union	
	+	+	+
Allowed foods	Foods in general	Foods in general	Foods in general
Maximum limit	*Quantum satis*	*Quantum satis*	*Quantum satis*
		United States	
Denomination	73,260	73,250	73,530
Allowed foods	Foods generally	Foods generally	Confections (including candy and chewing gum), frostings, ice cream and frozen desserts, dessert coatings and toppings, beverage mixes and powders, yogurts, custards, puddings, cottage cheese, gelatin, breadcrumbs, ready-to-eat cereals (excluding extruded cereals), coating formulations applied to dietary supplement tablets and capsules
Maximum limit	GMP [1]	GMP	GMP

[1] GMP: good manufacturing practices.

6. Stabilization of Green Color

Chlorophylls are stable pigments in their natural environment in physiological conditions. However, once extracted or processed with changes in pH value and temperature during the processing and storage of green foods, chlorophylls are prone to experience modifications in their structure, which consequently change their chromatic properties. Probably, the main reaction that affects chlorophylls is the substitution of the central magnesium ion by two hydrogens. The significance of the reaction for the food industry is related with the drastic change in color, because magnesium-derivatives are green, while magnesium-free derivatives (mainly pheophytins and pheorphorbides) are brown. Consequently, this easy and fast reaction is the principal cause of the loss of the original green color during the processing and storage of green foods. As has been stated before, the reduction or even withdrawal of the original green color is associated by the consumer with a decrease in the quality of the product. Therefore, the food industry has developed several strategies to preserve the initial green coloration. The early attempts consisted of the addition (even in-the-can coating) of alkalinizing agents, but the appearance of negative secondary effects such as a softening of the edible material or an off-flavor release [6] made them to fall into disuse.

A second historical approach has been the substitution of the central hydrogen atoms by zinc or copper ions to form the more stable green metallochlorophylls which consequently "re-green" the corresponding food product. The conditions for the production of zinc and copper complexes of chlorophyll derivatives in vegetables during processing have been optimized through the years [66–68]. Due to the industrial and commercial importance of this process, numerous patents have been published, with the most known being the so-called "Veri-green" method [69]. This patent was developed by the former Continental Can Company and consisted of the blanching of vegetables in brine solutions with some amounts of Zn^{+2} or Cu^{+2} salts to form mainly zinc or copper pheophytins [70] and make the processed vegetables greener. However, the patent was unproductive because the maximum limit for zinc concentrations established by the FDA is 75 ppm, and higher amounts of zinc are required to yield an acceptable and desirable green color. To overcome this limitation, new strategies for the encapsulation of zinc-chlorophylls are currently under development using different matrices such as gum Arabic, maltodextrin and OSA modified starch [71] or whey proteins [72]. Indeed, different techniques are being set up to microencapsulate raw chlorophyll extractions from edible vegetables without any salt treatment. For example, spray drying is a dehydration process which has been successfully applied to encapsulate other food compounds, and recently, the physicochemical properties of the process have been optimized to form whey protein isolate–kale leaf chlorophyll microcapsules [73]. A further step in the use of this procedure as an available alternative for the food industry is the enhancement of the stability of green pigments for heated foods. For example, the stability of alfalfa leaves microencapsulated towards temperature regimes and pH conditions has been determined [74], obtaining optimal conditions.

7. Conclusions

The market of the food colorants is evolving to more natural formulations, which means the appearance of new definitions and consequently, the necessity of new legislations. Green food colorants are one of the most affected. First, because natural green colorants are very labile and secondly, because it is not easy to reproduce green hues naturally. Consequently, in the near future the consumer will face new compounds responsible of the green color in foods. At present, independently of the existence of serious problems due to different legislations and different definitions between countries, authorized green natural colorants are incompletely characterized. This review has summarized the present knowledge of the compounds comprised the authorized green colorants, highlighting the common errors in their daily management. Finally, the scarce research developed on the ADME of green authorized food colorants is also considered but with the confidence that new advances will be shortly achieved.

Author Contributions: I.V., A.P.-G. and M.R. have contributed equally to the conceptualization, investigation, writing, review, and editing of this manuscript.

Funding: This research was funded by MINECO-CICYT, grant numbers ALI2018-R. The APC was partially funded by CSIC.

Conflicts of Interest: The authors declare no conflict of interest.

References

1. First Moment of Truth by Procter and Gamble. Available online: https://news.pg.com/video/first-moment-truth-30-s-reel (accessed on 5 November 2018).
2. Singh, S. Impact of color on marketing. *Manag. Decis.* **2006**, *44*, 783–789. [CrossRef]
3. Food Colors Market Size, Share & Trend Analysis Report by Product (Synthetic, Natural), by Application (Non-Dairy Food, CSD & Non-Alcoholic Beverages), and Segment Forecasts, 2018–2025. Available online: https://www.grandviewresearch.com/industry-analysis/food-colorants-market (accessed on 24 September 2018).
4. Martins, N.; Roriz, C.L.; Morales, P.; Barros, L.; Ferreira, I.C.F.R. Food colorants: Challenges, opportunities and current desires of agroindustries to ensure consumer expectations and regulatory practices. *Trends Food Sci.* **2016**, *52*, 1–15. [CrossRef]
5. Delgado-Vargas, F.; Paredes-López, O. *Natural Colorants for Food and Nutraceultical Uses*; CRC Press: Boca Raton, FL, USA, 2003; pp. 27–34.
6. Schwartz, S.J.; Cooperstone, J.L.; Cichon, M.J.; Von Elbe, J.H.; Giusti, M.M. Colorants. In *Fennema's Food Chemistry*; CRC Press: Boca Raton, FL, USA, 2017; pp. 681–752.
7. US FDA Electronic Code of Federal Regulations (eCFR) Listing of Color Additives Exempt from Certification Title 21, Chapter I, Subchapter A, Part 73. Available online: https://www.ecfr.gov/cgi-bin/retrieveECFR?gp=&SID=3463c48f55ae08efd099682901bb9500&r=PART&n=pt21.1.73 (accessed on 25 September 2018).
8. Regulation (EC) No 1333/2008 of the European Parliament and of the Council of 16 December 2008 on Food Additives (OJ L 354 31.12.2008, p. 16). Available online: https://eur-lex.europa.eu/eli/reg/2008/1333/2016-05-25 (accessed on 25 September 2018).
9. Wrolstad, R.E.; Culver, C.A. Alternatives to Those Artificial FD&C Food Colorants. *Annu. Rev. Food Sci. Technol.* **2012**, *3*, 59–77.

10. Carocho, M.; Morales, P.; Ferreira, I.C.F.R. Natural food additives: Quo vadis? *Trends Food Sci. Technol.* **2015**, *45*, 284–295. [CrossRef]
11. Lehto, S.; Buchweitz, M.; Klimm, A.; Straßburger, R.; Bechtold, C.; Ulberth, F. Comparison of food colour regulations in the EU and the US: A review of current provisions. *Food Addit. Contam.* **2017**, *34*, 335–355. [CrossRef] [PubMed]
12. McCann, D.; Barrett, A.; Cooper, A.; Crumpler, D.; Dalen, L.; Grimshaw, K.; Kitchen, E.; Lok, K.; Porteous, L.; Prince, E.; et al. Food additives and hyperactive behaviour in 3-year-old and 8/9-year-old children in the community: A randomised, double-blinded, placebo controlled trial. *Lancet* **2007**, *370*, 1560–1567. [CrossRef]
13. Simon, J.E.; Decker, E.A.; Ferruzzi, M.G.; Giusti, M.; Mejia, C.D.; Goldschmidt, M.; Talcott, S.T. Establishing standards on color from natural sources. *J. Food Sci.* **2017**, *82*, 2539–2553. [CrossRef] [PubMed]
14. Rodriguez-Amaya, D. Natural food pigments and colorants. *Curr. Opin. Food Sci.* **2016**, *7*, 20–26. [CrossRef]
15. Ngamwonglumlert, L.; Devahastin, S.; Chiewchan, N. Natural colorants: Pigment stability and extraction yield enhancement via utilization of appropriate pretreatment and extraction methods. *Crit. Rev. Food Sci. Nutr.* **2017**, *57*, 3243–3259. [CrossRef] [PubMed]
16. Oplatowska-Stachowiak, M.; Elliott, C.T. Food colors: Existing and emerging food safety concerns. *Crit. Rev. Food Sci. Nutr.* **2017**, *57*, 524–548. [CrossRef] [PubMed]
17. Sigurdson, G.T.; Tang, P.; Giusti, M.M. Natural Colorants: Food Colorants from Natural Sources. *Annu. Rev. Food Sci. Technol.* **2017**, *8*, 261–280. [CrossRef] [PubMed]
18. Gandul-Rojas, B.; Roca, M.; Gallardo-Guerrero, L. Detection of the color adulteration of green table olives with copper chlorophyllin complexes (E-141ii colorant). *LWT Food Sci. Technol.* **2012**, *1*, 311–318. [CrossRef]
19. China Food Additive Regulation. Available online: http://www.cfsa.net.cn/Article/News.aspx?id=4486BF3709A8E32BE89BDD2A36733D3CD2A97E8CE2A3F2E2 (accessed on 26 September 2018).
20. Japanese Food Additives Regulations. Available online: https://www.mhlw.go.jp/english/topics/foodsafety/foodadditives/index.html (accessed on 25 September 2018).
21. The Indian Food Safety and Standards Regulations. Available online: https://www.fssai.gov.in/home/fss-legislation/fss-regulations.html (accessed on 25 September 2018).
22. The Codex Alimentarius General Standard of Food Additives. Available online: http://www.fao.org/gsfaonline/additives/index.html#S (accessed on 25 September 2018).
23. Roca, M.; Chen, K.; Pérez-Gálvez, A. Chlorophylls. In *Handbook on Natural Pigments in Food and Beverages: Industrial Applications for Improving Food Color*; Woodhead Publishing: Cambridge, UK, 2016; pp. 125–158.
24. Scotter, M.J.; Castle, L.; Roberts, D. Method development and HPLC analysis of retail foods and beverages for copper chlorophyll (E141[i]) and chlorophyllin (E141[ii]) food colouring materials. *Food Addit. Contam. A* **2005**, *12*, 1163–1175. [CrossRef] [PubMed]
25. Inoue, H.; Furuya, K.; Watanabe, K.; Tanaka, K.; Shirai, T.; Miyoshi, E. Separation and determination of copper (II) chlorophylls by reversed-phase high performance liquid chromatography. *Anal. Sci.* **1988**, *4*, 599–603. [CrossRef]
26. Lian, W.; Shiuec, J.; Wang, H.; Hong, W.; Shih, P.H.; Hsu, C.K.; Huang, C.Y.; Hsing, C.R.; Wei, C.M.; Wang, J.K.; et al. Rapid detection of copper chlorophyll in vegetable oils based on surface-enhanced Raman spectroscopy. *Food Addit. Contam.* **2015**, *32*, 627–634.
27. Mortensen, A.; Geppel, A. HPLC–MS analysis of the green food colorant sodium copper chlorophyllin. *Innov. Food Sci. Emerg. Technol.* **2007**, *8*, 419–425. [CrossRef]
28. Pérez-Gálvez, A.; Ríos, J.J.; Roca, M. A new probe for tracking the presence of E141i food colorant. *Food Control* **2015**, *51*, 240–243. [CrossRef]
29. Willstätter, R. Chlorophyll. *J. Am. Chem. Soc.* **1915**, *37*, 323–345. [CrossRef]
30. Humphrey, A.M. Chlorophyll. *Food Chem.* **1980**, *5*, 57–67. [CrossRef]
31. Del Giovine, L.; Fabietti, F. Copper chlorophyll in olive oils: Identification and determination by LIF capillary electrophoresis. *Food Control* **2005**, *16*, 267–272. [CrossRef]
32. Roca, M.; Gallardo-Guerrero, L.; Mínguez-Mosquera, M.I.; Gandul-Rojas, B. Control of Olive Oil Adulteration with Copper-Chlorophyll Derivatives. *J. Agric. Food Chem.* **2010**, *58*, 51–56. [CrossRef] [PubMed]
33. Yasuda, K.; Tadano, K.; Ushiyama, H.; Ogawa, H.; kawai, Y.; Nishima, T. Investigation to find an indicator substance for the analysis of sodium copper chlorophyllin in foods. *J. Food Hyg. Soc. Jpn.* **1995**, *36*, 710–716. [CrossRef]

34. Inoue, H.; Yamashita, H.; Furuya, K.; Nonomura, Y.; Yoshioka, N.; Li, S. Determination of copper (II) chlorophyllin by reversed-phase high-performance liquid chromatography. *J. Chromatogr. A* **1994**, *679*, 99–104. [CrossRef]
35. Chernomorsky, S.; Rancourt, R.; Sahai, D.; Poretz, R. Evaluation of commercial chlorophyllin copper complex preparations by liquid chromatography with photodiode array detection. *J. AOAC Int.* **1997**, *80*, 433–435.
36. Aparicio-Ruiz, R.; Riedl, K.M.; Schwartz, S.J. Identification and quantification of metallo-chlorophyll complexes in bright green table olives by high-performance liquid chromatography-mass spectrometry quadrupole/time-of-flight. *J. Agric. Food Chem.* **2011**, *59*, 11100–11108. [CrossRef] [PubMed]
37. Suzuki, I.; Kubota, H.; Terami, S.; Hara, T.; Hirakawa, Y.; Iizuka, T.; Tatebe, C.; Ohtsuki, T.; Yano, T.; Sato, K.; et al. Development of an analytical method for copper chlorophyll and sodium copper chlorophyllin in processed foods. *Jpn. J. Food Chem. Saf.* **2016**, *23*, 55–62.
38. Negro, C.; De Bellis, L.; Sabella, E.; Nutricati, E.; Luvisi, A.; Micelli, A. Detection of not allowed food-coloring additives (copper chlorophyllin, copper-sulphate) in Green table olives sold on the Italian market. *Adv. Hort. Sci.* **2017**, *31*, 225–233.
39. Delpino-Rius, A.; Cosovanu, D.; Eras, J.; Vilaró, F.; Balcells, M.; Canela-Garayoa, R. A fast and reliable ultrahigh-performance liquid chromatography method to assess the fate of chlorophylls in teas and processed vegetable foodstuff. *J. Chromatogr.* **2018**, *1568*, 69–79. [CrossRef] [PubMed]
40. Ferruzzi, M.G.; Failla, M.L.; Schwartz, S.J. Assessment of degradation and intestinal cell uptake of carotenoids and chlorophyll derivatives from spinach puree using an in vitro digestion and Caco-2 human cell model. *J. Agric. Food Chem.* **2001**, *49*, 2082–2089. [CrossRef] [PubMed]
41. Gallardo-Guerrero, L.; Gandul-Rojas, B.; Minguez-Mosquera, M.I. Digestive stability, micellarization, and uptake by Caco-2 human intestinal cells of chlorophyll derivatives from different preparations of pea (Pisum sativum). *J. Agric. Food Chem.* **2008**, *56*, 8379–8386. [CrossRef] [PubMed]
42. Gandul-Rojas, B.; Gallardo-Guerrero, L.; Minguez-Mosquera, M.I. Influence of the chlorophyll pigment structure on its transfer from an oily food matrix to intestinal epithelium cells. *J. Agric. Food Chem.* **2009**, *57*, 5306–5314. [CrossRef] [PubMed]
43. Chen, K.; Roca, M. In vitro bioavailability of chlorophyll pigments from edible seaweeds. *J. Funct. Foods* **2018**, *41*, 25–33. [CrossRef]
44. Viera, I.; Chen, K.; Ríos, J.J.; Benito, I.; Pérez-Gálvez, A.; Roca, M. First-Pass Metabolism of Chlorophylls in Mice. *Mol. Nutr. Food Res.* **2018**, *62*. [CrossRef] [PubMed]
45. Brugsch, J.T.; Sheard, C. Chlorophyll, determination and quantitative estimation of decomposition of chlorophyll in the human body. *J. Lab. Clin. Med.* **1938**, *24*, 230–234.
46. Fisher, H.; Hendschel, A. Isolation of chlorophyll derivatives from the feces of elephants and human beings. *Hoppe Seylers Z. Physiol. Chem.* **1993**, *216*, 56–59.
47. Baxter, H. Absorption of chlorophyll phytol in normal man and in patients with Refsum's disease. *J. Lipid Res.* **1968**, *9*, 636–641. [PubMed]
48. Fernandes, T.M.; Gomes, B.B.; Lanfer-Marquez, U.M. Apparent absorption of chlorophyll from spinach in an assay with dogs. *Innov. Food Sci. Emerg. Technol.* **2007**, *8*, 426–432. [CrossRef]
49. Brink, D.M.; Wanders, R.J.A. Phytanic acid: Production from phytol, its breakdown and role in human disease. *Cell Mol. Life Sci.* **2006**, *63*, 1752–1765. [CrossRef]
50. EFSA Panel on Food Additives and Nutrient Sources added to Food (ANS). Scientific Opinion on the Re-Evaluation of Chlorophylls (E 140 (i)) as Food Additives. *EFSA J.* **2015**, *13*, 4089. Available online: https://www.efsa.europa.eu/en/efsajournal/pub/4089 (accessed on 2 November 2018).
51. EFSA Panel on Food Additives and Nutrient Sources added to Food (ANS). Scientific Opinion on the Re-Evaluation of Chlorophylls (E 140 (ii)) as Food Additives. *EFSA J.* **2015**, *13*, 4081. Available online: https://www.efsa.europa.eu/en/efsajournal/pub/4085 (accessed on 2 November 2018).
52. EFSA Panel on Food Additives and Nutrient Sources added to Food (ANS). Scientific Opinion on the Re-Evaluation of Chlorophylls (E 141 (i and ii)) as Food Additives. *EFSA J.* **2015**, *13*, 4151. Available online: https://www.efsa.europa.eu/en/efsajournal/pub/4151 (accessed on 2 November 2018).
53. EU Call for Scientific and Technical Data on the Permitted Food Additives E 140 (i) Chlorophylls, E 140 (ii) Chlorophyllins, E 141 (i) Copper Complexes of Chlorophylls and E 141 (ii) Copper Complexes of Chlorophyllins. Available online: https://ec.europa.eu/food/safety/food_improvement_agents/additives/re-evaluation_en. (accessed on 2 November 2018).

54. Ferruzzi, M.G.; Failla, M.L.; Schwartz, S.J. Sodium copper chlorophyllin: In vitro digestive stability and accumulation by Caco-2 human intestinal cells. *J. Agric. Food Chem.* **2002**, *50*, 2173–2179. [CrossRef] [PubMed]
55. Henderson, H.J.; Long, E.R. Accumulation of Chlorophyll Pigments in Visceral Organs and their Elimination. *Exp. Biol. Med.* **1941**, *48*, 438–439. [CrossRef]
56. Reber, E.F.; Willigan, D.A. The effects of a chlorophyll derivative when included in a ration fed rats. II. Reproduction, blood, and tissue studies. *Am. J. Vet. Res.* **1954**, *15*, 643–646. [PubMed]
57. Harrisson, J.W.E.; Levin, S.E.; Trabin, B. The safety and fate of potassium sodium copper chlorophyllin and other copper compounds. *J. Am. Pharm. Assoc.* **1954**, *43*, 722–737. [CrossRef]
58. Park, K.K.; Surh, Y.J. Chemopreventive activity of chlorophyllin against mouse skin carcinogenesis by benzoapyrene and benzoapyrene-7,8-dihydrodiol-9,10-epoxide. *Cancer Lett.* **1996**, *102*, 143–149. [CrossRef]
59. Gomes, B.; Barro, S.; Andrade-Wartha, E.; Silva, A.; Silva, V.V.; Lanfer-Marquez, U.L. Bioavailability of dietary sodium copper chlorophyllin and its effect on antioxidant defense parameters of Wistar rats. *J. Sci. Food Agric.* **2009**, *89*, 2003–2010. [CrossRef]
60. Chernomorsky, S.A.; Segelman, A.B. Biological activities of chlorophyll derivatives. *J. Med.* **1988**, *85*, 669–673.
61. Young, R.W.; Beregi, J.S. Use of chlorophyllin in the treatment of geriatric patients. *J. Am. Geriatr. Soc.* **1980**, *28*, 46–47. [CrossRef]
62. Egner, P.A.; Stanbubury, K.H.; Snyder, E.P.; Rogers, M.E.; Hintz, P.A.; Kensler, T.W. Identification and characterisation of chlorine e4 ethyl ester in sera of individuals participating in chlorophyllin chemoprevention trial. *Chem. Res. Toxicol.* **2000**, *13*, 900–906. [CrossRef]
63. Ferruzzi, M.G.; Blakeslee, J. Digestion, absorption and cancer preventive activity of dietary chlorophyll derivatives. *Nutr. Res.* **2007**, *27*, 1–12. [CrossRef]
64. Guidance Notes on the Classification of Food Extracts with Colouring Properties. Available online: https://ec.europa.eu/food/sites/food/files/safety/docs/fs_food-improvement-agents_guidance_additive-eu-rules.pdf. (accessed on 6 November 2018).
65. Palabiyik, I.; Durmaz, Y.; Öner, B.; Toker, O.S.; Coksari, G.; Konar, N.; Tamtürk, F. Using spray-dried microalgae as a natural coloring agent in chewing gum: Effects on color, sensory, and textural properties. *J. Appl. Phycol.* **2018**, *30*, 1031–1039. [CrossRef]
66. Jones, I.; White, R.; Gibbs, E.; Butler, L.; Nelson, L. Experimental formation of zinc and copper complexes of chlorophyll derivatives in vegetable tissue by thermal processing. *J. Agric. Food Chem.* **1977**, *25*, 149–153. [CrossRef]
67. Schanderl, S.H.; Marsh, G.; Chichester, C. Color reversion in processed I. Studies on regreened vegetables pea purges. *J. Food Sci.* **1965**, *30*, 312–316. [CrossRef]
68. LaBorde, F.F.; von Elbe, J.H. Zinc complex formation in heated vegetables purees. *J. Agric. Food Chem.* **1990**, *38*, 484–487. [CrossRef]
69. Segner, W.P.; Ragusa, T.J.; Nank, W.K.; Hoyle, W.C. Process for the Preservation of Green Color in Canned Vegetables. U.S. Patent No. 4473591, 25 September 1984.
70. Von Elbe, J.H.; Huang, A.S.; Attoe, E.L.; Nank, W.K. Pigment composition and color of conventional and Veri-Green canned beans. *J. Agric. Food Chem.* **1986**, *34*, 52–54. [CrossRef]
71. Porraud, S.; Pranee, A. Microencapsulation of Zn-chlorophyll pigment from Pandan leaf. *Int. Food Res.* **2010**, *17*, 1031–1042.
72. Bilek, S.E.; Özkan, G. Encapsulation of zinc-chlorophyll derivatives in whey protein matrix by emulsion/cold-set gelation. *J. Food* **2018**, *43*, 174–183.
73. Zhang, Z.; Peng, H.; Ma, H.; Zeng, X. Effect of inlet air drying temperatures on the physicochemical properties and antioxidant activity of whey protein isolate-kale leaves chlorophyll (WPI-CH) microcapsules. *J. Food Eng.* **2019**, *245*, 149–156. [CrossRef]
74. Afsaneh Raei, A.; Ali Yasini Ardakani, S.; Daneshi, M. Microencapsulation of the green pigment of alfalfa and its applications on heated food. *J. Food Process. Eng.* **2017**, *40*. [CrossRef]

molecules

MDPI

Review

Urtica spp.: Ordinary Plants with Extraordinary Properties

Dorota Kregiel, Ewelina Pawlikowska and Hubert Antolak *

Institute of Fermentation Technology and Microbiology, Faculty of Biotechnology and Food Science, Lodz University of Technology, 171/173 Wolczanska, 90-924 Lodz, Poland; dorota.kregiel@p.lodz.pl (D.K.); ewelina.pawlikowska@edu.p.lodz.pl (E.P.)

* Correspondence: hubert.antolak@gmail.com; Tel.: +48-42-631-3475

Received: 5 June 2018; Accepted: 6 July 2018; Published: 9 July 2018

Abstract: Nettles (genus *Urtica*, family Urticaceae) are of considerable interest as preservatives in foods for both human and animal consumption. They have also been used for centuries in traditional medicine. This paper reviews the properties of nettles that make them suitable for wider applications in the food and pharmaceutical industries. Nettles contain a significant number of biologically-active compounds. For example, the leaves are rich sources of terpenoids, carotenoids and fatty acids, as well as of various essential amino acids, chlorophyll, vitamins, tannins, carbohydrates, sterols, polysaccharides, isolectins and minerals. Extracts from the aerial parts of nettles are rich sources of polyphenols, while the roots contain oleanol acid, sterols and steryl glycosides. Due to the variety of phytochemicals and their proportions they contain, nettles show noticeable activity against both Gram-positive and Gram-negative bacteria. These properties make nettles suitable for a range of possible applications, including functional food, dietary supplements and pharmacological formulations. Despite these benefits, the nettle is still an underestimated plant source. This paper provides a unique overview of the latest research on nettle plants focusing on the possibilities for transforming a common weed into a commercial plant with a wide range of applications. Special attention is paid to the antimicrobial activity of the active compounds in nettles and to possible uses of these valuable plants in food and feed formulations.

Keywords: *Urtica* spp.; bioactive compounds; antioxidant activity; antimicrobial activity; traditional medicine; food industry; animal breeding

1. Habitats of *Urtica* spp. Plants

The genus *Urtica* belongs to the family Urticaceae in the major group Angiosperms (flowering plants). There are 46 species of flowering plant of the genus *Urtica* [1] (Table 1). The most prominent members of the genus are the stinging nettle *Urtica dioica* L. and the small nettle *U. urens* L., which are native to Europe, Africa, Asia and North America. Plants belonging to the genus *Urtica* are herbaceous perennials and can grow up to 2 m tall. Serrated leaves are attached in pairs opposite each other to the stem. The soft leaves and the rest of the plant are coated in hairs, some of which sting. The serrated, hairy leaves and sting are generally recognized characteristics of this plant [2]. However, the European variety *U. galeopsifolia* does not have stinging hairs [3]. The underground roots by which the plant spreads are noticeably yellow. Small flowers, each with four greenish-white petals, sit in dense clusters on elongated inflorescences towards the top of the stem.

The word "nettle" is said to derive from the Anglo-Saxon word "noedl" meaning "needle", while its Latin name "urtica" means "to burn". This refers to the stinging effects of the tiny hairs on the stems and leaves, which when rubbed against the skin cause a burning sensation and temporary rash. The hairs, so small they are almost invisible to the naked eye, are referred to as "trichomes".

When touched by the skin, the bulbous tips of the trichomes break off, leaving sharp, needle-like tubes [4]. These can pierce the skin and inject a fluid-containing substances including formic acid, the same chemical in ant and bee stings. Other compounds contained in the fluid include histamine, acetylcholine and serotonin. When these toxins are delivered into the skin, a painful itching and burning sensation occurs that may last up to 12 h. The hairs are naturally designed to protect the plant from insects.

Nettles grow all over the world in mild to temperate climates. They prefer open or partly shady habitats with plenty of moisture and are often found in forests, by rivers or streams and on roadsides. *Urtica* spp. are widespread throughout Europe and North America, North Africa and in parts of Asia. Both species of stinging nettle (*U. dioica* L. and *U. urens* L.) prefer to grow in nitrogen-rich soil and are commonly found in soils high in inorganic nitrates and heavy metals. Heavy metals are poorly processed by the plant and tend to accumulate in the leaves. Long vegetation seasons lead to on-going growth, while harsh winters cause destruction of the plants [5].

Nettles are considered weeds due to their rapid growth and soil coverage. However, there are economic and ecological reasons for cultivating stinging nettles. According to Dreyer and Müssing, nettles can improve soils over-fertilized with nitrogen and phosphate [5]. They can also promote the biodiversity of local flora and fauna [6,7]. Over 40 species of insect are supported by nettles [8]. *U. dioica* can reduce heavy metal content in soil [9]. *Urtica* spp. can be used to produce new high-quality agricultural raw materials for the dyeing, textile and energy sectors [7]. Due to their content of tough fibers, nettles were used in Germany and Austria to make textiles during the First World War [10]. Despite these benefits, most stinging nettles are wild harvested. Primary producers of stinging nettles include Eastern Germany, the former USSR, Bulgaria, the former Yugoslavia, Hungary and Albania [5].

Table 1. The list of species belonging to the *Urtica* genus [1,11,12].

No.	Name	Synonyms	Habitats
1	*U. angustifolia* Fisch. ex Hornem.	*U. dioica* var. *angustifolia* (Fisch. ex Hornem.) Ledeb. *U. foliosa* Blume	China, Japan, Korea
2	*U. ardens* Link	*U. himalayensis* Kunth & Bouché *U. mairei* var. *oblongifolia* C.J. Chen *U. zayuensis* C.J. Chen	Bhutan, India, Nepal, Sikkim
3	*U. atrichocaulis* (Hand.-Mazz.) C.J. Chen	*U. dioica* var. *atrichocaulis* Hand.-Mazz.	China -Guizhou, Sichuan, Yunnan
4	*U. atrovirens* Req. ex Loisel.	-	France, Italy, Spain
5	*U. ballotifolia* Wedd.	-	Colombia, Ecuador
6	*U. berteroana* Phil.	-	Chile, Bolivia, Argentina, Colombia
7	*U. buchtienii* Ross	-	Chile, Argentina
8	*U. cannabina* L.	*U. cannabina* f. *angustiloba* Chu	Russia, Sweden, Netherlands, China
9	*U. chamaedryoides* Pursh	*U. chamaedryoides* var. *runyonii* Correll	United States, Mexico
10	*U. circularis* Sorarú	*U. chamaedryoides* var. *circularis* Hauman *U. spathulata* var. *circularis* Hicken	Brazil, Argentina, Paraguay, Uruguay
11	*U. dioica* L.	*U. galeopsifolia* Wierzb. ex Opiz	United States, New Zealand, Turkey, Europe
12	*U. echinata* Benth.	*U. andicola* Wedd.	Bolivia, Peru, Argentina, Ecuador
13	*U. fissa* E. Pritz.	*U. pinfaensis* H. Lév. & Blin in H. Lév.	China, Taiwan, Egypt, Vietnam
14	*U. flabellata* Kunth	-	Bolivia, Peru, Ecuador, Chile, Colombia
15	*U. galeopsifolia* J. Jacq. ex Blume	-	Russia, Ukraine, Belarus
16	*U. glomeruliflora* Steud.	*U. fernandeziana* Ross ex Skottsb.	Chile
17	*U. haussknechtii* Boiss.	-	Turkey
18	*U. hyperborea* Jacq. ex Wedd.	*U. kunlunshanica* Chang Y. Yang	Nepal, India, China
19	*U. kioviensis* Rogow.	*U. dioica* var. *kioviensis* (Rogow.) Wedd. *U. dioica* subsp. *koviensis* Buia	Europe, Israel, Russia
20	*U. laetevirens* Maxim.	-	China, Japan, Korea
21	*U. leptophylla* Kunth	*U. copeyana* Killip *U. nicaraguensis* Liebm.	Costa Rica, Colombia, Peru, Bolivia, Ecuador
22	*U. lilloi* (Hauman) Geltman	*U. magellanica* var. *lilloi* Hauman	Argentina
23	*U. longispica* Killip	-	Ecuador, Peru, Colombia
24	*U. macbridei* Killip	-	Ecuador, Peru

Table 1. *Cont.*

No.	Name	Synonyms	Habitats
25	*U. magellanica* Juss. ex Poir.	*U. bracteata* Steud. *U. darwinii* Hook. *U. dioica* var. *pycnantha* Wedd. & DC. *U. dioica* var. *steudelii* Wedd. *U. magellanica* subsp. *bracteata* (Steud.) Geltman *U. magellanica* var. *bracteata* (Steud.) Wedd. *U. pseudodioica* Steud.	Chile, Peru, Bolivia, Argentina, Ecuador
26	*U. mairei* H. Lév.	-	China, India, Bhutan
27	*U. masafuerae* Phil.	-	Chile
28	*U. membranacea* Poir. ex Savigny	*Dubrueilia membranacea* Gaudich. *U. caudata* Vahl *U. dubia* Forssk.	Europe, Algeria
29	*U. mexicana* Liebm.	-	Mexico, Guatemala
30	*U. mollis* Steud.	*U. dioica* var. *mollis* (Steud.) Wedd. *U. diplotricha* Phil.	Peru, Chile, Argentina
31	*U. morifolia* Poir.	-	Europe
32	*U. orizabae* Liebm.	-	Mexico, United States, Cuba
33	*U. parviflora* Roxb.	-	Nepal, India, United States, China, Bhutan
34	*U. pilulifera* L.	*U. dodartii* L.	Tunisia, Israel, Cyprus, Costa Rica, Turkey
35	*U. platyphylla* Wedd.	*U. dioica* subsp. *platyphylla* P. Medvedev *U. takedana* Ohwi	Japan, Russia
36	*U. praetermissa* V.W. Steinm.	-	Mexico
37	*U. pubescens* Ledeb.	-	Mexico
38	*U. rupestris* Guss.	-	Italy
39	*U. sondenii* (Simmons) Avrorin ex Geltman	*U. dioica* subsp. *sondenii* (Simmons) Hyl. *U. dioica* var. *sondenii* Simmons	Canada
40	*U. spiralis* Blume	-	Mexico
41	*U. stachyoides* Webb & Benth.	-	Spain, Mexico
42	*U. taiwaniana* S.S. Ying	-	Taiwan
43	*U. thunbergiana* Siebold & Zucc.	*U. macrorrhiza* Hand.-Mazz.	Japan, Korea, China
44	*U. triangularis* Hand.-Mazz.	-	China
45	*U. trichantha* (Wedd.) Acevedo & NAVAS	*U. echinata* var. *trichantha* Wedd.	Chile, Bolivia, Peru
46	*U. urens* L.	*U. trianae* Rusby	Unite States, Mexico, Europe, Israel, New Zealand

2. Phytochemical Composition of *Urtica* spp.

Different factors affect the chemical composition of nettle plants, such as the variety, genotype, climate, soil, vegetative stage, harvest time, storage, processing and treatment [11–13]. Stinging nettles are a rich source of nutrients. A comprehensive proximate analysis showed that harvested upgrowths contained approximately 90% moisture, up to 3.7% proteins, 0.6% fat, 2.1% ash, 6.4% dietary fiber and 7.1% carbohydrates [10]. On the other hand, nettle leaf powders contain on average 30% proteins, 4% fats, 40% non-nitrogen compounds, 10% fiber and 15% ash (Table 2).

Table 2. Chemical composition of nettle leaf powders [10].

Parameter	Content
Moisture (%)	7.04 ± 0.77
Crude protein (%)	33.77 ± 0.35
Crude fiber (%)	9.08 ± 0.14
Crude fat (%)	3.55 ± 0.06
Total ash (%)	16.21 ± 0.54
Carbohydrate (%)	37.39 ± 0.72
Calcium (mg/100 g)	168.77 ± 1.47
Iron (mg/100 g)	227.89 ± 0.21
Tannins (%)	0.93 ± 0.01
Polyphenols (mg GAE/g)	128.75 ± 0.21
Carotenoids (μg/g, db)	3496.67 ± 0.56
Caloric value (kcal/100 g)	307.24 ± 0.13

In a study of nettles by Rafajlovska et al., higher quantities of proteins were found in the leaves than in the stems and roots. The content of proteins in the leaves ranged from 16.08 ± 0.38–26.89 ± 0.39%, depending on the source of the sample. The highest protein contents in the stem and roots were 14.54 ± 0.27% and 10.89 ± 0.11%, respectively [14]. Other studies of nettle composition have found that the plants contain a significant number of biologically-active compounds. The nettle leaves contain terpenoids [15], carotenoids [16] including β-carotene, neoxanthin, violaxanthin, lutein and lycopene [17,18], fatty acids, especially palmitic, *cis*-9,12-linoleic and α-linolenic acids [17,19], different polyphenolic compounds [20–23], essential amino acids, chlorophyll, vitamins, tannins, carbohydrates, sterols, polysaccharides, isolectins [16,17,24] and minerals [25,26], the most important of which is iron.

The leaves of *Urtica* spp. contain around 4.8 mg/g DM of chlorophyll, depending on the climate and environmental conditions. Interestingly, more chlorophyll and carotenoids are usually found in plants that have been harvested from shady places. Kukrić and co-workers noted that there were differences in the content of chlorophyll and carotenoids in leaves of different ages [16]. The concentration of chlorophyll increases in growing leaves and decreases during plant aging. The fresh leaves contain high concentrations of vitamins A, C, D, E, F, K and P, as well as of vitamin B-complexes [23]. The leaves are also known to contain particularly large amounts of the metals selenium, zinc, iron and magnesium. Rafajlovska et al. noted that stinging nettle leaves, stems and roots contained larger amounts of calcium than magnesium. These two elements were present at quantities almost three-times higher in the leaves than in the stems and roots. The calcium content expressed in relation to the dry mass ranged from 2.63–5.09% in leaves, from 0.76–1.42% in stems and from 0.61–0.92% in roots. Zinc was found in the highest concentrations in the leaves (27.44 mg/kg of dry mass), followed by copper (17.47 mg/kg) and manganese (17.17 mg/kg). The mean values for cobalt content were significantly higher in leaves than in stems and roots. The contents of cobalt in the leaves, stems and roots, respectively, were in the ranges of 0.11–0.21 mg/kg, 0.10–0.18 mg/kg, and 0.08–0.16 mg/kg, relative to the corresponding dry mass. *Urtica* leaves in addition contain boron, sodium, iodine, chromium, copper and sulfur [14].

The total phenolic content of one gram of nettle powder has been reported as 129 mg GAE (Gaelic Acid Equivalent), which is two-times higher than the phenolic content in 100 mL of cranberry juice (66.61 mg GAE) [27]. Stinging nettles have been shown to be richer in individual polyphenols than other wild plants [13]. Ghaima and co-workers found that the content of phenolic compounds in stinging nettle leaves was significantly higher than in dandelion leaves [28]. Vajić et al. reported that the predominant phenolic compound in stinging nettle leaves is rutin [29]. Đurović and co-workers studied the chemical composition of stinging nettle leaves using different analytical approaches. Soxhlet extraction was performed and qualitative analysis of Ultrasound-Assisted (UA) extracts using the UHPLC-DAD technique with MS/MS. Differences in the chemical profiles were found. For example, after Soxhlet extraction, syringic, cinnamic and protocatechuic acids were detected in the products, which was not the case with the UA extract. On the other hand, ferulic, caffeic, chlorogenic and sinapic acids were detected only after ultrasound-assisted extraction [30].

Orčić and co-workers quantified various plant phenolics in methanol extracts of *U. dioica*, for flowers, roots, stems and leaves separately, harvested from different locations in Serbia [20]. The polyphenol profiles were dependent not only on the parts of the plants, but also on the location of their acquisition. The researchers found that inflorescence extracts were the richest in phenolics. The most abundant compound was 5-*O*-caffeoylquinic acid (chlorogenic acid), followed by quercetin 3-*O*-rhamnosylglucoside (rutin) and 3-*O*-glucoside (isoquercetin).

The quantitative and qualitative composition of the roots differs from that of the aerial parts of the plants [22]. The content of the majority of the phenols in the root extracts is significantly lower, and the only prominent compound is secoisolariciresinol. Therefore, the roots of stinging nettles are considered to be the poorest part in terms of bioactive compounds. The root extracts contain only a few acids and derivatives in significant amounts (Table 3) [20]. They contain starch, gum, albumen, sugars and resins, as well as neurotransmitters and receptors, such as histamine, acetylcholine, choline or serotonin. Methanolic extracts of nettle roots have an inhibitory effect on aromatase, a key enzyme in the biosynthesis of estrogens. The lipophilic fractions usually contain phytosterols, pentacyclic triterpenoids, coumarins, ceramides and hydroxyl fatty acids. In turn, isolectins and some polysaccharides have been isolated from the hydrophilic fractions [31]. Krauss and Spitteler identified eighteen phenolic compounds (including homovanillyl alcohol, vanillin, vanillic acid and phenylpropanes) and nineteen ligands (including isolaric, iresinol, secoisolariciresinol and neoolivil) in root extracts [32]. Scopoletin, a coumarin derivative, has also been identified in nettle roots [24]. All these compounds are considered to be very important in medicine and pharmacology. For example, homovanillyl alcohol has been shown to protect against cardiovascular disease [33], while histamine influences the complex physiology of brain systems, affecting cognitive processes, including learning and memory [34], as well as neurotransmitters involved in neuromodulation processes [35]. Phytosterols reduce the absorption of cholesterol in the gut and thereby lower blood cholesterol levels [36]. Scopoletin is a stimulator of lipoprotein lipase activity and protects against cardiovascular diseases [37]. Lignans improve immune responses [38].

Table 3. The phenolic profiles in *U. dioica* extracts (mg per g of dry extract) [20].

Group of Compounds	Compound	Origin							
		Fruska Gora I/II				*Stara Planina*			
		Flowers	Leaves	Steams	Roots	Flowers	Leaves	Steams	Roots
Phenolic acids	*p*-Hydroxybenzoic acid	0.064/0.017	0.037/0.021	0.021/0.023	0.032/0.029	0.036	0.051	0.014	0.048
	Gentisic acid	0.0096/0.0044	0.0034/0.0082	not det/det	not det/0.036	not det	det	not det	not det
	Protocatechuic acid	0.070/0.032	0.48/0.16	not det/0.014	not det/0.015	0.022	0.072	0.0069	0.0106
	Vanillic acid	det */not det **	not det/det	not det/not det	not det/det	not det	not det	not det	0.09
	Quinic acid	**1.6/0.27**	**0.3/0.36**	**0.047/0.039**	**0.1/0.31**	**0.86**	**0.66**	**0.088**	**0.36**
	Ferulic acid	0.071/0.09	0.009/0.013	0.031/0.061	0.011/0.028	0.05	0.052	0.024	0.024
	***p*-Coumaric acid**	**not det/0.0105**	**not det/not det**	**0.24/0.38**	**0.12/02**	**0.022**	**0.026**	**0.18**	**0.23**
	Caffeic acid	**0.48/0.41**	**0.21/0.29**	**0.0053/0.033**	**det/0.0118**	**0.64**	**0.93**	**0.031**	**0.0039**
	5-*O*-Caffeolylquinic acid	**36/15.8**	**1.23/2.7**	**0.29/1.87**	**0.056/0.029**	**35**	**28**	**2.3**	**0.025**
Coumarins	Esculetin	0.041/0.0078	0.0120/0.0125	0.015/0.019	det/0.0047	0.0095	0.0074	det	det
	Scopoletin	0.103/0.018	0.12/0.21	0.026/0.054	0.076/0.11	0.04	0.073	0.048	0.18
Lignans	Secoisolariciresinol	not det/not det	not det/not det	not det/not det	det/0.2	not det	not det	not det	0.009
Flavones	Chrysoeriol	det/det	det/det	det/det	det/det	0.0027	det	det	det
Flavonols	Kaempferol	det/0.007	not det/not det	not det/det	not det/not det	0.019	not det	not det	not det
	Kaempferol 3-*O*-glucoside	**0.074/0.7**	**not det/not det**	**not det/0.0068**	**not det/not det**	**0.6**	**0.07**	**0.017**	**not det**
	Quercitrin	0.0124/not det	not det/not det	not det/not det	not det/not det	not det	not det	not det	not det
	Quercetin 3-*O*-glucoside	**0.63/3.64**	**not det/0.0024**	**0.0316/0.38**	**not det/0.0054**	**2.82**	**1.08**	**0.48**	**not det**
	Quercetin 3-*O*-rutinoside	**6.1/4.6**	**0.0018/0.0206**	**0.40/1.35**	**0.0023/0.0186**	**9.5**	**4.6**	**2.25**	**0.0054**
	Isorhamnetin	det/0.036	not det/not det	not det/det	not det/not det	0.047	not det	not det	not det
Biflavonoids	Amentoflavone	det/det	det/det	det/det	det/det	det	det	det	det
Flavan-3-ols	Catechin	not det/0.076	not det/not det	not det/not det	not det/not det	1.0	not det	not det	not det

* det: peak observed, but the concentration was too low to evaluate it; ** not det: peak not observed. Bolded compounds are those that occur at the highest concentration.

Fresh nettle leaves contain smaller amounts of sterols and higher concentrations of flavonol glycosides. The leaves of the plant also contain carotenoids, mainly β-carotene, violaxanthin, xanthophylls, zeaxanthin, luteoxanthin and lutein epoxide [5]. Terpene diols, terpene diol glucosides, α-tocopherol, as well as five monoterpenoid components have also been detected in nettle leaves [39]. Weglarz and Roslon studied the content of polyphenolic acids in leaves and rhizomes. They found that the level of these compounds was higher in the male forms, but the chemical profiles of polyphenolic acids from the female plants were much more diverse [40,41]. Moreover *U. dioica* is considered the only plant that contains choline acetyl-transferase, an acetylcholine-synthesizing enzyme [42]. Fifteen hydroxycinnamic acid derivatives and sixteen flavonoids, flavones and flavonol-type glycosides were identified in hydroalcoholic extracts from the aerial parts of *U. dioica*, *U. urens* and *U. membranacea* using HPLC-PDA-ESI/MS. Of these, 4-caffeoyl-5-*p*-coumaroylquinic acid and three statin-like 3-hydroxy-3-methylglutaroyl flavone derivatives were identified for the first time in *U. urens* and *U. membranacea*, respectively. *U. membranacea* showed a higher content of flavonoids, mainly luteolin and apigenin glycosides, which are almost absent in the other species studied [43].

The hairs of *Urtica* plants contain an acrid fluid with the active components: acetylcholine, histamine and formic acid, as well as silica, serotonin and 5-hydroxy tryptamine. Many of these chemicals are smooth muscle stimulants [44]. The fresh hairs of *U. dioica* also contain a high level of acetylcholine [45]. The results of numerous experiments suggest that each species of nettle, a well as each part of the plant (root, stalk or leaves) have a different content and profile of bioactive compounds. Therefore, different species of nettle may have different uses, depending on their chemical characteristics [21].

Generally, the phenolic composition of plants is affected by different factors, including the variety, genotype, climate, soil, vegetative stage of the plant, harvest time, storage, processing and treatment [46,47]. When and how nettles are harvested strongly determines the final product. For example, for fiber production, stinging nettles should be harvested when the seeds are mature or when the stalks reach 80% of the aboveground biomass, from the second year of planting. During the first year, the stalks are too thin, too ramified and have too many leaves. If the main product is to be the leaves, younger plants are harvested. The time of year for nettle harvesting depends on the purpose. Plants collected in April are used for fodder, medicine or chlorophyll production. Nettles harvested at the end of June are used for fiber production. The second harvest in September may be used for the collection of leaves [7].

3. Antimicrobial Activities of *Urtica* spp.

Nettles possess noticeable antimicrobial activity against Gram-positive and Gram-negative bacteria when compared with standard and strong antimicrobial compounds, such as miconazole nitrate, amoxicillin-clavulanic acid, ofloxacin and netilmicin [48]. Different fractions of various *Urtica* species have been studied to determine their antimicrobial activity. The results indicate the great potential of this plant for the discovery of novel effective compounds (Tables 4 and 5) [49–56].

Table 4. In vitro activity of *Urtica* spp. against microorganisms.

Urtica spp. Extract	Microorganisms	Location	Reference
U. dioica L. water extract	*Escherichia coli* *Proteus mirabilis* *Citrobacter koseri* *Staphylococcus aureus* *Streptococcus pneumoniae* *Enterobacter aerogenes* *Micrococcus luteus* *Staphylococcus epidermidis* *Candida albicans*	Turkey	[48]

Table 4. *Cont.*

Urtica spp. Extract	Microorganisms	Location	Reference
U. dioica L. ethyl acetate extract	*Aeromonas hydrophila* *Salmonella typhi* *Staphylococcus aureus* *Bacillus cereus* *Escherichia coli*	Iraq	[28]
U. dioica L. water extract	*Salmonella* spp. *Proteus* spp. *Bacillus subtilis* *Staphylococcus aureus* *Pseudomonas aeruginosa* *Escherichia coli*	Iraq	[50]
U. dioica L. leaves aqueous extract	*Escherichia coli* *Enterococcus gallinarum* *Enterococcus faecalis* *Streptococcus pyogenes* *Clavibacter michiganensis* *Pseudomonas tomato* *Xanthomonas vesicatoria*	Turkey	[51]
U. dioica L. root aqueous extract	*Enterococcus faecalis* *Streptococcus pyogenes* *Klebsiella pneumoniae*		
U. dioica L. seeds aqueous extract	*Enterococcus faecalis*		
U. dioica L. aqueous extract 0.45 mg/100 mL	*Staphylococcus epidermidis* 3615 *Staphylococcus aureus* 740 *Escherichia coli* 443 *Salmonella typhimurium* 98 *Serratia marcescens* 97	India	[56]
U. dioica L. supercritical CO_2 extract	*Bacillus subtilis* *Saccharomyces cerevisiae* *Aspergillus niger* *Botrytis cinerea* *Geotrichum candidum*	Macedonia	[49]
U. dioica L. ethanol extract	*Salmonella* spp. *Bacillus subtilis* *Staphylococcus aureus* *Pseudomonas aeruginosa*	Iraq	[50]
U. dioica L. methanol extract	15 different strains of *Staphylococcus aureus* MRSA	Iran	[54]
U. dioica L. methanol extract	*Shigella dysenteriae* *Salmonella enteritidis*	Iran	[55]
U. dioica L. leaves methanol extract	*Escherichia coli* *Streptococcus pyogenes* *Listeria monocytogenes* *Pseudomonas aeruginosa* *Klebsiella pneumoniae* *Proteus vulgaris* *Erwinia carotovora*	Turkey	[51]
U. dioica L. root methanol extract	*Enterococcus gallinarum* *Xanthomonas vesicatoria*		
U. dioica L. seeds methanol extract	*Enterococcus gallinarum* *Enterococcus faecalis* *Streptococcus pyogenes* *Staphylococcus aureus* *Listeria monocytogenes* *Pseudomonas aeruginosa* *Proteus vulgaris* *Shigella* spp. *Bacillus pumilus* *Clavibacter michiganensis* *Xanthomonas vesicatoria*		

Table 4. *Cont.*

Urtica spp. Extract	Microorganisms	Location	Reference
U. dioica L. flowers methanol extract	*Escherichia coli* *Bacillus subtilis* *Staphylococcus aureus* *Pseudomonas aeruginosa* *Candida albicans* *Aspergillus niger*	Iran	[52]
U. pilulifera L. leaves aqueous extract	*Enterococcus faecalis* *Streptococcus pyogenes*	Turkey	[51]
U. pilulifera L. root aqueous extract	*Xanthomonas vesicatoria*		
U. pilulifera L. seeds aqueous extract	*Enterococcus faecalis* *Proteus vulgaris* *Shigella* spp.		
U. pilulifera L. leaves methanol extract	*Enterococcus gallinarum* *Enterococcus faecalis* *Streptococcus pyogenes* *Listeria monocytogenes* *Pseudomonas aeruginosa* *Klebsiella pneumoniae* *Proteus vulgaris* *Shigella* spp. *Bacillus pumilus* *Clavibacter michiganensis* *Pseudomonas tomato* *Erwinia carotovora*		
U. pilulifera L. roots methanol extract	*Enterococcus gallinarum* *Enterococcus faecalis* *Streptococcus pyogenes* *Clavibacter michiganensis* *Pseudomonas tomato* *Erwinia carotovora*		
U. pilulifera L. seeds methanol extract	*Enterococcus gallinarum* *Enterococcus faecalis* *Streptococcus pyogenes* *Staphylococcus aureus* *Listeria monocytogenes* *Pseudomonas aeruginosa* *Proteus vulgaris* *Shigella* spp. *Bacillus pumilus* *Clavibacter michiganensis*		
U. urens L. ethanol extract	*Staphylococcus aureus* ATCC 6538 *Pseudomonas aeruginosa* ATCC 27893 *Bacillus subtilis* JN 934392 *Salmonella enteritidis* *Escherichia coli* ATCC 25922 *Staphylococcus epidermidis* MTCC 3615 *Micrococcus luteus* ATCC 4698 *Enterococcus faecalis* ATCC 29212	Tunisia	[53]

Table 5. In vitro activity of *Urtica* spp. against microorganisms: minimal inhibitory concentrations.

Urtica spp. Extract	Microorganisms	Minimal Inhibitory Concentration (MIC)	Location	Reference
U. dioica L. ethanol extract	*Bacillus subtilis* *Escherichia coli* (food-origin) *Escherichia coli* (urine-origin) *Pseudomonas aeruginosa* *Lactobacillus plantarum*	36.21 mg/mL 36.21 mg/mL 72.43 mg/mL 72.43 mg/mL 72.43 mg/mL	Serbia	[16]
U. dioica L. hexane extract	*Staphylococcus aureus* MRSA *Bacillus cereus* *Bacillus spizizenii* ATCC 663 *Vibrio parahaemolyticus*	66.66 mg/mL 16.66 mg/mL 16.66 mg/mL 66.66 mg/mL	Iran	[57]
U. dioica L. chloroform extract	*Bacillus cereus* *Vibrio parahaemolyticus*	33.33 mg/mL 4.16 mg/mL		
U. dioica L.	*Acinetobacter calcoaceticus*	33.33 mg/mL		
U. dioica L. ethyl acetate extract I	*Bacillus spizizenii* ATCC 663 *Vibrio parahaemolyticus* *Saccharomyces cerevisiae*	8.33 mg/mL 16.66 mg/mL 2.08 mg/mL		
U. dioica L. ethyl acetate extract II	*Vibrio parahaemolyticus*	0.13 mg/mL		
U. dioica L. methanol extract	*Acinetobacter calcoaceticus*	16.66 mg/mL		
U. dioica L. butanol extract	*Escherichia coli* *Bacillus subtilis* *Staphylococcus aureus* MRSA	66.66 mg/mL 8.33 mg/mL 16.66 mg/mL		
U. dioica L. hexane fraction	*Aeromonas hydrophila* *Aeromonas salmonicida* subsp. *salmonicida* *Flavobacterium columnare* *Vibrio salmonicida* *Yersinia ruckeri* *Pseudomonas aeruginosa* *Staphylococcus aureus* *Salmonella typhi* *Klebsiella pneumoniae* *Enterococcus faecalis*	125 µg/mL 125 µg/mL 250 µg/mL 62.5 µg/mL 31.25 µg/mL 250 µg/mL 31.25 µg/mL 7.81 µg/mL 31.25 µg/mL 125 µg/mL	India	[58]
U. dioica L. leaves hexane fraction	*Staphylococcus aureus* *Pseudomonas aeruginosa* *Salmonella typhi* *Klebsiella pneumoniae* *Shigella flexneri*	31.25 µg/mL 250 µg/mL 7.81 µg/mL 31.25 µg/mL 125 µg/mL	India	[59]
U dioica L. essential oils hydrodistillation method	*Bacillus cereus* PTCC1565 *Staphylococcus aureus* *Pseudomonas aeruginosa* *Klebsiella pneumoniae* *Enterococcus faecalis* PTCC1239 *Escherichia coli* ATCC1533	1.8 µg/mL 3.75 µg/mL 3.75 µg/mL 3.75 µg/mL 7.5 µg/mL 7.5 µg/mL	Iran	[60]
U. dioica L. ethanol extract	Methicillin-sensitive strains of *S. aureus* Methicillin-resistant strains of *S. aureus*	0.188–0.500 mg/mL 0.063–0.500 mg/mL	Portugal	[61]

The results presented in Table 5 show the antimicrobial activities of various nettle extracts obtained by different researchers. As can be seen, some nettle extracts show activity at a concentration of 72 mg/mL and others at 1 µg/mL. These differences appear excessive, and the results should therefore be viewed with caution. Such variations may be associated with the location of the plant habitat and climactic conditions, as well as being due to the use of different extraction techniques and evaluation methods. Despite their significant differences, however, the results of these studies show that nettle plants exhibit antimicrobial activity against a wide spectrum of microbial strains, often isolated from foods of low microbiological quality. A study by Kukrić et al. revealed that nettle extracts had inhibitory effects on various Gram-positive and Gram-negative bacteria including *Bacillus subtilis*, *Lactobacillus plantarum*, *Pseudomonas aeruginosa* and *Escherichia coli* [16]. Mahmoudi et al. reported that all microorganisms tested in their research, Gram-negative and Gram-positive bacteria, as well as *Candida albicans* yeast, were sensitive to alcoholic extract from the nettle stem [62]. In recent studies conducted by Antolak et al., the ethanol extract of *U. dioica* showed inhibitory activity against the growth of acetic acid bacteria belonging to the genus *Asaia*, a beverage spoilage bacteria found in functional drinks [63]. On the other hand, Shale et al. noted that *E. coli* and *P. aeruginosa* were completely resistant to the ethanol and methanol extracts from stems and leaves of *U. dioica* [64]. Different antimicrobial properties may be the result of the isolation of different compounds in different solvents, of different extraction efficiencies and possibly of chemical degradation by polar and non-polar solvents. The extraction method, the plant type, the geographical and ecological status, the climate, seasonal and experimental conditions, the age of the plant, environmental stress factors, as well as inter-species differences all play a role and may explain the diversity of results in different studies [21,65,66].

4. *Urtica* spp. in Traditional and Modern Medicine

Nettles are one of the most commonly-used medicinal plants in the world, due to their health-enhancing qualities. Because of their high content of nutritive substances, nettles are also used in folk veterinary medicine [48,67,68]. There are many dietary supplements based on *Urtica* spp. now on the market. Their popularity can be explained by their non-toxic chemical composition, relatively low cost and wide availability. The most recognized health benefit of using stinging nettles is activity against Benign Prostatic Hyperplasia (BPH), also known as an enlarged prostate, as well as urinary tract infections. Clinical studies suggest that *Urtica* spp. contain compounds that affect the hormones responsible for BPH. In addition, nettle root extract shows activity against prostate cancer cells. In therapy, nettles are usually used in combination with saw palmetto (*Serenoa repens*) [69,70]. They are also used as a home remedy for bladder infections.

Nettles can help alleviate the symptoms of osteoarthritis and joint pain, typically in the case of hands, knees, hips and spine. Nettles can work in combination with nonsteroidal anti-inflammatory drugs (NSAIDs), allowing patients to decrease their use of NSAIDs. The prolonged use of NSAIDs can increase the risk of heart attack or stroke. In a study by Randall and co-workers, nettles were able to decrease osteoarthritic pain in the base of the thumb when applied to the painful area. In a clinical trial of 37 people with acute arthritis, 50 g of stewed nettle leaves consumed daily, combined with 50 mg of diclofenac, were shown to be as effective as the full 200-mg dose of diclofenac over a two-week period [71]. Studies have also shown that applying nettle leaves directly decreases joint pain and can treat arthritis. In a study by Christensen and Bliddal, it was found that a combination of nettles, fish oil and vitamin E reduced the need for analgesics and other drugs for the symptoms of osteoarthritis [72].

Another study conducted by Klingelhoefer et al. showed the anti-inflammatory benefits of stinging nettles against other autoimmune diseases, such as rheumatoid arthritis [73]. Nettle leaves contain histamine, which may seem inadvisable for allergy medication. However, histamine has been already used to treat strong allergy symptoms [74]. Histamine production causes unwanted allergic reactions, associated with unpleasant nasal congestion, sneezing or itching. Stinging nettles affect numerous receptors and/or enzymes involved in allergic reactions [75]. In addition, because of their anti-histamine and anti-inflammatory properties, stinging nettles can be used as a natural component in eczema medications. Infusions of the plant can be used for nasal and menstrual hemorrhage, diabetes, anemia, asthma, hair loss and to promote lactation [76]. Terpenes and phenols are major groups associated with the inhibition of cancers, as well as with the treatment of headache, rheumatism and some skin diseases [58,77,78]. Phenols also have been associated with the inhibition of atherosclerosis and cancer, as well as age-related degenerative brain disorders [79,80].

The combination of *U. dioica* with common thyme (*Thymus vulgaris*), liquorice (*Glycyrrhiza glabra*), common grape (*Vitis vinifera*) and lesser galangal (*Alpinia officinarum*) has been known in Turkey as an Ankaferd Blood Stopper (ABS). This traditional medicine works on endothelium, blood cells, angiogenesis, cellular proliferation, vascular dynamics and cell mediators to stop bleeding [81]. In a study conducted by Bourgeois et al., nettles were used for cosmetic applications as an anti-aging complex, involving the inhibition of collagenase and elastase activities. These properties could be ascribed to the ursolic acid and quercetin present in the nettle extracts [82].

Herb extract of *Urtica* plants is useful for bladder disorders, reduces postoperative blood loss and prevents hemorrhagic and purulent inflammation following adenomectomy. Aqueous infusions of *U. dioica* exhibit antioxidant activity towards iron-promoted oxidation of phospholipids, linoleic acid and deoxyribose [83]. For a long time, the hypoglycemic effects of *U. dioica* were only speculative. Recent studies show that nettles possess anti-diabetic properties [84]. Thus, nettles could serve as good adjuvant to other oral hypoglycemic agents and seem promising for the development of phytomedicines for diabetes mellitus. In addition, as organic nitrogenous compounds, amino acids from nettles are building blocks in the process of protein biosynthesis [85]. The safety of aqueous extracts of *U. dioica* and their antidiabetic effects have been confirmed with mice models [86].

5. Food and Feed Applications

Nettles have traditionally been used as a nutritious food, particularly in spring time in rural areas. The Romans are said to have consumed nettles, and a recipe for Saint Columba's broth survives to this day. In Greek and Roman times, nettle roots were used for meat tenderization. Nowadays, nettles are used in a large number of recipes. *Urtica* spp. are blended with fromage blanc, potatoes and nutmeg to make nettle nouvelle [39]. Stinging nettles remain popular, especially in poor countries and among the lower socioeconomic classes [87]. For example, amino acids from dehydrated nettle meal are nutritionally better in comparison to those of alfalfa meal [88]. These plants may be consumed primarily as a boiled or cooked fresh vegetable, which is added to soups, cooked as a pot herb or used in vegetable salads [10]. In the temperate region of the Himalayas, *U. plaviflora* leaves are cooked and eaten as a green vegetable. Upgrowths and leaves are collected with the help of bamboo or iron pincers and cooked as soup. The plants are boiled with maize, millet or wheat flour, with the addition of salt and chili to make a kind of porridge. Due to their seasonality, *U. plaviflora* plants are preserved by lactic acid fermentation [4].

In European countries, nettles are used in soup or as a steamed or wilted vegetable. Since it has a similar flavor and texture, cooked nettle can be used as a substitute for spinach. Raw nettles after blending can be also used in pesto sauces, salad dressings or dips. Boiled nettles with walnuts is a common dish in Georgia, while Romanians make sour soup using fermented wheat bran, vegetables and young nettle leaves [89]. Mature leaves are used in the production of semi-hard Cornish cheese, made from grass-rich milk and wrapped in stinging nettles. The nettle changes the acidity of the outside of the cheese, affecting the way the curd breaks down and matures. It has also been documented that nettle leaves can be used to coagulate milk in the process of fresh cheese making [90]. In some European countries, for example in Serbia and Poland, bread with nettle leaves (up to 1%) is sold as a commercial product [30]. Compared to barley and wheat flour, nettle flour has a much higher content of proteins, crude fibers, fats, ash, calcium and iron and has a low glycemic index. As compared to barley and wheat, nettles have much higher levels of tannins and total polyphenols [10].

Nettle leaves can also be used to make an herbal tea, which is rich in vitamins and minerals. Depending on the amount used, nettle tea has a mild to strong flavor and tastes similar to vegetable broth. Concentrated nettle tea can be used as a soup base or as a component in drinks or green cocktails. Nettle tea can also be used as a nutritional replacement for water. Nettle roots can be used as liquid or powdered extracts, as well as in special decoctions. Nettles are also used in herbal liquors [91]. In the British Isles, *Urtica* plants are used in an alcoholic beverage, which is similar to ginger beer and brewed in the same way. Nettle and oat extracts are the subject of a U.S. Patent describing the use of plant powders as additives in beverages or fruit juice to provide nutritional drinks [92]. Aqueous infusions of *U. dioica* exhibit antioxidant activity towards iron-promoted oxidation of phospholipids, linoleic acid and deoxyribose [83]. The use of such antioxidant and antimicrobial compounds is of considerable interest for the preservation of foods, as well as for improving the shelf-life of food products [53,93].

Despite their beneficial properties, the consumption of nettle teas or juices may cause a skin rash in individual cases. Although it is rare, there have been reports of allergic reactions after ingesting raw nettle leaves in the form of puree or juice [94]. Therefore, stinging nettles need to be correctly prepared by hot water infusion, maceration, drying or tincturing. This pretreatment deactivates the formic acid, allowing safe consumption of this valuable plant.

Oxidation is a serious problem in the food industry. Meat products are particularly susceptible to changes in the oxidation of lipids and heme pigments during storage. Along with meat spoilage bacteria, the oxidation of lipids and myoglobin has a major effect on reducing the shelf-life of meat and meat products. Lipid oxidation decreases the nutritional value of meat through the deterioration of essential fatty acids, causing an unacceptable flavor, generating potentially toxic products and promoting the oxidation of other important molecules, such as myoglobin [95]. Although synthetic antioxidants are available, in recent years, the demand for natural antioxidants has increased, mainly because of the adverse effects of synthetic antioxidants [96]. Iron, the most abundant ion in meat, is released from heme pigments and ferritin and may be an important catalyst in the oxidation of lipids and proteins. Studies have demonstrated that commercially-available polyphenols and extracts rich in polyphenols can inhibit myoglobin oxidation.

Inai et al. studied the inhibition of myoglobin oxidation by some plant polyphenols with activity for flavonols (kaempferol, myricetin and quercetin) [97]. Slightly weaker activity was observed for other polyphenols: sinapic acid, catechin, nordihydroguaiaretic acid, taxifolin, morin and ferulic acid. The use of natural antioxidants from *U. dioica* water extract and dried leaves as a functional ingredient significantly decreased the level of lipid deterioration and increased color stability during storage. Therefore, nettle water extract can be successfully used to reduce lipid oxidation and to enhance the functionality of the final products. Other studies have investigated the use of *U. dioica* water extract in sucuk, a Turkish dry-fermented sausage [98–100], in ground beef [101], meatballs [102], in super-chilled minced meat [103], vacuum-packed beef steaks [104] and in cooked pork sausage [105].

The use of *Urtica* spp. as a dietary supplement may positively affect the health and productivity of poultry and cows. The large number of active compounds in this plant may show stronger antibacterial activities than synthetic antimicrobials. It has been suggested that extracts of *Urtica* spp. may also possess appetite- and digestion-stimulating properties [106–110], which stimulate the growth of beneficial bacteria in the gastrointestinal tract of poultry [111]. In some parts of Europe, fresh leaves are traditionally fed to pigs and poultry. Safamehr and co-workers showed that nettles had no effect on the percentage of breast, thigh and abdominal fat [68]. However, chickens fed with 1% nettles had the highest carcass yield. In contrast, Alçiçek, Bozkurt and Çabuk observed an improvement in the carcass yield of broilers supplemented with a combination of essential oils [109]. This can be explained by more intense protein anabolism [112]. The relative weights of most organs were not affected by the inclusion of nettles to the diet, which is in agreement with the findings by Nobakht [113].

It has been suggested that the antioxidant activity of nettles may induce a decrease in the relative weight of the liver [114]. A study conducted by Safamehr and co-workers showed that 1%–2% supplementation with nettles had a positive effect on the performance, carcass traits and blood biochemical parameters of broilers [68]. A study by Humphries and Reynolds confirmed the usefulness of nettles as a forage crop for cows. Production of milk was maintained when nettles were used to replace dry grass silage in the diet of lactating dairy cows [115]. The addition of nettle haylage to the diet caused changes in rumen pH that were potentially beneficial to lactating dairy cows on high grain diets. It is worth noting that diets based on plant material, rich in immune-promoting bioactive compounds, can avoid the need for antibiotic growth promoters. There is increasing public and government pressure in several countries of the EU and some non-EU nations to find natural alternatives to antibiotics [116,117].

Şandru and co-workers noted that *U. dioica* alcoholic extract increased cell-mediated innate immune potential in chickens. Alcoholic nettle plant extract significantly increased total leukocyte numbers, from 15,400–17,125 cells/mm^3 [118]. Similarly, nettle extract treatment significantly enhanced the in vitro functional capacity of phagocytes. This could lead to a higher resistance to diseases and improve the post-vaccination response of broilers, thus reducing economic losses. It was noted that the glucose and total protein content, as well as the heterophil and lymphocyte levels in chickens were not influenced by different quantities of nettle, while the concentrations of cholesterol and triglycerides in the blood were affected significantly. The effect on lowering cholesterol in blood serum may be because of the presence of plant sterols, such as stigmasterol and campesterol. These decrease the cholesterol concentration in micelles [119,120]. Fermont and co-workers and Visioli and co-workers report that the cholesterol levels in blood serum and meat are probably lowered by phenolic compounds [121,122]. However, Khosravi et al. found that the addition of nettle extract to the diet of boilers had no significant positive effect on their total cholesterol [123].

Urtica spp. provide animals with nutrients and bioactive components, which support antimicrobial activity, immune enhancement and stress reduction. However, the phytochemical composition is complex, and the mode of action is unclear [124]. Further studies are needed to investigate the bioactive components of nettles and their modes of action. It is worth noting that the World Health Organization (WHO), in its monographs on 'Selected medicinal plants', describes *Urticae* as valuable herbs for many medicinal uses [125]. The European Commission Directorate-General for Health and Food Safety showed that *Urtica* spp. fulfils the criteria of a foodstuff, as defined in Regulation (EC) No. 178/2002. This opinion is supported by the European Food Safety Authority (EFSA). It concluded that *Urtica* spp. has neither an immediate nor delayed harmful effect on human or animal health and has no negative effect on the environment [126].

6. Conclusions

Stinging nettles can be found all over the world. Plant hairs located on the leaves and stems contain a number of chemicals, which can cause a stinging reaction and uncomfortable irritation when brought into contact with human skin. Nevertheless, stinging nettles have a number of health benefits and have been used medicinally since at least the times of Ancient Greece. Studies have shown that all parts of the nettle have antioxidant, antimicrobial and pro-health capabilities. Most nettle medicines are made from the flowers, stems and leaves, but roots are also used in pharmacology. This valuable plant has been used most commonly as a diuretic and for treating painful muscles and joints, eczema, gout and anemia. Nettles may be used as a vegetable, in juice, tea and as an ingredient in many dishes. The use of *Urtica* spp. as a feed component could also positively affect the health of poultry and animal productivity. However, despite these proven benefits, the nettle is still an undervalued plant.

Research is continuing into this ordinary plant with unique pharmacological and dietary properties. It is worth investigating the possible wider inclusion of nettles in the daily diet to promote well-being and prevent diseases.

Funding: This research received no external funding.

Conflicts of Interest: The authors declare no conflicts of interest.

References

1. The Plant List. A Working List of All Plant Species. Available online: http://www.theplantlist.org/tpl1.1/search?q=Urtica+ (accessed on 5 April 2018).
2. Jiarui, C.; Qi, L.; Friis, I.; Wilmot-Dear, C.M.; Monro, A.K. Urticaceae. In *Flora of China*; Missouri Botanical Garden Press: St. Louis, MO, USA, 2003; pp. 76–189.
3. Geltman, D.V. *Urtica galeopsifolia* Wierzb. ex Opiz (*Urticaceae*) in Wicken Fen (E. England). *Watsonia* **1992**, *19*, 127–129.
4. Mishra, A.; Kharel, G.P. Preservation and Quality Evaluation of Sisnu (*Urtica plaviflora*) by making Gundruk like fermented product. *J. Food Sci. Technol. Nepal.* **2010**, *6*, 114–117. [CrossRef]
5. Upton, R. Stinging nettles leaf (*Urtica dioica* L.): Extraordinary vegetable medicine. *J. Herb. Med.* **2013**, *3*, 9–38.
6. Dreyer, J.; Müssig, J. New Horizons in Natural Fibre Production: Separation of Hemp and Nettle with Enzymes. Presented at the 3th International Symposium Biorohstoff Hanf & Andere Faserpflanzen, Wolfsburg, Germany, 13–16 September 2000.
7. Di Virgilio, N.; Papazoglou, E.G.; Jankauskiene, Z.; Di Lonardo, S.; Pralczyk, M.; Wilegusz, K. The potential of stinging nettle (*Urtica dioica* L.) as a crop with multiple uses. *Ind. Crops Prod.* **2015**, *68*, 42–49. [CrossRef]

8. Lang, A.; Otto, M. Feeding behaviour on host plants may influence potential exposure to *Bt* maize pollen of aglais *Urticae larvae* (Lepidoptera, Nymphalidae). *Insects* **2015**, *6*, 760–771. [CrossRef] [PubMed]
9. Viktorova, J.; Jandova, Z.; Madlenakova, M.; Prouzova, P.; Bartunek, V.; Vrchotova, B.; Lovecka, P.; Musilova, L.; Macek, T. Native phytoremediation potential of *Urtica dioica* for removal of PCBs and heavy metals can be improved by genetic manipulations using constitutive CaMV 35S promoter. *PLoS ONE* **2016**, *11*, e0167927. [CrossRef] [PubMed]
10. Adhikari, B.M.; Bajracharya, A.; Shrestha, A.K. Comparison of nutritional properties of stinging nettle (*Urtica dioica*) flour with wheat and barley flours. *Food Sci. Nutr.* **2016**, *4*, 119–124. [CrossRef] [PubMed]
11. eFloras.org. Available online: http://www.efloras.org/browse.aspx?flora_id=0&name_str=urtica&btnSearch=Search (accessed on 5 April 2018).
12. Plants for a Future. Earth. Plants. People. Available online: https://www.pfaf.org/user/DatabaseSearhResult.aspx (accessed on 5 April 2018).
13. Augspole, I.; Duma, M.; Ozola, B.; Cinkmanis, I. Phenolic Profile of Fresh and Frozen Nettle, Goutweed, Dandelion and Chickweed Leaves. In Proceedings of the 11th Baltic Conference on Food Science and Technology "Food Science and Technology in a Changing World", Jelgava, Latvia, 27–28 April 2017.
14. Rafajlovska, V.; Kavrakovski, Z.; Simonovska, J.; Srbinoska, M. Determination of protein and mineral contents in stinging nettle. *Qual. Life* **2013**, *4*, 26–30. [CrossRef]
15. Gül, S.; Demirci, B.; Başer, K.H.C.; Akpulat, H.A.; Aksu, P. Chemical composition and *in vitro* cytotoxic, genotoxic effects of essential oil from *Urtica dioica* L. *Bull. Environ. Contam. Toxicol.* **2012**, *88*, 666–671. [CrossRef] [PubMed]
16. Kukrić, Z.Z.; Topalić-Trivunović, L.N.; Kukavica, B.M.; Matoš, S.B.; Pavičić, S.S.; Boroja, M.M.; Savić, A.V. Characterization of antioxidant and antimicrobial activities of nettle leaves (*Urtica dioica* L.). *APPTEF* **2012**, *43*, 1–342. [CrossRef]
17. Guil-Guerrero, J.L.; Rebolloso-Fuentes, M.M.; Isasa, M.E.T. Fatty acids and carotenoids from stinging nettle (*Urtica dioica* L.). *J. Food. Compos. Anal.* **2003**, *16*, 111–119. [CrossRef]
18. Kudritsata, S.E.; Filman, G.M.; Zagorodskaya, L.M.; Chikovanii, D.M. Carotenoids of *Urtica dioica*. *Chem. Nat. Compd.* **1986**, *22*, 604–605. [CrossRef]
19. Bağci, E. Fatty Acid Composition of the Aerial Parts of *Urtica Dioica* (Stinging Nettle) L. (*Urticaceae*). In *Biodiversity*; Şener, B., Ed.; Springer: Boston, MA, USA, 2002; pp. 323–327.
20. Orčić, D.; Francišković, M.; Bekvalac, K.; Svirčev, E.; Beara, I.; Lesjak, M.; Mimica-Dukić, N. Quantitative determination of plant phenolics in *Urtica dioica* extracts by high-performance liquid chromatography coupled with tandem mass spectrometric detection. *Food Chem.* **2014**, *143*, 48–53. [CrossRef] [PubMed]
21. Otles, S.; Yalcin, B. Phenolic compounds analysis of root, stalk, and leaves of nettle. *Sci. World J.* **2012**, 564367. [CrossRef] [PubMed]
22. Pinelli, P.; Ieri, F.; Vignolini, P.; Bacci, L.; Baroni, S.; Romani, A. Extraction and HPLC analysis of phenolic compounds in leaves, stalks, and textile fibers of *Urtica dioica* L. *J Agric. Food Chem.* **2008**, *56*, 9127–9132. [CrossRef] [PubMed]
23. Rutto, L.K.; Xu, Y.; Ramirez, E.; Brandt, M. Mineral properties and dietary value of raw and processed stinging nettle (*Urtica dioica* L.). *Int. J. Food Sci.* **2013**, 857120. [CrossRef]
24. Sajfrtová, M.; Sovová, H.; Opletal, L.; Bártlová, M. Near-critical extraction of β-sitosterol and scopoletin from stinging nettle roots. *J. Supercrit. Fluid* **2005**, *35*, 111–118. [CrossRef]
25. Kara, D. Evaluation of trace metal concentrations in some herbs and herbal teas by principal component analysis. *Food Chem.* **2009**, *114*, 347–354. [CrossRef]
26. Tack, F.M.; Verloo, M.G. Metal contents in stinging nettle (*Urtica dioica* L.) as affected by soil characteristics. *Sci. Total Environ.* **1996**, *192*, 31–39. [CrossRef]
27. Keskin-Šašić, I.; Tahirović, A.; Topčagić, A.; Klepo, L.; Salihović, M.; Ibragić, S.; Toromanović, J.; Ajanović, A.; Velispahić, E. Total phenolic content and antioxidant capacity of fruit juices. *Bull. Chem. Technol. Bosnia Herzeg.* **2012**, *39*, 25–28.
28. Ghaima, K.K.; Hashim, N.M.; Ali, S.A. Antibacterial and antioxidant activities of ethyl acetate extract of nettle (*Urtica dioica*) and dandelion (*Taraxacum officinale*). *J. Pharm. Sci.* **2013**, *3*, 96–99.
29. Vajić, U.J.; Grujić-Milanović, J.; Živković, J.; Šavikin, K.; Gođevac, D.; Miloradović, Z.; Bugarski, B.; Mihailović-Stanojević, N. Optimization of extraction of stinging nettle leaf phenolic compounds using response surface methodology. *Ind. Crop. Prod.* **2015**, *74*, 912–917. [CrossRef]

30. Đurović, S.; Pavlić, B.; Šorgić, S.; Popov, S.; Savić, S.; Pertonijević, M.; Radojković, M.; Cvetanović, A.; Zeković, Z. Chemical composition of stinging nettle leaves obtained by different analytical approaches. *J. Funct. Food.* **2017**, *32*, 18–26. [CrossRef]
31. Van Damme, E.J.M.; Peumans, W.J. Isolectin composition of individual clones of *Urtica dioica*: Evidence for phenotypic differences. *Physiol. Plant.* **1987**, *71*, 328–334. [CrossRef]
32. Krauss, R.; Spitteler, G. Phenolic compounds from roots of *Urtica dioica*. *Phytochemistry* **1990**, *29*, 1653–1659. [CrossRef]
33. De la Torre, R.; Corella, D.; Castañer, O.; Martínez-González, M.A.; Salas-Salvador, J.; Vila, J.; Estruch, R.; Sorli, J.V.; Arós, F.; Fiol, M.; et al. Protective effect of homovanillyl alcohol on cardiovascular disease and total mortality: Virgin olive oil, wine, and catechol-methylathion. *Am. J. Clin. Nutr.* **2017**, *105*, 1297–1304. [CrossRef] [PubMed]
34. Blandina, P.; Efoudebe, M.; Cenni, G.; Mannaioni, P.; Passani, M.B. Acetylcholine, histamine, and cognition: Two sides of the same coin. *Learn. Mem.* **2004**, *11*, 1–8. [CrossRef] [PubMed]
35. Aquino, G.; Arias-Montano, J.-A. Neuromodulation and histamine: Regulation of neurotransmitter release by H 3 receptors. *Salud Ment.* **2012**, *35*, 345–352.
36. Salo, P.; Hopia, A.; Ekblom, J.; Lahtinen, R.; Laakso, P. Plant Stanol Ester as a Cholesterol Lowering Ingredient of Benecol® Foods. In *Healthful Lipids*; Akoh, C., Lai, O.-M., Eds.; AOCS Press: Urbana, IL, USA, 2005; pp. 335–360.
37. Yang, J.-Y.; Koo, J.-H.; Yoon, H.-Y.; Lee, J.-H.; Park, B.-H.; Kim, J.-S.; Chi, M.S.; Park, J.-W. Effect of scopoletin on lipoprotein lipase activity in 3T3-L1 adipocytes. *Int. J. Mol. Med.* **2007**, *20*, 527–531. [CrossRef] [PubMed]
38. Francišković, M.; Gonzalez-Pérez, R.; Orčić, D.; Sánchez de Medina, F.; Martínez-Augustin, O.; Svirčev, E.; Simin, N.; Mimica-Dukić, N. Chemical composition and immuno-modulatory effects of *Urtica dioica* L. (stinging nettle) extracts. *Phytother. Res.* **2017**, *31*, 1183–1191. [CrossRef] [PubMed]
39. Kavalali, G.M. *Urtica: The Genus Urtica*; CRC Press: Boca Raton, FL, USA, 2004.
40. Rosłon, W.; Węglarz, Z. Polyphenolic acids of female and male forms of *Urtica dioica*. Presented at the International Conference on Medicinal and Aromatic Plants (Part II), Budapest, Hungary, 8 July 2001; pp. 101–104.
41. Węglarz, Z.; Rosłon, W. Developmental and chemical variation in aboveground organs in the male and female forms of common nettle (*Urtica dioica* L.). *Herb. Polonica* **2000**, *46*, 324–331.
42. Nasiri, S.; Nobakht, A.; Safamehr, A. The effect of different levels of nettle *Urtica dioica* L. (*Urticaceae*) medical plant in starter and grower feeds on performance, carcass traits, blood biochemical and immunity parameters of broilers. *Iran. J. Appl. Anim. Sci.* **2011**, *1*, 177–181.
43. Carvalho, A.R.; Costab, G.; Figueirinha, A.; Liberal, J.; Prior, J.A.V.; Lopes, M.C.; Cruz, M.T.; Batista, M.T. *Urtica* spp.: Phenolic composition, safety, antioxidant and anti-inflammatory activities. *Food Res. Int.* **2017**, *99*, 485–494. [CrossRef] [PubMed]
44. Oliver, F.; Amon, E.U.; Breathnach, A.; Francis, D.M.; Sarathchandra, P.; Black, A.K.; Greaves, M.W. Contact urticaria due to the common stinging nettle (*Urtica dioica*)-histological, ultrastructural and pharmacological studies. *Clin. Exp. Dermatol.* **1991**, *16*, 1–7. [CrossRef] [PubMed]
45. Emmelin, N.; Feldberg, W. Systemic effect of adenosine triphosphate. *Br. J. Pharmacol.* **1948**, *3*, 273–284. [CrossRef]
46. Angela, M.; Meireles, A. *Extracting Bioactive Compounds for Food Products: Theory and Applications*; CRC Press: Boca Raton, FL, USA, 2009.
47. Marrelli, M.; Menichini, F.; Statti, G.A.; Bonesi, M.; Duez, P.; Menichini, F.; Conforti, F. Changes in the phenolic and lipophilic composition, in the enzyme inhibition and antiproliferative activity of *Ficus carica* L. cultivar Dottato fruits during maturation. *Food Chem. Toxicol.* **2012**, *50*, 726–733. [CrossRef] [PubMed]
48. Gülçin, I.; Küfrevioglu, O.I.; Oktay, M.; Büyükokuroglu, M.E. Antioxidant, antimicrobial, antiulcer and analgesic activities of nettle (*Urtica dioica* L.). *J. Ethnopharmacol.* **2004**, *90*, 205–215. [CrossRef] [PubMed]
49. Rafajlovska, V.; Djarmati, Z.; Najdenova, V.; Cvetkov, L. Extraction of stinging nettle (Urtica dioica L.) with supercritical carbon dioxide. Presented at the 1st Black Sea Basin Conference on Analytical Chemistry, Odessa, Ukraine, 11–15 September 2001.
50. Salih, N.A.; Arif Edh, D.J. Antibacterial effect of nettle (*Urtica dioica*). *AL-Qadisiyah J. Vet. Med. Sci.* **2014**, *13*, 1–6.

51. Körpe, D.A.; Işeri, O.D.; Sahin, F.I.; Cabi, E.; Haberal, M. High-antibacterial activity of *Urtica* spp. seed extracts on food and plant pathogenic bacteria. *Int. J. Food Sci. Nutr.* **2013**, *64*, 355–362. [CrossRef] [PubMed]
52. Ghaedi, M.; Naghiha, R.; Jannesar, R.; Dehghanian, N.; Mirtamizdoust, B.; Pezeshkpour, V. Antimicrobial and antifungal activity of flower extract of *Urtica dioica*, *Chamaemelum nobile* and *Salvia officinalis*: Effect of $Zn[OH]_2$ nanoparticles and Hp-2-minh on their property. *J. Ind. Eng. Chem.* **2015**, *32*, 353–359. [CrossRef]
53. Mzid, M.; Khedir, S.B.; Salem, M.B.; Regaieg, W.; Rebai, T. Antioxidant and antimicrobial activities of ethanol and aqueous extracts from *Urtica urens*. *Pharm. Biol.* **2017**, *55*, 775–781. [CrossRef] [PubMed]
54. Salehzadeh, A.; Asadpour, L.; Naeemi, A.S.; Houshmand, E. Antimicrobial activity of methanolic extracts of *Sambucus ebulus* and *Urtica dioica* against clinical isolates of methicillin resistant *Staphylococcus aureus*. *Afr. J. Tradit. Complement. Altern. Med.* **2014**, *11*, 38–40. [CrossRef] [PubMed]
55. Ranjbar, R.; Shamami, K.S. Evaluation of antimicrobial activity of *Eucalyptus* essential oil and *Urtica* alcoholic extract on *Salmonella enteritidis* and *Shigella dysenteriae in vitro* conditions. *Bull. Environ. Pharmacol. Life Sci.* **2015**, *4*, 56–59.
56. Jyoti, K.; Baunthiyal, M.; Singh, A. Characterization of silver nanoparticles synthesized using *Urtica dioica* Linn. leaves and their synergistic effects with antibiotics. *J. Radiat. Res. Appl. Sci.* **2016**, *9*, 217–227. [CrossRef]
57. Modarresi-Chahardehi, A.; Ibrahim, D.; Sulaiman, S.F.; Mousavi, L. Screening antimicrobial activity of various extracts of *Urtica dioica*. *Rev. Biol. Trop.* **2012**, *60*, 1567–1576. [CrossRef] [PubMed]
58. Dar, S.A.; Ganai, F.A.; Yousuf, A.R.; Balkhi, M.U.; Bhat, T.M.; Sharma, P. Pharmacological and toxicological evaluation of *Urtica dioica*. *Pharm. Biol.* **2013**, *51*, 170–180. [CrossRef] [PubMed]
59. Singh, R.; Dar, S.A.; Sharma, P. Antibacterial activity and toxicological evolution of semi purified hexane extract of *Urtica dioica* leaves. *Res. J. Med. Plant.* **2012**, *6*, 123–135.
60. Ramtin, M.; Massiha, A.; Khoshkholgh-Pahlaviani, M.R.M.; Issazadeh, K.; Assmar, M.; Zarrabi, S. In vitro antimicrobial activity of *Iris pseudacorus* and *Urtica dioica*. *Zahedan J. Res. Med. Sci.* **2013**, *16*, 35–39.
61. Zenão, S.; Aires, A.; Dias, C.; Saavedra, M.J.; Fernandes, C. Antibacterial potential of *Urtica dioica* and *Lavandula angustifolia* extracts against methicillin resistant *Staphylococcus aureus* isolated from diabetic foot ulcers. *J. Herb. Med.* **2017**, *10*, 53–58. [CrossRef]
62. Mahmoudi, R.; Amini, K.; Fakhri, O.; Alem, M. Aroma profile and antimicrobial properties of alcoholic and aqueous extracts from root, leaf and stalk of nettle (*Urtica dioica* L.). *J. Microbiol. Biotechnol. Food Sci.* **2014**, *4*, 220–224. [CrossRef]
63. Antolak, H.; Czyżowska, A.; Kręgiel, D. Antibacterial and antiadhesive activities of extracts from edible plants against soft drink spoilage by *Asaia* spp. *J. Food Protect.* **2017**, *80*, 5–34. [CrossRef] [PubMed]
64. Shale, T.; Stirk, W.; Van Standen, J. Screening of medicinal plants used in Lesotho for anti-bacterial and anti-inflammatory activity. *J. Ethnopharmacol.* **1999**, *67*, 347–354. [CrossRef]
65. Özkan, A.; Yumrutaş, Ö.; Saygideğer, S.D.; Kulak, M. Evaluation of antioxidant activities and phenolic contents of some edible and medicinal plants from Turkey's flora. *Adv. Environ. Biol.* **2011**, *5*, 231–236.
66. Pourmorad, F.; Hosseinimehr, S.J.; Shahabimajd, N. Antioxidant activity, phenol and flavonoid contents of some selected Iranian medicinal plants. *Afr. J. Biotechnol.* **2006**, *5*, 1142–1145.
67. Viegi, L.; Pieroni, A.; Guarrera, P.M.; Vangelisti, R. A review of plants used in folk veterinary medicine in Italy as basis for a databank. *J. Ethnopharmacol.* **2003**, *89*, 221–244. [CrossRef] [PubMed]
68. Safamehr, A.; Mirahmadi, M.; Nobakht, A. Effect of nettle (*Urtica dioica*) medicinal plant on growth performance, immune responses, and serum biochemical parameters of broiler chickens. *Int. Res. J. Appl. Basic Sci.* **2012**, *3*, 721–728.
69. Sökeland, J. Combined sabal and *Urtica extract* compared with finasteride in men with benign prostatic hyperplasia: Analysis of prostate volume and therapeutic outcome. *BJU Int.* **2000**, *86*, 439–442. [CrossRef] [PubMed]
70. Koch, E. Extracts from fruits of saw palmetto (*Sabal serrulata*) and roots of stinging nettle (*Urtica dioica*): Viable alternatives in the medical treatment of benign prostatic hyperplasia and associated lower urinary tracts symptoms. *Plant Med.* **2001**, *67*, 489–500. [CrossRef] [PubMed]
71. Randall, C.; Randall, H.; Dobbs, F.; Hutton, C.; Sanders, H. Randomized controlled trial of nettle sting for treatment of base-of-thumb pain. *J. R. Soc. Med.* **2000**, *93*, 305–309. [CrossRef] [PubMed]
72. Christensen, R.; Bliddal, H. Is Phytalgic® a goldmine for osteoarthritis patients or is there something fishy about this nutraceutical? A summary of findings and risk-of-bias assessment. *Arthritis Res. Ther.* **2010**, *12*, 105. [CrossRef] [PubMed]

73. Klingelhoefer, S.; Obertreis, B.; Quast, S.; Behnke, B. Antirheumatic effect of IDS 23, a stinging nettle leaf extract, on in vitro expression of T helper cytokines. *J. Rheumatol.* **1999**, *26*, 2517–2522. [PubMed]
74. Roschek, B., Jr.; Fink, R.C.; McMichael, M.; Alberte, R.S. Nettle extract (*Urtica dioica*) affects key receptors and enzymes associated with allergic rhinitis. *Phytother. Res.* **2009**, *23*, 920–926. [CrossRef] [PubMed]
75. D'Amato, G. Pollen Allergy in Europe. The UCB Institute of Allergy. Available online: https://www.ucb.com/_up/tuioa_com/images/PollenAllergy-DAmato-simplified-V2-070910_PP.pdf (accessed on 5 April 2018).
76. Khare, V.; Kushwaha, P.; Verma, S.; Gupta, A.; Srivastava, S.; Rawat, A.K.S. Pharmacognostic evaluation and antioxidant activity of *Urtica dioica* L. *Chin. Med. UK* **2012**, *3*, 128–135. [CrossRef]
77. Lalitharani, S.; Mohan, V.R.; Regini, G.S. GC-MS analysis of ethanolic extract of *Zanthoxylum rhetsa* (roxb.) dc spines. *J. Herb. Med. Toxicol.* **2010**, *4*, 191–192.
78. Taylor, R.S.; Edel, F.; Manandhar, N.P.; Towers, G.H. Antimicrobial activities of southern Nepalese medicinal plants. *J. Ethnopharmacol.* **1996**, *50*, 97–102. [CrossRef]
79. Cheung, L.M.; Cheung, P.C.K.; Ooi, V.E.C. Antioxidant activity and total phenolics of edible mushroom extracts. *Food Chem.* **2003**, *81*, 249–255. [CrossRef]
80. Wang, T.; Jónsdóttir, R.; Ólafsdóttir, G. Total phenolic compounds, radical scavenging and metal chelation of extracts from Icelandic seaweeds. *Food Chem.* **2009**, *116*, 240–248. [CrossRef]
81. Fisgin, N.S.; Cayci, Y.T.; Coban, A.Y.; Ozatli, D.; Tanyel, E.; Durupinar, B.; Tulek, N. Antimicrobial activity of plant extract Ankaferd Blood Stopper®. *Fiterapia* **2009**, *80*, 48–50. [CrossRef] [PubMed]
82. Bourgeois, C.; Leclerc, É.A.; Corbin, C.; Doussot, J.; Serrano, V.; Vanier, J.R.; Seigneuret, J.-M.; Auguina, D.; Pichon, C.; Lainé, É.; Hano, C. Nettle (*Urtica dioica* L.) as a source of antioxidant and anti-aging phytochemicals for cosmetic applications. *C. R. Chim.* **2016**, *16*, 1090–1100. [CrossRef]
83. Matsingou, T.C.; Kapsokefalou, M.; Salifoglou, A. Aqueous infusions of Mediterranean herbs exhibit antioxidant activity towards iron promoted oxidation of phospholipids, linoleic acid, and deoxyribose. *Free Radic. Res.* **2001**, *35*, 593–605. [CrossRef] [PubMed]
84. Sharafetdinov, Kh.Kh; Kiseleva, T.L; Kochetkova, A.A.; Mazo, V.K. Promising plant sources of anti-diabetic micronutrients. *J. Diabetes Metab.* **2017**, *8*, 778.
85. Wahba, H.E.; Motawe, H.M.; Ibrahim, A.Y. Growth and chemical composition of *Urtica pilulifera* L. plant as influenced by foliar application of some amino acids. *J. Mater. Environ. Sci.* **2015**, *6*, 499–506.
86. Mukundi, M.J.; Mwaniki, N.E.N.; Piero, N.M.; Murugi, N.J.; Kelvin, J.K.; Yusuf, A.A.; Mwonjoria, K.J.; Ngetich, K.A.; Agyirifo, S.D.; Gathumbi, K.P.; et al. Potential anti-diabetic effects and safety of aqueous extracts of *Urtica dioica* collected from Narok County, Kenya. *Pharm. Anal. Acta* **2017**, *7*, 548. [CrossRef]
87. Uprety, Y.; Poudel, R.C.; Shrestha, K.K.; Rajbhandary, S.; Tiwari, N.N.; Shrestha, U.B.; Asselin, H. Diversity of use and local knowledge of wild edible plant resources in Nepal. *J. Ethnobiol. Ethnomed.* **2012**, *8*, 16. [CrossRef] [PubMed]
88. Bekele, B.; Melesse, A.; Beyan, M.; Berihun, K. The effect of feeding stinging nettle (*Urtica simensis* S.) leaf meal on feed intake, growth performance and carcass characteristics of hubbard broiler chickens. *Glob. J. Sci. Front. Res.* **2015**, *15*. Available online: https://globaljournals.org/GJSFR_Volume15/1-The-Effect-of-Feeding.pdf (accessed on 15 June 2018).
89. Costa, H.S.; Albuquerque, T.G.; Sanches-Silva, A.; Vasilopoulou, E.; Trichopoulou, A.; D'Antuono, L.F.; Alexieva, I.; Boyko, N.; Costea, C.; Fedosova, K.; et al. New nutritional composition data on selected traditional foods consumed in Black Sea Area countries. *J. Sci. Food Agric.* **2013**, *93*, 3524–3534. [CrossRef] [PubMed]
90. Fiola, C.; Prado, D.; Mora, M.; Alava, J.I. Nettle cheese: Using nettle leaves (*Urtica dioica*) to coagulate milk in the fresh cheese making process. *Int. J. Gastronomy Food Sci.* **2016**, *4*, 19–24. [CrossRef]
91. Sánchez-Mata, M.; Tardio, J. *Mediterranean Wild Edible Plants*; Springer: New York, NY, USA, 2016.
92. Kovacs, J. Compositions of Oats and Nettle Extracts to be Used as a Food Additive or Pharmaceutical Preparation in Human Health Care. U.S. Patent 4886665, 12 December 1989.
93. Antolak, H.; Kręgiel, D. Food preservatives from plants. In *Food Additives*; Karunaratne, D.N., Pamunuwa, G., Eds.; InTech: Berlin, Germany, 2017; pp. 45–85.
94. Johnson, A.; Johnson, S. Garden plants poisonous to people. *Primefact* **2006**, *359*. Available online: https://www.dpi.nsw.gov.au/__data/assets/pdf_file/0008/112796/garden-plants-poisonous-to-people.pdf (accessed on 25 June 2018).

95. Papuc, C.; Goran, G.V.; Predescu, C.N.; Nicorescu, V. Mechanisms of oxidative processes in meat and toxicity induced by postprandial degradation products: A review. *Compr. Rev. Food Sci. Food Saf.* **2017**, *16*, 96–123. [CrossRef]
96. Shah, M.A.; Bosco, S.J.D.; Mir, S.A. Plant extracts as natural antioxidants in meat and meat products. *Meat Sci.* **2014**, *98*, 21–33. [CrossRef] [PubMed]
97. Inai, M.; Miura, Y.; Honda, S.; Masuda, A.; Masuda, T. Metmyoglobin reduction by polyphenols and mechanism of the conversion of metmyoglobin to oxymyoglobin by quercetin. *J. Agric. Food Chem.* **2014**, *62*, 893–901. [CrossRef] [PubMed]
98. Aksu, M.I.; Kaya, M. Effect of usage *Urtica dioica* L. on microbiological properties of sucuk, a Turkish dry-fermented sausage. *Food Control* **2004**, *8*, 591–595. [CrossRef]
99. Kaban, G.; Aksu, M.İ.; Kaya, M. Behaviour of *Staphylococcus aureus* in sucuk with nettle (*Urtica dioica* L.). *J. Food Saf.* **2007**, *27*, 400–410. [CrossRef]
100. Karabacak, S.; Bozkurt, H. Effects of *Urtica dioica* and *Hibiscus sabdariffa* on the quality and safety of sucuk (Turkish dry-fermented sausage). *Meat Sci.* **2007**, *78*, 288–296. [CrossRef] [PubMed]
101. Alp-Erbay, E.; Aksu, M.I. Effects of water extract of *Urtica dioica* L. and modified atmosphere packaging on the shelf life of ground beef. *Meat Sci.* **2010**, *86*, 468–473.
102. Oz, F. Effects of water extract of *Urtica dioica* L. on the quality of meatballs. *J. Food Process. Pres.* **2014**, *38*, 1356–1363. [CrossRef]
103. Ahmadi, M.; Razavilar, V.; Motallebi, A.A.; Esmailzadeh Kenari, R.; Khanipour, A.A. Effects of hydroalcoholic and water extracts of nettle leaf (*Urtica dioica* L.) on chemical properties of superchilled minced meat of common Kilka (Clupeonella cultriventris caspia). *J. Food Qual. Hazards Control* **2014**, *1*, 85–88.
104. Aksu, M.I.; Alinezhad, H.; Erdemir, E. Effect of lyophilized water extract of *Urtica dioica* L. on the shelf life of vacuum-packaged beef steaks. *J. Food Process Pres.* **2015**, *39*, 3059–3066. [CrossRef]
105. Latoch, A.; Stasiak, D.M. Effect of water extract of *Urtica dioica* L. on lipid oxidation and color of cooked pork sausage. *J. Food Process Pres.* **2017**, *41*, e12818. [CrossRef]
106. Gill, C. Herbs and plant extracts as growth enhancers. *Feed Int.* **1999**, *20*, 20–23.
107. Langhout, P. New additives for broiler chickens. *J. World's Poult. Sci. J.* **2000**, *16*, 22–27.
108. Madrid, J.; Hernández, F.; García, V.; Orengo, J.; Megías, M.D.; Sevilla, V. Effect of plant extracts on ileal apparent digestibility and carcass yield in broilers at level of farm. Presented at the 14th European Symp on Poultry Nutrition, Lillehammer, Norway, 10–14 August 2003; p. 187.
109. Alçiçek, A.; Bozkurt, M.; Çabuk, M. The effects of an essential oil combination derived from selected herbs growing wild in Turkey on broiler performance. *S. Afr. J. Anim. Sci.* **2004**, *33*, 89–94.
110. Zhang, K.Y.; Yan, F.; Keen, C.A.; Waldroup, P.W. Evaluation of microencapsulated essential oils and organic acids in diets for broiler chickens. *Int. J. Poult. Sci.* **2005**, *4*, 612–619.
111. Wenk, C. Why all the discussion about herbs? In *Feed Industry, Proceedings of the Alltech's 16th Anniversary Symposium Biotechnology*; Lyons, T.P., Ed.; Alltech Tech.: Nottingham, UK; University Press: Nicholasvile, KY, USA, 2000; pp. 79–96.
112. Szewczyk, A.; Hanczakowska, E.; Świątkiewicz, M. The effect of nettle (*Urtica dioica*) extract on fattening performance and fatty acid profile in the meat and serum lipids of pigs. *J. Anim. Feed Sci.* **2006**, *15*, 81–84. [CrossRef]
113. Nobakht, A. Effects of different levels of chicory (*Cichorium intybus* L.), zizaphora (*Zizaphora tenuior* L.), nettle (*Urtica dioica* L.) and savoury (*Satureja hortensis* L.) medicinal plants on carcass characteristics of male broilers. *J. Med. Plants Res.* **2011**, *5*, 4354–4359.
114. Toldy, A.; Stadler, K.; Sasvari, M.; Jakus, J.; Jung, K.J.; Chung, H.Y.; Berkes, I.; Nyakas, C.; Radák, Z. The effect of exercise and nettle supplementation on oxidative stress markers in the rat brain. *Brain Res. Bull.* **2005**, *65*, 487–493. [CrossRef] [PubMed]
115. Humphries, D.J.; Reynolds, C.K. The effect of adding stinging nettle (*Urtica dioica*) haylage to a total mixed ration on performance and rumen function of lactating dairy cows. *Anim. Feed Sci. Technol.* **2014**, *189*, 72–81. [CrossRef]
116. Williams, P.; Losa, R. The use of essential oils and their compounds in poultry nutrition. *J. World's Poult. Sci. J.* **2001**, *17*, 14–15.
117. McCartney, E. The natural empire strikes back. *Poult. Int.* **2002**, *41*, 36–42.

118. Şandru, C.D.; Niculae, M.; Popescu, S.; Paştiu, A.I.; Páll, E.; Spînu, M. *Urtica dioica* alcoholic extract increases the cell-mediated innate. *Ind. Crop Prod.* **2016**, *88*, 48–50. [CrossRef]
119. Avci, G.; Kupeli, E.; Eryavuz, A.; Yesilada, E.; Kucukkurt, I. Antihypercholesterolaemic and antioxidant activity assess ment of some plants used as remedy in Turkish folk medicine. *J. Ethnopharmacol.* **2006**, *107*, 418–423. [CrossRef] [PubMed]
120. Mavi, A.; Terzi, Z.; Ozgen, U.; Yildirin, A.; Coskun, M. Antioxidant properties of some medicinal plants: *Prangos ferulacea* (*Apiaceae*), *Sedum* (*Crassulaceae*), *Malva neglecta* (*Malvaceae*), *Cruciata taurica* (*Rubiaceae*), *Rosa pimpinellifolia* (*Rosaceae*), *Galium verum* subsp. *verum* (*Rubiaceae*), *Urtica dioica* (*Urticaceae*). *Biol. Pharm. Bull.* **2004**, *27*, 702–705. [CrossRef] [PubMed]
121. Fermont, L.; Gozzelino, M.T.; Linard, A. Response of plasma lipids to dietary cholesterol and wine polyphenols in rats fed polyunsaturated fat diets. *Lipids* **2000**, *35*, 991–999. [CrossRef]
122. Visioli, F.; Galli, C.; Galli, G.; Caruso, D. Biological activities and metabolic fate of olive oil phenols. *J. Lipid Sci. Technol.* **2002**, *104*, 677–684. [CrossRef]
123. Khosravi, A.; Boldaji, F.; Dastar, B.; Hasani, S. The use of some feed additives as growth promoter in broilers nutrition. *Int. J. Poult. Sci.* **2008**, *7*, 1095–1099. [CrossRef]
124. Wang, R.J.; Li, D.F.; Bourne, S.; Wang, R.J.; Li, D.F. Can 2000 years of herbal medicine history help us solve problems in the year 2000? Presented at the Alltech's 14th Annual Symposium on Biotechnology in the Feed Industry, Nicholasville, KY, USA, 1998; pp. 273–292.
125. World Health Organization. *WHO Monographs on Selected Medicinal Plants*; WHO: Geneva, Switzerland, 2004; Volume 2, pp. 1–358.
126. European Commission. *Directorate—General for Health and Food Safety*; Final Review Report for the Basic Substance *Urtica* spp.; European Commission: Brussels, Belgium, 2017.

molecules

MDPI

Article

Olive Leaf Addition Increases Olive Oil Nutraceutical Properties

Imen Tarchoune [1], Cristina Sgherri [2,*], Jamel Eddouzi [3], Angela Zinnai [2,4], Mike Frank Quartacci [2,4,*] and Mokhtar Zarrouk [1]

1 Laboratoire de Biotechnologie d'Olivier, Centre de Biotechnologie de Borj-Cedria, B.P. 901, 2050 Hammam-Lif, Tunisia; imen_tarchoune@yahoo.fr (I.T.); zarrouk.mokhtar@gmail.com (M.Z.)

2 Department of Agriculture, Food and Environment (DAFE), Via del Borghetto 80, 56124 Pisa, Italy; angela.zinnai@unipi.it

3 Laboratoire de Biotechnologie Appliquée à l'Agriculture, INRA Tunis, Université de Carthage, El Menzah, 1004 Tunis, Tunisia; douzi.jamel@gmail.com

4 Interdepartmental Research Centre "Nutraceuticals and Food for Health", University of Pisa, Via del Borghetto 80, 56124 Pisa, Italy

* Correspondence: cristina.sgherri@unipi.it (C.S.); mike.frank.quartacci@unipi.it (M.F.Q.); Tel.: +39-050-2216609 (C.S.); +39-050-2216612 (M.F.Q.)

Academic Editor: Lillian Barros

Received: 17 January 2019; Accepted: 1 February 2019; Published: 2 February 2019

Abstract: The aim of the present research was to study the effects of olive leaf addition (0 and 3%) on the major antioxidants and the antioxidant activity of Neb Jmel and Oueslati olive oils. Olives and leaves of the two Tunisian varieties were harvested during the 2016/2017 crop season. Both leaves and oils were characterised for their concentrations in phenolics, tocopherols and antioxidant power. Other parameters such as free acidity, peroxide value, chlorophyll and carotenoid concentrations were also taken into consideration. Compared to Oueslati, the Neb Jmel oil showed a lower free acidity (50%) and peroxide value (5.6-fold), and higher chlorophyll (1.6-fold), total phenolics (1.3-fold), flavonoid (3-fold) and oleuropein derivative (1.5-fold) concentrations, in addition to an increased antioxidant activity (1.6-fold). Leaf addition promoted a significant increment in total chlorophyll, α-tocopherol and phenolics in both varieties, above all in Oueslati oil, due to a higher abundance of bioactive constituents in the corresponding leaves. In particular, chlorophyll and carotenoid concentrations reached values twice higher than in Neb Jmel leaves, and flavonoids and oleouperin derivatives were three-fold higher. This prevented the oxidation and the formation of peroxides, reducing the peroxide value of the fortified oil to the half. The results provide evidence on the performance of the Tunisian Neb Jmel and Oueslati varieties, showing that their oils present a chemical profile corresponding to the extra virgin olive oil category and that, after leaf addition, their nutritional value was improved.

Keywords: extra virgin olive oil; leaf addition; Tunisian varieties; phenolics; tocopherols; antioxidant activity

1. Introduction

In the Mediterranean area, the olive tree is so economically important that oliviculture is one of the most widespread agricultural activities. It is estimated that about 8 million ha are cultivated with olive trees [1,2]. In particular, in Tunisia, an area of 1.7 million ha is covered by olive trees, which results in the production of more than 4% of the global olive oil amount. Due to the 170,000 tons produced per year, Tunisia is the fourth largest producer and exporter of olive oil in the world [3].

The nutritional and health-promoting effects of olives and olive oils are more and more recognized [4]. Indeed, it is well known that olive oil is a potential antioxidant [5–7] showing anti-inflammatory [6,8,9], cardioprotective [6,10], anticancer [6,11], antidiabetic [12] and neuroprotective effects [13].

The importance of olive oil is related to its high amounts of monounsaturated fatty acids and to the presence of low-represented components such as α-tocopherol, phenolics, chlorophyll and carotenoids. Phenols are among the most important nutraceutical compounds because of their nutritional and sensorial characteristics [14]. Phenolic compounds and tocopherols play a protective role against oxidative stress [15] and are able to extend the extra virgin olive oil shelf-life due to their antioxidative properties [16]. Olive oil colour is determined by its pigment composition and concentration, especially with reference to chlorophyll. In the dark, this pigment is also endowed with antioxidant activity, even if under light conditions it can act as a prooxidant, reacting with triplet oxygen to form the excited-state singlet oxygen [17]. Oxidation is the most important process that causes quality deterioration of olive oil because of its high concentration of unsaturated fatty acids. Consequently, synthetic antioxidants have been used as food additives to improve oil stability. As many studies suggest that the use of synthetic antioxidants may lead to health risks [18], considerable attention has been recently focused on plant phenolics due to their healthful and nutritional effects [19]. Recent papers have provided evidence that olive leaves have a high antioxidant activity, originated by the presence of phenolics, thus exhibiting strong preventive effects against oxidation [18,20,21]. Olive leaves are a by-product of olive cultivation and can be easily obtained either from pruning or olive oil industry as a waste product [1].

The present research was designed to verify a possible improvement of the nutritional value of two Tunisian olive oils by adding olive leaves, as a source of natural antioxidants, during oil extraction.

2. Results and Discussion

2.1. Quality Parameters

According to the definitions and standards established by the International Olive Oil Council [22], the classification of virgin olive oil into different categories depends on its chemical, physical and sensory parameters, among which the degree of acidity and the peroxide value (Table 1) are the most used. In this study, the differences observed between the two oils could be due, besides variety, to the different geographical location of the plants in terms of climate and soil composition.

Table 1. Quality parameters and chlorophyll and carotenoid contents of Neb Jmel and Oueslati olive oils extracted with and without addition of olive leaves (3%).

Parameter	Neb Jmel		Oueslati	
	Oil	Oil + Leaves	Oil	Oil + Leaves
Free acidity (%C18:1)	0.56 ± 0.05 [b]	0.57 ± 0.03 [b]	1.00 ± 0.00 [a]	0.60 ± 0.01 [b]
Peroxide value (meq O_2/kg)	6.00 ± 0.71 [c]	6.00 ± 0.50 [c]	34.00 ± 9.53 [a]	15.33 ± 3.21 [b]
Chlorophyll (mg/kg)	8.37 ± 1.32 [b]	14.91 ± 2.91 [a]	5.33 ± 1.22 [c]	16.51 ± 2.95 [a]
Carotenoids (mg/kg)	2.95 ± 0.13 [c]	5.21 ± 0.54 [b]	2.76 ± 0.60 [c]	7.26 ± 0.61 [a]

Data are means of three independent experiments ± SE (n = 3). Means followed by different letters are significantly different at $p \leq 0.05$ as determined by Duncan's multiple range test. SE, standard error; meq, milliequivalent.

2.2. Free Acidity

Oil acidity is a simple and effective parameter to evaluate and classify a commercial-grade olive oil [23]. The extraction of olive oil from fresh undamaged fruits carried out following a correct crushing procedure gives oils with very low acidity [23]. However, during extraction and storage the olive oil can be altered by the release of free fatty acids as a consequence of the hydrolysis of triglycerides, thus increasing the free acidity.

The acidity values of Neb Jmel and Oueslati olive oils, extracted without (control) and with the addition of 3% olive leaves, are reported in Table 1. The Neb Jmel oils showed very low values (0.6%).

The addition of leaves during oil extraction did not affect the free acidity in this variety. Regarding Oueslati, acidity of oils (1%) was higher in comparison with the Neb Jmel ones. However, the addition of leaves influenced Oueslati oil quality, decreasing free acidity to 0.6%. Our results agree with those reported by Ben Mansour et al. [24] for the Neb Jmel variety, whereas the free acidity of Oueslati oils was higher in comparison with the findings of Ouni et al. [25] on the same variety. This behaviour could be related to the use of olives at an advanced stage of maturation. Indeed, a late harvest of olives may alter oil acidity by increasing the lipolytic enzyme activities [26]. However, in our case the addition of olive leaves was responsible for the decrease of free acidity likely due to the presence of antioxidant compounds. Consequently, fortified Oueslati oils could still be classified as extra virgin olive oils as the free acidity value was lower than 0.8%.

2.3. Peroxide Value

Peroxides are intermediate products of oil oxidation which originate a complex mixture of volatile compounds such as aldehydes, ketones, hydrocarbons, alcohols and esters. These compounds are responsible for the alteration of the organoleptic characteristics [27], dramatically reducing oil shelf-life as well as consumer acceptance. Furthermore, also light and high temperatures are well-known factors generally promoting peroxide formation [28].

Table 1 shows that the peroxide value of Neb Jmel oils was well below the established limit (<20 milliequivalents (meq) O_2/kg) for all the categories of olive oil [22]. In contrast, a high peroxide value was observed in the Oueslati control oil (34 meq O_2/kg), so that it could no longer be classified as an extra virgin olive oil. However, the addition of leaves during oil extraction and processing prevented the oxidation and the formation of peroxides, reducing the peroxide value to half. Contrary to our results, Malheiro et al. [17] reported that leaf addition increased the peroxide value. Such discrepancy could be explained with differences in the relative presence of additional antioxidants and of leaf residues. These constituents, through gas exchanges occurring during the respiration process, may have increased the availability of oxygen, thus inducing peroxidation.

2.4. Chlorophylls and Carotenoids

Chlorophylls and carotenoids play important roles in olive oils. They interfere with the oxidative stability, acting as antioxidants in the dark or as prooxidants when exposed to light [29]. Furthermore, these compounds are responsible for the yellow-green pigmentation of olive oils, increasing consumer acceptability. Chlorophyll and carotenoid concentrations of olive oils and leaves are reported in Tables 1 and 2, respectively.

Table 2. Chlorophyll, carotenoid and total phenols, flavonoids and tocopherol (α and γ) contents, and ABTS radical cation ($ABTS^{\bullet+}$) scavenging activity of Neb Jmel and Oueslati olive leaves.

Parameter	Neb Jmel	Oueslati
Chlorophyll (μg/g DW)	506.08 ± 10.22 ^{b}	829.29 ± 80.06 ^{a}
Carotenoids (μg/g DW)	26.90 ± 4.45 ^{b}	44.33 ± 4.38 ^{a}
Total phenolics (mg GA eq/g DW)	62.84 ± 1.29 ^{b}	67.91 ± 3.18 ^{a}
Total flavonoids (mg CA eq/g DW)	5.85 ± 1.16 ^{b}	13.61 ± 0.81 ^{a}
$ABTS^{\bullet+}$ (μmol TE/g DW)	69.05 ± 3.26 ^{b}	113.84 ± 5.11 ^{a}
γ (μg/g DW)	tr	tr
α (μg/g DW)	82.37 ± 9.95 ^{a}	10.12 ± 0.84 ^{b}

Data are means of six independent experiments ± SE (n = 6). Means followed by different letters are significantly different at $p \leq 0.05$ as determined by Duncan's multiple range test. DW, dry weight; GA, gallic acid; CA, catechin; TE, trolox equivalent; α, alpha tocopherol; γ, gamma tocopherol, tr, trace.

The Neb Jmel oil showed a higher chlorophyll value (Table 1) in comparison with Oueslati. However, Oueslati leaves (Table 2) displayed the highest chlorophyll (829.29 μg/g) and carotenoid (44.3 μg/g) concentrations, highlighting the role that the addition of chlorophyll could have had in the oxidative stability of the olive oil. This can further explain the difference with the results reported by Malheiro et al. [17] concerning the peroxide value, as in their study olive leaf addition did not affect the chlorophyll content.

In the present work, the addition of leaves (3%) to Neb Jmel and Oueslati oils enhanced chlorophyll and carotenoid concentrations. In fact, chlorophyll reached the same level in both fortified oils, even if Oueslati oil was the one that showed the highest carotenoid concentration. Therefore, significant effects of leaf addition on oil pigments were observed in both varieties (Tables 1 and 2). The addition of leaves also turned the olive oils greener, this visual observation being clear in both oils, and was very likely associated with the increased pigment concentration. The increase in chlorophyll concentrations makes the fortified oils interesting from a nutritional point of view due to the antioxidant activity of chlorophyll and its potential to exert chemopreventive actions against carcinogens [30].

2.5. Total Phenolics

Phenolics are important components for olive oil quality and organoleptic characteristics. Moreover, they are very effective antioxidants playing an important role in human diet and health [3]. Current evidences strongly support the contribution of phenols to the prevention of cancer, cardiovascular and neurodegenerative diseases. The shelf-life of an oil is also correlated with its natural antioxidant amount [16]. Indeed, phenolics delay the oxidative degradation process, thus extending the shelf-life of the product [3,15].

Figure 1 shows that total phenolic concentration of the Neb Jmel oil (736 mg GA eq/kg oil) was significantly higher than that of Oueslati (528 mg GA eq/kg oil). In the literature, it was reported that the total phenolic concentration of Neb Jmel olive oils varies from 562 to 1167 mg GA eq/kg oil [24], whereas for Oueslati it changes from 100 to 859 mg GA eq/kg oil [25,31]. The effect of the geographic location on phenols can be evidenced from the different behaviour showed by each variety [24,25,31]. Indeed, Neb Jmel oil (from the north of Tunisia) showed a higher phenol concentration than Oueslati (from central Tunisia). The present data confirm previous findings on the effect of the variety on phenol concentration in oils [24,25,31].

Oueslati leaves showed a higher phenolic concentration than Neb Jmel ones (Table 2). It is worth noting that leaf addition, increasing total phenolic concentration by 44 and 10% in Oueslati and Neb Jmel oils, respectively, (Figure 1), determined the same mean value in the fortified products. Similar findings were also confirmed by other studies [32,33].

2.6. Total Flavonoids

Flavonoids are plant secondary metabolites with different phenolic structures. These compounds are used mostly to generate pigments which play an important role in the colours of plants. During the past decade, many studies have reported their beneficial effects on human health [34,35]. Indeed, flavonoids display important anti-inflammatory, antiallergic and anticancer activities as well as antiviral properties [34,35]. In this study, total flavonoid concentration was determined in both the oils and leaves of the two Tunisian olive cultivars Neb Jmel and Oueslati (Table 2, Figure 1). The results point out that Oueslati leaves showed a concentration of total flavonoids two-fold higher than Neb Jmel. However, in Oueslati oil the total flavonoid content was about half than that found in Neb Jmel one (Figure 1). The effect of leaf addition during oil extraction on these compounds was remarkable. Our results showed an increase in total flavonoids by 22% in Neb Jmel oils and by 160% in Oueslati, thus determining a not significant difference between the two fortified oils. This was likely due to the highest level of total flavonoids of Oueslati leaves. According to the present research, the findings of Ebrahimi et al. [36] reported values ranging from 156 to 361 mg rutin eq/kg for refined and crude olive oils, respectively.

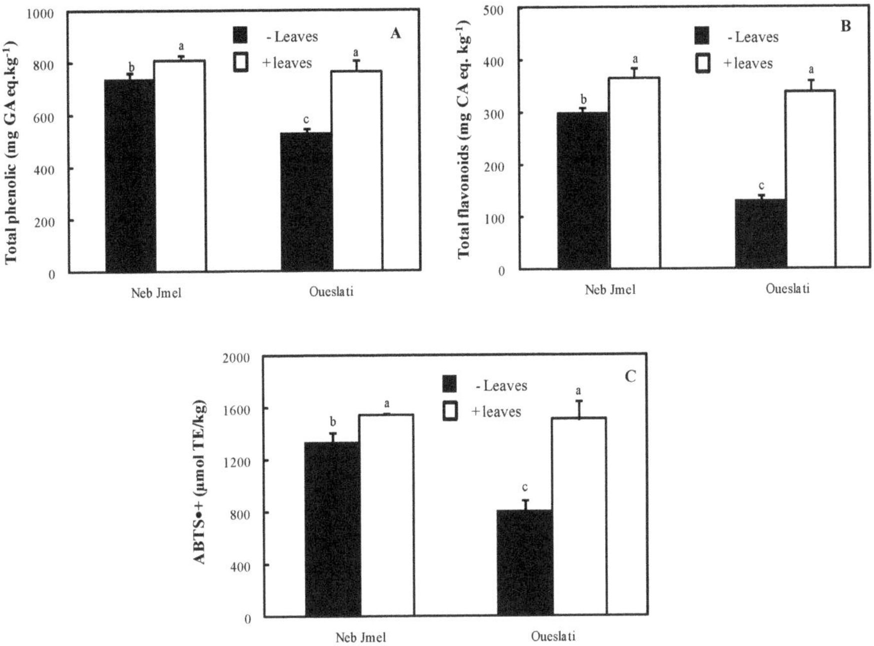

Figure 1. Total phenol (**A**) and total flavonoid (**B**) contents and ABTS$^{•+}$ scavenging activity (**C**) of Neb Jmel and Oueslati olive oils extracted with and without addition of olive leaves (3%). Data are means of three independent experiments ± SE (n = 3). Means followed by different letters are significantly different at $p \leq 0.05$ as determined by Duncan's multiple range test.

2.7. ABTS$^{•+}$ Scavenging Activity

ABTS$^{•+}$ scavenging activity of the Neb Jmel oil was 66% higher than that of Oueslati (Figure 1). In contrast, Oueslati leaves showed a higher free-radical scavenging activity compared to the Neb Jmel variety (Table 2). Following 3% leaf addition, the antioxidant capacity was increased by 15% and 87% in Neb Jmel and Oueslati oils, respectively. This wide change was probably due to the increases in chlorophyll, carotenoid, total phenolic and flavonoid concentrations. Indeed, some authors found a good correlation between the total phenolic amount and the radical scavenging power [37,38].

2.8. Phenolic Compounds

The analysis of the phenolic profile of olive leaf extracts and oils from the two varieties is reported in Tables 3 and 4, respectively. Thirteen phenolic compounds, among which phenolic acids, phenolic alcohols and secoiridoids, were identified and quantified. As regards olive leaf extracts, oleuropein derivatives were the most represented, followed by phenolic acids and phenolic alcohols (Table 3).

With minor changes compared to leaves, oleuropein derivatives were still the most abundant compounds in oils, followed by phenolic alcohols and phenolic acids (Table 4). In particular, Oueslati leaves showed a three-fold higher content in oleuropein derivatives than Neb Jmel ones, whereas Neb Jmel oil exhibited the highest concentration of these compounds. With leaf addition, oleuropein derivative concentration increased by 9 and 48% in Neb Jmel and Oueslati oils, respectively, confirming that olive leaves are a source of oleuropein derivatives [39]. Consistent with other reports, we found that oleuropein derivative amounts changed depending on the variety.

Table 3. Phenolic compounds (μg/g DW) of Neb Jmel and Oueslati olive leaves.

Parameter	Neb Jmel	Oueslati
	Secoiridoids	
Oleuropein derivatives	648.6 ± 92.2 [b]	1948.6 ± 96.8 [a]
	Phenolic Acids	
Gallic	1.76 ± 0.65 [a]	0.31 ± 0.25 [b]
Protocatechuic	37.39 ± 8.68 [a]	32.39 ± 12.80 [a]
p-Hydroxybenzoic	0.93 ± 0.05 [b]	4.16 ± 0.41 [a]
Chlorogenic	1.80 ± 0.00	Tr.
Vanillic	33.84 ± 2.66 [a]	12.11 ± 1.36 [b]
Caffeic	1.01 ± 0.05 [a]	1.19 ± 0.61 [a]
Syringic	2.28 ± 0.55 [b]	8.91 ± 2.65 [a]
Vanillin	2.90 ± 0.11 [b]	11.32 ± 3.84 [a]
p-Coumaric	5.66 ± 0.52 [a]	0.87 ± 0.57 [b]
Ferulic	15.07 ± 0.40 [b]	37.18 ± 4.71 [a]
	Phenolic Alcohols	
Hydroxytyrosol	5.75 ± 1.40 [b]	24.03 ± 2.43 [a]
Tyrosol	3.17 ± 0.76 [b]	14.46 ± 2.23 [a]

Data are means of three independent experiments ± SE (n = 3). Means followed by different letters are significantly different at $p \leq 0.05$ as determined by Duncan's multiple range test.

Table 4. Phenolic compounds (mg/kg) of Neb Jmel and Oueslati olive oils extracted with and without addition of olive leaves (3%).

Parameter	Neb Jmel		Oueslati	
	Oil	Oil + Leaves	Oil	Oil + Leaves
		Secoiridoids		
Oleuropein derivatives	194.4 ± 12.1 [b]	211.0 ± 1.9 [a]	126.4 ± 2.6 [c]	187.6 ± 10.5 [b]
		Phenolic Acids		
Gallic	-	-	-	-
Protocatechuic	-	-	-	-
***p*-Hydroxybenzoic**	-	-	-	-
Chlorogenic	-	-	-	-
Vanillic	1.52 ± 0.02 [a]	1.44 ± 0.05 [a]	0.77 ± 0.14 [b]	0.27 ± 0.03 [c]
Caffeic	0.007 ± 0.003 [b]	0.02 ± 0.0001 [a]	0.004 ± 0.001 [b]	0.015 ± 0.0001 [a]
Syringic	0.005 ± 0.004 [b]	Tr.	0.02 ± 0.01 [a]	0.02 ± 0.005 [a]
Vanillin	0.15 ± 0.0008 [a]	0.17 ± 0.01 [a]	0.17 ± 0.05 [a]	0.11 ± 0.01 [a]
***p*-Coumaric**	0.11 ± 0.008 [b]	0.12 ± 0.0004 [b]	0.29 ± 0.01 [a]	0.32 ± 0.04 [a]
Ferulic	0.10 ± 0.005 [a]	0.11 ± 0.005 [a]	0.11 ± 0.01 [a]	0.17 ± 0.01 [a]
		Phenolic Alcohols		
Tyrosol	17.97 ± 0.018 [a]	14.67 ± 0.59 [b]	1.57 ± 0.05 [c]	1.07 ± 0.13 [c]
Hydroxytyrosol	3.57 ± 0.02 [b]	4.13 ± 0.11 [a]	1.38 ± 0.20 [c]	1.17 ± 0.12 [c]

Data are means of three independent experiments ± SE (n = 3). Means followed by different letters are significantly different at $p \leq 0.05$ as determined by Duncan's multiple range test.

During the past few years, the biological activities of olive oil phenolics, namely oleuropein derivatives, have been thoroughly investigated. Several studies have attempted to elucidate the performance of oleuropein derivatives as antioxidant compounds. As the excessive presence of reactive oxygen species has been suggested to participate in the aetiology of several diseases [40], the focus on powerful antioxidants able to counteract the free-radical attack has become increasingly important. The antioxidant actions of oleuropein have been mostly assigned to its free-radical scavenging activity. Considering all together, the addition of olive leaves during oil extraction process suggests that it could be a means for improving oil quality.

Although olive fruits are rich in secoiridoids, hydroxytyrosol and tyrosol represent the two most important phenolic alcohols of both olive leaves [41] and oils [42]. Leaf composition of the Oueslati variety was characterised by a four-fold higher amount of hydroxytyrosol and tyrosol compared to Neb Jmel (Table 3). However, the oils, and likely the fruits, were mostly endowed with these phenolic alcohols in the Neb Jmel variety, showing concentrations of hydroxytyrosol and tyrosol of 3.57 and 17.97 mg/kg, respectively (Table 4). Obviously, the difference in the amounts of phenolic alcohols (hydroxytyrosol and tyrosol) depends not only on the variety, but also on the organ—leaf or fruit—considered. Our findings are in agreement with previous studies on phenolics in olive oil [17,21]. Oueslati oil, after 3% leaf addition, did not show any significant change in the hydroxytyrosol and tyrosol contents in comparison with the control (Table 4). These results suggest that a very high content of phenolics and related compounds could have ended up in olive mill wastewater. Indeed, many investigations clearly showed the occurrence of a high content of phenolic compounds in olive mill wastewater [43–46]. The hydrophilic character of polyphenols was likely responsible for the solubilisation of the most part of phenols into the water phase during oil extraction.

Phenolic acids found at considerable concentrations in Neb Jmel and Oueslati leaves (Table 3) were represented by gallic, protocatechuic, *p*-hydroxybenzoic, chlorogenic, vanillic, caffeic, syringic, vanillin, *p*-coumaric and ferulic acids. In the oils of the two varieties, the identified phenolic acids did not overcome the value of 2 mg/kg (Table 4), which agrees with the results reported by Kelebek et al. [47]. In both olive varieties, no significant effect of leaf addition was registered.

2.9. Tocopherols

Tocopherols occur in vegetable oils, playing an important role during oxidative processes. Two identified compounds (α- and γ-tocopherol) were quantified in this study. The quantitative profile of α- and γ-tocopherol is shown in Table 2 and Figure 2.

Tocopherol concentration in olive leaves is reported in Table 2. α-Tocopherol was present at higher amounts in Neb Jmel leaves (82.37 µg/g DW) in comparison with Oueslati, which showed an eight-fold lower amount (10.12 µg/g DW); γ-tocopherol was found in trace amounts in the leaves of both varieties.

In control oils (Figure 2), the α-isomer reached values of 257.8 and 283.6 mg/kg in Oueslati and Neb Jmal, respectively, not showing any significant difference between the two cultivars. Following leaf addition, α-tocopherol concentration showed a slight increase in the Neb Jmel oil, where it reached the value of 328.08 mg/kg. This result could be related to the higher presence of this compound in the corresponding leaves (Table 2). In contrast, the lower amount of α-tocopherol detected in Oueslati leaves (10.12 µg/g) did not determine any improvement in the oil following leaf addition (Figure 2). According to Malheiro et al. [17], the amount of α-tocopherol in oils was not significantly influenced when leaf addition was less than 5%.

Concerning γ-tocopherol (Figure 2), the amount was very low compared to the α-isomer (1.17–28.73 mg/kg), being it more represented in the Neb Jmel oil. Leaf addition (3%) did not significantly affect γ-tocopherol concentrations due to the fact that in the leaves of both varieties this isomer was detected in trace amounts (Table 2).

Consistent with previous reports [48], we found that olive leaves can be used as an alternative source to improve the chemical composition of olive oils, mainly the α-tocopherol concentration. Likewise, the present results agree with previous studies on the influence of the cultivar on α-tocopherol concentration. Franco et al. [49] reported very high levels of α-tocopherol in seven varieties of Spanish oils (217–345 mg/kg). In contrast, in the Portuguese olive oil studied by Cunha et al. [50], values ranging from 93 to 260 mg/kg were found. Similar values to those reported in the present experiment were found in some studies performed on different Tunisian oils [51–53].

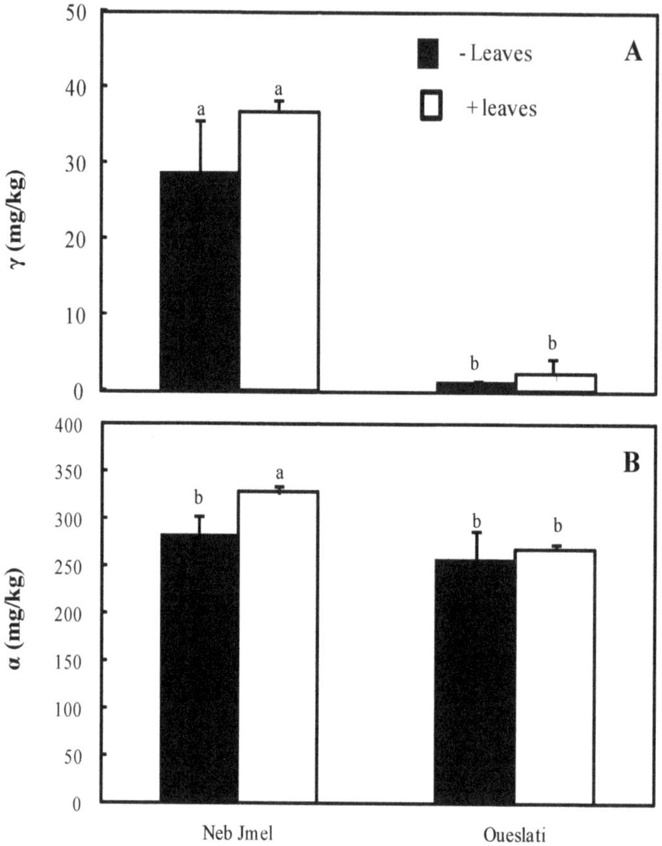

Figure 2. γ-Tocopherol (**A**) and α-tocopherol (**B**) content of Neb Jmel and Oueslati olive oils extracted with and without addition of olive leaves (3%). Data are means of three independent experiments ± SE ($n = 3$). Means followed by different letters are significantly different at $p \leq 0.05$ as determined by Duncan's multiple range test.

Antioxidants, such as vitamin E (tocopherols), may prevent the detrimental effects of free radicals. In the Mediterranean diet, olive oil substantially contributes to the daily intake of these antioxidants. The health benefits of vitamin E are evidenced by the fact that the ingestion of fresh fruits and vegetables is inversely related to the extent of some cancers as well as to plasma lipid peroxidation [54]. It should be highlighted that the two vitamin E isoforms have different health-related properties. In fact, γ-tocopherol is the less powerful antioxidant, although being capable of trapping peroxynitrites. For this reason, γ-tocopherol has been acknowledged as the "other" vitamin E important for human health [54].

3. Materials and Methods

3.1. Olive Leaves and Fruit Sampling

The olives of two varieties were collected from different regions: Oueslati, in the centre (Khit El Oued), and Neb Jmel, in the north (Borj El Amri) of Tunisia. The harvest was performed at the same stage of maturity, considering a colour maturity index of about three. The maturity index was evaluated taking into consideration the changes in skin and pulp colours. Samples of 100 fruits were taken randomly and classified into eight groups or categories: green intense (category 0), yellow or yellowish green (category 1), green with reddish spots (category 2), reddish or light violet (category 3), black with white pulp (category 4), black with <50% purple flesh (category 5), black with ≥50% purple flesh (category 6) and black with 100% purple flesh (category 7). The maturity index was calculated as $A \times 0 + B \times 1 + C \times 2 + D \times 3 + E \times 4 + F \times 5 + G \times 6 + H \times 7/100$, where A, B, C, D, E, F, G and H are the number of fruits in each class.

The olives were picked by hand from three trees during the 2016/2017 crop season (November). Olive leaves were also collected from the same trees at harvest. The following percentages of olive leaves were added to fruits (*w*/*w*) prior to crushing: 0% (control) and 3% (fortified). The choice of the percentage of olive leaves was based on previous reports [32,33]. According the above studies, 3% was the optimal percentage of olive leaves which can be added to improve oil quality without any negative effects. Only healthy fruits without any kind of infection or physical damage were processed. After harvesting, fresh olives (2.5–3.0 kg) were washed and crushed with a hammer crusher, and the paste, mixed at 25 °C for 30 min, was centrifuged without the addition of warm water. The oil yield from each extraction was 200–250 mL/kg. The oil produced was then transferred into dark glass bottles and stored in the dark at 4 °C until analysis.

3.2. Methods

3.2.1. Quality Parameters

Determination of physicochemical quality parameters (free acidity and peroxide values) was carried out following the analytical methods described by Regulation EEC/2568/91 and EEC/1429/92 of the European Union Commission (European Union Commission Regulation, 1991, 1992).

Free acidity, given as percentage of oleic acid, was determined by titration of the oil dissolved in an ethanol–ether solution (1:1, *v*/*v*) with a 0.1 M potassium hydroxide ethanolic solution. The peroxide value, expressed in milliequivalents of active oxygen per kg oil (meq/kg), was determined as follows: a mixture of oil and chloroform–acetic acid was left to react with a solution of potassium iodide (10.5 M) in the darkness. The free iodine was then titrated with a sodium thiosulfate solution (0.01 N).

3.2.2. Pigment Concentration

Oil Pigment Determination

One millilitre of oil was diluted ten-fold in *n*-hexane. Chlorophylls and carotenoids were determined colourimetrically as previously described [55]. The maximum absorption at 670 nm was related to the chlorophyll fraction and that at 470 nm to the carotenoid one. The specific extinction coefficients considered for calculation were 613 for pheophytin, as a major component of the chlorophyll fraction, and 2000 for lutein, as a major component of the carotenoid fraction. The pigment concentrations were calculated as follows:

$$\text{Chlorophyll (mg/kg)} = (A_{670} \times 10^6)/(613 \times 100 \times d)$$

$$\text{Carotenoids (mg/kg)} = (A_{470} \times 10^6)/(2000 \times 100 \times d)$$

where A is the absorbance and d is the spectrophotometer cell thickness (1 cm).

Leaf Pigment Determination

Fresh leaf tissue (0.1 g) was ground in a mortar with sand and 70% ethanol solution. The homogenates were then filtered and washed with 70% ethanol (up to 5 mL). After centrifugation for 10 min at 12,100× *g*, absorbance was read at 646.6 and 663.6 nm for chlorophylls and at 480 nm for carotenoids. Concentrations of total chlorophylls and total carotenoids (μg/g DW) were calculated according to Porra et al. [56].

3.3. *Extraction of Phenolic Compounds*

3.3.1. Fresh Leaves

Leaf samples (0.2 g) were ground in a mortar at room temperature with 70% methanol containing 1% HCl. The homogenates were sonicated for 30 min and centrifuged at 12,100× *g* for 30 min at 4 °C. The supernatants were stored at −20 °C and used to determine both phenolic compounds and antioxidant activity.

3.3.2. Olive Oils

Phenolic compounds of olive oils were extracted according to Rotondi et al. [57]. Two grams of oil were added to 1 mL of *n*-hexane and 2 mL of a methanol/water (70:30, *v*/*v*) solution in a 10 mL centrifuge tube. After vigorous mixing, tubes were centrifuged for 10 min. The hydroalcoholic phase was collected, and the hexane phase was re-extracted twice with 2 mL of a methanol/water (70:30, *v*/*v*) solution. The hydroalcoholic fractions were combined, washed with 2 mL of *n*-hexane to remove the residual oil and vacuum-dried.

3.4. *Total Polyphenol and Flavonoid Concentrations*

Total phenolic content was estimated by the Folin Ciocalteu method as described by Singleton and Rossi [58]. To the extract, diluted with distilled water, 1 mL of sodium carbonate (20%) and 1 mL of Folin Ciocalteu reagent were added. The mixture was allowed to stand in a water bath for 30 min at 40 °C. The concentration of the total phenolic compounds was expressed as mg of gallic acid equivalents. The absorbance was measured at 765 nm using a UV–vis spectrophotometer (VARIAN, Milan, Italy). The experiments were performed in triplicate, and mean values and standard deviations were calculated using the Microsoft Excel software (Microsoft Corporation, Redmond, WA, USA).

The total flavonoid concentration was determined by the aluminium trichloride method using catechin as reference compound [59]. A 5% $NaNO_2$ solution was added to the extract, followed after 6 min by 10% aluminium trichloride. The mixture was incubated for further 5 min and then 1 M NaOH was added. The final volume was 2.5 mL. After 15 min of incubation, the absorbance at 510 nm was detected. Total flavonoid concentration was expressed as mg of catechin equivalents.

3.5. *Free-Radical Scavenging Ability*

The free-radical scavenging activity of samples was determined by the ABTS radical cation decolourisation assay described by Pellegrini et al. [60]. The radical solution was generated by adding 7 mM ABTS solution to 4.9 mM potassium persulfate. Before use, the radical solution was diluted with ethanol to obtain an absorbance of 0.700 at 734 nm. A control containing ethanol and $ABTS^{\bullet+}$ solution was also prepared, and the absorbance was taken as the initial. After a 15 min incubation period at room temperature, the final absorbance was read at 734 nm. Calculations were performed by percentage of inhibition of the ABTS cation radical as follows:

$$\% \text{ of inhibition} = ((\text{initial Abs} - \text{final Abs})/\text{initial Abs}) \times 100$$

To quantify antioxidant capacity, a calibration curve of the percentage of inhibition against Trolox in the range 2–20 nmol was used.

3.6. *HPLC Analysis of Phenolic Compounds*

Qualitative and quantitative analysis were performed by reverse-phase HPLC (RP-HPLC) [61]. Twenty microlitres of extract were injected into a Waters model 515 HPLC system fitted with a 3.9 mm × 150 mm Nova-Pak C18 column (Waters, Milford, MA, USA). Detection was conducted at 280 nm using a Waters 2487 dual λ UV–visible detector. Mobile phase A contained 98% water and 2% acetic acid, and mobile phase B contained 68% water, 30% acetonitrile and 2% acetic acid. A linear gradient of 10 to 95% mobile phase B was run for 90 min at 1 mL/min. The identity of the phenolic acids was confirmed by cochromatography on HPLC with authentic standards (Sigma Chemical Co., St. Louis, MO, USA), and quantification was performed using a standard curve in the range of 20 to 200 ng of standard mixtures containing gallic, protocatechuic, *p*-hydroxybenzoic, chorogenic, vanillic, caffeic, syringic, *p*-coumaric, ferulic, tyrosol, hydroxytyrosol, vanillin and oleuropein. Chromatogram analysis was performed by the software Millennium 32 (Waters).

3.7. Extraction and Detection of Tocopherols (Vitamin E)

Tocopherols were determined in the lipid extracts from olive leaves and in oils. Extractions were performed in the dark as previously reported [54] and according to the method of Gimeno et al. [62]. Tocopherol isoforms (α and γ) were determined by isocratic RP-HPLC using a Shimadzu apparatus (model LC-20AD) with an electrochemical detector (Metrohm model 791, Varese, Italy) equipped with a glassy carbon electrode and LC Solution software (Shimadzu) for the integration of peaks. Detection was performed according to Galatro et al. [63] at +0.6 V at 25 °C with a Nova Pak C-18 4 μm column (3.9 × 150 mm). The extracts were eluted with 95% methanol containing 20 mM $LiClO_4$ at a flow rate of 1 mL min^{-1}. For identification and quantification of peaks, a calibration curve was prepared using standard mixtures of α-, β-, γ- and δ- tocopherol provided by Sigma (Milan, Italy) in the range of 25 to 75 ng.

3.8. Statistical Analysis

The results are means from three replicates. All data are reported as mean values ± SE. The significance of differences among mean values was determined by one-way ANOVA. Comparisons among means were performed using Duncan's multiple range test. Means in tables and figures accompanied by different letters are significantly different at $p \leq 0.05$.

4. Conclusions

The present study confirms the dependence of olive oil quality on the cultivar, besides geographical location, climate and soil characteristics. The Neb Jmel oil showed the best chemical composition with the lowest free acidity and peroxide values, the highest chlorophyll, total phenol and total flavonoid concentrations as well as antioxidant activity. The addition of olive leaves (3%) to Neb Jmel and Oueslati oils affected both their quality and chemical composition, mainly conferring an increased resistance to oxidation as well as improving the nutritional qualities. In particular, a remarkable increase in oleuropein derivatives was observed, which was responsible for the enhancement in total phenolic amounts. As Oueslati leaves are particularly enriched in chlorophyll, carotenoids, flavonoids and oleouperin derivatives, their addition during oil extraction may have prevented the oxidation and the formation of peroxides. Thus, in the Oueslati variety leaf addition reduced both peroxide values and free acidity of oils, allowing them to be still classified as extra virgin olive oils. The enrichment of the oils with antioxidant compounds from the leaves also led to a remarkable increase in the nutritional quality of the Oueslati oil, which became similar to the Neb Jmel one. In conclusion, the addition of a small percentage of olive leaves could improve the nutraceutical properties of extra virgin olive oil by increasing the phenolic compound content. These compounds, together with tocopherols, play a protective role against oxidative stress, being also able to extend the extra virgin olive oil shelf-life due to their antioxidative properties.

Author Contributions: I.T. conceived the experiments. I.T., C.S., and M.F.Q. performed the experiments and wrote the paper. A.Z. analysed and interpreted the data. A.Z., M.F.Q., M.Z. and J.E. revised the paper.

Funding: This study was financially supported by the University of Pisa (Fondi di Ateneo, 2017).

Conflicts of Interest: The authors declare no conflict of interest.

References

1. Boudhrioua, N.; Bahloul, N.; Ben Slimen, I.; Kechaou, N. Comparison on the total phenol concentrations and the color of fresh and infrared dried olive leaves. *Ind. Crops Prod.* **2009**, *29*, 412–419. [CrossRef]
2. Zinnai, A.; Venturi, F.; Andrich, L.; Silvestri, S.; Andrich, G. A kinetic method to evaluate the effect of environmental variability on the quality of an extra virgin olive oil. *Agrochimica* **2014**, *58*, 35–50.
3. Bouaziz, M.; Chamkha, M.; Sayadi, S. Comparative study on phenolic concentration and antioxidant activity during maturation of the olive cultivar Chemlali from Tunisia. *J. Agric. Food Chem.* **2004**, *52*, 5476–5481. [CrossRef] [PubMed]
4. Harwood, J.L.; Yaqoob, P. Nutritional and health aspects of olive oil. *Eur. J. Lipid Sci. Technol.* **2002**, *104*, 685–697. [CrossRef]
5. Turner, R.; Etiene, N.; Garcia-Alonso, M.; de Pascual-Teresa, S.; Minihane, A.M.; Weinberg, P.D.; Rimbach, G. Antioxidant and anti-atherogenic activities of olive oil phenolics. *Int. J. Vitaminol. Nutr. Res.* **2010**, *75*, 61–70. [CrossRef] [PubMed]
6. Boss, A.; Bishop, K.S.; Marlow, G.; Barnett, M.P.G.; Ferguson, L.R. Evidence to Support the Anti-Cancer Effect of Olive Leaf Extract and Future Directions. *Nutrients* **2016**, *8*, 513. [CrossRef] [PubMed]
7. Zinnai, A.; Venturi, F.; Quartacci, M.F.; Sanmartin, C.; Favati, F.; Andrich, G. Solid carbon dioxide to promote the extraction of extra-virgin olive oil. *Grasas y Aceites* **2016**, *67*, e121. [CrossRef]
8. Zhang, X.; Cao, J.; Zhong, L. Hydroxytyrosol inhibits pro-inflammatory cytokines, iNOS, and COX-2 expression in human monocytic cells. *Naunyn Schmiedebergs Arch. Pharmacol.* **2009**, *379*, 581–586. [CrossRef]
9. Venturi, F.; Sanmartin, C.; Taglieri, I.; Nari, A.; Andrich, G.; Terzuoli, E.; Donnini, S.; Nicolella, C.; Zinnai, A. Development of Phenol-Enriched Olive Oil with Phenolic Compounds Extracted from Wastewater Produced by Physical Refining. *Nutrients* **2017**, *9*, 916. [CrossRef]
10. Bayram, B.; Ozcelik, B.; Grimm, S.; Roeder, T.; Schrader, C.; Ernst, I.M.; Wagner, A.E.; Grune, T.; Frank, J.; Rimbach, G. A diet rich in olive oil phenolics reduces oxidative stress in the heart of SAMP8 mice by induction of Nrf 2-dependent gene expression. *Rejuvenation Res.* **2012**, *15*, 71–81. [CrossRef]
11. Gill, C.I.R.; Boyd, A.; McDermott, E.; McCann, M.; Servili, M.; Selvaggini, R.; Taticchi, A.; Esposto, S.; Montedoro, G.; McGlynn, H.; et al. Potential anti-cancer effects of virgin olive oil phenols on colorectal carcinogenesis models in vitro. *Int. J. Cancer* **2005**, *117*, 1–7. [CrossRef] [PubMed]
12. Rigacci, S.; Guidotti, V.; Bucciantina, M.; Parri, M.; Nediani, C.; Cerbai, E.; Stefani, M.; Berti, A. Oleuropein aglycon prevents cytotoxic amyloid aggregation of human amylin. *J. Nutr. Biochem.* **2010**, *21*, 725–726. [CrossRef] [PubMed]
13. Schaffer, S.; Podstawa, M.; Visioli, F.; Bogani, P.; Muller, W.E.; Eckert, G.P. Hydroxytyrosol-rich olive mill wastewater extract protects brain cells in vitro and ex vivo. *J. Agric. Food Chem.* **2007**, *55*, 5043–5049. [CrossRef] [PubMed]
14. Beltran, G.; Aguilera, M.P.; Del Rio, C.; Sanchez, S.; Martinez, L. Influence of fruit ripening process on the natural antioxidant concentration of Hojiblanca virgin olive oils. *Food Chem.* **2005**, *89*, 207–215. [CrossRef]
15. Gambacorta, G.; Faccia, M.; Previtali, M.A.; Pati, S.; La Notte, E.; Bioano, A. Effects of olive maturation and stoning on quality indices and antioxidant concentration of extra virgin oils (cv. *Coratina*) during storage. *J. Food Sci.* **2010**, *75*, 229–235. [CrossRef] [PubMed]
16. Cinquanta, L.; Esti, M.; Notte, E. Evaluation of phenolic compounds in virgin olive oil during storage. *J. Am. Oil Chem. Soc.* **1997**, *74*, 1259–1264. [CrossRef]
17. Malheiro, R.; Casal, S.; Teixeira, H.; Bento, A.; Pereira, J.A. Effect of Olive Leaves Addition during the Extraction Process of Overmature Fruits on Olive Oil Quality. *Food Bioprocess Technol.* **2013**, *6*, 509–521. [CrossRef]
18. Bouaziz, M.; Fki, I.; Jemai, H.; Ayadi, M.; Sayadi, S. Effect of storage on refined and husk olive oils composition: Stabilization by addition of natural antioxidants from Chemlali olive leaves. *Food Chem.* **2008**, *108*, 253–262. [CrossRef]
19. Bubonja-Sonje, M.; Giacometti, J.; Abram, M. Antioxidant and antilisterial activity of olive oil, cocoa and rosemary extract polyphenols. *Food Chem.* **2011**, *127*, 1821–1827. [CrossRef]
20. Kiritsakis, K.; Kontominas, M.G.; Kontogiorgis, C.; Litina, D.H.; Moustakas, A.; Kiritsakis, A. Composition and Antioxidant Activity of Olive Leaf Extracts from Greek Olive Cultivars. *J. Am. Oil Chem. Soc.* **2010**, *87*, 369–376. [CrossRef]

21. Paiva-Martins, F.; Correia, R.; Félix, S.; Ferreira, P.; Gordon, M.H. Effects of enrichment of refined olive oil with phenolic compounds from olive leaves. *J. Agric. Food Chem.* **2007**, *55*, 4139–4143. [CrossRef] [PubMed]
22. IOC. *Norme Commercial Applicable Aux Huiles D'olive Et Aux Huiles De Grignon D'olive*; COI/NCn°3 Rev.4; Conseil Olèicole International: Madrid, Spain, 2009.
23. Gutierrez, F.; Varona, I.; Albi, M.A. Relation of acidity and sensory quality with sterol concentration of olive oil from stored fruit. *J. Agric. Food Chem.* **2000**, *48*, 1106–1110. [CrossRef] [PubMed]
24. Ben Mansour, A.; Gargouri, B.; Flamini, G.; Bouaziz, M. Effect of Agricultural Sites on Differentiation between Chemlali and Neb Jmel Olive Oils. *J. Oleo Sci.* **2015**, *64*, 381–392. [CrossRef] [PubMed]
25. Ouni, Y.; Flamini, G.; Issaoui, M.; Ben Youssef, N.; Luigi, C.P.; Hammami, M.; Daoud, D.; Mokhta, Z. Volatile compounds and compositional quality of virgin olive oil from Oueslati variety: Influence of geographical origin. *Food Chem.* **2011**, *124*, 1770–1776.
26. Martinéz-Suárez, J.M. Recientes estúdios de la almazara experimental del instituto de la grasa. *Rivista Italiana delle Sostanze Grasse* **1973**, *50*, 325–330.
27. Matthaüs, B.; Guillaume, D.; Gharby, S.; Haddad, A.; Harhar, H.; Charrouf, Z. Effect of processing on the quality of edible argan oil. *Food Chem.* **2010**, *120*, 426–432. [CrossRef]
28. Gharby, S.; Harhar, H.; Guillaume, D.; Haddad, A.; Matthäus, B.; Charrouf, Z. Oxidative Stability of Edible Argan Oil: A TwoYear Period Study. *LWT Food Sci. Technol.* **2011**, *44*, 1–8. [CrossRef]
29. Gutiérrez, F.; Perdiguero, S.; Gutiérrez, R.; Olías, J.M. Evaluation of the bitter taste in virgin olive oil. *J. Am. Oil Chem. Soc.* **1992**, *69*, 394–395. [CrossRef]
30. Sgherri, C.; Perez-Lopez, U.; Pinzino, C. Antioxidant properties of food products containing lycopene are increased by the presence of chlorophyll. In *Lycopene: Food Sources, Potential Role in Human Health and Antioxidant Effects*; Baily, J.R., Ed.; Nova Publisher: Hauppauge, NY, USA, 2015; pp. 39–89, ISBN 978-1-63117-927-3.
31. Abaza, L.; Taamalli, W.; Ben Temime, S.; Daoud, D.; Gutierrez, F.; Zarrouk, M. Natural antioxidant composition as correlated to stability of some Tunisian virgin olive oils. *Rivista Italiano delle Sostanze Gresse* **2005**, *132*, 12–18.
32. Giovacchino, L.D.; Angerosa, F.; Giacinto, L.D. Effect of mixing leaves with olives on organoleptic quality of oil obtained by centrifugation. *J. Am. Oil Chem. Soc.* **1996**, *73*, 371–374. [CrossRef]
33. Sevim, D.; Tuncay, O. Effect of Olive Leaves Addition before Extraction of Turkish Olive Cultivars on Olive Oil Minor Components and Antioxidant Activity. *Open Access Sci. Rep.* **2013**, *2*, 2–8.
34. Kandaswami, C.; Lee, L.T.; Lee, P.P.; Hwang, J.J.; Ke, F.C.; Huang, Y.T.; Lee, M.T. The antitumor activities of flavonoids. *In Vivo* **2005**, *19*, 895–909. [PubMed]
35. Crozier, A.; Ashihara, H. *Plant Secondary Metabolites and the Human Diet*; Blackwell Publishing: Oxford, UK, 2006.
36. Ebrahimi, N.; Khanikia, G.J.; Keshavarz, S.A.; Shariatifar, N.; Nabizadeh, R.; Sadighara, P. Spectrophotometric method for quantification of flavonoids in olive oil supplied from Tehran market of Iran. *J. Food Saf. Hyg.* **2015**, *1*, 21–24.
37. Gorinstein, S.; Martin-Belloso, O.; Katrich, E.; Lojek, A.; Cíz, M.; Gligelmo-Miguel, N.; Haruenkit, R.; Park, Y.S.; Jung, S.T.; Trakhtenberg, S. Comparison of the concentrations of the main biochemical compounds and the antioxidant activity of some Spanish olive oils as determined by four different radical scavenging tests. *J. Nutr. Biochem.* **2003**, *14*, 154–159. [CrossRef]
38. Samaniego, S.C.; Troncoso González, A.M.; García-Parrilla, M.C.; Quesada Granados, J.J.; López García de la Serrana, H.; López Martínez, M.C. Different radical scavenging tests in virgin olive oil and their relation to the total phenol content. *Anal. Chim. Acta* **2007**, *593*, 103–107. [CrossRef] [PubMed]
39. Ben Salah, M.; Abdelmelek, H.; Abderraba, M. Study of Phenolic Composition and Biological Activities Assessment of Olive Leaves from different Varieties Grown in Tunisia. *J. Med. Chem.* **2012**, *2*, 107–111. [CrossRef]
40. Chin, K.Y.; Ima-Nirwana, S. The Role of Vitamin E in Preventing and Treating Osteoarthritis—A Review of the Current Evidence. *Front. Pharmacol.* **2018**, *9*, 946. [CrossRef]
41. Sedef, N.; Karakaya, S. Olive tree (*Olea europaea*) leaves: Potential beneficial effects on human health. *Nutr. Rev.* **2009**, *67*, 632–638.

42. Cortesi, N.; Rovellini, P.; Fusari, P. Dosaggio dei biofenoli degli oli vergini di oliva: Idrossitirosolo, tirosolo, agliconi secoiridoidi, acidi secoridoidi, lignani e flavonoidi. *Rivista Italiana delle Sostanze Grasse* **2002**, *79*, 145–150.
43. De Leonardis, A.; Macciola, V.; Lembo, G.; Aretini, A.; Nag, A. Studies on oxidative stabilisation of lard by natural antioxidants recovered from olive-oil mill wastewater. *Food Chem.* **2007**, *100*, 998–1004. [CrossRef]
44. Khoufi, S.; Aloui, F.; Sayadi, S. Extraction of antioxidants from olive mill wastewater and electro-coagulation of exhausted fraction to reduce its toxicity on anaerobic digestion. *J. Hazard. Mater.* **2008**, *151*, 531–539. [CrossRef] [PubMed]
45. El-Abbassi, A.; Kiai, H.; Hafidi, A. Phenolic profile and antioxidant activities of olive mill wastewater. *Food Chem.* **2012**, *132*, 406–412. [CrossRef] [PubMed]
46. Azaizeh, H.; Halahlih, F.; Najami, N.; Brunner, D.; Faulstich, M.; Tafesh, A. Antioxidant activity of phenolic fractions in olive mill wastewater. *Food Chem.* **2012**, *134*, 2226–2234. [CrossRef] [PubMed]
47. Kelebek, H.; Selli, S.; Kol, O. Quantitative determination of phenolic compounds using LC-DAD-ESI-MS/MS in cv. Ayvalik olive oils as affected by harvest time. *J. Food Meas. Charact.* **2017**, *11*, 226–235. [CrossRef]
48. Lucas, A.; Martinez, E.; Rincón, J.; Blanco, M.A.; Garcia, I. Supercritical fluid extraction of tocopherol concentrates from olive tree leaves. *J. Supercrit. Fluids* **2002**, *22*, 221–228. [CrossRef]
49. Franco, M.N.; Galeano-Diaz, T.; Sanchez, J.; De Miguel, C.; Martin-Vertedor, D. Total phenolic compounds and tocopherols profiles of seven olive oil varieties grown in the south-west of Spain. *J. Oleo Sci.* **2014**, *63*, 115–125. [CrossRef] [PubMed]
50. Cunha, S.S.; Fernandes, J.O.; Oliveira, M.B. Quantification of free and esterified sterols in Portuguese olive oils by solid-phase extraction and gas chromatography-mass spectrometry. *J. Chromatogr. A* **2006**, *1128*, 220–227. [CrossRef] [PubMed]
51. Baccouri, O.; Guerfel, M.; Baccouri, B.; Cerretani, L.; Bendini, A.; Lercker, G.; Zarrouk, M.; Daoud Ben Miled, D. Chemical composition and oxidative stability of Tunisian monovarietal virgin olive oils with regard to fruit ripening. *Food Chem.* **2008**, *109*, 743–754. [CrossRef]
52. Krichene, D.; Taamalli, W.; Daoud, D.; Salvador, M.D.; Fregapane, G.; Zarrouk, M. Phenolic Compounds, Tocopherols and others minor components in virgin olive oils of some Tunisian Varieties. *J. Food Biochem.* **2007**, *31*, 179–194. [CrossRef]
53. Manai-Djebali, H.; Krichène, D.; Ouni, Y.; Gallardo, L.; Sánchez, J.; Osorio, E.; Daoud, D.; Guido, F.; Zarrouk, M. Chemical profiles of five minor olive oil varieties grown in central Tunisia. *J. Food Compos. Anal.* **2012**, *27*, 109–119. [CrossRef]
54. Sgherri, C.; Kadlecova, Z.; Prdossi, A.; Navari-Izzo, F.; Izzo, R. Irrigation with Diluted Seawater Improves the Nutritional Value of Cherry Tomatoes. *J. Agric. Food Chem.* **2008**, *56*, 3391–3397. [CrossRef] [PubMed]
55. Mínguez-Mosquera, M.I.; Rejano-Navarro, L.; Gandul-Rojas, B.; Sánchez-Gómez, A.H.; Garrido-Fernández, J. Color–Pigment Correlation in Virgin Olive Oil. *J. Am. Oil Chem. Soc.* **1991**, *69*, 332–336. [CrossRef]
56. Porra, R.J.; Thompson, W.A.; Kriedemann, P.E. Determination of accurate extinction coefficients and simultaneous equations for assaying chlorophylls a and b extracted with four different solvents: Verification of the concentration of chlorophyll standards by atomic absorption spectrometry. *Biochim. Biophys. Acta* **1989**, *975*, 384–394. [CrossRef]
57. Rotondi, A.; Bendini, A.; Cerretani, L.; Mari, M.; Lercker, G.; Gallina-Toschi, G.T. Effect of Olive Ripening Degree on the Oxidative Stability and Organoleptic Properties of Cv. Nostrana di Brisighella Extra Virgin Olive Oil. *J. Agric. Food Chem.* **2004**, *52*, 3649–3654. [CrossRef]
58. Singleton, V.L.; Rossi, J.A., Jr. Colorimentry of total phenolics with phosphomolybdic phosphotungstic acid reagents. *Am. J. Enol. Vitic.* **1965**, *16*, 144–158.
59. Zhishen, J.; Mengcheng, T.; Jianming, W. The determination of flavonoid contents in mulberry and their scavenging effects on superoxide radicals. *Food Chem.* **1999**, *64*, 555–559. [CrossRef]
60. Pellegrini, N.; Re, R.; Yang, M.; Rice-Evans, C. Screening of dietary carotenoids and carotenoid-rich fruit extracts for antioxidant activities applying 2,2′-azinobis (3-ethylbenzothiazolyne-6-sulfonic acid) radical cation decolorization assay. *Methods Enzymol.* **1999**, *299*, 379–389.
61. Talcott, S.T.; Howard, L.R. Phenolic autoxidation is responsible for color degradation in processed carrot puree. *J. Agric. Food Chem.* **1999**, *47*, 2109–2115. [CrossRef]

62. Gimeno, E.; Castellote, A.I.; Lamuela-Raventós, R.M.; de la Torre, M.C.; López-Sabater, M.C. Rapid determination of vitamin E in vegetable oils by reversed-phase high-performance liquid chromatography. *J. Chromatogr. A* **2000**, *9*, 251–254. [CrossRef]
63. Galatro, A.; Simontacchi, M.; Puntarulo, S. Free radical generation and antioxidant content in chloroplasts from soybean leaves exposed to ultraviolet-B. *Physiol. Plant.* **2001**, *113*, 564–570. [CrossRef]

Sample Availability: Not available.

molecules

Article

Anti-Oxidant and Anti-Melanogenic Properties of Essential Oil from Peel of Pomelo cv. Guan Xi

Wanying He [1,2], Xiaoyan Li [1,2], Ying Peng [1,2], Xiaoyan He [1,2] and Siyi Pan [1,2,*]

1 College of Food Science and Technology, Huazhong Agricultural University, Wuhan 430070, Hubei, China; wanyinghe@webmail.hzau.edu.cn (W.H.); lixiaoyanhzau@sina.com (X.L.); pengkangying@webmail.hzau.edu.cn (Y.P.); hexiaoyanhzau@126.com (X.H.)

2 Key Laboratory of Environment Correlative Dietology, Ministry of Education, Huazhong Agricultural University, Wuhan 430070, Hubei, China

* Correspondence: pansiyi@mail.hzau.edu.cn

Received: 20 November 2018; Accepted: 7 January 2019; Published: 10 January 2019

Abstract: Here, we investigated the anti-oxidant and anti-melanogenic effects of pomelo peel essential oil (PPEO) from pomelo cv. Guan Xi. The volatile chemical composition of PPEO was analyzed with gas chromatography–mass spectrometry (GC/MS). The most abundant component of PPEO was limonene (55.92%), followed by β-myrcene (31.17%), and β-pinene (3.16%). PPEO showed strong anti-oxidant activities against 2,2-diphenyl-2-picryhydrazyl (DPPH), 2,2′-azinobis-(3-ethylbenzthiazoline-6-sulphonate (ABTS) and superoxide anion free radicals. Based on the B16 melanoma cell system, the effects of PPEO on the viability and morphology of B16 cells and the production of melanin were evaluated. The results revealed that PPEO at concentrations below 50 μg/mL could decrease the melanin content without affecting cell viability and morphology. Intracellular tyrosinase (TYR) activity and Western blot analysis showed that PPEO could down-regulate the expression level of TYR in B16 cells and dose-dependently inhibit TYR activity (by a maximum of 64.54%). In conclusion, PPEO has good anti-oxidant and anti-melanogenic activity, and thus can be widely used as a natural antioxidant in the food, pharmaceutical, and cosmetic industries.

Keywords: pomelo peel; essential oil; anti-oxidant; anti-melanogenic; B16 melanoma cell

1. Introduction

Pomelo, a citrus fruit belonging to the genus Rutaceae, is grown and consumed worldwide due to its unique flavor and high nutritional value [1]. In the production of pomelo juice, jam and other products, pomelo peel (PP) is a major by-product accounting for about 50% of the total weight of the fruit. However, most of the PP is disposed of in landfills, resulting in environmental pollution and loss of economic value [2,3]. Pomelo peel contains many natural chemical ingredients, making it a good source of valuable extracts. Compared with the peel of other fruits, PP has a higher concentration of essential oil (EO). Citrus EO is generally considered to be safe with a broad spectrum of biological activities such as anti-inflammatory and anxiolytic effects [4]. Due to its high content of active substances such as terpenes, sesquiterpene, aldehydes, ketones, and esters, pomelo EO has strong aromatic, antioxidant, bacteriostatic, and antiviral properties [4,5], and thus can be used as a functional ingredient and premium fragrance in the food, cosmetic and pharmaceutical industries. Therefore, the high value-added utilization of PP has important research significance and economic prospects.

Oxidative stress can produce reactive oxygen such as superoxide (O_2), hydrogen peroxide (H_2O_2), and hydroxyl radical (HO). These reactive oxygen species can disrupt the balance of normal metabolic activity in the human body and are associated with many chronic diseases such as aging, cancer, atherosclerosis, and inflammation [6]. As a result, research of natural antioxidants has gradually become a hot spot. These antioxidants can remove excess free radicals from the body and relieve conditions caused by excessive free radicals [7,8]. There has been extensive research on the antioxidant activities of plant EOs. Most current research on antioxidants is focused on citrus EOs. For example, the antioxidant properties of 34 kinds of citrus EOs have been tested. The results showed that most of the EOs have good inhibitory effects on 2,2-diphenyl-2-picryhydrazyl (DPPH), which are significantly better than the effects of water-soluble vitamin E [9]. Extract of sweet orange peel have significant inhibitory effects on DPPH, with 50% inhibitory concentrations (IC_{50}) of 600 μL/mL [10]. Moreover, the EOs extracted with a cold pressing method also have high antioxidant activity, because cold pressing can better preserve the active ingredients in EOs [11]. However, the above studies are limited to the antioxidant activities of EO mixtures as a whole, while there have been few studies of the antioxidant activities of specific components in EOs.

Melanin is a pigment widely distributed on the surface of skin, hair, retina, and adrenal medulla. It is synthesized from tyrosine under the enzymatic oxidation of tyrosinase (TYR) [12]. However, excessive production and accumulation of melanin can cause pigmentation spots and skin discoloration such as chloasma, freckles, and age spots. Tyrosinase is associated with a variety of diseases and may be a key factor of dopamine neurotoxicity and neurodegeneration associated with Parkinson's disease [13]. Besides, it is also a key rate-limiting enzyme in the initial reaction of melanin production. At present, the application of EOs as a natural enzyme inhibitor has become a research hotspot. Plant EOs have strong biological activities and great application potentials in biology and medicine [14]. Cinnamon EO and clove EO can inhibit the TYR activity in B16 cells by 37% and 10%, respectively [7]. Lemon EO was observed to have significant inhibitory effects on TYR activity, and its major components were determined to be monoterpenoids and oxindoles [15], which are also the main components of pomelo EO. Hence, it can be speculated that pomelo EO might also have inhibitory effects on TYR activity. Mint leaf EO could reduce the synthesis rate of melanin in B16-F10 cells, and β-caryophyllene (main component) can also decrease melanin production by down-regulating the expression of microphthalmia-associated transcription factor protein (MITF), tyrosinase-related protein-1 (Trp-1), tyrosinase-related protein-2 (Trp-2), and TYR [16]. Aromatic or aliphatic compounds such as anisaldehyde and cuminaldehyde are effective TYR inhibitors [13]. Citrus EOs contain a large amount of fatty aldehyde compounds, indicating their potential inhibitory activity against TYR. However, the current research on pomelo mostly focuses on the storage and preservation of fresh fruit and the extraction of pectin from the peel, while little research attention has been paid to the antioxidant and anti-melanogenic effects of pomelo peel EO (PPEO). Therefore, clarifying the effect of PPEO on melanin synthesis is of great significance for improving the high value-added utilization of PP and expanding its application in the health food, pharmaceutical and cosmetics fields.

In this study, we extracted PPEO from pomelo cv. Guan Xi by a cold pressing method and analyzed its main components using gas chromatography–mass spectrometry (GC/MS). The anti-oxidant activities were examined with 2,2-diphenyl-1-picrylhydrazyl (DPPH); 2,2′-azino-bis(3-ethylbenzothiazoline-6-sulfonic acid) diammonium salt (ABTS), and superoxide anion radical scavenging. Furthermore, based on the B16 melanoma cell system, the effect of PPEO on the viability and morphology of B16 cells and the production of melanin was evaluated. Finally, intracellular TYR activity assay and Western blot analysis were performed to validate the inhibitory effect of PPEO on TYR.

2. Results and Discussion

2.1. Chemical Composition of Pomelo Peel Essential Oil

The extraction rate of PPEO by the method of cold pressing was 0.82%, which was higher than that in a previous study of citrus essential oils (0.25%) [17]. The obtained PPEO was pale yellow and a clear liquid with a natural aroma. Figure 1 shows the total ion chromatogram obtained from GC/MS analysis of the PPEO 10-fold diluted with absolute ethanol. A total of 21 compounds were detected (Table 1). Among these compounds, 13 were terpenes, which accounted for 94.15% of the total. The highest content of limonene in PPEO was 55.92%, followed by β-myrcene (31.17%), β-pinene (3.16%), ocimene (1.42%), and β-copaene (1.24%) (Table 1).

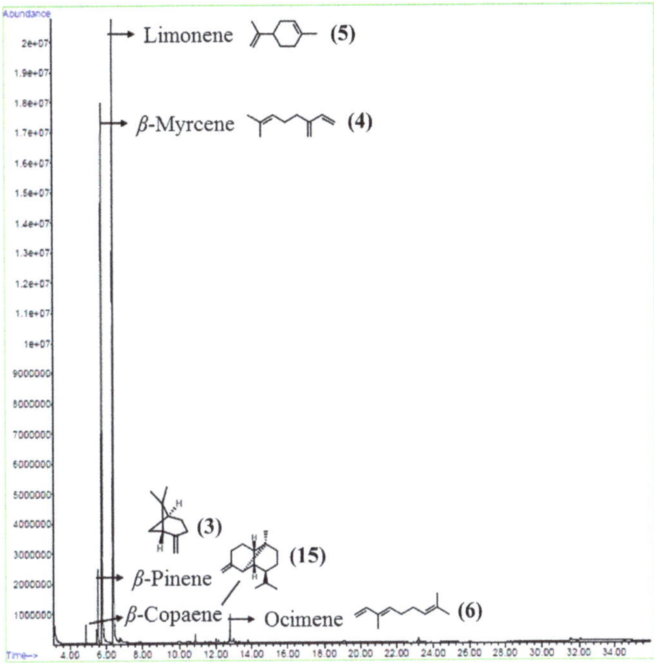

Figure 1. Total ion chromatogram of aroma components from pomelo peel essential oil (PPEO).

Table 1. Chemical components of essential oils.

No.	Compound	Retention Index	Molecular Formula	Peak Area (%)
1	α-Pinene	951	$C_{10}H_{16}$	0.15
2	Artemisia triene	966	$C_{10}H_6$	0.05
3	β-Pinene	994	$C_{10}H_{16}$	3.16
4	β-Myrcene	1009	$C_{10}H_{16}$	31.17
5	Limonene	1050	$C_{10}H_{16}$	55.92
6	Ocimene	1073	$C_{10}H_{16}$	1.42
7	Propionamide	1149	$C_9H_{11}NO$	0.40
8	Metaraminol	1295	$C_9H_{13}NO_2$	0.21
9	Citral	1348	$C_{10}H_{16}O$	0.73
10	4-Carene	1366	$C_{10}H_{16}$	0.38
11	Norephedrine	1382	$C_9H_{13}NO$	0.04
12	Caryophyllene	1418	$C_{15}H_{24}$	0.13
13	Cubebene	1463	$C_{15}H_{24}$	0.15
14	Cathinone	1498	$C_9H_{11}NO$	0.05
15	β-Copaene	1516	$C_{15}H_{24}$	1.24
16	Bicyclogermacrene	1523	$C_{15}H_{24}$	0.24
17	γ-Elemene	1596	$C_{15}H_{24}$	0.10
18	2,6,11,15-Tetramethyl-hexadeca-2,6,8,10,14-pentaene	1979	$C_{20}H_{30}O_2$	0.50
19	β-Farnesene	2014	$C_{15}H_{24}$	0.18
20	7-Methoxy-6-(3-methyl-2-oxobutyl)-2*H*-1-benzopyran-2-one	2302	$C_{16}H_{16}O_4$	0.61
21	2-(Methylamino)-1-phenylethanol	2329	$C_9H_{13}NO$	0.04
	Total			96.87

2.2. Antioxidant Activities of Pomelo Peel Essential Oil

Figure 2 shows the DPPH free radical scavenging rate, superoxide anion radical scavenging rate, and total antioxidant activity of PPEO. PPEO exhibited significant effects on the free radical scavenging rate in a concentration dependent manner. At a low concentration (5 mg/mL), PPEO had no significant effect on DPPH free radical scavenging (Figure 2A). However, with increasing concentration, the DPPH free radical scavenging rate of PPEO reached 68.13% at 150 mg/mL, which was statistically significantly different from that of the control group ($p < 0.01$). TheIC_{50} of PPEO was 70.12 mg/mL. The positive control butylated hydroxytoluene (BHT) showed good DPPH free radical scavenging ability at low concentrations. Superoxide anion free radicals can induce lipid peroxidation in the body, thereby accelerating the aging of human skin and even internal organs [18]. As shown in Figure 2B, when the concentration of PPEO was 0.2 mg/mL, the superoxide anion clearance rate was only 11.93% ($p < 0.05$). With increasing PPEO concentration, the clearance rate increased to 44.74% at 1.0 mg/mL ($p < 0.01$). It seems that PPEO could effectively remove the superoxide anion radicals, but with a lower scavenging ability than L-ascorbic acid at the same concentration. Figure 2C shows that the total antioxidant activity of PPEO was slightly lower than that of BHT at the concentrations lower than 0.4 mg/mL. However, as the concentration increased, the antioxidant activity of PPEO exceeded that of BHT. At the concentration of 1.0 mg/mL, the total antioxidant activity of PPEO was 20.35% higher than that of BHT. These results indicated that low-concentration PPEO has no obvious antioxidant effect while high-concentration PPEO has good antioxidant effect. Butylated hydroxytoluene is an industrially synthesized antioxidant that facilitates fast oxidation resistance at low concentrations, while PPEO is a mixture of various compounds with antioxidant activities, which may have complex interactions with each other [19].

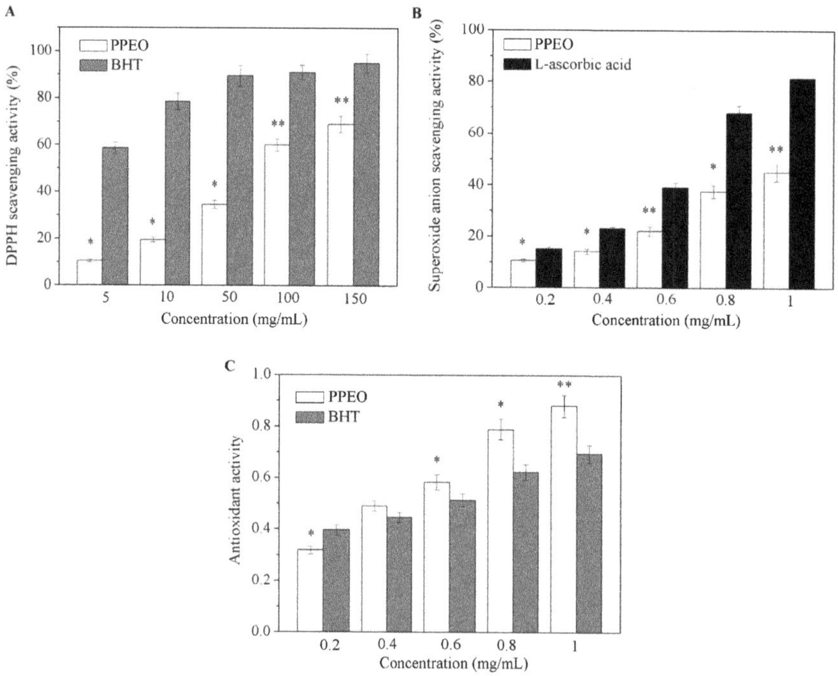

Figure 2. Antioxidant activities of PPEO. (**A**) 1,1-Dipheny-2-Picryhydrazyl (DPPH) radical scavenging assay; (**B**) superoxide anion radical scavenging activity assay; (**C**) 2,2′-azinobis-(3-ethylbenzthiazoline-6-sulphonate) (ABTS) total antioxidant activity. * Indicates samples that are significantly different ($n = 3$; * $p < 0.05$ and ** $p < 0.01$ compared with the positive control group).

2.3. Anti-Melanogenic Effects of Pomelo Peel Essential Oil

2.3.1. Effect of Pomelo Peel Essential Oil on Cell Viability

B16 melanoma cells were treated with different concentrations of PPEO for 24 h, and the viability of each group was detected by MTT assay (Figure 3). The survival rate of B16 cells treated with a low concentration of PPEO (5 μg/mL) was higher than 100%, indicating that low concentration of PPEO may facilitate the proliferation of B16 cells, which might be related to the active volatile components in PPEO [20]. The cell viability decreased significantly along with increasing PPEO concentration. When the concentration of PPEO was 150 μg/mL, the cell survival rate significantly decreased to 29.35% ($p < 0.01$). The results show that at concentrations lower than 50 μg/mL, PPEO did not affect cell viability. However, high concentrations of PPEO (>50 μg/mL) significantly inhibited cell viability.

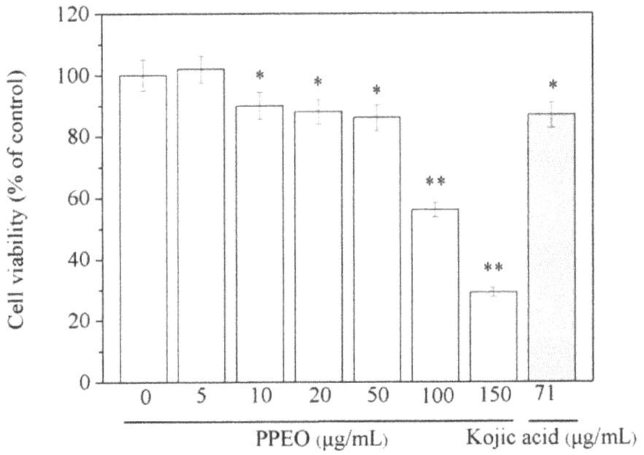

Figure 3. Effect of PPEO on B16 melanoma cell viability. * Indicates samples that are significantly different ($n = 3$; * $p < 0.05$ and ** $p < 0.01$ compared with the blank control group).

2.3.2. Effect of Pomelo Peel Essential Oil on Cell Morphology

B16 cells are adherent and mostly fusiform cells with dendrites and relatively more divisions. They are tightly connected monolayers with high transparency (Figure 4). The cells in the control group showed uniform fluorescence, clear cell boundaries, and normal dendritic morphology. When the concentration of PPEO was 10–50 μg/mL, the number of cell deaths was small, the cell boundary was clear, and the fluorescence was relatively uniform. However, at a concentration of 100 μg/mL, the number of cell deaths increased and the boundaries between cells became blurred, accompanied by the appearance of obvious fluorescent spots. At the same time, cells were dispersed and the dendrites were reduced. In the high-concentration PPEO group (100 μg/mL), the cells showed enhanced fluorescence, and were swollen and separated from each other, presenting a typical state of apoptosis in the cells.

2.3.3. Inhibition of Pomelo Peel Essential Oil on Intracellular Tyrosinase Activity and Melanin Content

To determine the anti-melanogenic activity of PPEO, we evaluated its effect on TYR activity and melanin content in B16 melanoma cells. The B16 melanoma cells were treated with various concentrations of PPEO and then co-cultured for 72 h. As shown in Figure 5, PPEO dose-dependently inhibited TYR activity and melanin content. At a concentration of 50 μg/mL, melanin synthesis and TYR activity were inhibited by 48.28% and 64.54%, respectively, and the IC_{50} of melanin synthesis in inhibition was 67.64 μg/mL. Kojic acid, the positive control, inhibited TYR activity by 62.09%, which is similar to the inhibitory effect of PPEO at 50 μg/mL (Figure 5B).

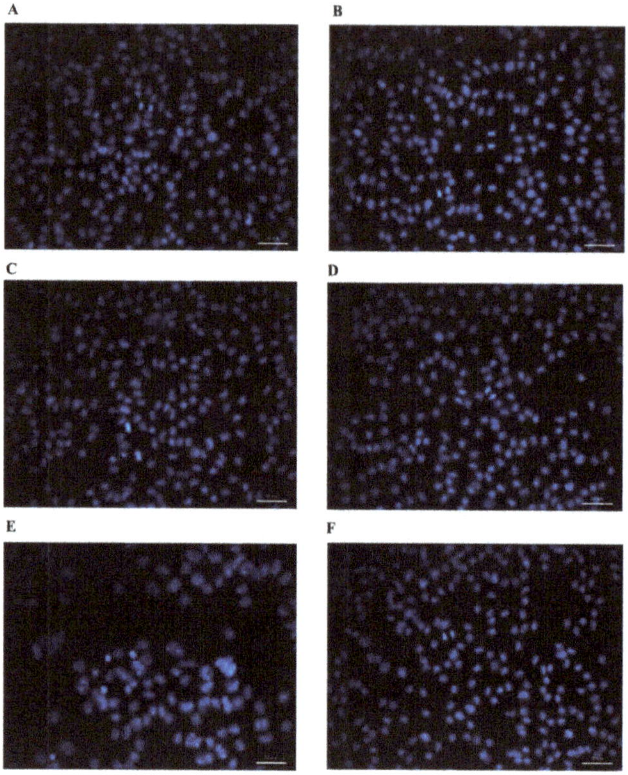

Figure 4. Optical microscopic morphology of B16 cells. PPEO at concentrations of 0, 10, 20, 50, 100 μg/mL were for (**A–E**), respectively, and kojic acid at a concentration of 71 μg/mL was for (**F**). Scale bar: 50 μm

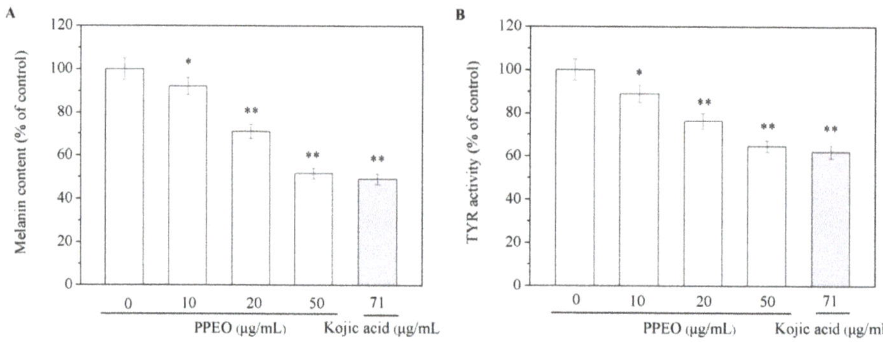

Figure 5. Effect of PPEO on melanin content (**A**) and tyrosinase activity (**B**) in B16 cells. * Indicates samples that are significantly different (n = 3; * $p < 0.05$ and ** $p < 0.01$ compared with the blank control group).

2.3.4. Effect of Pomelo Peel Essential Oil on Tyrosinase Expression in B16 Cells

The expression of TYR protein was detected by Western blotting, and the intensity of protein expression was determined by the ratio of the target band to the internal reference band (Figure 6). PPEO down-regulated the expression level of TYR in B16 cells in a concentration-dependent manner.

Compared with that in the blank control group, the TYR expression gradually decreased with increasing PPEO concentration. When the PPEO concentration was 50 μg/mL, the TYR expression was 60.38% lower than that of the blank group and was close to that of the positive group (kojic acid). This result is consistent with the tyrosinase enzyme linked immunosorbent assays (ELISA) test results.

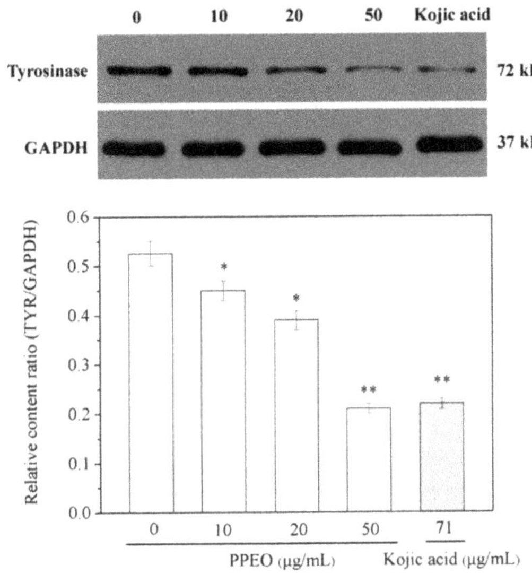

Figure 6. Effect of PPEO on tyrosinase expression in B16 cells. * Indicates samples that are significantly different ($n = 3$; * $p < 0.05$ and ** $p < 0.01$ compared with the blank control group). GADPH: glyceraldehyde-3-phosphate dehydrogenase.

2.4. Discussion

Previous studies have shown that the characteristic aroma of pomelo is mainly attributable to a variety of compounds [2]. In this study, we first extracted PPEO from pomelo cv. Guan Xi and analyzed its main components using GC/MS. Our study demonstrated that the fresh and natural fruit aromas were mainly due to the presence of aldehydes and terpenoids in PPEO. The safety of natural products used in health foods, drug, and cosmetic ingredients is a major concern. Several studies have explored the use of extracts from pulp of guava [21], waste of citrus [6], earthworm [18], oil of *Aquilaria crassna* [8], and oil from *Alpinia zerumbet* [22]. We first determined the antioxidant activities of PPEO. Previous studies have shown that plant EOs have universal antioxidant activities [14]. Meanwhile, it has been reported that terpenes such as limonene, β-myrcene, β-pinene, ocimene, β-copaene, and citral showed anti-oxidant activities [21,23,24]. The strong anti-oxidant activity of PPEOs may be attributable to these components. In our experiment, PPEO was extracted using a cold pressing method under normal temperature conditions, which could better preserve the active components with antioxidant function, resulting in a high total antioxidant capacity. It is known that ultra violet (UV) radiation induces free radical formation in the skin, which is linked directly to the onset of skin photodamage and biological damage. Thus, our results suggest that PPEO may be a useful anti-oxidant source and have the potential to prevent UV-induced damage.

Next, we highlighted the anti-melanogenic effects of PPEO through cell viability, cell morphology, intracellular melanin content, intracellular TYR activity and expression, and compared these results with those of the positive control (kojic acid). Previous studies have shown that the action mechanism of terpenoids in cells is related to the destruction of lipophilic compounds in biofilms [25,26]. Due to the high hydrophobicity of terpenoids, their toxic effects lead to swelling and enhanced fluidity and permeability of the cell membranes [27]. Therefore, terpenoids might cause death of cells at high concentrations of PPEO. However, when the concentration of PPEO was below 50 μg/mL, the state of the cell was normal. On considering this possibility, concentrations below 50 μg/mL of PPEO were used to evaluate its effects on melanin content and intracellular TYR activity.

To our knowledge, TYR catalyzes the first two steps of mammalian melanogenesis [18], namely the hydroxylation of monophenol to *o*-diphenol and the oxidation of diphenol to *o*-quinones, both of which use molecular oxygen, followed by a series of nonenzymatic steps to finally result in the formation of melanin [12]. Therefore, inhibition of TYR activity may help to avoid abnormal melanin pigmentation in skin. The results of this study preliminarily demonstrated the good antioxidant performance of PPEO. In the co-culture of PPEO and B16 cells, PPEO acted as an antioxidant to inhibit the catalytic reaction of TYR and block the synthesis pathway of melanin, resulting in a decrease in melanin production. Meanwhile, the effect of PPEO on the dendritic morphology of cells might also destroy the normal physiological functions of cells, which in turn affects the formation of melanin in cells. The above results demonstrate that PPEO can decrease the melanin content without affecting the cell viability.

In addition, previous study showed that citral, myrcene, (2E)-alkenal, and terpinolene were popular tyrosinase inhibitors [13]. In another study, citral and myrcene were found to have significant inhibitory effects on TYR, and trans-citral has better inhibitory effect than *cis*-citral. In plant EOs, the content of *trans*-citral is higher than that of *cis*-citral. Meanwhile, citral and myrcene are the main active substances that inhibit TYR in EOs [28]. Therefore, it is most likely that the inhibition of TYR expression by the components (limonene, β-myrcene, β-pinene, ocimene, β-copaene) in PPEO is a synergistic effect, and the main active components could be partly attributed to citral and myrcene.

3. Materials and Methods

3.1. Materials

Fully ripe fruits of pomelo cv. Guan Xi were harvested from Fujian province of China. Reagents 2,2-diphenyl-1-picrylhydrazyl (DPPH), dimethyl sulfoxide (DMSO), Hoechst 33258 staining solution, fetal bovine serum (FBS), *n*-paraffins (C7–C30), and tyrosinase enzyme linked immunosorbent assays (ELISA) kit were purchased from Sigma-Aldrich Co. (St. Louis, MO, USA). Dulbecco's Modified Eagle Medium (DMEM) was from Gibco Chemical Co. (Grand Island, NY, USA). Penicillin-streptomycin double antibody, trypsin, and methyl thiazolyl tetrazolium (MTT) were purchased from Genivew Co. (El Monte, CA, USA). Cell Lysates, tert-butyl hydroxytoluene (BHT), kojic acid, and protein quantification test kit were purchased from Shanghai Yuye biotechnology Co., Ltd. (Shanghai, China). RPMI 1640 medium was purchased from Hyclone Co. (Logan, UT, USA). ECL chemiluminescence detection kit and sodium dodecyl sulfate polyacrylamide gels (SDS-PAGE) Gel Preparation Kit were purchased from Aspentech Co. (Bedford, MA, USA). B16 melanoma cells were purchased from Shanghai Tongpai Biotechnology Co., Ltd (Shanghai, China). All other analytical grade chemicals were bought from Sinopharm chemical reagent Co., Ltd (Shanghai, China).

3.2. Extraction of Essential Oil

Fresh pomelo was picked and the white peel from the fresh pomelo peel was removed, and then the exocarp chopped in 1.5 cm × 0.5 cm × 0.2 cm sized pieces. Pomelo peel (100 g) was crushed and pressed twice with a cold hydraulic press (Model 6YL-190; Changbai Mountain Technology Limited company, Changchun, China) [28] and then the peel slurry was collected. After filtering through a steel sieve (0.15 mm), saturated sodium chloride (NaCl) solution was added to the sample for the extraction of PPEO for 3 h. Then, the sample was centrifuged at 10,000× g for 30 min at 4 °C. Supernatants were stored in separate amber bottles at −20 °C until use. The extracted PPEO was weighed to determine the extraction yield as follows: extraction yield (%) = [(weight of extracted oil)/(weight of pomelo peel)] × 100%.

3.3. *Gas Chromatography-Mass Spectrometry Analysis*

The extracted PPEO was filtered through a 0.45 μm microporous organic membrane. Volatile compounds were analyzed using an Agilent 7890A GC coupled to an Agilent 5975C mass spectrometer (Palo Alto, CA, USA). The components of PPEO were identified using HP-5Ms phenylmethylsiloxane capillary column (30 m × 0.25 mm i.d., 0.25 μm; Agilent Technologies, J & W Scientific Products, Folsom, CA, USA) [24]. The helium was used as a carrier gas with a flow rate of 1 mL/min. Injector temperature was 250 °C. The split ratio was 10:1. The temperature program was 45 °C (hold for 1 min), increase at 10 °C/min to 165 °C (hold for 2 min), increase at 1.5 °C/min to 180 °C (hold for 2 min), and then increase at 10 °C/min to 250 °C (hold for 2 min). The temperature of both injector and detector was set at 250 °C. Mass spectra were scanned from *m/z* 35–350 amu. The electron impact ionization energy was 70 eV. Identification of compounds detected by GC/MS analysis was performed by comparing mass spectra and retention indices (RIs) with published data obtained under similar conditions, as well as by comparing their mass spectra with the MS library of Wiley 7.0 and Nist 05 [29]. A mixture of *n*-paraffins (C7–C30) as standards was used for calculating RIs. Samples were analyzed and identified using an available Retention Time Locked (RTL) database with Deconvolution Reporting Software (DRS) and a database of 926 DRS compounds.

3.4. *Antioxidant Activities*

3.4.1. DPPH Radical Scavenging Assay

The DPPH radical scavenging assay was performed according to Boskou et al. [23]. The sample PPEO was prepared with different concentrations with absolute ethanol. A mixture of 50 μL sample and 150 μL 0.1 mmol/L DPPH free radical ethanol solution was taken and placed in a 96-well plate. After vigorous shaking, the mixture was incubated at room temperature in the dark for 30 min. The absorbance at 517 nm was measured. Pomelo peel essential oil was replaced with absolute ethanol to serve as the blank control, and BHT was used as positive control.

3.4.2. Superoxide Anion Radical Scavenging Activity Assay

The superoxide anion radical scavenging activity was measured as described by Zhang et al. [30]. The reaction mixture consisted of 4.5 mL 50 mM Tris-HCl buffer (pH 8.2) and 1 mL sample solution at different concentrations. The mixed solution was pre-incubated at 25 °C for 10 min, and then initiated by the addition of 0.45 mL 2.5 mM pyrogallol. After vigorous shaking for 5 min, the reaction was terminated by the addition of 8 mol/L HCl. The absorbance was read at 517 nm.

3.4.3. ABTS Radical Scavenging Assay

The ABTS radical scavenging assay was conducted following the method previously described by Re et al. [31]. Diluted radical solution was prepared by mixing 7 mM ABTS and 2 mM $K_2S_2O_8$ in equal amounts, followed by reaction in the dark overnight at room temperature. The samples were prepared in different concentrations with ultrapure water. Aliquots of 10 μL of samples were mixed with 200 μL of the diluted radical solution in 96-well plate and the absorbance was measured at 734 nm after 5 min using an M200 pro enzyme-labeled instrument (Tecan, Ltd., Männedorf, Switzerland). BHT was used as the positive control.

3.5. *Cell Culture and Treatment*

The murine metastatic melanoma cell line B16 was cultured in sterile cell culture flasks with RPMI 1640 medium supplemented with 100 U/mL penicillin, 100 U/mL streptomycin, and 10% heat inactivated FBS at 37 °C in a humidified incubator containing 5% CO_2. Cells in logarithmic growth phase were selected for subsequent experiments [32]. The extracted PPEO was dissolved in Tween 80 and filtered through a 0.45 μm microporous organic membrane. The PPEO was diluted to different concentrations and then was added to the medium.

3.6. MTT Assay for Cell Viability

Cell viability was evaluated by 3-[4, 5-dimethylthiazol-2-yl]-2, 5-diphenyl tetrazolium bromide (MTT) assay according to the method of Satooka et al. [33]. The cell density was 7×10^4 cells/mL and the cells were seeded on 96-well cell culture plates at 100 μL per well. After 24 h of culture, the original culture solution was aspirated. Cells were exposed to various concentrations of PPEO or kojic acid (71 μg/mL), with 6 replicates for each concentration (500 μmol/L of kojic acid is equal to 71 μg/mL). After culture for another 24 h, 100 μL of 0.5 mg/mL MTT was added to each well, followed by inoculation for 4 h at 37 °C. The liquid was carefully aspirated from the well, and then 150 μL of DMSO was added to each well. After 10 min of shaking, the absorbance was measured at 490 nm using an M200 pro enzyme-labeled instrument (Tecan, Ltd.).

3.7. Immunofluorescence Analysis and Hoechst Staining

A sterilized coverslip was placed into each hole of the six-hole plate on an ultra-clean workbench. The cell suspension was added to each coverslip and placed in an incubator with a CO_2 concentration of 5% at 37 °C until cell fixation (2 h). After the addition of 2 mL culture medium, the culture was continued for about 6 h. The medium was decanted and the cells were washed for 5 min with PBS for 3 times. The cells were fixed with 4% paraformaldehyde for 30 min, and then paraformaldehyde was removed by PBS buffer washing. After the addition of appropriate amounts of Hoechst stain, the coverslips were incubated at room temperature for 15 min in the dark. The coverslips were then rinsed 3 times with PBS for 5 min each time, and the side with the cells was observed under a confocal laser scanning microscope [34].

3.8. Determination of Melanin Content

Melanin content was determined as described by Huang et al. [35]. B16 cells were plated at a density of 7×10^4 cells/well in a 6-well plate. The experimental group was added with 0.2 μmol/L α-MSH (melanocyte-stimulating hormone) to construct a cell model with high melanin expression. After 12 h of culture, cells were exposed to various concentrations of PPEO (10–100 μg/mL). Kojic acid at a concentration of 71 μg/mL was used as a positive control. After 48 h of culture, the supernatant was discarded and the cells were washed 3 times with PBS buffer. After the addition of 200 μmol/L NaOH solution (containing 10% DMSO) to each well, the cells were fully lysed at 80 °C for 1 h, and the absorbance was measured at 492 nm. The amount of protein was measured by Micro BCA protein assay kit (Shanghai Yuye biotechnology Co., Ltd, Shanghai, China). The melanin content was calculated by normalization to the total cellular protein (1 g of melanin/mg of protein) and reported as a percentage of the control.

3.9. Intracellular Tyrosinase Activity

B16 cells were plated at a density of 7×10^4 cells/well in a 6-well plate. After 24 h of culture, cells were exposed to various concentrations of PPEO (10–100 μg/mL) or kojic acid (71 μg/mL), and incubated for additional 48 h. The cells were then washed with ice-cold phosphate buffer. The plates were frozen at −80 °C for 30 min. After thawing and mixing, the tyrosinase activity was measured by ELISA kit [36].

3.10. Protein Extraction and Western Blot Analysis

Total protein was extracted from cells lysed by RIPA Lysis buffer with 1% phenylmethylsulfonyl fluoride (PMSF). The amount of protein was measured by Micro BCA protein assay kit (Shanghai Yuye biotechnology Co., Ltd.). The samples were loaded on 10% sodium dodecyl sulfate polyacrylamide gels (SDS-PAGE) and transferred onto polyvinylidene difluoride (PVDF) membranes. The membranes were then placed in blocking solution and blocked at room temperature for 1 h. The membranes were incubated overnight at 4 °C with appropriate concentrations of specific antibodies, including rabbit monoclonal antibodies glyceraldehyde-3-phosphate dehydrogenase (GAPDH, 1:10000 dilution) and rabbit monoclonal antibodies TYR (1:1000 dilution). After five or six times of washing, the blots were then incubated with secondary antibody (HRP-Goat anti Rabbit). BandScan was used to analyze the integrated density of bands.

3.11. Statistical Analysis

All the experiments were performed with freshly prepared samples in triplicate. The results were expressed as means ± standard deviation (SD) and analyzed by one-way analysis of variance (ANOVA) test using SPSS 19.0 (IBM Corporation, Armonk, NY, USA). Differences were considered as statistically significant at the level of $p < 0.05$.

4. Conclusions

Our study is the first to extract essential oil from the peel of pomelo cv. Guan Xi by a cold pressing method and analyze its main components of limonene, β-myrcene, β-pinene, ocimene, and β-copaene. Our results reveal that PPEO has strong antioxidant activities against DPPH, ABTS, and superoxide anion radicals and the main active components responsible for the effect are terpenes. Besides, the effects of PPEO on the viability of B16 cells and the production of melanin were evaluated based on the B16 melanoma cell system. The results indicate that PPEO down-regulates the expression level of TYR in B16 cells, which inhibits the catalytic reaction of TYR and blocks the synthesis pathway of melanin; and it further reduces melanin production without affecting the cell viability. This study provides data support for expanding the potential application of essential oil from pomelo peel as a natural antioxidant in the food, pharmaceutical and cosmetic industries. Further research in vivo is needed to fully evaluate the potential anti-melanogenic effect of PPEO.

Author Contributions: S.P. provided the initial idea for research. W.H. and Y.P. designed the research. X.L. conducted the experimental work and W.H. drafted the manuscript. X.H. edited the paper. All authors discussed and approved the final manuscript.

Funding: This work was financially supported by National Natural Science Foundation of China (No. 31571847) and Modern Agricultural Industry Technology System Post Scientist Project of China (No. CARS-26).

Conflicts of Interest: The authors declare no conflict of interest.

References

1. Liu, Y.; Liu, A.; Ibrahim, S.A.; Yang, H.; Huang, W. Isolation and characterization of microcrystalline cellulose from pomelo peel. *Int. J. Biol. Macromol.* **2018**, *111*, 717–721. [CrossRef] [PubMed]
2. Guo, X.; Zhao, W.; Liao, X.; Hu, X.; Wu, J.; Wang, X. Extraction of pectin from the peels of pomelo by high-speed shearing homogenization and its characteristics. *LWT Food Sci. Technol.* **2017**, *79*, 640–646. [CrossRef]
3. Zhu, L.; Wang, Y.; Wang, Y.; You, L.; Shen, X.; Li, S. An environmentally friendly carbon aerogels derived from waste pomelo peels for the removal of organic pollutants/oils. *Microporous Mesoporous Mater.* **2018**, *241*, 285–292. [CrossRef]
4. Hosni, K.; Zahed, N.; Chrif, R.; Abid, I.; Medfei, W.; Kallel, M.; Brahim, N.B.; Sebei, H. Composition of peel essential oils from four selected Tunisian Citrus species: Evidence for the genotypic influence. *Food Chem.* **2010**, *123*, 1098–1104. [CrossRef]
5. Wu, F.; Jin, Y.; Xu, X.; Yang, N. Electrofluidic pretreatment for enhancing essential oil extraction from citrus fruit peel waste. *J. Clean. Prod.* **2017**, *159*, 85–94. [CrossRef]
6. Akhtar, M.; Iqbal, L.; Lateef, M.; Nawab, B.; Saleem, M.; Afza, N. Bio-reactive properties of citrus waste: An investigation of antioxidant and tyrosinase inhibitory activities. *Pak. J. Bot.* **2011**, *43*, 2881–2883.

7. Fiocco, D.; Arciuli, M.; Arena, M.P.; Benvenuti, S.; Gallone, A. Chemical composition and the anti-melanogenic potential of different essential oils. *Flavour Fragr. J.* **2016**, *31*, 255–261. [CrossRef]
8. Dahham, S.S.; Tabana, Y.M.; Iqbal, M.A.; Ahamed, M.B.; Ezzat, M.O.; Majid, A.S.; Majid, A.M. The Anticancer, Antioxidant and Antimicrobial Properties of the Sesquiterpene β-Caryophyllene from the Essential Oil of *Aquilaria crassna*. *Molecules* **2015**, *20*, 11808–11829. [CrossRef]
9. Choi, H.S.; Song, H.S.; Ukeda, H.; Sawamura, M. Radical-Scavenging Activities of Citrus Essential Oils and Their Components: Detection Using 1,1-Diphenyl-2-picrylhydrazyl. *J. Agric. Food Chem.* **2000**, *48*, 4156–4161. [CrossRef]
10. Dayanand, D.; Balamurugan, K.; Agrawal, R.C.; Rahul, V.; Rahi, J. Evalution of antioxidant activity of methanolic and hydromethanolic extract of sweet orange peels. *Recent Res. Sci. Technol.* **2011**, *3*, 22–25.
11. Asikin, Y.; Maeda, G.; Tamaki, H.; Mizu, M.; Oku, H.; Wada, K. Cultivation line and fruit ripening discriminations of Shiikuwasha (*Citrus depressa* Hayata) peel oils using aroma compositional, electronic nose, and antioxidant analyses. *Food Res. Int.* **2015**, *67*, 102–110. [CrossRef]
12. Seo, S.Y.; Sharma, V.K.; Sharma, N. Mushroom tyrosinase: Recent Prospects. *J. Agric. Food Chem.* **2003**, *51*, 2837–2853. [CrossRef] [PubMed]
13. Ritaro, M.; Hiroyuki, U.; Masayoshi, S. Tyrosinase inhibitory activity of citrus essential oils. *J. Agric. Food Chem.* **2006**, *54*, 2309–2313.
14. Proestos, C.; Lytoudi, K.; Mavromelanidou, O.K.; Zoumpoulakis, P.; Sinanoglou, V.J. Antioxidant Capacity of Selected Plant Extracts and Their Essential Oils. *Antioxidants* **2013**, *2*, 11–22. [CrossRef] [PubMed]
15. Hu, J.J.; Li, X.; Liu, X.H. Inhibitory effect of lemon essential oil on mushroom tyrosinase activity in vitro. *Mod. Food Sci. Technol.* **2015**, *31*, 97–105.
16. Yang, C.H.; Huang, Y.C.; Tsai, M.L.; Chen, Y.C.; Li, L.L.; Ya, W.Y. Inhibition of melanogenesis by β-caryophyllene from lime mint essential oil in mouse B16 melanoma cells. *Int. J. Cosmet. Sci.* **2015**, *37*, 550–554. [CrossRef] [PubMed]
17. Phi, N.T.L.; Hung, P.V.; Chi, P.T.L.; Dung, N.H. Impact of Extraction Methods on Antioxidant and Antimicrobial Activities of Citrus Essential Oils. *J. Essent. Oil Bear. Plants* **2015**, *18*, 806–817. [CrossRef]
18. Azmi, N.; Hashim, P.; Hashim, D.M.; Halimoon, N.; Majid, N.M. Anti-elastase, anti-tyrosinase and matrix metalloproteinase-1 inhibitory activity of earthworm extracts as potential new anti-aging agent. *Asian Pac. J. Trop. Biomed.* **2014**, *4* (Suppl. 1), S348–S352. [CrossRef]
19. Dong, H.; Zhang, Q.; Li, L.; Liu, J.; Shen, L.; Li, H.; Qin, W. Antioxidant activity and chemical compositions of essential oil and ethanol extract of *Chuanminshen violaceum*. *Ind. Crops Prod.* **2015**, *76*, 290–297. [CrossRef]
20. Guo, J.J.; Gao, Z.P.; Xia, J.L.; Ritenour, M.A.; Li, G.Y.; Shan, Y. Comparative analysis of chemical composition, antimicrobial and antioxidant activity of citrus essential oils from the main cultivated varieties in China. *LWT Food Sci. Technol.* **2018**, *97*, 825–839. [CrossRef]
21. Almulaiky, Y.; Zeyadi, M.; Saleh, R.; Baothman, O.; Al-shawafi, W.; Al-Talhi, H. Assessment of antioxidant and antibacterial properties in two types of Yemeni guava cultivars. *Biocatal. Agric. Biotechnol.* **2018**, *16*, 90–97. [CrossRef]
22. Tu, P.T.; Tawata, S. Anti-oxidant, anti-aging, and anti-melanogenic properties of the essential oils from two varieties of *Alpinia zerumbet*. *Molecules* **2015**, *20*, 16723–16740. [CrossRef] [PubMed]
23. Boskou, G.; Salta, F.N.; Chrysostomou, S.; Mylona, A.; Chiou, A.; Andrikopoulos, N.K. Antioxidant capacity and phenolic profile of table olives from the Greek market. *Food Chem.* **2006**, *94*, 558–564. [CrossRef]
24. Roberto, D.; Micucci, P.; Sebastian, T.; Graciela, F.; Anesini, C. Antioxidant activity of limonene on normal murine lymphocytes: Relation to H_2O_2 modulation and cell proliferation. *Basic Clin. Pharmacol. Toxicol.* **2010**, *106*, 38–44. [CrossRef] [PubMed]
25. Sall, C.; Dombrowsky, L.; Bottzeck, O.; Praud-Tabaries, A.; Blache, Y. Targeting bacterial biofilms: Design of a terpenoid-like library as non-toxic anti-biofilm compounds. *Bioorg. Med. Chem. Lett.* **2011**, *21*, 1493–1497. [CrossRef] [PubMed]
26. Tepe, B.; Donmez, E.; Unlu, M.; Candan, F.; Daferera, D.; Vardar-Unlu, G.; Polissiou, M.; Sokmen, A. Antimicrobial and antioxidative activities of the essential oils and methanol extracts of *Salvia cryptantha* (Montbret et Aucher ex Benth.) and *Salvia multicaulis* (Vahl). *Food Chem.* **2004**, *84*, 519–525. [CrossRef]
27. Turina, A.V.; Nolan, M.V.; Zygadlo, J.A.; Perillo, M.A. Natural terpenes: Self-assembly and membrane partitioning. *Biophys. Chem.* **2006**, *122*, 101–113. [CrossRef]

28. Zhang, Y.; Wu, L.; Yao, Z.; Ma, Z.; Liu, J. Hypolipidemic effects of hickory nut oil using cold pressure extraction. *Food Sci. Biotechnol.* **2016**, *25* (Suppl. 1), 41–46. [CrossRef]
29. Yao, Y.Y.; Pan, S.Y.; Fan, G.; Dong, L.; Ren, J.N.; Zhu, Y. Evaluation of volatile profile of Sichuan dongcai, a traditional salted vegetable, by SPME-GC-MS and E-nose. *LWT Food Sci. Technol.* **2015**, *64*, 528–535. [CrossRef]
30. Zhang, Q.A.; Wang, X.; Song, Y.; Fan, X.H.; Garcia Martin, J.F. Optimization of Pyrogallol Autoxidation Conditions and Its Application in Evaluation of Superoxide Anion Radical Scavenging Capacity for Four Antioxidants. *J. AOAC Int.* **2016**, *99*, 504–511. [CrossRef]
31. Roberta, E.N.; Annap, R.; Ananth, P.; Annala, M.; Catherine, R.E. Antioxidant activity applying an improved ABTS radical cation decolorization assay. *Free Radic. Biol. Med.* **1999**, *26*, 1231–1237.
32. Meira, W.V.; Heinrich, T.A.; Cadena, S.; Martinez, G.R. Melanogenesis inhibits respiration in B16-F10 melanoma cells whereas enhances mitochondrial cell content. *Exp. Cell Res.* **2017**, *350*, 62–72. [CrossRef] [PubMed]
33. Satooka, H.; Cerda, P.; Kim, H.J.; Wood, W.F.; Kubo, I. Effects of matsutake mushroom scent compounds on tyrosinase and murine B16-F10 melanoma cells. *Biochem. Biophys. Res. Commun.* **2017**, *487*, 840–846. [CrossRef] [PubMed]
34. Monika, P.; Claus, R.; Gudrun, M.; Donat, K.; Jochen, H.M. Ceramide-induced apoptosis of D283 medulloblastoma cells requires mitochondrial respiratory chain activity but occurs independently of caspases and is not sensitive to Bcl-Xl overexpression. *J. Neurochem.* **2002**, *82*, 424–494.
35. Huang, Y.C.; Liu, K.C.; Chiou, Y.L.; Yang, C.H.; Chen, T.H.; Li, T.T.; Liu, L.L. Fenofibrate suppresses melanogenesis in B16-F10 melanoma cells via activation of the p38 mitogen-activated protein kinase pathway. *Chem. Biol. Interact.* **2013**, *205*, 157–164. [CrossRef] [PubMed]
36. Li, X.; Guo, L.; Sun, Y.; Zhou, J.; Gu, Y.; Li, Y. Baicalein inhibits melanogenesis through activation of the ERK signaling pathway. *Int. J. Mol. Med.* **2010**, *25*, 923–927. [CrossRef] [PubMed]

Sample Availability: Not available.

molecules

Article

Influence of Temperature, Solvent and pH on the Selective Extraction of Phenolic Compounds from Tiger Nuts by-Products: Triple-TOF-LC-MS-MS Characterization

Elena Roselló-Soto [1], Francisco J. Martí-Quijal [1], Antonio Cilla [1], Paulo E. S. Munekata [2], Jose M. Lorenzo [2], Fabienne Remize [3] and Francisco J. Barba [1,*]

1 Nutrition and Food Science Area, Preventive Medicine and Public Health, Food Science, Toxicology and Forensic Medicine Department, Universitat de València, Faculty of Pharmacy, Avda; Vicent Andrés Estellés, s/n, Burjassot, 46100 València, Spain; eroso2@alumni.uv.es (E.R.-S.); francisco.j.marti@uv.es (F.J.M.-Q.); antonio.cilla@uv.es (A.C.)

2 Centro Tecnológico de la Carne de Galicia, Avda. Galicia No. 4, Parque Tecnológico de Galicia, San Cibrao das Viñas, 32900 Ourense, Spain; pmunekata@gmail.com (P.E.S.M.); jmlorenzo@ceteca.net (J.M.L.)

3 QualiSud, Université de La Réunion, CIRAD, Université Montpellier, Montpellier SupAgro, Université d'Avignon, 2 rue J. Wetzell, F-97490 Sainte Clotilde, France; fabienne.remize@univ-reunion.fr

* Correspondence: francisco.barba@uv.es; Tel.: +34-963-54-4972; Fax: +34-963-54-4954

Academic Editors: Lillian Barros and Isabel C.F.R. Ferreira
Received: 26 January 2019; Accepted: 18 February 2019; Published: 22 February 2019

Abstract: The aim of this study was to assess the effect of temperature, solvent (hydroethanolic mixtures) and pH on the recovery of individual phenolic compounds from "horchata" by-products. These parameters were optimized by response surface methodology and triple-TOF-LC-MS-MS was selected as the analytical tool to identify and quantify the individual compounds. The optimum extraction conditions were 50% ethanol, 35 °C and pH 2.5, which resulted in values of 222.6 mg gallic acid equivalents (GAE)/100 g dry matter and 1948.1 μM trolox equivalent (TE)/g of dry matter for total phenolic content (TPC) and trolox equivalent antioxidant capacity (TEAC), respectively. The extraction of phenolic compounds by the conventional solvent method with agitation was influenced by temperature (p = 0.0073), and more strongly, by the content of ethanol in the extraction solution (p = 0.0007) while the pH did not show a great impact (p = 0.7961). On the other hand, the extraction of phenolic acids was affected by temperature (p = 0.0003) and by ethanol amount (p < 0.0001) but not by the pH values (p = 0.53). In addition, the percentage of ethanol influenced notably the extraction of both 4-vinylphenol (p = 0.0002) and the hydroxycinnamic acids (p = 0.0039). Finally, the main individual phenolic extracted with hydroethanolic mixtures was 4-vinylphenol (303.3 μg/kg DW) followed by spinacetin3-*O*-glucosyl-(1→6)-glucoside (86.2 μg/kg DW) and sinensetin (77.8 μg/kg DW).

Keywords: polyphenols; tiger nut; by-products; solvent extraction; horchata de chufa; triple TOF-LC-MS-MS

1. Introduction

"Horchata de chufa" is a typical beverage from the Valencian Community. It is obtained from tiger nuts (*Cyperus esculentus*), which are tuberous rhizomes that protrude from the tips of the plant's roots under the ground [1]. During "horchata" preparation a large amount of waste and by-products are obtained, representing "horchata" by-products ~60% of the total amount of the raw material used to obtain the beverage [2]. These by-products are a source of polysaccharides, fiber, oil (rich in oleic acid) and antioxidant compounds (e.g., vitamin E and polyphenols), among others [3].

Some previous studies have evaluated the potential application of "horchata" by-products for the preparation of new meat products, due to their high content in fiber [4–6]. However, the exploitation of "horchata" by-products as a source of phenolic compounds for food industries, nutraceuticals and cosmetics has not been widely explored. Some existing studies have evaluated the impact of the use of enzyme pre-treatments alone or combined with high-pressure to extract phenolic compounds from tiger nuts [7]. In this line, a previous study conducted by our research group evaluated the impact of conventional solvent extraction using a combination of binary mixtures consisting of ethanol and water at different percentages, at different temperatures and extraction times to recover total phenolic compounds and total flavonoids with antioxidant capacity from "horchata" by-products, obtaining promising results, particularly with the use of mild heating (up to 60 °C) and hydroethanolic solvents (0%–50% ethanol) [8]. Another relevant aspect of phenolic compound extraction is the selection of an appropriate pH that can influence the yield and stability of phenolic compounds. Acidic conditions are associated with higher extraction yields (higher interaction of phenolic compounds with the solvent) on different vegetable sources of phenolic compounds [9,10].

However, only spectrophotometric methods were used, as it was a preliminary study. It is well known that it is not only important to evaluate the total amount of polyphenols but also to characterize their profile as the biological activity differs according to the targeted compound. Accordingly, the use of chromatographic techniques is currently encouraged to establish the structure and activity of bioactive compounds (that can be complemented by less specific but informative spectrophotometric methodologies) to estimate the impact of conventional and non-conventional extraction processes, processing and bioaccessibility outcomes [11]. Therefore, in the present study, the impact of temperature, solvent (hydroethanolic mixtures) and pH on the recovery of individual phenolic compounds from "horchata" by-products was evaluated. For this purpose, a response surface methodology (RSM) approach was used to optimize the extraction. Moreover, triple-TOF-LC-MS-MS was selected as the analytical tool to identify and quantify the individual compounds. In addition, the results were compared to those obtained after using supercritical carbon dioxide and Folch extraction methodology.

2. Results and Discussion

2.1. Impact of Temperature, Solvent and pH on the Selective Extraction of Total Phenolic Compounds (TPC) and Trolox Equivalent Antioxidant Capacity (TEAC) from Tiger Nuts by-Products

The conventional extraction with hydroethanolic mixtures was optimized according to a Box-Behnken design in order to maximize the TEAC and TPC values. The TPC and TEAC values for each extraction are shown in Table 1.

The TPC and TEAC values ranged from 186.52 to 222.58 mg GAE/100 g of dry matter and 617.80 to 1948.07 μM TE/g of dry matter, respectively. The best condition according to the factorial design for TPC and TEAC was 35 °C, 50% ethanol and pH 2.5. Our TPC values were higher than those obtained by Ogunlade et al. [12] who found values of 115.70 mg GAE/100 g of tiger nut in roasted tubers. In addition, Oladele et al. [13] noticed TPC values of 351 and 134 mg/100 g for yellow and brown tiger nuts, respectively. Koubaa et al. [14] obtained TPC values of 4.53–6.21 mg GAE/100 g of oil and 4.71–5.29 GAE/100 g of oil, for supercritical fluids (SC-CO_2) and mechanical expression (ME) extractions, respectively. Parker et al. [15] found TPC values ranging from 5.63 to 64.9 mg/100 g for tiger nut, whereas Badejo et al. [16] obtained TPC values of 21.67 mg/100 mL for a tiger nut aqueous extract drink.

Moreover, Roselló-Soto et al. [8] reported that TPC values obtained from "horchata" by-products according to the solvent used, temperature and extraction time, showing that the highest TPC values were obtained using 25% ethanol (*v*/*v*), at 60 °C with an extraction time of 3 h. The difference in the TPC values could be due to raw material studied but also to the methods of extraction and analysis.

Table 1. Total phenolic content (TPC) and total antioxidant capacity (TEAC) from "horchata" by-products obtained after conventional extraction with hydroethanolic mixtures.

Run#	T (°C)	Ethanol (%)	pH	TPC (mg GAE/100 g of Dry Matter)	TEAC Values (μM TE/g of Dry Matter)
1	25	0	7.25	186.52 ± 4.58	705.31 ± 18.45
2	25	25	2.5	197.74 ± 1.38	996.68 ± 5.26
3	25	25	12	196.74 ± 2.09	1325.68 ± 109.73
4	25	50	7.25	200.06 ± 2.34	1185.43 ± 284.17
5	35	0	2.5	203.95 ± 2.64	617.80 ± 24.15
6	35	0	12	207.20 ± 13.06	862.10 ± 50.04
7	35	25	7.25	211.79 ± 3.75	1110.00 ± 216.41
8	35	25	7.25	206.92 ± 8.28	1094.15 ± 171.80
9	35	25	7.25	203.64 ± 2.39	1091.04 ± 50.88
10	35	50	2.5	222.58 ± 2.16	1948.07 ± 434.18
11	35	50	12	215.72 ± 1.51	1785.94 ± 84.12
12	50	0	7.25	186.78 ± 0.94	983.24 ± 101.28
13	50	25	2.5	209.17 ± 3.32	1101.18 ± 79.84
14	50	25	12	210.38 ± 4.31	1328.06 ± 76.90
15	50	50	7.25	220.48 ± 2.47	1644.27 ± 28.53

The influence of pH (2.5–12), temperature (25–50 °C) and volume of ethanol (0%–50%) to obtain phenolic compounds by conventional extraction with an ethanol:water mixtures was analyzed using a response surface methodology (RSM). As can be seen in Figure 1A, the extraction of phenolic compounds by the conventional method was influenced by temperature (p = 0.0073), and more strongly, by the content of ethanol in the extraction solution (p = 0.0007). On the contrary, the pH did not show a great impact (p = 0.7961). Regarding temperature, we appreciated that at pH = 7, the optimum value in an extraction without ethanol was 37 °C. However, at the highest studied ethanol concentration (50%), the optimum temperature was increased up to 43.5 °C. Furthermore, increasing the ethanol percentage led to a clear increase in extraction yield at temperatures above 40 °C, but on the contrary, this improvement was less clear at room temperature (25 °C). As proposed elsewhere, when temperature increases, the integrity of the cell wall is altered, and therefore there is a greater contact of the cellular components, among them the polyphenols, with the extraction solution [17]. However, over a certain threshold, despite the increased extraction of polyphenols due to matrix degradation, a decrease of several of these bioactive compounds can happen due to thermal lability, as observed in Figure 1A.

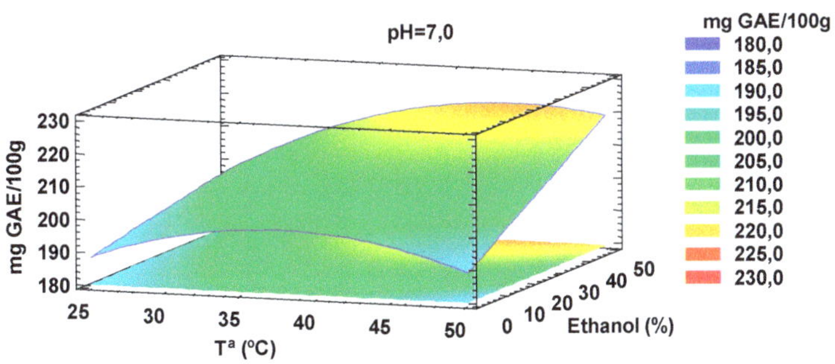

(A)

Figure 1. *Cont.*

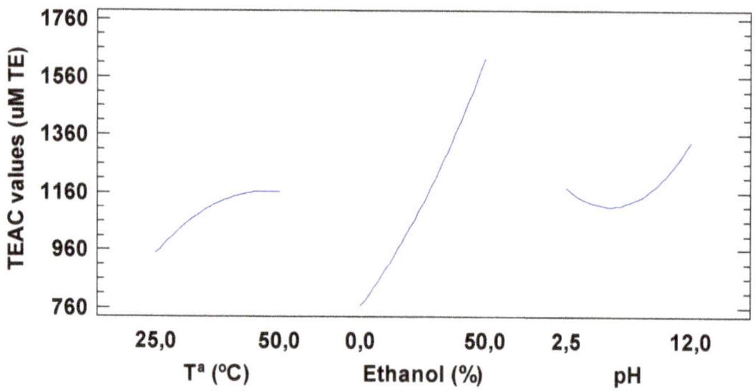

(**B**)

Figure 1. Plot for the influence of extraction condition parameters in total phenolic content (mg gallic acid equivalents ((GAE)/100 g of dry matter) (**A**) and the main effects chart for antioxidant activity (**B**) using solid-liquid extraction. One gram of tiger nuts by-products was extracted using different temperatures (25–50 °C), ethanol:water mixture (0%–50% ethanol) and pH (2.5–12).

These values fully agree with those obtained by Moreira et al. [18], who observed that using a concentration of 50% ethanol, a 2-fold increase in the values of TPC from apple tree wood was obtained, compared to control samples without ethanol. In addition, they also found that temperature had an important impact on the extraction, getting a greater extraction at 55 °C compared to 20 °C. Vatai et al. [19] also reached the same conclusions, obtaining the maximum TPC values in Refosk (red grape marc) and lyophilized elderberry with an optimum percentage of 50% ethanol at 60 °C. On the contrary, Roselló-Soto et al. [8] observed an 82% increase in the TPC value from tiger nuts by-products with an ethanol volume of 50% at 60 °C during 2 h of extraction, but this changed at 3 h, obtaining the maximum TPC values with an ethanol concentration of 25%.

Figure 1B shows the main effects observed for antioxidant capacity at different temperatures, ethanol concentration and pH. It is clearly observed how an increase in the ethanolic fraction during the extraction of the polyphenols significantly increased the antioxidant capacity of the samples ($p = 0.0029$). Besides, neither the extraction temperature nor the pH influenced the antioxidant capacity ($p = 0.2328$ and 0.3635, respectively). Roselló-Soto et al. [8] also obtained an increase in the antioxidant capacity (TEAC assay) of extracts from "horchata" by-products when they used an ethanol concentration of 50% and an increase in temperature up to 60 °C. In our case, there was also an increasing trend ($p > 0.05$) in the antioxidant capacity when the temperature was augmented, as well as when highly acidic and alkaline conditions were used. However, as it was mentioned above, these increases are not significant. Moreover, Li et al. [20] also indicated that the TEAC values of extracts obtained from *Gordonia axillaris* increased with an ethanol volume of 40% and a temperature of 40 °C, but decreased when these values were higher. According to the authors, this could be due to the degradation of some thermolabile antioxidant compounds. Moreira et al. [18] studied the extraction of polyphenols and antioxidant capacity from apple tree wood, observing the highest antioxidant capacity (measured using FRAP assay) after using a temperature of 55 °C and 50% ethanol volume. Similarly, Rusu et al. [21] indicated that increasing ethanol percentage in the solvent (from 50% to 95%) and extraction temperature (from 20 to 40 °C) were associated with higher TPC and TEAC values in walnut septum extract. Interestingly, the authors also obtained higher TPC and TEAC values by carrying out the extraction with 50% ethanol solution at 20 °C. In addition, Bamba et al. [22] obtained an increase in the antioxidant capacity (measured using DPPH assay) with augmented temperature, but since the content of phenolic compounds decreased, this increase could be due to the presence of other antioxidant compounds.

The ANOVA results (Table 2) show that not only these parameters by themselves can affect the extraction yield, but the combination of them also can produce significant changes. This is the case of TPC, in which the combination of temperature and ethanol can modify TPC extraction ($p = 0.0318$).

Table 2. Analysis of variance (ANOVA) results for each effect in response surface methodology (RSM) of total phenolic compounds (TPC) and trolox equivalent antioxidant capacity (TEAC) values.

	TPC		TEAC	
Source	***p*-Value**	**Sig.**	***p*-Value**	**Sig.**
A: T^{a} (°C)	0.0073	**	0.2328	n.s.
B: Ethanol (%)	0.0007	***	0.0029	***
C: pH	0.7961	n.s.	0.3635	n.s.
AA	0.0036	***	0.6071	n.s.
AB	0.0318	*	0.8815	n.s.
AC	0.7439	n.s.	0.9067	n.s.
BB	0.9735	n.s.	0.5495	n.s.
BC	0.2318	n.s.	0.3976	n.s.
CC	0.0497	*	0.3016	n.s.

Sig: significance; ns: not significant, * ($p < 0.05$); ** ($p < 0.01$); *** ($p < 0.001$).

2.2. Impact of Temperature, Solvent and pH on the Selective Extraction of Individual Phenolic Compounds from Tiger Nuts by-Products

The profile and content of specific phenolic compounds extracted with ethanol:water mixtures from tiger nut by-products using the RSM methodology for optimization and after analyzing the extracts by Triple-TOF-LC-MS-MS are shown in Table 3 and Figure 2.

Specifically, as can be seen in Figure 2A, both pH and temperature had a great influence on the extraction of lignans ($p = 0.0256$ and 0.0251, respectively). The maximum yield was observed at pH 10.62 and 43.3 °C. However, at room temperature (25 °C) the optimum pH dropped to 6.2. This outcome agrees with the data previously obtained by Tu et al. [23] who found an optimum pH in the range 5.5–6 for the extraction of lignans at room temperature for *Fructus forsythiae*.

Phenolic acid extraction was also influenced in a very large amount by temperature ($p = 0.0003$) and by the volume of ethanol ($p < 0.0001$) but not by the pH ($p = 0.53$) (Figure 2B). In this case, it should be noted how the maximum extraction yield was obtained for a temperature of 50 °C and an ethanol content of 41.4%. Elez-Garofulić et al. [24] also observed an increase in the extraction of phenolic acids by increasing the temperature using the microwave-assisted extraction on sour cherry Marasca, obtaining an optimum temperature of 70 °C. A similar trend was also reported by Waszkowiak et al. [25], who studied the influence of the percentage of ethanol used for the extraction of phenolic compounds of flaxseeds extracts, ranging from 60% and 90%, and they observed that the best ratio for phenolic acids corresponded to 60% ethanol in water.

Table 3. Individual phenolic compounds (µg/kg) (lignans, flavones, flavonoids, dihydroxybenzoic acids, hydroxycinnamic and phenolic acids) determined by Triple-TOF-LC-MS-MS from "horchata" by-products obtained after conventional extraction with hydroethanolic mixtures.

Tª (°C)	Ethanol (%)	pH	1-AP	7-HS	Cyanidin	Ethyl Vanillin	4-Vinylphenol	Sinensetin	SGG	Cinnamic Acid	DPC	PC	FG+FG	Ferulaldehyde	4-HB+BA
25	0	7.25	ND	ND	ND	10.40	85.00	ND	ND	3.30	10.40	ND	4.40	ND	ND
25	25	2.5	ND	ND	ND	10.90	209.2	ND	ND	4.00	31.50	ND	ND	1.80	23.00
25	25	12	ND	ND	ND	10.60	258.8	ND	ND	ND	34.90	ND	ND	3.10	21.40
25	50	7.25	ND	ND	ND	11.00	264.3	ND	ND	ND	38.30	ND	ND	3.70	24.50
35	0	2.5	1.70	ND	ND	14.90	126.2	ND	35.50	6.10	13.50	ND	8.80	0.50	ND
35	0	12	ND	3.70	ND	21.50	20.10	ND	25.80	ND	ND	0.20	7.40	ND	0.60
35	25	7.25	4.30	3.00	2.30	10.60	229.9	50.10	85.00	ND	ND	28.00	8.70	3.00	21.30
35	25	7.25	4.40	4.10	2.40	11.60	248.6	50.00	86.00	ND	ND	33.50	8.00	3.90	22.80
35	25	7.25	4.30	4.50	2.00	12.20	234.7	50.00	81.90	ND	ND	31.90	8.50	3.80	22.60
35	50	2.5	3.90	ND	4.10	13.20	303.3	77.80	86.20	ND	ND	41.80	8.10	6.60	23.20
35	50	12	4.00	ND	5.50	11.10	297.4	76.10	95.40	ND	ND	38.90	ND	6.10	22.10
50	0	7.25	1.40	4.20	ND	26.00	22.30	ND	23.00	11.00	ND	ND	4.90	ND	11.40
50	25	2.5	ND	ND	ND	12.20	242.5	57.30	92.50	ND	ND	32.90	ND	5.30	25.00
50	25	12	4.60	4.40	3.50	11.90	302.6	60.40	92.80	ND	ND	42.10	9.30	5.20	28.80
50	50	7.25	ND	ND	5.30	18.70	352.7	ND	ND	ND	ND	48.00	ND	3.60	34.20

Lignans: 1-Acetoxypinoresinol (AP); 7-Hydroxysecoisolariciresinol (7-HS). Dihydroxybenzoic acid: Ethylvanillin. Hydroxycinnamic acids: Cinnamic acid; Dihydro-p-coumaric acid (DPC); p-coumaric acid (PC); Ferulicacid 4-*O*-glucoside + Feruloyl glucose (FG+FG). Phenolicacids: 4-Hydroxybenzaldehyde+Benzoic acid (4-HB+BA); Ferulaldehyde. Flavones: Sinensetin. Flavonoids: Cyanidin; Spinacetin 3-*O*-glucosyl-(1→6)-glucoside (SGG). ND: not detected.

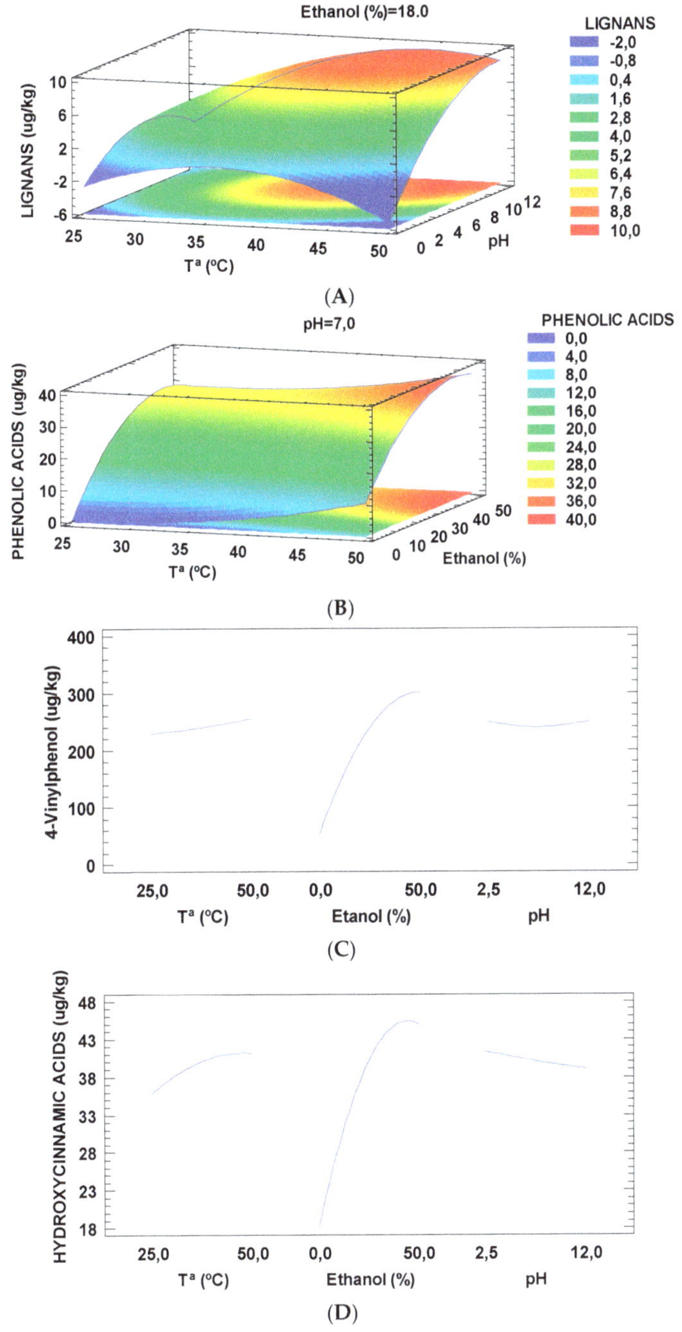

Figure 2. Response surface plot for the influence of extraction condition parameters in lignans (**A**) and phenolic acids (**B**). Main effects chart for 4-vinylphenol (**C**) and hydroxycinnamic acids (**D**) using solid-liquid extraction. 1 g of tiger nuts by-products was extracted by varying temperature (25–50 °C), ethanol:water mixture (0%–50%) and pH (2.5–12).

Furthermore, the percentage of ethanol influenced notably the extraction of both 4-vinylphenol , and the hydroxycinnamic acids ($p = 0.0039$) (Figure 2C,D). This finding can be explained by the polarity of the solvent, since increasing the volume of ethanol in water increases the polarity. As it is known, these compounds are polar, so a more polar solvent will extract them better. Other authors such as Woźniak et al. [26], Chew et al. [27] and Paini et al. [28] also used mixtures with ethanol to improve the yield of polyphenols extraction. In the case of flavonoids, no statistically significant differences were found on the extraction yield of dihydroxybenzoic acids and flavones when varying these parameters.

In Table 4 the ANOVA results are shown for the influence of temperature, ethanol and pH in the extraction of the compounds explained in Figure 2. As can be seen, for the lignans extraction the combination of temperature and pH had a significant effect ($p = 0.0344$). This could be explained by the increase of solubility, which improved the extraction of these compounds.

Table 4. ANOVA results for each effect in RSM of TPC and TEAC values.

Source	Lignans		Phenolic Acids		4-Vinylphenol		Hydroxycinnamic Acids	
	p-Value	Sig.	*p*-Value	Sig.	*p*-Value	Sig.	*p*-Value	Sig.
A: T[a] (°C)	0.0251	*	0.0003	***	0.3599	n.s.	0.3558	n.s.
B: Ethanol (%)	0.5893	n.s.	0.0000	***	0.0002	***	0.0039	***
C: pH	0.0256	*	0.5304	n.s.	0.9698	n.s.	0.6837	n.s.
AA	0.0050	**	0.0165	*	0.9056	n.s.	0.7121	n.s.
AB	0.1028	n.s.	0.5178	n.s.	0.0843	n.s.	0.4661	n.s.
AC	0.0344	*	0.1846	n.s.	0.6726	n.s.	0.1701	n.s.
BB	0.0153	*	0.0000	***	0.0219	*	0.0804	n.s.
BC	0.3209	n.s.	0.5820	n.s.	0.2235	n.s.	0.5392	n.s.
CC	0.0473	*	0.2462	n.s.	0.5964	n.s.	0,9739	n.s.

Sig: significance; ns: not significant, * ($p < 0.05$); ** ($p < 0.01$); *** ($p < 0.001$).

Possible beneficial effects of polyphenols on human health are the subject of increasing scientific interest. For example, phenolic acids and lignans have been shown to have a positive hepatoprotective action [29]. The anti-inflammatory action of ferulaldehyde was also found in mice by other authors [30]. For all these reasons, it is necessary to consider the most appropriate extraction conditions depending on the compounds desired to obtain.

2.3. *Optimization and Validation of the Extraction Conditions*

The combination of critical parameters (temperature, ethanol and pH), which allowed to obtain the highest TPC yield and TEAC value. To do this, an optimization based on desirability was used. Theoretically, in the case of TPC, optimum values of 229.29 mg GAE/100 g of dry matter were obtained with conditions of 43.7 °C, 50% ethanol and pH = 2.5. For the antioxidant capacity, the optimum value obtained was 1846.34 μM TE/g of dry matter at a temperature of 50 °C, a volume of ethanol of 50% and a pH = 2.5. As can be seen in Table 1, the maximum extraction of TPC and the maximum TEAC values were obtained experimentally with the conditions of 35 °C, 50% ethanol and pH = 2.5 (222.58 ± 2.16 mg GAE/100 g of dry matter and 1948.07 ± 434.18 μM TE/g of dry matter respectively). These results are close to those expected theoretically, so we can affirm that the method has been validated for TPC and TEAC.

2.4. *Comparison of Hydroethanolic Extraction of Individual Phenolics Compounds from Horchata by-Products with Those of Folchand Supercritical-CO_2 Extraction*

The data on individual phenolics obtained in the present study were compared to those obtained after conventional extraction method (Folch) and an innovative extraction method (supercritical CO_2 extraction). The results for the comparison were obtained from a previously published article about the extraction of phenolic compounds in the oil fraction of "horchata" by-products [31].

It can be noticed that there was a great difference in the compounds obtained with the three different extraction techniques. As for conventional extraction with Folch method, the major compound obtained by far was 4-vinylphenol (216.9 μg/kg DW), followed by p-coumaric acid (25.35 μg/kg DW) and benzoic acid (13.54 μg/kg DW). In contrast, the main compounds extracted by SC-CO_2, especially at 40 MPa, were isohydroxymatairesinol (399.44 μg/kg DW), scopoletin (93.24 μg/kg DW) and caffeic acid (30.66 μg/kg DW). In the present study (see Table 2), the main individual phenolic extracted with hydroethanolic mixtures was 4-vinylphenol (303.3 μg/kg DW) followed by spinacetin3-O-glucosyl-(1→6)-glucoside (86.2 μg/kg DW) and sinensetin (77.8 μg/kg DW).

It is worth noting that 4-vinylphenol was the predominant phenolic compound obtained, after using both conventional methods, despite the different polarity of solvents employed, though higher extraction is accomplished with hydroethanolic mixtures. Moreover, a great difference was also observed between the compounds obtained after conventional extraction (Folch and hydroethanolic mixtures) and those obtained by SC-CO_2. As already mentioned, a more polar solvent facilitates the extraction of more phenolic compounds, therefore the low yield obtained in the SC-CO_2 extraction is not surprising. We should also keep in mind that the way to prepare the sample is different. In the case of conventional extraction, homogenization is carried out at 10,000× rpm, while for SC-CO_2 extraction the sample is only milled. This protocol differences greatly modify the accessibility of the solvents to the intracellular compounds, since in the case of conventional extraction the samples were homogenized with the solvent, breaking the cellular structures, whereas in the SC-CO_2 extraction only the particle size of the samples was reduced.

These findings agree with the data previously reported by Parker et al. [15] who demonstrated that tiger nuts skins are richer in *p*-coumaric acid than tiger nuts tubers, being this compound the fourth most important (48 μg/kg DW) in "horchata" by-products after hydroethanolic extraction. In addition, Ezeh et al. [7] studied the polyphenols present in tiger nuts, finding mainly *trans*-cinnamic acid. Oladele et al. [13] also determined the phenolic profile of tiger nuts, considering the yellow and brown varieties. They obtained differences between both species, since in yellow tiger nuts the main compounds were ferulic acid (~58 mg/100 g), *p*-hydroxybenzoic acid (~29 mg/100 g), *p*-hydroxybenzaldehyde (~16 mg/100 g) and vanillic acid (~6 mg/100 g), whereas in brown tiger nut, vanillic (~15 mg/100 g), *p*-coumaric (~17 mg/100 g), caffeic (~15 mg/100 g), ferulic (~34 mg/100 g) and sinapinic acids (~21 mg/100 g) were predominant.

3. Material and Methods

3.1. Chemicals and Reagents

Sodium hydroxide (NaOH), sodium carbonate (Na_2CO_3), and acetone were obtained from J.T.Baker (Deventer, Holland). ABTS radical (2,2′-azino-bis(3-ethylbenzothiazoline-6-sulphonic acid), Trolox (6-hydroxy-2,5,7,8-tetramethylchroman-2-carboxylic acid), Folin-Ciocalteau phenol reagent 1N, gallic acid, sodium nitrite ($NaNO_2$), formic acid (HPLC grade), potassium persulfate ($K_2S_2O_8$) and ethanol p.a. (99.5%) were purchased from Sigma-Aldrich (Steinheim, Germany). Sodium hydrogen carbonate (reagent grade; 99.7%) and methanol (reagent grade; 99.9%) were purchased from Scharlau (Barcelona, Spain). Deionized water was obtained from Millipore (Bedfore, MA, USA).

3.2. Samples

A conventional process was used for obtaining "horchata" from tiger nuts with a denomination of origin "Chufa de Valencia" (*Cyperus esculentus*). Then "horchata" by-products were taken and provided by the "Consejo Regulador D.O. Chufa de Valencia" (Valencia, Spain). Afterwards, they were dried at 60 °C for 72 h using a Memmert UFP 600 air-circulating oven (Schwabach, Germany) and ground for obtaining uniform particle size. Finally, they were vacuum packed until needed.

3.3. Extraction at Different Temperatures, Ethanol:Water Mixtures and pH

The conditions for solid–liquid extraction were selected based on a previous study [32]. First, one gram of dehydrated "horchata" by-product was weighed and fifteen milliliter of the hydroethanolic mixtures at different ethanol concentrations (0%, 25% and 50%, *v*/*v*) and different pH (2.5, 7.25 and 12), adjusted with NaOH or HCl was used for the extraction. The beakers with the samples were then placed and stirred on a plate with a magnetic stirrer. To avoid the evaporation of the solvent during the extraction, the samples were covered with aluminum foil. The temperature was adjusted to 25, 35 and 50 °C in each of the stirring plate's rows. The extraction was carried out for 3 h. The obtained samples were filtered through Whatman No. 1 and used for the determination of phenolic compounds.

3.4. Total Antioxidant Capacity

TEAC (Trolox equivalent antioxidant capacity) assay was used for the determination of the total antioxidant capacity [8]. Twenty-five milliliter of ABTS (7 mM) was mixed with 440 μL of $K_2S_2O_8$ (140 mM) and kept at room temperature for 12–16 h under darkness. For the determination, the absorbance of $ABTS^{\bullet+}$ working solution was measured at a wavelength of 734 nm on a Perkin-Elmer UV/Vis Lambda 2 spectrophotometer (Perkin-Elmer, Jügesheim, Germany) to obtain the initial absorbance (A_0). When the absorbance of the mixture was 0.700 ± 0.020, 100 μL of the extracts appropriately diluted were added, and the absorbance was measured at 20 min (A_f). The following equation was used to calculate the inhibition percentage (%) of the samples:

$$\% \text{ Inhibition} = (1 - A_f/A_0) \times 100,$$

where A_0 is the absorbance at the initial time and A_f is the absorbance obtained after 20 min.

The results were expressed as micromolar Trolox equivalent (μM TE)/g of dry matter.

3.5. Determination of Total Phenolic Content (TPC)

The method previously reported [33], with some modifications [34] was used. To sum up, 500 μL of extract was mixed with 4.5 mL of distilled water and then 1 mL of the 2% Na_2CO_3 solution (*w*/*v*), and 0.25 mL of the Folin-Ciocalteau reagent (1N) were added. The mixture was left to stand for one hour under darkness at room temperature. Afterwards, the absorbance was measured at 765 nm. TPC was determined by interpolating the absorbance values in a calibration curve using gallic acid standard (10 μg/mL) at different concentrations between 0 and 5 μg/mL. The results were expressed as mg equivalents of gallic acid (GAE)/100 g of dry matter.

3.6. Triple TOF–LC–MS–MS Characterization of Phenolic Compounds

The phenolic profile characterization and quantification was performed according to the previously described method [31], using an Agilent 1260 Infinity (Agilent, Waldbronn, Germany) with a Waters UPLC C18 column 1.7 μm (2.1 × 50 mm) Acquity UPLC BEH.C18 (Waters, Cerdanyola del Vallès, Spain) for the separation of the main phenolic compounds in the samples. Moreover, a TripleTOF™ 5600 LC/MS/MS system (AB SCIEX, Foster City, CA, USA) was utilized for the identification. For that purpose, a mobile phase consisting of solvent A (water, 0.1% formic acid) and solvent B (methanol, 0.1% formic acid) was used as follows: 0 min 90% A; 13 min 100% (B); 15 min 90% A. Five microliter and 0.4 mL/min were the injection volume and flow rate, respectively.

MS data were obtained between 80 and 1200 *m*/*z* on negative mode, and the IDA acquisition method was carried out in the survey scan type (TOF-MS) using the dependent scan type (product ion). Ion spray voltage (−4500 V); declustering potential (90 V); collision energy (−50 V); temperature with 25 psi curtain gas (400 °C); 50 psi for both ion source gas 1 (GC1) and ion source gas 2 (GS2) were used as the main parameters for the MS analysis.

The IDA MS/MS analysis was carried out with ion tolerance of 50 mDa, 25 V collision energy and activated dynamic background subtract. The software analyst PeakView1.1 (AB SCIEX, Foster

City, CA, USA) and its applications (XIC Manager and Formula Finder) were used for data acquisition and processing. Finally, an external calibration curve using resveratrol as standard was prepared for the quantification of phenolic compounds.

3.7. Experimental Design and Statistical Analyses

Box-Behnken design with three levels (maximum, minimum, and central) of each independent variable, temperature (25–50 °C), the concentration of ethanol (0%–50%), and pH (2.5–12), leading to 15 combinations of these variables. Independent variable levels were selected accounting for the sample and the potential degradation of thermolabile antioxidant compounds after high temperatures (>50 °C). The combinations included temperature–ethanol–pH conditions with an intermediate level (central point) of the three variables replicated three times, which was used to check the reproducibility and stability of the results. Experiments were randomized to minimize the systematic bias in the observed responses due to extraneous factors and for higher precision. In addition, we studied whether there were correlations between a pair of variables. The significant differences ($p < 0.05$) between the results were calculated by analysis of variance (ANOVA), using the least significant difference (LSD) test to indicate the samples between which there were differences. All statistical analyses were performed using the software Statgraphics® Centurion XV (Statpoint Technologies, Inc., The Plains, VA, USA).

4. Conclusions

The optimization of phenolic compounds extraction by response surface methodology, followed by Triple-TOF-LC-MS-MS characterization, indicated that temperature and ethanol content are more important variables than pH. The optimum extraction conditions for total phenolic content and antioxidant activity were 50% ethanol, 35 °C and pH 2.5, which could obtain values of 222.6 mg GAE/100 g of dry matter and 1948.1 μM TE/g of dry matter for TPC and TEAC, respectively. The optimized extraction condition also positively influenced the extraction of the main individual phenolic compounds of tiger nuts by-products, particularly 4-vinylphenol and hydroxycinnamic acids. Therefore, tiger nuts by-products can be explored as a valuable source of phenolic compounds with potential applications in food industries, nutraceuticals and cosmetics.

Author Contributions: E.R.-S., F.J.B. and J.M.L. conceived, designed and carried out the experiments. E.R.-S., F.J.B., J.M.L., and P.E.S.M. analysed the data and wrote the paper. E.R.-S., F.J.M.-Q., A.C., F.J.B., J.M.L., P.E.S.M. and F.R. wrote and reviewed the paper before submitting.

Funding: This work was supported by the project GV/2018/040 "Implementación y optimización de procesos innovadores para la valorización de los subproductos obtenidos a partir del proceso de elaboración de la horchata" for emerging research groups from the Generalitat Valenciana.

Acknowledgments: Paulo E. Munekata acknowledges the postdoctoral fellowship support from the Ministry of Economy and Competitiveness (MINECO, Spain) "Juan de la Cierva" program (FJCI-2016-29486). The authors would like to thank the Regulatory Council D.O. Tiger nut of Valencia (Valencia, Spain) for providing "horchata" by-products. Moreover, we are also grateful for the technical support from the "Central Service for Experimental Research (SCSIE)" at the Universitat de València for the help provided in the analysis using the Triple TOF-LC-MS-MS.

Conflicts of Interest: The authors have no conflict of interest to disclose.

References

1. Roselló-Soto, E.; Poojary, M.M.; Barba, F.J.; Koubaa, M.; Lorenzo, J.M.; Mañes, J.; Moltó, J.C. Thermal and non-thermal preservation techniques of tiger nuts' beverage "horchata de chufa". Implications for food safety, nutritional and quality properties. *Food Res. Int.* **2018**, *105*, 945–951. [CrossRef]
2. Roselló-Soto, E.; Poojary, M.M.; Barba, F.J.; Lorenzo, J.M.; Mañes, J.; Moltó, J.C. Tiger nut and its by-products valorization: From extraction of oil and valuable compounds to development of new healthy products. *Innov. Food Sci. Emerg. Technol.* **2018**, *45*, 306–312. [CrossRef]

3. Sánchez-Zapata, E.; Fernández-López, J.; Angel Pérez-Alvarez, J. Tiger nut (*Cyperus esculentus*) commercialization: Health aspects, composition, properties, and food applications. *Compr. Rev. Food Sci. Food Saf.* **2012**, *11*, 366–377. [CrossRef]
4. Sánchez-Zapata, E.; Díaz-Vela, J.; Pérez-Chabela, M.L.; Pérez-Alvarez, J.A.; Fernández-López, J. Evaluation of the effect of tiger nut fibre as a carrier of unsaturated fatty acids rich oil on the quality of dry-cured sausages. *Food Bioprocess Technol.* **2013**, *6*, 1181–1190. [CrossRef]
5. Sánchez-Zapata, E.; Fuentes-Zaragoza, E.; Viuda-Martos, M.; Fernández-López, J.; Sendra, E.; Sayas, E.; Pérez-Alvarez, J.A. Reclaim of the by-products from "Horchata" elaboration process. *Food Bioprocess Technol.* **2012**, *5*, 954–963. [CrossRef]
6. Sánchez-Zapata, E.; Zunino, V.; Pérez-Alvarez, J.A.; Fernández-López, J. Effect of tiger nut fibre addition on the quality and safety of a dry-cured pork sausage "Chorizo" during the dry-curing process. *Meat Sci.* **2013**, *95*, 562–568. [CrossRef] [PubMed]
7. Ezeh, O.; Niranjan, K.; Gordon, M.H. Effect of enzyme pre-treatments on bioactive compounds in extracted tiger nut oil and sugars in residual meals. *JAOCS J. Am. Oil Chem. Soc.* **2016**, *93*, 1541–1549. [CrossRef] [PubMed]
8. Roselló-Soto, E.; Barba, F.J.; Putnik, P.; Bursać Kovačević, D.; Lorenzo, J.M.; Cantavella-Ferrero, Y. Enhancing bioactive antioxidants' extraction from "Horchata de Chufa" by-products. *Foods* **2018**, *7*, 161. [CrossRef] [PubMed]
9. Wang, C.; Shi, L.; Fan, L.; Ding, Y.; Zhao, S.; Liu, Y.; Ma, C. Optimization of extraction and enrichment of phenolics from pomegranate (*Punica granatum* L.) leaves. *Ind. Crop. Prod.* **2013**, *42*, 587–594. [CrossRef]
10. Friedman, M.; Jürgens, H.S. Effect of pH on the stability of plant phenolic compounds. *J. Agric. Food Chem.* **2000**, *48*, 2101–2110. [CrossRef] [PubMed]
11. Granato, D.; Shahidi, F.; Wrolstad, R.; Kilmartin, P.; Melton, L.D.; Hidalgo, F.J.; Miyashita, K.; Camp, J.V.; Alasalvar, C.; Ismail, A.B.; et al. Antioxidant activity, total phenolics and flavonoids contents: Should we ban in vitro screening methods? *Food Chem.* **2018**, *264*, 471–475. [CrossRef] [PubMed]
12. Ogunlade, I.; Adeyemi Bilikis, A.; Aluko Olanrewaju, G. Chemical compositions, antioxidant capacity of Tigernut (*Cyperus esculentus*) and potential health benefits. *Eur. Sci. J.* **2015**, *11*.
13. Oladele, A.; Adebowale, J.; Bamidele, O. Phenolic Profile and Antioxidant Activity of Brown and Yellow Varieties of Tigernut (*Cyperus esculentus* L.). *Niger. Food J.* **2017**, *35*, 51–59.
14. Koubaa, M.; Barba, F.J.; Mhemdi, H.; Grimi, N.; Koubaa, W.; Vorobiev, E. Gas assisted mechanical expression (GAME) as a promising technology for oil and phenolic compound recovery from tiger nuts. *Innov. Food Sci. Emerg. Technol.* **2015**, *32*, 172–180. [CrossRef]
15. Parker, M.L.; Ng, A.; Smith, A.C.; Waldron, K.W. Esterified phenolics of the cell walls of chufa (*Cyperus esculentus* L.) tubers and their role in texture. *J. Agric. Food Chem.* **2000**, *48*, 6284–6291. [CrossRef] [PubMed]
16. Badejo, A.A.; Damilare, A.; Ojuade, T.D. Processing effects on the antioxidant activities of beverage blends developed from *Cyperus esculentus*, *Hibiscus sabdariffa*, and *Moringa oleifera* extracts. *Prev. Nutr. Food Sci.* **2014**, *19*, 227–233. [CrossRef] [PubMed]
17. Yang, L.; Jiang, J.-G.; Li, W.-F.; Chen, J.; Wang, D.-Y.; Zhu, L. Optimum extraction process of polyphenols from the bark of *Phyllanthus emblica* L. based on the response surface methodology. *J. Sep. Sci.* **2009**, *32*, 1437–1444. [CrossRef] [PubMed]
18. Moreira, M.M.; Barroso, M.F.; Boeykens, A.; Withouck, H.; Morais, S.; Delerue-Matos, C. Valorization of apple tree wood residues by polyphenols extraction: Comparison between conventional and microwave-assisted extraction. *Ind. Crop. Prod.* **2017**, *104*, 210–220. [CrossRef]
19. Vatai, T.; Škerget, M.; Knez, Ž. Extraction of phenolic compounds from elder berry and different grape marc varieties using organic solvents and/or supercritical carbon dioxide. *J. Food Eng.* **2009**, *90*, 246–254. [CrossRef]
20. Li, Y.; Li, S.; Lin, S.J.; Zhang, J.J.; Zhao, C.N.; Li, H. Bin Microwave-assisted extraction of natural antioxidants from the exotic *Gordonia axillaris* fruit: Optimization and identification of phenolic compounds. *Molecules* **2017**, *22*, 1481. [CrossRef] [PubMed]
21. Rusu, M.E.; Gheldiu, A.-M.; Mocan, A.; Moldovan, C.; Popa, D.-S.; Tomuta, I.; Vlase, L. Process optimization for improved phenolic compounds recovery from walnut (*Juglans regia* L.) septum: Phytochemical profile and biological activities. *Molecules* **2018**, *23*, 2814. [CrossRef] [PubMed]

22. Bamba, B.S.B.; Shi, J.; Tranchant, C.C.; Xue, S.J.; Forney, C.F.; Lim, L.T. Influence of extraction conditions on ultrasound-assisted recovery of bioactive phenolics from blueberry pomace and their antioxidant activity. *Molecules* **2018**, *23*, 1685. [CrossRef] [PubMed]
23. Tu, Q.-Y.; Zhou, C.-S.; Tang, J.-P. Microwave assisted-semi bionic extraction of lignan compounds from *Fructus Forsythiae* by orthogonal design. *J. Cent. South Univ. Technol.* **2008**, *15*, 59–63. [CrossRef]
24. Elez Garofulić, I.; Dragović-Uzelac, V.; Režek Jambrak, A.; Jukić, M. The effect of microwave assisted extraction on the isolation of anthocyanins and phenolic acids from sour cherry Marasca (*Prunus cerasus* var. Marasca). *J. Food Eng.* **2013**, *117*, 437–442. [CrossRef]
25. Waszkowiak, K.; Gliszczyńska-Świgło, A. Binary ethanol–water solvents affect phenolic profile and antioxidant capacity of flaxseed extracts. *Eur. Food Res. Technol.* **2016**, *242*, 777–786. [CrossRef]
26. Woźniak, Ł.; Marszałek, K.; Skąpska, S.; Jędrzejczak, R.; Woźniak, Ł.; Marszałek, K.; Skąpska, S.; Jędrzejczak, R. The application of supercritical carbon dioxide and ethanol for the extraction of phenolic compounds from chokeberry pomace. *Appl. Sci.* **2017**, *7*, 322. [CrossRef]
27. Chew, K.K.; Khoo, M.Z.; Ng, S.Y.; Thoo, Y.Y.; Wan Aida, W.M.; Ho, C.W. Effect of ethanol concentration, extraction time and extraction temperature on the recovery of phenolic compounds and antioxidant capacity of *Orthosiphon stamineus* extracts. *Int. Food Res. J.* **2011**, *18*, 1427–1435.
28. Paini, M.; Casazza, A.A.; Aliakbarian, B.; Perego, P.; Binello, A.; Cravotto, G. Influence of ethanol/water ratio in ultrasound and high-pressure/high-temperature phenolic compound extraction from agri-food waste. *Int. J. Food Sci. Technol.* **2016**, *51*, 349–358. [CrossRef]
29. Pereira, C.; Barros, L.; Ferreira, I.C. Extraction, identification, fractionation and isolation of phenolic compounds in plants with hepatoprotective effects. *J. Sci. Food Agric.* **2016**, *96*, 1068–1084. [CrossRef]
30. Radnai, B.; Tucsek, Z.; Bognar, Z.; Antus, C.; Mark, L.; Berente, Z.; Gallyas, F.; Sumegi, B.; Veres, B. Ferulaldehyde, a water-soluble degradation product of polyphenols, inhibits the lipopolysaccharide-induced inflammatory response in mice. *J. Nutr.* **2009**, *139*, 291–297. [CrossRef]
31. Roselló-Soto, E.; Barba, F.J.; Lorenzo, J.M.; Munekata, P.E.S.; Gómez, B.; Moltó, J.C. Phenolic profile of oils obtained from "horchata" by-products assisted by supercritical-CO_2 and its relationship with antioxidant and lipid oxidation parameters: Triple TOF-LC-MS-MS characterization. *Food Chem.* **2019**, *274*, 865–871. [CrossRef] [PubMed]
32. Franco, D.; Rodríguez-Amado, I.; Agregán, R.; Munekata, P.E.S.; Vázquez, J.A.; Barba, F.J.; Lorenzo, J.M. Optimization of antioxidants extraction from peanut skin to prevent oxidative processes during soybean oil storage. *LWT* **2018**, *88*, 1–8. [CrossRef]
33. Roselló-Soto, E.; Barba, F.J.; Parniakov, O.; Galanakis, C.M.; Lebovka, N.; Grimi, N.; Vorobiev, E. High voltage electrical discharges, pulsed electric field, and ultrasound assisted extraction of protein and phenolic compounds from olive kernel. *Food Bioprocess Technol.* **2015**, *8*, 885–894. [CrossRef]
34. Singleton, V.L.; Orthofer, R.; Lamuela-Raventós, R.M. Analysis of total phenols and other oxidation substrates and antioxidants by means of Folin-Ciocalteu reagent. In *Methods in Enzymology*; Packer, L., Ed.; Oxidants and Antioxidants Part A Academic Press: Cambridge, MA, USA, 1999; Volume 299, pp. 152–178.

Sample Availability: Samples of the compounds are available from the authors.

molecules

MDPI

Article

Ultrasound as a Rapid and Low-Cost Extraction Procedure to Obtain Anthocyanin-Based Colorants from *Prunus spinosa* L. Fruit Epicarp: Comparative Study with Conventional Heat-Based Extraction

Maria G. Leichtweis [1], Carla Pereira [1], M.A. Prieto [1,2], Maria Filomena Barreiro [1,3], Ilton José Baraldi [4], Lillian Barros [1,*] and Isabel C.F.R. Ferreira [1,*]

1. Centro de Investigação de Montanha (CIMO), Instituto Politécnico de Bragança, Campus de Santa Apolónia, 5300-253 Bragança, Portugal; mg.leichtweis@hotmail.com (M.G.L.); carlap@ipb.pt (C.P.); michaelumangelum@gmail.com (M.A.P.); barreiro@ipb.pt (M.F.B.)
2. Nutrition and Food Science Group, Department of Analytical and Food Chemistry, CITACA, CACTI, University of Vigo-Vigo Campus, 36310 Vigo, Spain
3. Laboratory of Separation and Reaction Engineering—Laboratory of Catalysis and Materials (LSRE-LCM), Polytechnic Institute of Bragança, Campus Santa Apolónia, 5300-253 Bragança, Portugal
4. Departamento Acadêmico de Alimentos (DAALM), Universidade Tecnológica Federal do Paraná, Campus Medianeira, 85884-000 Paraná, Brasil; ijbaraldi@gmail.com

* Correspondence: lillian@ipb.pt (L.B.); iferreira@ipb.pt (I.C.F.R.F); Tel.: +351-273-303285 (L.B.); +351-273-303219 (I.C.F.R.F); Fax: +351-273-325405 (L.B.); +351-273-325405 (I.C.F.R.F)

Academic Editor: Marcello Locatelli
Received: 13 December 2018; Accepted: 1 February 2019; Published: 5 February 2019

Abstract: An ultrasound rapid and low-cost procedure for anthocyanin-based colorants from *Prunus spinosa* L. fruit epicarp was developed, and the advantages were compared with conventional heat-based extraction. To obtain the conditions that maximize anthocyanins' extraction, a response surface methodology was applied using the variables of time, temperature, and ethanol content, in the case of heat extraction, whereas for ultrasound assisted extraction, temperature was replaced by ultrasound power. Two anthocyanin compounds were identified by HPLC-DAD-ESI/MS—namely, cyanidin 3-rutinoside and peonidin 3-rutinoside. The responses used were the extraction yield and the content of the identified anthocyanins. Ultrasound extraction was the most effective method at 5.00 ± 0.15 min, 400.00 ± 32.00 W, and 47.98% ± 2.88% of ethanol obtaining 68.60% ± 2.06% of extracted residue, with an anthocyanin content of 18.17 mg/g (extract-basis) and 11.76 mg/g (epicarp-basis). Overall, a viable green process was achieved that could be used to support pilot-scale studies for industrial production of anthocyanin-based colorants from *P. spinosa* fruit epicarp.

Keywords: *Prunus spinosa* L. fruit epicarp; wild fruit valorization; cyanidin 3-rutinoside; peonidin 3-rutinoside; heat and ultrasound assisted extraction; response surface methodology

1. Introduction

Prunus spinosa L. (blackthorn) is a spontaneous wild shrub found in Portugal, Spain, and other European countries. Its fruits are commonly used for liqueur and jam preparations, as well as for medicinal purposes [1]. Nevertheless, no reports were found regarding the industrial, or large scale, use of these fruits, probably because of their bitter and astringent taste.

The valorization of agricultural products has gained much attention in the late years as a mean for a sustainable management, which can concomitantly increase the profit of local economies. In this regard, *P. spinosa* constitutes an underexploited source and can serve as a raw material for the recovery/production of compounds for food applications [2]. As with other *Prunus* species, anthocyanin compounds can be found in blackthorn fruits at high levels, being responsible for their typical coloration [3,4]. In fact, a complex profile of anthocyanins was previously identified in *P. spinosa* fruits, among which cyanidin 3-rutinoside and peonidin 3-rutinoside were found to be predominant [5,6].

Anthocyanins are natural pigments belonging to the phenolic compounds group and, within that, to the flavonoids class, presenting a range of colors between red, blue, and violet that are characteristic of various fruits and vegetables [7]. Beyond their various physiological benefits, which include effects against cardiovascular diseases, atherosclerosis, and cancer, recently, an increasing interest in these compounds began to arise because of their colorant properties [8,9].

The industrial production of natural-based colorants has been established for years and consists mainly of obtaining colorant-rich extracts through conventional heat assisted extraction (HAE, or maceration) using water as a solvent followed by several isolation/drying steps. This type of conventional process, although used at large-scale, is known for requiring high-energy consumption and long extraction times [10–12]. Alternative extraction processes, able to replace traditional ones, have been established to shorten the needed time, decrease energy requirements, and reduce solvent consumption. Among the non-conventional procedures applied to anthocyanins' extraction, ultrasound, microwave, and supercritical fluid assisted extraction techniques have attracted, in the recent years, the attention of industrials and researchers [10,13]. Regarding ultrasound assisted extraction (UAE), it is considered an inexpensive, simple, and efficient alternative to conventional techniques [14]. The extraction capability of UAE is attributed to mechanical and cavitation phenomena, which lead to cells' disruption, particle size reduction, and enhanced mass transfer across the cell membrane [11,13].

To obtain anthocyanin-rich extracts, it is crucial to consider the factors affecting the stability of these compounds, including structure and concentration, pH, temperature, light exposure, oxygen levels, and used extraction solvents [15]. Thus, the choice of the extraction method, along with the optimization of relevant extraction variables, are essential to guarantee a maximum recovery efficiency [16]. Additionally, the efficiency is also strongly affected by the variability observed among different matrices [17]. Through response surface methodology (RSM), it is possible to optimize the relevant variables simultaneously, obtaining polynomial models capable of describing, within the tested experimental interval, the ideal conditions that maximize the used response criteria [13].

In the present study, the goal was to explore blackthorn anthocyanin composition and promote a higher commercial value of these wild fruits through the development of an anthocyanin-based coloring extract. For that purpose, the fruit epicarp was used because it has a much more intense color than the pulp, and thus a higher concentration of anthocyanins and less interfering compounds in the extraction process (e.g., sugars). To the best of our knowledge, and according to a thorough literature survey, no reference or report on the optimization of anthocyanin compounds extraction from fruit epicarps of *P. spinose* was found. Therefore, the present study aimed to optimize the extraction of these compounds from *P. spinosa* fruit epicarps through HAE and UAE techniques, evaluating the following variables: i) type of solvent (water and green organic solvents); ii) extraction time; iii) solid-to-liquid ratio; and iv) temperature (for HAE) or pressure (for UAE). The most efficient parameters were obtained by response surface methodology (RSM). The identification and quantification of the anthocyanin compounds present in the extracts was assessed by HPLC-DAD-ESI/MS.

2. Results

2.1. Development of RSM Models to Optimize Responses and Conditions

The RSM is a valuable instrument to assess the impact of the main extraction factors and their interactions on one or more responses. The technique uses fixed experimental designs with the major goals of minimizing the experimental labor and finding optimal solutions. In this regard, the work presented here uses the *circumscribed central composite design (CCCD)* design plan, which is a popular design among researchers when trying to optimize food processing methods [18], such as the extraction of anthocyanin compounds.

In a previous study [5], authors identified, using HPLC-DAD-ESI/MS, the anthocyanin compounds of cyanidin 3-rutinoside ($[M + H]^+$ at m/z 595) and peonidin 3-rutinoside ($[M + H]^+$ at m/z 609) in *P. spinosa* fruits, and highlighted that the colorant capacity of these fruits is mainly attributed to these compounds. Although authors quantify the content of those anthocyanin compounds in *P. spinosa* fruits, the conditions of extraction were not optimized. Therefore, based on those preliminary findings, it seems logical to continue to explore the potential of *P. spinosa* fruits as a source of anthocyanin compounds. In this regard, the study applies a RSM technique under a *CCCD* to optimize the operating conditions of the extraction of two common techniques in the industrial environment (HAE and UAE) with the intention of maximizing their extraction. However, because the major quantity of anthocyanin compounds in *P. spinosa* fruits is located in the fruit epicarp, in this study, we ignored the inside parts of the fruit and focused the attention on the fruit epicarps. Additionally, by focusing on the epicarps, we are avoiding a high content of interfering compounds in the extraction process (e.g., sugars) that would require further purification steps. Figure S1 (supplemental material) shows a complete summary of all the steps used for the optimization procedure in order to recover the anthocyanin compounds from the epicarps of *P. spinosa* fruits. Figure S2 (supplementary material) shows a chromatographic example of HPLC-DAD-ESI/MS results for the quantification of anthocyanin compounds in the epicarps of *P. spinosa* fruits.

Table 1 shows the experimental results derived from the *CCCD* used to optimize the extraction of anthocyanins from the fruits epicarps (Y_1, mg C/g R; Y_2, mg C/g E dw; and *Yield*, %) for each one of the computed extraction techniques (HAE and UAE). As described, the *CCCD* experimental results are subjected to the mathematical analysis of Equation (1), by applying a fitting procedure coupled with non-linear least-squares estimations. The parametric values of Equation (1) derived from this analytical procedure, the corresponding confidence interval of the parameters ($\alpha = 0.05$) found after modelling the extraction response values, and basic statistical information of the mathematical procedure are presented in Table 2. The parametric values considered non-significant (*ns*) values were excluded from model construction and the final equations for describing the responses assessed using significant terms are presented in Table S1 (supplementary material).

The significant parametric values in Table 2 are presented as a function of the codification criteria of the *CCCD*. Although they could be presented as the real numerical ranges of the variables assessed (X_1 to X_3), such information would not provide any additional insights of the regression analysis performed or the possible effects that may occur. The key information is the weight of the numerical values of the significant parameters; therefore, it seems logical to present them under a codification mode that allows us to compare the values between them effortlessly. Therefore, based on the numerical values derived, some global conclusions can be deduced as follows:

- For the HAE technique: In global terms, the significant parametric values within the linear effect (LE) group have a far more relevant contribution to the description of the responses than the interactive effect (IE), with the quadratic effect (QE) group being the less representative one (LE > IE >> QE). In the extraction *Yield* response, the three variables assessed (t, T, and S) showed similar contributions to its description. Regarding the response values of Y_1 (mg CT/g R) and Y_2 (mg CT/g E dw), the contribution of the variables is $S >> T > t$.
- For the UAE technique: The contribution to the description of effects of the responses by the significant parametric values is distributed as LE > QE >> IE. In all the responses assessed (extraction *yield*, Y_1 and Y_2 values), the contribution of the variables is $S >> P > t$.

Table 1. Experimental results of the *CCCD* used for the response surface methodology (RSM) optimization of the three main variables involved (X_1, X_2, and X_3) in the heat assisted extraction (HAE) and ultrasound assisted extraction (UAE). Responses comprised three format values assessed (Y_1, mg C/g R; Y_2, mg C/g E dw; and *Yield*, %).

	Experimental Design									HAE							UAE						
	Coded Values			HAE			UAE			Residue		Individual Content				Total Content		Residue		Individual Content				Total Content	
	X_1	X_2	X_3	X_1: *t*	X_2: *T*	X_3: *S*	X_1: *t*	X_2: *P*	X_3: *S*	*Yield*	Y_1C1	Y_1C2	Y_2C1	Y_2C2	Y_1CT	Y_2CT	*Yield*	Y_1C1	Y_1C2	Y_2C1	Y_2C2	Y_1CT	Y_2CT
				min	°C	%	min	W	%	%	mg/g R	mg/g R	mg/g E	mg/g E	mg/g R	mg/g E	%	mg/g R	mg/g R	mg/g E	mg/g E	mg/g R	mg/g E
1	−1	−1	−1	21.2	34.2	20.3	9.1	160.8	20.3	53.75	6.65	3.04	3.58	1.64	9.73	5.21	61.24	9.26	3.65	5.67	2.24	12.91	7.91
2	−1	−1	1	21.2	34.2	79.7	9.1	160.8	79.7	45.43	7.54	4.65	3.42	2.11	12.22	5.54	52.87	9.45	4.44	5.00	2.35	13.89	7.34
3	−1	1	−1	21.2	75.8	20.3	9.1	339.2	20.3	57.26	6.08	2.81	3.48	1.61	8.78	5.09	70.17	8.96	3.44	6.29	2.41	12.40	8.70
4	−1	1	1	21.2	75.8	79.7	9.1	339.2	79.7	48.85	8.35	4.44	4.08	2.17	12.74	6.25	58.80	9.36	4.51	5.50	2.65	13.87	8.15
5	1	−1	−1	68.8	34.2	20.3	20.9	160.8	20.3	55.45	6.88	2.90	3.82	1.61	9.66	5.42	59.12	10.66	3.69	6.30	2.18	14.35	8.48
6	1	−1	1	68.8	34.2	79.7	20.9	160.8	79.7	50.00	10.05	4.05	5.03	2.02	14.03	7.05	55.22	8.99	4.14	4.97	2.29	13.14	7.25
7	1	1	−1	68.8	75.8	20.3	20.9	339.2	20.3	60.18	6.72	2.81	4.04	1.69	9.37	5.73	67.05	8.79	2.86	5.90	1.92	11.65	7.82
8	1	1	1	68.8	75.8	79.7	20.9	339.2	79.7	53.09	11.01	4.39	5.84	2.33	15.26	8.17	57.19	8.69	4.24	4.97	2.42	12.92	7.39
9	−1.68	0	0	5	55	50	5	250	50	53.88	8.15	3.80	4.39	2.05	12.01	6.44	68.58	10.53	4.77	7.22	3.27	15.30	10.49
10	1.68	0	0	85	55	50	25	250	50	56.51	9.09	3.70	5.14	2.09	12.74	7.23	56.13	10.14	4.56	5.69	2.56	14.71	8.25
11	0	−1.68	0	45	20	50	15	100	50	49.49	11.09	4.70	5.49	2.33	15.59	7.82	55.99	12.41	5.45	6.95	3.05	17.86	10.00
12	0	1.68	0	45	90	50	15	400	50	60.78	8.68	3.42	5.27	2.08	11.96	7.36	76.95	10.60	4.40	8.16	3.38	15.00	11.54
13	0	0	−1.68	45	55	0	15	250	0	54.73	3.81	1.63	2.09	0.89	5.46	2.98	50.18	8.22	1.92	4.12	1.19	10.14	5.31
14	0	0	1.68	45	55	100	15	250	100	47.62	5.68	2.69	2.70	1.28	8.23	3.99	34.40	11.06	1.15	3.81	0.39	12.21	4.20
15	−1.68	−1.68	−1.68	5	20	0	5	100	0	54.39	4.24	1.85	2.30	1.01	6.01	3.31	47.94	8.15	3.73	3.91	1.79	11.88	5.69
16	−1.68	−1.68	1.68	5	20	100	5	100	100	36.34	2.45	1.65	0.89	0.60	4.12	1.49	33.12	6.25	1.50	2.07	0.50	7.75	2.57
17	−1.68	1.68	−1.68	5	90	0	5	400	0	56.79	3.53	1.59	2.00	0.90	5.13	2.90	61.16	10.89	4.46	6.66	2.73	15.36	9.39
18	−1.68	1.68	1.68	5	90	100	5	400	100	47.24	4.51	2.39	2.13	1.13	6.88	3.26	28.92	11.57	4.64	3.34	1.34	16.21	4.69
19	1.68	−1.68	−1.68	85	20	0	25	100	0	51.72	4.45	2.00	2.30	1.03	6.40	3.33	45.30	9.99	4.55	4.53	2.06	14.54	6.59
20	1.68	−1.68	1.68	85	20	100	25	100	100	39.88	4.46	1.75	1.78	0.70	6.21	2.48	28.91	9.60	1.71	2.78	0.50	11.31	3.27
21	1.68	1.68	−1.68	85	90	0	25	400	0	64.04	2.14	0.78	1.37	0.50	2.95	1.87	51.96	9.85	3.88	5.12	2.02	13.73	7.13
22	1.68	1.68	1.68	85	90	100	25	400	100	61.84	6.75	2.82	4.17	1.74	9.54	5.91	23.34	10.08	3.72	2.35	0.87	13.80	3.22
23	0	0	0	45	55	50	15	250	50	56.07	8.65	3.78	4.85	2.12	12.53	6.97	65.72	10.85	4.50	7.13	2.96	15.35	10.08
24	0	0	0	45	55	50	15	250	50	56.55	8.79	4.13	4.97	2.34	13.00	7.31	65.90	10.93	4.34	7.20	2.86	15.27	10.06
25	0	0	0	45	55	50	15	250	50	54.57	8.99	4.09	4.91	2.23	13.22	7.14	66.06	10.44	4.04	6.89	2.67	14.47	9.56
26	0	0	0	45	55	50	15	250	50	54.35	8.65	3.78	4.70	2.05	12.36	6.76	67.94	11.08	4.35	7.53	2.96	15.43	10.48
27	0	0	0	45	55	50	15	250	50	54.57	9.26	4.20	5.02	2.27	13.33	7.14	67.80	10.27	4.09	6.96	2.77	14.36	9.74
28	0	0	0	45	55	50	15	250	50	54.35	9.12	4.18	5.02	2.30	13.37	6.76	68.10	10.26	3.92	6.99	2.67	14.18	9.66

In general, positive and highly significant effects of LE, QE, and IE are found to moderately affect the studied responses. In both techniques assessed (HAE and UAE), the variable *S* is the most relevant one. Initial increases of *S* cause an increase of the extraction efficiency until it reaches a maximum, in which case the increase will cause a decrease in the extraction, but its interactive effect with the variable *t* and *T* or *P* causes a more favorable influence.

Additionally, using the significant parametric values of Table 2 coupled with a simplex methodology, it is possible to determine the absolute/relative optimal values of conditions to maximize the responses individually or globally, in order to obtain the most efficient extraction process. Table 3 shows the individual and global optimal response values and the corresponding conditions that produced them. In consequence, the extracting techniques (HAE and UAE) according to the three response value formats (Y_1, mg C/g R; Y_2, mg C/g E dw; and *Yield*, %) for each assessed anthocyanin (C1 and C2), as well as for the total anthocyanin content (CT = C1 + C2), are depicted.

Table 2. Parametric results of the second-order polynomial equation of Equation (1) for the HAE and UAE techniques assessed and for the three response value formats (Y_1, mg C/g R; Y_2, mg C/g E dw; and *Yield*, %). The parametric subscript 1, 2, and 3 stands for the variables involving *t* (X_1), *T* or *P* (X_2), and *S* (X_3), respectively. Analyses of significance of the parameters (α = 0.05) are presented in coded values. Additionally, the statistical information of the fitting procedure to the model is presented.

Parameters		Residue	Individual Content				Total Content	
		Yield	Y_1C1	Y_1C2	Y_2C1	Y_2C2	Y_1CT	Y_2CT
HAE								
Intercept	b_0	54.86 ± 0.72	9.35 ± 0.38	4.15 ± 0.19	5.07 ± 0.17	2.26 ± 0.08	13.48 ± 0.55	7.29 ± 0.23
Linear effect	b_1	1.54 ± 0.43	0.33 ± 0.21	ns	0.25 ± 0.09	ns	0.28 ± 0.21	0.26 ± 0.13
	b_2	3.12 ± 0.43	ns	ns	0.15 ± 0.09	0.05 ± 0.02	ns	0.19 ± 0.13
	b_3	−3.07 ± 0.43	0.56 ± 0.21	0.33 ± 0.11	0.17 ± 0.09	0.11 ± 0.05	0.88 ± 0.31	0.27 ± 0.13
Quadratic effect	b_{11}	ns	−0.26 ± 0.21	−0.09 ± 0.07	−0.10 ± 0.05	−0.05 ± 0.03	−0.35 ± 0.26	−0.16 ± 0.15
	b_{22}	ns	ns	ns	ns	ns	ns	ns
	b_{33}	−1.24 ± 0.42	−1.60 ± 0.26	−0.69 ± 0.13	−0.94 ± 0.11	−0.40 ± 0.06	−2.26 ± 0.37	−1.33 ± 0.15
Interactive effect	b_{12}	0.78 ± 0.31	0.00 ± 0.00	ns	ns	ns	ns	ns
	b_{13}	0.54 ± 0.31	0.26 ± 0.15	0.04 ± 0.02	0.18 ± 0.07	0.04 ± 0.03	0.31 ± 0.22	0.22 ± 0.09
	b_{23}	0.69 ± 0.31	0.31 ± 0.15	0.14 ± 0.08	0.21 ± 0.07	0.09 ± 0.03	0.45 ± 0.22	0.29 ± 0.09
Statistics (R^2)		0.9375	0.9100	0.8755	0.9443	0.9272	0.9046	0.9489
UAE								
Intercept	b_0	68.11 ± 1.70	10.42 ± 0.47	4.10 ± 0.28	6.98 ± 0.22	2.83 ± 0.13	14.46 ± 0.62	9.75 ± 0.31
Linear effect	b_1	−1.70 ± 0.96	ns	ns	−0.13 ± 0.12	−0.10 ± 0.07	ns	−0.22 ± 0.17
	b_2	2.12 ± 0.96	0.23 ± 0.21	0.18 ± 0.15	0.25 ± 0.12	0.14 ± 0.07	0.37 ± 0.35	0.36 ± 0.17
	b_3	−6.46 ± 0.96	ns	−0.16 ± 0.15	−0.55 ± 0.12	−0.26 ± 0.07	ns	−0.82 ± 0.17
Quadratic effect	b_{11}	−2.29 ± 1.16	ns	0.26 ± 0.20	−0.13 ± 0.11	ns	ns	ns
	b_{22}	ns	ns	0.35 ± 0.20	ns	0.19 ± 0.09	0.66 ± 0.42	0.27 ± 0.21
	b_{33}	−7.33 ± 1.16	−0.36 ± 0.27	−0.84 ± 0.20	−1.02 ± 0.15	−0.68 ± 0.09	−1.22 ± 0.42	−1.84 ± 0.21
Interactive effect	b_{12}	ns	−0.34 ± 0.20	−0.11 ± 0.11	−0.18 ± 0.09	−0.07 ± 0.05	−0.46 ± 0.25	−0.23 ± 0.12
	b_{13}	ns	ns	ns	ns	ns	ns	ns
	b_{23}	−1.29 ± 0.69	0.17 ± 0.10	0.22 ± 0.11	−0.08 ± 0.05	ns	0.41 ± 0.25	ns
Statistics (R^2)		0.9431	0.7825	0.9032	0.9316	0.9035	0.8986	0.9380

Table 3. Variable conditions in natural values that lead to optimal global and individual response values for RSM for each of the extracting techniques assessed (HAE and UAE), for the three response value formats (Y_1, mg C/g R; Y_2, mg C/g E dw; and *Yield*, %), for each compound assessed (C1 and C2), and for the total compounds (CT = C1 + C2).

Criteria			Optimal Variable Conditions			Optimum Response	
			X_1: t (min)	X_2: T (°C) or P(W)	X_3: S (%)		
(A) Individual optimal variable conditions							
HAE	*Yield*		85.00 ± 8.50	90.00 ± 4.50	38.01 ± 3.04	65.10 ± 3.91	%
	Y_1	C1	64.89 ± 5.19	90.00 ± 8.10	62.01 ± 3.10	9.71 ± 0.49	mg C1/g R
		C2	47.18 ± 3.30	90.00 ± 7.20	62.22 ± 6.22	4.27 ± 0.34	mg C2/g R
		CT	58.85 ± 5.89	90.00 ± 9.00	61.97 ± 6.20	13.89 ± 0.14	mg CT/g R
	Y_2	C1	84.27 ± 0.84	90.00 ± 3.60	63.11 ± 3.79	5.64 ± 0.51	mg C1/g E dw
		C2	48.06 ± 1.92	90.00 ± 0.90	59.98 ± 3.60	2.38 ± 0.21	mg C2/g E dw
		CT	70.76 ± 5.66	90.00 ± 4.50	60.82 ± 3.04	7.89 ± 0.55	mg CT/g E dw
UAE	*Yield*		12.79 ± 0.51	400.00 ± 32.00	32.51 ± 1.95	74.53 ± 2.24	%
	Y_1	C1	5.00 ± 0.10	400.00 ± 28.00	61.80 ± 1.85	11.82 ± 0.71	mg C1/g R
		C2	5.00 ± 0.10	400.00 ± 28.00	53.58 ± 1.07	6.45 ± 0.45	mg C2/g R
		CT	5.00 ± 0.10	400.00 ± 20.00	58.39 ± 4.09	18.32 ± 1.47	mg CT/g R
	Y_2	C1	5.00 ± 0.50	400.00 ± 4.00	40.11 ± 2.81	7.88 ± 0.16	mg C1/g E dw
		C2	5.00 ± 0.40	400.00 ± 28.00	44.35 ± 4.43	3.96 ± 0.28	mg C2/g E dw
		CT	5.00 ± 0.35	400.00 ± 20.00	43.37 ± 4.34	12.23 ± 0.86	mg CT/g E dw
(B) Global optimal variable conditions							
HAE	*Yield*		49.02 ± 2.94	90.00 ± 7.20	50.00 ± 0.50	50.89 ± 3.05	%
	Y_1	C1				9.71 ± 0.29	mg C1/g R
		C2				4.22 ± 0.13	mg C2/g R
		CT				13.93 ± 0.42	mg CT/g R
	Y_2	C1				5.57 ± 0.11	mg C1/g E dw
		C2				2.36 ± 0.05	mg C2/g E dw
		CT				7.93 ± 0.08	mg CT/g E dw
UAE	*Yield*		5.00 ± 0.15	400.00 ± 32.00	47.98 ± 2.88	68.60 ± 2.06	%
	Y_1	C1				11.74 ± 0.23	mg C1/g R
		C2				6.43 ± 0.32	mg C2/g R
		CT				18.17 ± 1.82	mg CT/g R
	Y_2	C1				7.81 ± 0.47	mg C1/g E dw
		C2				3.95 ± 0.24	mg C2/g E dw
		CT				11.76 ± 0.82	mg CT/g E dw

2.2. Alternative Visual Illustration of the Effects of the Extraction Variables on the Target Responses Used

Although the parametric values show the behavior of the responses, and could be used to understand their patterns, a more visual way to express the effects of variables on the extraction of any type of response is to generate 3D surface and/or contour plots, by varying two variables in the experimental range under investigation and holding the other one at a fixed level. In this regard, Figures 1 and 2 show the 3D surface and 2D contour plots, respectively, representing the influence of the investigated effects of HAE and UAE parameters on the extraction behavior. The plots enable one to visualize the influence and interaction between the variables. Visual analyses of 3D surface and 2D contour plots are in accordance with parametric values derived from the multiple regression analysis, as described in Table 2 (parametric values) and Table S1 (full mathematical models, supplementary material).

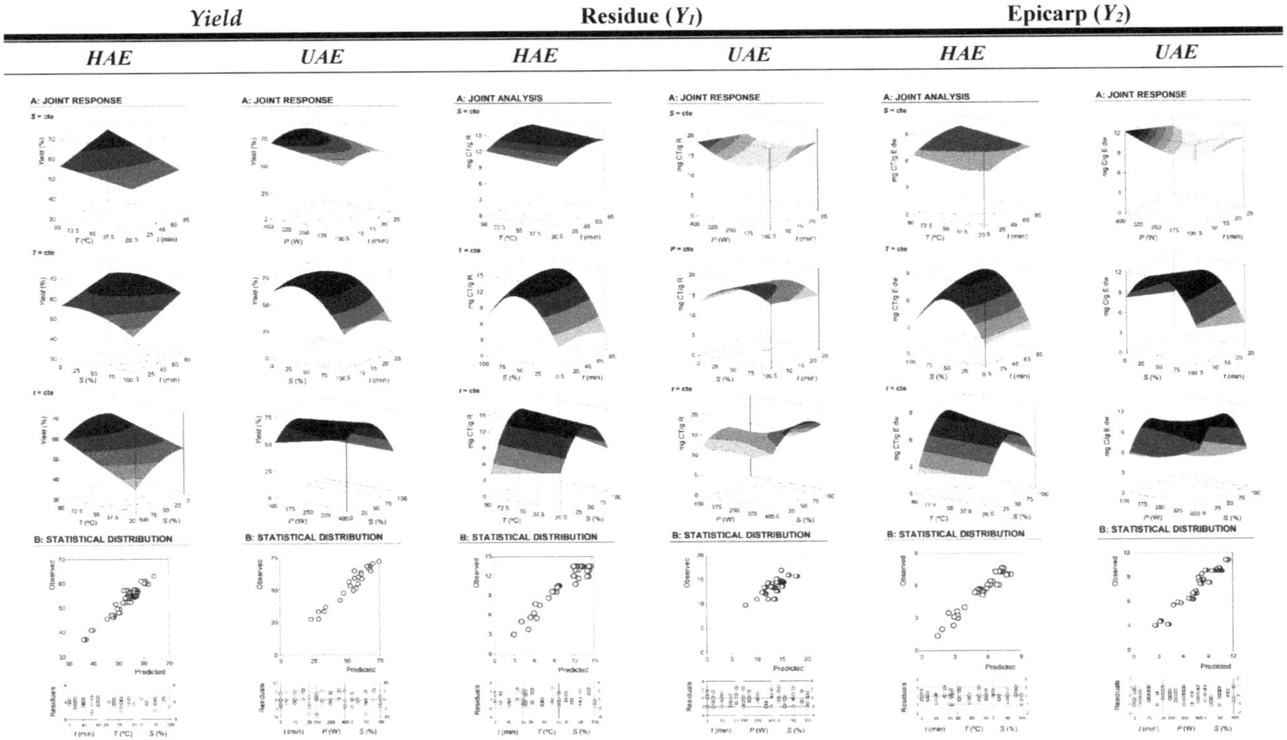

Figure 1. Illustration of the graphical results obtained by heat assisted extraction (HAE) and ultrasound assisted extraction (UAE) for the extraction *yield* of the residual content material (R) and the total detected anthocyanin compounds (cyanidin 3-rutinoside and peonidin 3-rutinoside, CT = C1 + C2) in terms of two response formats (Y_1, mg C/g R and Y_2, mg C/g E dw). Full results are described in Table 1. Every figure is presented in two parts. Part A shows the 3D net surfaces predicted by Equation (1) when the excluded variable is positioned at the individual optimum (Table 3). Part B describes the statistical analysis in a graphical form to show the goodness of fit of the models applied.

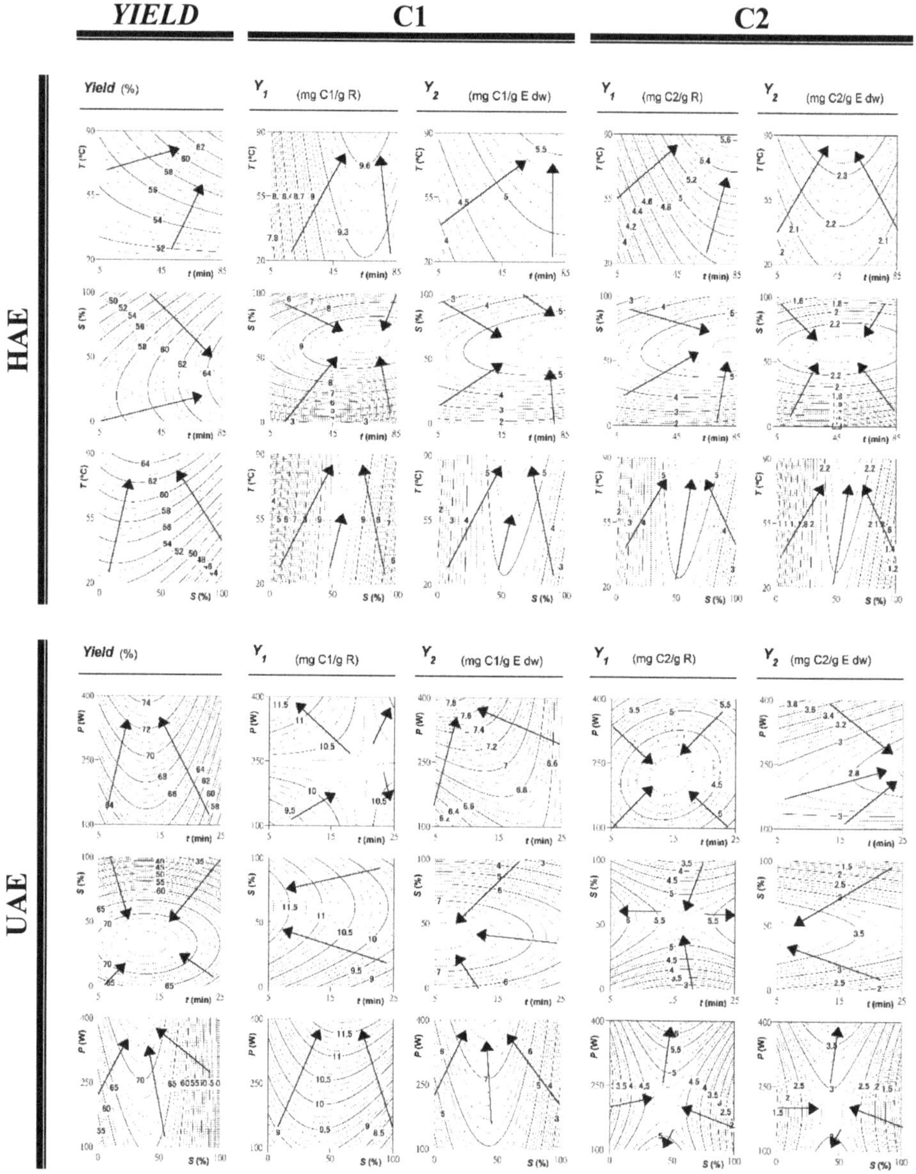

Figure 2. The optimized isolines projections for the extraction of C1 (cyanidin 3-rutinoside) and C2 (peonidin 3-rutinoside) as a function of the combination of the three main variables involved (X_1, X_2, and X_3) in the HAE and UAE. For each compound, the two response value formats (Y_1, mg C/g R and Y_2, mg C/g E dw) are presented to describe the most favorable conditions. Furthermore, the response projections of the *yield* of the extracted residual material are presented. All the contour graphs were built by the second order polynomial models generated by Equation (1) (Table S1) when the excluded variable is positioned at the individual optimum (Table 3).

The extraction results for HAE and UAE, as function of the combination of the three main involved variables ($X_{1\text{-}3}$: t, T or P, and S), can be observed in Figures 1 and 2. In this regard, Figure 1 shows the 3D surface plots of the extracted R (*Yield*, %), and CT, in two response formats (Y_1, mg CT/g R and Y_2, mg CT/g E dw). On the other hand, Figure 2 shows the optimized isolines projections for C1 (cyanidin 3-rutinoside) and C2 (peonidin 3-rutinoside) extraction, in the two response value formats (Y_1, mg C/g R and Y_2, mg C/g E dw). These figures show, respectively, optimized 3D graphical and 2D isolines projection results for the extracted anthocyanins (C1 or C2) as function of the three combined variables (t, T or P, and S) in HAE and UAE. The total anthocyanins (C1+C2) are accounted together (CT) in Figure 1, and individually in Figure 2. They are helpful to visualize the tendencies of each response and lead to define of the maximum favorable conditions, considering all together all responses.

Additionally, Figure 1B exemplifies the competence to predict the obtained results. In statistical terms, the distribution of residues (Figure 1) presents, for the majority of the cases, more than 90% of reliability, showing a good agreement between experimental and predicted values. This is also verified by the achieved high R^2 values (Table 2), which indicates the percentage of variability explained by the model.

In HAE, small differences between the extraction behavior of the two considered anthocyanins (when comparing C1 and C2, or Y_1 and Y_2) were clearly distinguished. The opposite occurred in UAE, the effects were distinct for each one of the detected anthocyanins, as well as according to the response format. However, for both extraction techniques, the S variable was the most significant one, producing a relevant impact on the level of extraction of all anthocyanins assessed. As described above, the LE and the QE of the significant parametric values of the variable S can be perceived in all figures. In almost all cases, the variable S indicates a maximum level at ~50% of hydroalcoholic mixture (water/ethanol, v/v). The negative impact of quadratic term of the variable S can be explained through the increase of water in the process, which expands the yield of extraction. Other negative effects such as those between T or P and S may suggest that the further use of lower P, in combination with higher S, will avoid the anthocyanin degradation. The results are in accordance with others recently reported by other authors [19–21], in which inclined surfaces to the side of T or P may increase the solubility of target compounds by using stronger energies, and consequently improve their release from the sample matrix, while destroying the integrity of connective and structural tissues.

2.3. Conditions That Maximize the Anthocyanins Extraction and Experimental Verification

The aim of this study was to maximize the extraction yield of targeted anthocyanin compounds from epicarps of *P. spinosa* fruits, in the applied HAE and UAE techniques, within pre-set variable conditions and ranges. The values of the variable conditions that lead to optimal response values for RSM using a *CCCD*, obtained with the aid of *simplex* procedure, for each of the assessed extracting techniques are shown in Table 3. Figure 3 part A shows the individual summary of the effects of all variables assessed for HAE and UAE systems in 2D illustrations, where the variables are positioned at the individual optimal values of the others (Table 3). The dots (⊙) presented alongside each line highlight the location of the optimum value, meanwhile lines are the predicted behavior found by the mathematical analysis of Equation (1) generated by the theoretical second-order polynomial models described in Table S1. Next, some relevant details of the results produced are highlighted:

- For the HAE: the global optimal variable conditions were found at 49.02 ± 2.94 min, 90.00 ± 7.20 °C, and 50.00% ± 0.50% of ethanol, producing maximum response values of 13.93 ± 0.42 mg CT/g R (Y_1), 7.93 ± 0.08 mg CT/g E dw (Y_2), and 50.89% ± 3.05% (*yield* of the extracted residue).
- For the UAE: the global optimal variable conditions were found at 5.00 ± 0.15 min, 400.00 ± 32.00 W, and 47.98% ± 2.88% of ethanol, producing maximum response values of 18.17 ± 1.82 mg CT/g R (Y_1), 11.76 ± 0.82 mg CT/g E dw (Y_2), and 68.60% ± 2.06 % (*yield* of the extracted residue).

Considering both the individual and global values, the higher amount of extracted compounds was obtained for the UAE technique. The ideal solvent composition was almost the same, and the two techniques required high energy values, where the highest values of T and P proposed by the experimental design were the optimal, but the UAE needed less t than HAE (~90% less). The obtained results are in accordance with similar conclusions found previously [10,17,22], in which UAE proved to consume less energy because of the lower t needed, and provide higher extraction values while increasing the purity and, additionally, aiding to meet the requirements of a green extraction concept.

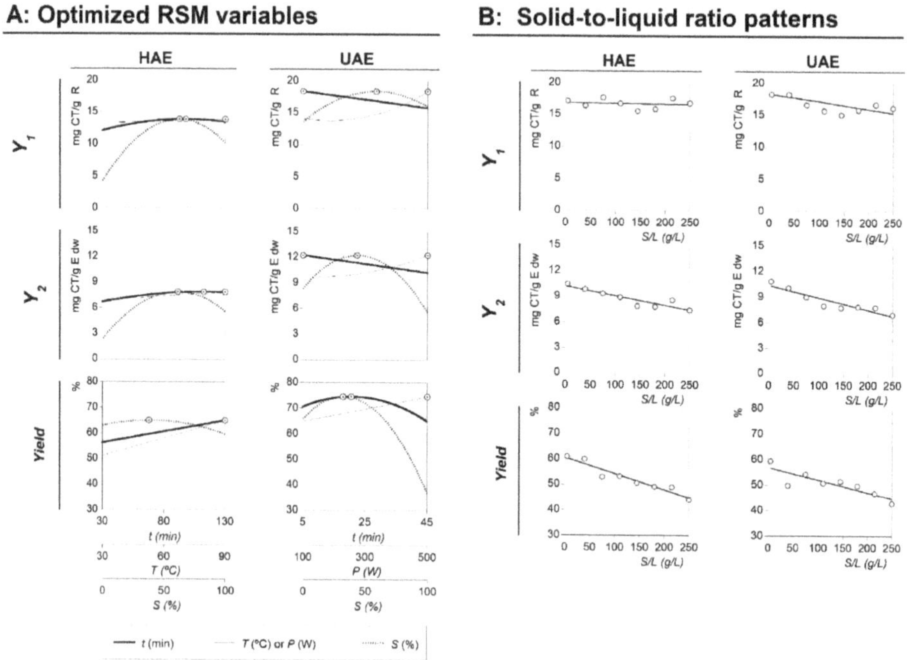

Figure 3. Final graphical effects of all variables assessed for HAE and UAE systems. Part A shows the individual 2D responses as a function of all the variables assessed that were positioned at the individual optimal values of the others (Table 2). The points (⊙) presented alongside each line highlight the location of the optimum value. Lines and dots are generated by the theoretical second order polynomial models generated by Equation (1) (Table S1). Part B shows the dose response of S/L at the global optimal values of the other three variables (Table 3). The limit value (~150 g/L) shows the maximum achievable experimental concentration until the sample cannot be physically stirred at laboratory scale. RSM—response surface methodology.

2.4. Dose-Response Analysis of the Solid-to-Liquid Ratio Effect at the Optimal Conditions

The study of S/L effect was performed at the optimal conditions (Table 3) predicted by the polynomial models obtained for each extraction technique (HAE and UAE) using the total anthocyanin content (CT), as quantified by HPLC analysis, as the response factor. The individual S/L study for each individual anthocyanin (C1 or C2) was not presented because the behavior was similar to the pattern of the total amount. In both processes, the S/L was designed to verify the behavior between 5 and 250 g/L. The maximum value of 250 g/L was used as a limit condition because of the impossibility of producing a homogenized extraction when higher values were used.

The obtained dose responses of the S/L were consistent for both HAE and UAE systems, and could be described by a simple linear relationship (shown in Part B of Figure 3). All experimental points are distributed around the linear equation; consequently, the dose response is explained by the slope (m) of the linear relationship. None of the cases showed positive m values (the extraction efficiency increases as the S/L rate increases), and two cases showed non-significant values or a zero value of m (the efficiency doesn't change as the S/L increases). In all the other cases, the m showed negative values (the efficiency decreases as the S/L increases). The responses from the Y_1 value format, for HAE and UAE, were the ones that showed non-significant m values, whereas all other responses showed significant negative values of m (Y_1 and *Yield* for HAE and UAE). The conclusions derived from this analysis are described below:

- For the Y_1 value format, the response of the parametric m value in HAE and UAE presents a non-significant interval of confidence, which means that the changes in the response are not statistically supported and, therefore, the parameter must be considered equal to zero. In other words, the amount of anthocyanins in the extracted residue does not vary as a function of the S/L increase. The extraction values were defined numerically by the intercept parametric value (b) of the linear equation as 14.85 ± 2.29 and 18.25 ± 3.95 mg CT/g R for HAE (R^2 = 0.9920) and UAE (R^2 = 0.9817), respectively.
- For the Y_2 value format, the parametric values for HAE were b = 9.21 ± 1.37 mg CT/g E dw and m = −0.0113 ± 0.0051, with R^2 = 0.9566; while for UAE, b = 10.32 ± 1.48 mg CT/g E dw and m = −0.0143 ± 0.0038, with R^2 = 0.9244. Negative m values show that the S/L increase leads to a decrease in the extraction ability, obtaining a maximum value of extraction at 5 g/L and a minimum at 250 g/L. However, the observed decrease is slight (less than −0.02), which means that the increase of 1 g/L implies the loss of 0.0113 ± 0.0051 mg CT/g E dw for the HAE process and 0.0143 ± 0.0038 mg CT/g E dw for UAE. Such values produce losses at the maximum tested experimental value (250 g/L) of ~15%, comparative with the one extracted at 5 g/L. Nevertheless, the economic advantages of working at 250 g/L are far more superior than the possible benefits of extracting at the optimal S/L value.
- For the *Yield* value format, the parametric values for HAE were b = 54.62% ± 4.87% and m = −0.0636 ± 0.0123, with R^2 = 0.9516; whereas for UAE, b = 58.90% ± 7.77% and m = −0.0491 ± 0.0116, with R^2 = 0.9618. Although, at the initial S/L values, the results obtained for HAE and UAE conducted to similar extraction yields, these values decreased as the S/L increased. The m parametric value is significantly lower for the UAE process, resulting in higher extraction yield values at 250 g/L. These results are in accordance with the conclusions highlighted in the literature, where UAE is reported as enhancing the extraction process by increasing the mass transfer between the plant material and the solvent [23]. The UAE leads to better cell disruption, facilitating the release of the extractable compounds by increasing the contact surface area between the solid and liquid phases [22,23].

2.5. Comparison with Other Studies Involving the Extraction of Anthocyanins

There are few works in the literature dealing with anthocyanins in *P. spinosa* fruits. In one of these studies with *P. spinosa* fruits, Guimarães et al. [5] performed the extraction using methanol with 0.5% TFA added as solvent, and identified eight different anthocyanins, predominantly peonidin 3-rutinoside and cyanidin 3-rutinoside, with 34.47 ± 0.03 µg/100 g fruits dw and 31.12 ± 0.11 µg/100 g fruits dw, respectively. Other authors [6], found 3.5 ± 0.5 mg of anthocyanins/100 g dw of *P. spinosa* fruits. Both authors used the whole fruit, while in this study, only the epicarp was used as the extraction material, a fact that may justify the significant differences between the encountered results, when compared with the present study. Compared with the pulp, the fruit epicarp presents a greater intensity of color and, therefore, a higher concentration of anthocyanins, in addition to less interfering compounds, is obtained. Moreover, another fact that aided the production of large amounts of anthocyanins from the extracted material was the optimization of the extraction process, which led to increased extraction efficiency and yield. In another study that used *P. spinosa* fruits as a source of anthocyanins [6], the total content was quantified by spectrophotometric methods, presenting values that cannot be compared with those found in the present study.

Some examples of other plant-based sources of anthocyanins are *Oryza sativa* L. (var. Glutinosa) bran, which shows 42.00 mg/g [24]; *Phaseolus vulgaris* L. (common beans) fruit coat, presenting 32.00 mg/g [25]; and *Rubus fruticosus* L. (blackberries) fruit, which possess 17.10 mg/g [26]. Although these values are slightly higher than those presented by *P. spinosa* fruits, in general, the referred fruits and vegetables already have a high commercial value and other industrial purposes, unlike *P. spinosa* fruits. On the other hand, residues such as grape vine (*Vitis vinifera* L.) pomace, and mango (*Mangifera indica*) skin presented lower anthocyanin amounts, that is, 6.33 mg/g [27] and 2.03 to 3.60 mg/g [28], respectively. Thus, these wild fruits revealed to be an excellent source of anthocyanins, serving as a base raw-material for the production of natural colorant additives for commercial purposes.

3. Discussion

The minimalism of using conventional methods (HAE or maceration) versus the compensations of new non-conventional technologies (microwave, ultrasound, cold pressing, squeezing, etc.) to recover compounds from plant materials, as well as by-products, is a principal topic in the list of many industries in order to increase profitability by decreasing energy costs and reducing greenhouse gas emissions to meet legal requirements. Additionally, non-conventional technologies favor the safety of processes and the quality of products, as well as the functionality and product standardization.

Scientific literature shows clear evidence that extraction procedures of target compounds from plant-based products must be assessed individually. Therefore, a nonstop effort needs to be performed, as agro-industrial and food sectors are looking for by-products' valorization into added-value products. However, in order to take full advantage of the technological advances, the extraction conditions need to be optimized. Mathematical solutions, such as RSM tools, could increase the efficiency and profitability of the process and help to change conventional extraction approaches.

Colorants are one of the most important additives in terms of marketing because their presence in food products is considered to influence customers' perceptions, choices, and preferences. *P. spinosa* fruit epicarps have been scarcely explored and, to the best of the authors knowledge, the potential industrial use of their anthocyanin compounds has not been previously investigated. In such a context, the present work presents a new rapid method to extract anthocyanin compounds from *P. spinosa* fruit epicarps. RSM and other mathematical strategies were successfully employed to optimize the extraction conditions that maximize the anthocyanin compounds' recovery to produce a rich extract with potential industrial application as a natural coloring additive.

4. Materials and Methods

4.1. *Plant Material*

Ripe *P. spinosa* fruits were harvested in Bragança (Trás-os-Montes, Northeast Portugal) in September 2017, the epicarp was separated from the rest of the fruit body, frozen, and lyophilized. They were then triturated, to be reduced to a fine powder (~20 mesh), and stored under refrigeration, protected from light until further use.

4.2. *Extraction Procedures for P. Spinosa Fruit Epicarps*

4.2.1. Heat Assisted Extraction (HAE)

HAE was performed in a water reactor agitated internally with a Cimarec^{TM} Magnetic Stirrer at a constant speed (~500 rpm, Thermo Scientific, San Jose, CA, USA), following a procedure previously performed [13]. The powdered epicarp samples of *P. spinosa* (1.0 g) were extracted with 20 mL of solvent (ethanol/water) acidified with citric acid (pH = 3), under diverse conditions, as previously defined by the established RSM plan (Table 1). The ranges of the experimental design were as follows: time (t or X_1, 5 to 85 min), temperature (T or X_2, 20 to 90 °C), and ethanol content (S or X_3, 0% to 100%). The solid-to-liquid ratio (S/L or X_4) was kept at 50 g/L for all conditions.

4.2.2. Ultrasound-Assisted Extraction (UAE)

An ultrasonic device (QSonica sonicators, model CL-334, Newtown, CT, USA) equipped with a water reactor (EUP540A, Euinstruments, France) at a fixed frequency (40 kHz) was used for UAE procedure. The variables considered were as follows: ultrasonic power (*P*, in watts), *S*, and *t*, which were programmed according to the defined RSM plan (Table 1), following a procedure previously performed [29]. The powdered epicarp samples (2.5 g) were placed in a reactor with 50 mL of solvent (ethanol/water) acidified with citric acid (pH = 3), and extracted under diverse conditions, maintaining the *S/L* constant at 50 g/L. The ranges of the experimental design were as follows: *t* (X_1, 5 to 25 min), *P* (or X_2, 100 to 400 W), and *S* (or X_3, 0% to 100%).

4.2.3. Post-Extraction Sample Processing

When all the individual extraction conditions were carried out (for HAE and UAE), the samples were immediately centrifuged (6000 rpm during 20 min at 10 °C) and filtered (paper filter Whatman n° 4) to eliminate the non-dissolved material. The supernatant was collected and divided into two portions for HPLC and extraction yield analysis. The portion separated for HPLC analysis (3 mL) was dried at 35 °C, re-dissolved in acidified water (citric acid solution with pH 3), and filtered through an LC filter disk (0.22 µm), whereas the portion for the extraction yield determination (5 mL) was dried at 105 °C during 48 h and thereafter weighted.

4.3. Identification and Quantification of Anthocyanins by HPLC

The extract was analyzed using an HPLC-DAD-ESI/MSn (Dionex Ultimate 3000 UPLC, Thermo Scientific, San Jose, CA, USA) system, previously described [30]. The detection was carried out using a DAD (520 nm as the preferred wavelength) and mass spectrometer (Linear Ion Trap LTQ XL mass spectrometer, Thermo Finnigan, San Jose, CA, USA) equipped with an ESI source. The anthocyanins present in the samples were characterized according to their UV and mass spectra. The anthocyanins cyanidin 3-rutinoside (C1) and peonidin 3-rutinoside (C2) were the most relevant compounds found, and were, therefore, quantified using a five-level calibration curve of known concentrations (200–20 µg/mL) of cyanidin 3-glucoside ($y = 243287\ x - 1000000$; $R^2 = 0.9953$, Polyphenols, Sandnes, Norway) and peonidin 3-glucoside ($y = 122417\ x - 447974$; $R^2 = 0.9965$, Polyphenols, Sandnes, Norway).

4.4. Response Value Formats for Results Presentation

The two anthocyanin compounds (C, either C1 or C2) and their sum (C total, CT) were used as responses in each applied technique. The results were presented according to two response formats (*Y*): Y_1, in mg of C per gram of extracted residue (mg C/g R), which was specifically used to evaluate the C purity in the extracts; and Y_2, in mg of C per g of fruit epicarp dry weight (mg C/g E dw), which was specifically used to analyses the C extraction yield. Both responses were equally analyzed, but additional considerations regarding the last one (Y_2, mg C/g E dw) were taken in the results presentation, becuse it is considered as an important guiding response when dealing in terms of optimization for industrial transference. Note that by dividing those responses, Y_2/Y_1, the extracted residue quantity (g R/g E dw) is obtained, which provides information regarding the third response criterion expressed (*Yield*, %).

4.5. Experimental Design, Model Analysis, and Statistical Evaluation

4.5.1. RSM Experimental Design

Trials based on one-at-the-time analysis (analysis of each of the variables for each one of the selected techniques) were conducted, the ranges originating significant changes were selected (Table 1). The joint effects of the three defined variables were studied using a *circumscribed central composite design* (CCCD), using five levels for each one with twenty eight response combinations, as described previously [31].

4.5.2. Mathematical Model

The experimental data produced by the RSM design were analyzed mathematically by means of least-squares calculation, using the following second-order polynomial equation with interactive terms [32]:

$$Y = b_0 + \sum_{i=1}^{n} b_i X_i + \sum_{\substack{i=1 \\ j>i}}^{n-1} \sum_{j=2}^{n} b_{ij} X_i X_j + \sum_{i=1}^{n} b_{ii} X_i^2 \quad (1)$$

where Y is the dependent variable (response variable) modelled, X_i and X_j define the independent variables, b_0 is the constant coefficient, b_i is the coefficient of linear effect, b_{ij} is the coefficient of interaction effect, b_{ii} the coefficient of quadratic effect, and n is the number of variables. As responses, the following three value formats were used: Y_1 (mg C/g R), Y_2 (C/g E dw), and *Yield* (%).

4.5.3. Procedure to Optimize the Variables to a Maximum Response

A simplex algorithm method was used to find the optimum values by solving nonlinear problems in order to maximize the extraction yield and the recovery of anthocyanin compounds, as explained previously [33]. Certain limitations were imposed (i.e., times cannot be lower than 0) to avoid variables with unnatural and unrealistic physical conditions.

4.5.4. Dose-Response Analysis of the Solid-to-Liquid Ratio

Once the optimal conditions (X_1, X_2, and X_3) were found, the following natural optimization step was used to describe the pattern of the S/L (or X_4, expressed in g/L). The objective was to achieve more productive conditions as required by industrial applications. In all cases, experimental points are distributed following linear patterns as the S/L increases, consequently, linear models with intercepts were used to evaluate the responses. The parametric value of the slope (m) was used to assess the dose response. Positive values will indicate an increase in the extraction responses, whereas negative values will designate a decrease in the extraction efficiency, as the S/L increases.

4.6. Mathematical Procedures

The analytical procedures to model the data, to determine the parametric values, confidence intervals, and statistical calculations, were obtained following the descriptions of other authors [34]. In brief, (a) the parametric values were obtained using the quasi-Newton algorithm (least-square) by running the integrated macro '*Solver*' in Microsoft Excel; (b) the coefficient significance of the parameters produced ($\alpha = 0.05$) was assessed using the '*SolverAid*' macro to conclude their confidence intervals; and c) the model consistency was proven by means of several statistical criteria, such as (i) the Fisher *F*-test ($\alpha = 0.05$); (ii) the '*SolverStat*' macro; and (iii) the R^2 coefficient.

5. Conclusions

The efficiency of the UAE was higher than that obtained with HAE. The main anthocyanins identified were cyanidin 3-rutinoside and peonidin 3-rutinoside, being the ones quantified. Through the optimization of the extraction process, it was possible to reach by UAE 18.17 ± 1.82 mg CT/g R (Y_1), 11.76 ± 0.82 mg CT/g E dw (Y_2), and 68.60% ± 2.06% (yield of the extracted residue), with the optimal parameters of extraction being 5.00 ± 0.15 min, 400.00 ± 32.00 W, and 47.98% ± 2.88% of ethanol. The used mathematical models (RSM and dose-response models) were statistically significant and allowed the optimization of the anthocyanins extraction. For the S/L effects, inspected at the optimum conditions, the responses for all assessed criteria followed a decreasing linear relation until 250 g/L.

In conclusion, the present study contributed to the valorization of the wild fruits of *P. spinosa* by exploring anthocyanin-rich extracts that can find potential application as natural colorants in different industrial fields. For that purpose, an optimized extraction method was obtained using advanced and efficient extraction systems.

Supplementary Materials: The supplementary materials are available online.

Author Contributions: Formal analysis, M.A.P.; Investigation, M.G.L.; Methodology, C.P. and L.B.; Resources, M.F.B.; Supervision, I.J.B. and I.C.F.R.F.

Funding: The authors are grateful to the Foundation for Science and Technology (FCT, Portugal) and FEDER under Programme PT2020 for financial support to CIMO (UID/AGR/00690/2013), L. Barros and C. Pereira research contract; to FEDER-Interreg España-Portugal programme for financial support through the project 0377_Iberphenol_6_E.; to European Regional Development Fund (ERDF) through the Regional Operational Program North 2020, within the scope of Project NORTE-01-0145-FEDER-023289: DeCodE and project Mobilizador Norte-01-0247-FEDER-024479: ValorNatural®. This work was also financially supported by the following: Project POCI-01-0145-FEDER-006984, Associate Laboratory LSRE-LCM funded by FEDER through COMPETE2020, Programa Operacional Competitividade e Internacionalização (POCI), and national funds through FCT. The authors thank the GAIN (Xunta de Galicia) for financial support (P.P. 0000 421S 140.08) to M.A. Prieto by a post-doctoral (modality B) grant.

Conflicts of Interest: The authors declare no conflict of interest.

References

1. Morales, P.; Ferreira, I.C.F.R.; Carvalho, A.M.; Fernández-Ruiz, V.; Sánchez-Mata, M.S.O.S.C.C.; Cámara, M.; Morales, R.; Tardío, J. Wild edible fruits as a potential source of phytochemicals with capacity to inhibit lipid peroxidation. *Eur. J. Lipid Sci. Technol.* **2013**, *115*, 176–185. [CrossRef]
2. Naziri, E.; Nenadis, N.; Mantzouridou, F.T.; Tsimidou, M.Z. Valorization of the major agrifood industrial by-products and waste from Central Macedonia (Greece) for the recovery of compounds for food applications. *Food Res. Int.* **2014**, *65*, 350–358. [CrossRef]
3. Usenik, V.; Fabčič, J.; Štampar, F. Sugars, organic acids, phenolic composition and antioxidant activity of sweet cherry (*Prunus avium* L.). *Food Chem.* **2008**, *107*, 185–192. [CrossRef]
4. Ieri, F.; Pinelli, P.; Romani, A. Simultaneous determination of anthocyanins, coumarins and phenolic acids in fruits, kernels and liqueur of *Prunus mahaleb* L. *Food Chem.* **2012**, *135*, 2157–2162. [CrossRef]
5. Guimarães, R.; Barros, L.; Dueñas, M.; Carvalho, A.M.; Queiroz, M.J.R.P.; Santos-Buelga, C.; Ferreira, I.C.F.R. Characterisation of phenolic compounds in wild fruits from Northeastern Portugal. *Food Chem.* **2013**, *141*, 3721–3730. [CrossRef]
6. Pinacho, R.; Cavero, R.Y.; Astiasarán, I.; Ansorena, D.; Calvo, M.I. Phenolic compounds of blackthorn (*Prunus spinosa* L.) and influence of in vitro digestion on their antioxidant capacity. *J. Funct. Foods* **2015**, *19*, 49–62. [CrossRef]
7. Hernández-Herrero, J.A.; Frutos, M.J. Degradation kinetics of pigment, colour and stability of the antioxidant capacity in juice model systems from six anthocyanin sources. *Int. J. Food Sci. Technol.* **2011**, *46*, 2550–2557. [CrossRef]
8. Olivas-Aguirre, F.J.; Rodrigo-García, J.; Martínez-Ruiz, N.D.R.; Cárdenas-Robles, A.I.; Mendoza-Dïaz, S.O.; Álvarez-Parrilla, E.; González-Aguilar, G.A.; De La Rosa, L.A.; Ramos-Jiménez, A.; Wall-Medrano, A. Cyanidin-3-O-glucoside: Physical-chemistry, foodomics and health effects. *Molecules* **2016**, *21*, 264. [CrossRef]
9. Dahmoune, F.; Nayak, B.; Moussi, K.; Remini, H.; Madani, K. Optimization of microwave-assisted extraction of polyphenols from *Myrtus communis* L. leaves. *Food Chem.* **2015**, *166*, 585–595. [CrossRef]
10. Zhu, Z.; He, J.; Liu, G.; Barba, F.J.; Koubaa, M.; Ding, L.; Bals, O.; Grimi, N.; Vorobiev, E. Recent insights for the green recovery of inulin from plant food materials using non-conventional extraction technologies: A review. *Innov. Food Sci. Emerg. Technol.* **2016**, *33*, 1–9. [CrossRef]
11. Wang, X.; Wu, Y.; Chen, G.; Yue, W.; Liang, Q.; Wu, Q. Optimisation of ultrasound assisted extraction of phenolic compounds from *Sparganii rhizoma* with response surface methodology. *Ultrason. Sonochem.* **2013**, *20*, 846–854. [CrossRef]

12. Wang, W.; Jung, J.; Tomasino, E.; Zhao, Y. Optimization of solvent and ultrasound-assisted extraction for different anthocyanin rich fruit and their effects on anthocyanin compositions. *LWT Food Sci. Technol.* **2016**, *72*, 229–238. [CrossRef]
13. Roriz, C.L.; Barros, L.; Prieto, M.A.; Morales, P.; Ferreira, I.C.F.R. Floral parts of *Gomphrena globosa* L. as a novel alternative source of betacyanins: Optimization of the extraction using response surface methodology. *Food Chem.* **2017**, *229*, 223–234. [CrossRef]
14. Agcam, E.; Akyıldız, A.; Balasubramaniam, V.M. Optimization of anthocyanins extraction from black carrot pomace with thermosonication. *Food Chem.* **2017**, *237*, 461–470. [CrossRef]
15. Rodriguez-Amaya, D.B. Natural food pigments and colorants. In *Current Opinion in Food Science*; Elsevier Ltd.: Amsterdam, The Netherlands, 2016; Volume 7, pp. 20–26. ISBN 2214-7993.
16. Jiménez, L.C.; Caleja, C.; Prieto, M.A.; Barreiro, M.F.; Barros, L.; Ferreira, I.C.F.R. Optimization and comparison of heat and ultrasound assisted extraction techniques to obtain anthocyanin compounds from *Arbutus unedo* L. fruits. *Food Chem.* **2018**, *264*, 81–91.
17. Montesano, D.; Fallarino, F.; Cossignani, L.; Simonetti, M.S.; Puccetti, P.; Damiani, P. Innovative extraction procedure for obtaining high pure lycopene from tomato. *Eur. Food Res. Technol.* **2008**, *226*, 327–335. [CrossRef]
18. Bezerra, M.A.; Santelli, R.E.; Oliveira, E.P.; Villar, L.S.; Escaleira, E.A.; Escaleira, L.A. Response surface methodology (RSM) as a tool for optimization in analytical chemistry. *Talanta* **2008**, *76*, 965–977. [CrossRef]
19. Pinela, J.; Prieto, M.A.; Carvalho, A.M.; Barreiro, M.F.; Oliveira, M.B.P.; Barros, L.; Ferreira, I.C.F.R. Microwave-assisted extraction of phenolic acids and flavonoids and production of antioxidant ingredients from tomato: A nutraceutical-oriented optimization study. *Sep. Purif. Technol.* **2016**, *164*, 114–124. [CrossRef]
20. Roriz, C.L.; Barros, L.; Prieto, M.A.; Barreiro, M.F.; Morales, P.; Ferreira, I.C.F.R. Modern extraction techniques optimized to extract betacyanins from *Gomphrena globosa* L. *Ind. Crops Prod.* **2017**, *105*, 29–40. [CrossRef]
21. Oludemi, T.; Barros, L.; Prieto, M.A.; Heleno, S.A.; Barreiro, M.F.; Ferreira, I.C.F.R. Extraction of triterpenoids and phenolic compounds from: Ganoderma lucidum: Optimization study using the response surface methodology. *Food Funct.* **2018**, *9*, 209–226. [CrossRef]
22. Chemat, F.; Rombaut, N.; Sicaire, A.G.; Meullemiestre, A.; Fabiano-Tixier, A.S.; Abert-Vian, M. Ultrasound assisted extraction of food and natural products. Mechanisms, techniques, combinations, protocols and applications. A review. *Ultrason. Sonochem.* **2017**, *34*, 540–560. [CrossRef]
23. Tomšik, A.; Pavlić, B.; Vladić, J.; Ramić, M.; Vidović, S. Optimization of ultrasound-assisted extraction of bioactive compounds from wild garlic (*Allium ursinum* L.). *Ultrason. Sonochem.* **2016**, *29*, 502–511.
24. Chen, M.H.; McClung, A.M.; Bergman, C.J. Phenolic content, anthocyanins and antiradical capacity of diverse purple bran rice genotypes as compared to other bran colors. *J. Cereal Sci.* **2017**, *77*, 110–119. [CrossRef]
25. Mojica, L.; Berhow, M.; Gonzalez de Mejia, E. Black bean anthocyanin-rich extracts as food colorants: Physicochemical stability and antidiabetes potential. *Food Chem.* **2017**, *229*, 628–639. [CrossRef]
26. Elisia, I.; Hu, C.; Popovich, D.G.; Kitts, D.D. Antioxidant assessment of an anthocyanin-enriched blackberry extract. *Food Chem.* **2006**, *101*, 1052–1058. [CrossRef]
27. Bosiljkov, T.; Dujmić, F.; Cvjetko Bubalo, M.; Hribar, J.; Vidrih, R.; Brnčić, M.; Zlatic, E.; Radojčić Redovniković, I.; Jokić, S. Natural deep eutectic solvents and ultrasound-assisted extraction: Green approaches for extraction of wine lees anthocyanins. *Food Bioprod. Process.* **2017**, *102*, 195–203. [CrossRef]
28. Ajila, C.M.; Naidu, K.A.; Bhat, S.G.; Rao, U.J.S.P. Bioactive compounds and antioxidant potential of mango peel extract. *Food Chem.* **2007**, *105*, 982–988. [CrossRef]
29. Backes, E.; Pereira, C.; Barros, L.; Prieto, M.A.; Kamal, A.; Filomena, M.; Ferreira, I.C.F.R. Recovery of bioactive anthocyanin pigments from *Ficus carica* L. peel by heat, microwave, and ultrasound based extraction techniques. *Food Res. Int.* **2018**, *113*, 197–209. [CrossRef]
30. Gonçalves, G.A.; Soares, A.A.; Correa, R.C.G.; Barros, L.; Haminiuk, C.W.I.; Peralta, R.M.; Ferreira, I.C.F.R.; Bracht, A. Merlot grape pomace hydroalcoholic extract improves the oxidative and inflammatory states of rats with adjuvant-induced arthritis. *J. Funct. Foods* **2017**, *33*, 408–418. [CrossRef]
31. Heleno, S.A.; Prieto, M.A.; Barros, L.; Rodrigues, A.A.; Barreiro, M.F.; Ferreira, I.C.F.R. Optimization of microwave-assisted extraction of ergosterol from *Agaricus bisporus* L. by-products using response surface methodology. *Food Bioprod. Process.* **2016**, *100*, 25–35. [CrossRef]

32. Pinela, J.; Prieto, M.A.; Barros, L.; Carvalho, A.M.; Oliveira, M.B.P.P.; Saraiva, J.A.; Ferreira, I.C.F.R. Cold extraction of phenolic compounds from watercress by high hydrostatic pressure: Process modelling and optimization. *Sep. Purif. Technol.* **2018**, *192*, 501–512. [CrossRef]
33. Vieira, V.; Prieto, M.A.; Barros, L.; Coutinho, J.A.P.; Ferreira, O.; Ferreira, I.C.F.R. Optimization and comparison of maceration and microwave extraction systems for the production of phenolic compounds from *Juglans regia* L. for the valorization of walnut leaves. *Ind. Crops Prod.* **2017**, *107*, 341–352. [CrossRef]
34. Caleja, C.; Barros, L.; Prieto, M.A.; Barreiro, F.M.F.; Oliveira, M.B.P.; Ferreira, I.C.F.R. Extraction of rosmarinic acid from *Melissa officinalis* L. by heat-, microwave- and ultrasound-assisted extraction techniques: A comparative study through response surface analysis. *Sep. Purif. Technol.* **2017**, *186*, 297–308. [CrossRef]

Sample Availability: Samples of the plant and extracts are available from the authors.

molecules

MDPI

Article

Optimization of the Extraction Process to Obtain a Colorant Ingredient from Leaves of *Ocimum basilicum* var. *purpurascens*

Filipa Fernandes [1], Eliana Pereira [1], Miguel A. Prieto [1,2], Ricardo C. Calhelha [1], Ana Ćirić [3], Marina Soković [3], Jesus Simal-Gandara [2], Lillian Barros [1,*] and Isabel C. F. R. Ferreira [1,*]

1 Centro de Investigação de Montanha (CIMO), Instituto Politécnico de Bragança, Campus de Santa Apolónia, 5300-253 Bragança, Portugal; filipafernandes1994@outlook.com (F.F.); eliana@ipb.pt (E.P.); mprieto@ipb.pt (M.A.P.); calhelha@ipb.pt (R.C.C.)

2 Nutrition and Bromatology Group, Department of Analytical and Food Chemistry, Faculty of Food Science and Technology, University of Vigo—Ourense Campus, E-32004 Ourense, Spain; jsimal@uvigo.es

3 University of Belgrade, Department of Plant Physiology, Institute for Biological Research "Siniša Stanković", Bulevar Despota Stefana 142, 11000 Belgrade, Serbia; rancic@ibiss.bg.ac.rs (A.Ć.); mris@ibiss.bg.ac.rs (M.S.)

* Correspondence: lillian@ipb.pt (L.B.); iferreira@ipb.pt (I.C.F.R.F.); Tel.: +351-273-303285 (L.B.); +351-273-303219 (I.C.F.R.F.)

Academic Editor: Derek J. McPhee

Received: 22 January 2019; Accepted: 13 February 2019; Published: 14 February 2019

Abstract: Heat-Assisted Extraction (HAE) was used for the optimized production of an extract rich in anthocyanin compounds from *Ocimum basilicum* var. *purpurascens* leaves. The optimization was performed using the response surface methodology employing a central composite experimental design with five-levels for each of the assessed variables. The independent variables studied were the extraction time (t, 20–120 min), temperature (T, 25–85 °C), and solvent (S, 0–100% of ethanol, *v/v*). Anthocyanin compounds were analysed by HPLC-DAD-ESI/MS and the extraction yields were used as response variables. Theoretical models were developed for the obtained experimental data, then the models were validated by a selected number of statistical tests, and finally, those models were used in the prediction and optimization steps. The optimal HAE conditions for the extraction of anthocyanin compounds were: $t = 65.37 \pm 3.62$ min, $T = 85.00 \pm 1.17$ °C and $S = 62.50 \pm 4.24\%$, and originated 114.74 ± 0.58 TA mg/g of extract. This study highlighted the red rubin basil leaves as a promising natural matrix to extract pigmented compounds, using green solvents and reduced extraction times. The extract rich in anthocyanins also showed antimicrobial and anti-proliferative properties against four human tumor cell lines, without any toxicity on a primary porcine liver cell line.

Keywords: natural colorants; anthocyanins; *Ocimum basilicum* var. *purpurascens* leaves; red rubin basil; Heat-Assisted Extraction; extraction optimization

1. Introduction

Consumers' interest in food quality has been increasing, selecting foods with health benefits. Colour is the main organoleptic attribute in the selection and acceptance of foods [1,2]. Some vegetable matrices are composed by natural pigments, attracting much attention from the scientific community and leading to studies to characterize these compounds and explore their subsequent application, not only in the food industry as natural colorants, but also in the pharmaceutical sector, as antioxidants [3–5].

Anthocyanins are natural pigment studied worldwide; however, when these compounds are incorporated in food products, there are several intrinsic and extrinsic factors that affect and influence their stability [6]. The pH is an important parameter, because it is crucial in the determination of the anthocyanin colour, which shows a significant pigmentation variability. In aqueous medium, they are red at pH = 1–3, colourless at pH = 4–5, purple at pH = 6–7, blue at pH = 7–8, and yellow at pH = 8–9 [3,7]. Other variables to be taken into account in anthocyanins' stability are the handling and storage temperature, the chemical composition of target products (presence of enzymes, proteins, metal ions and even other flavonoids), and exposure to light and oxygen [3].

The use of anthocyanins as food colorants is approved in several countries [8] and according to the Regulation (EU) nr 1129/2011 of the Commission of 11 November 2011, their application is authorised in numerous food products and processes, such as cured cheeses and cheese products of red marbled paste, vegetables in vinegar, oil or brine (except olives), jams, jellies and marmalades, fruit-flavored breakfast cereals, fish pastes and crustaceans, pre-cooked crustaceans, and smoked fish, among other products. The acceptable daily intake (ADI) is not regulated, which means that sufficient quantity can be added to food products to achieve the desired coloration effect [9].

Anthocyanins can be found in numerous natural sources, especially in fruits, cereals, leaves, flowers, and roots, such as in the leaves of *Ocimum basilicum* var. *purpurascens* (red rubin basil) [10]. Red rubin basil belongs to the *Lamiaceae* family, being a variety of *Ocimum basilicum*, and is used not only as an ornamental plant, but also in traditional medicine [10–12].

In order to apply sustainable extraction methodologies at an industrial level, mathematical studies are performed to maximize the extraction of compounds from natural matrices [13,14]. The patterns of the response variables of the extraction method, such as processing temperature, time and solvent [13,14] can be evaluated using the response surface methodology (RSM). This technique allows to save time, reagents and reduce the operational costs, meanwhile increases the efficiency of the optimization process. Aiming to promote the applicability of natural pigments present in the *Ocimum basilicum* var. *pupurascens* leaves at an industrial level, this work optimized the HAE extraction of anthocyanin compounds, particularly cyanidin and pelargonidin derivatives using Response Surface Methodology (RSM).

2. Results

2.1. Response Criteria for the RSM Analysis

The HPLC anthocyanin profile of the red rubin basil leaves extract from experimental run number 18 is shown in Figure 1. Up to 13 anthocyanin compounds were identified (Table 1) based their chromatographic characteristics (UV-Vis, mass spectral fragmentation patterns) and literature information [15,16].

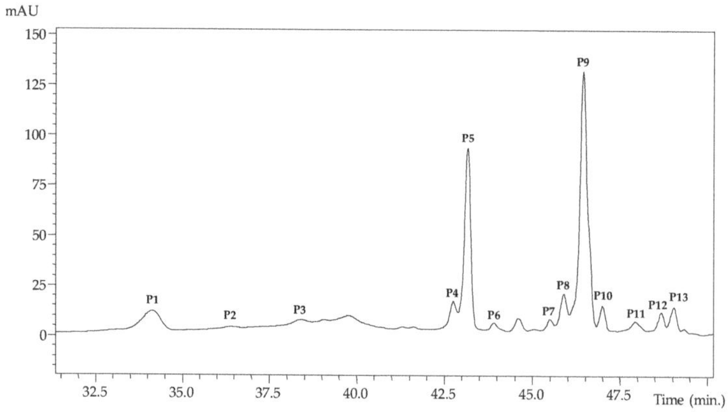

Figure 1. HPLC profile of anthocyanin molecules found in red rubin basil leaves extract obtained in the data set number 18 (as described in Table 2).

Table 1. Retention time (Rt), wavelengths of maximum absorption in the UV-Vis region (λ_{max}), and tentative identification of anthocyanin compounds in *O. basilicum* var. *purpurascens* (mean ± SD).

Peak	Rt (min)	λ_{max} (nm)	$[M + H]^+$	Main Fragment ESI-MSn [Intensity (%)]	Tentative Identification
P1	34.1	520	919	757(49),449(6),287(13)	Cyanidin-3-(*p*-coumaroyl-6′-caffeoyl)sophoroside isomer 1 [A]
P2	36.4	520	919	757(49),449(6),287(13)	Cyanidin-3-(*p*-coumaroyl-6′-caffeoyl)sophoroside isomer 2 [A]
P3	38.4	522	1005	757(6),535(11),287(11)	Cyanidin-3-(6-*p*-coumaroyl)sophoroside-5-(6-malonyl) glucoside [A]
P4	42.8	522	757	595(100),449(11),287(61)	Cyanidin-3-(6-*p*-coumaroyl)glucoside-5-glucoside [A]
P5	43.2	530	1081	919(15),449(6),287(6)	Cyanidin-3-(6-*p*-coumaroyl-6′-caffeoyl)-5-glucoside isomer1 [A]
P6	43.9	532	1167	919(44),757(5),287(20)	Cyanidin-3-(6-*p*-coumaroyl-6′-caffeoyl)sophoroside-5-(6-malonyl)glucoside isomer 1 [A]
P7	44.6	530	1167	919(27),757(5),287(6)	Cyanidin-3-(6 *p*-coumaroyl-6′-caffeoyl)sophoroside-5-(6-malonyl)glucoside isomer 2 [A]
P8	45.5	530	1081	919(100),449(11),287(20)	Cyanidin-3-(6-*p*-coumaroyl-6′-caffeoyl)sophoroside-5-glucoside isomer2 [A]
P9	45.9	530	1065	903(20),449(5),287(3)	Cyanidin-3-(6,6′-di *p*-coumaroyl)sophoroside-5-glucoside [A]
P10	46.4	526	1151	989(10),903(5),287(5)	Cyanidin-3-(6,6′-di *p*-coumaroyl)sophoroside-5-(6-malonyl)glucoside [A]
P11	47.0	514	1049	887(33),433(9),271(5)	Pelargonidin-3-(6,6′-di *p*-coumaroyl)sophoroside-5-glucoside [B]
P12	48.0	526	1167	1005(63),919(23), 449(8),287(13)	Cyanidin-3-(6-*p*-coumaroyl-X-malonyl-6′-caffeoyl)sophoroside-5-glucoside [A]
P13	48.7	530	1151	989(28),449(17),287(5)	Cyanidin-3-(6-*p*-coumaroyl-X-malonyl-6′-*p*-coumaroyl)sophoroside-5-glucoside [A]

Calibration curves used: [A]- cyanidin-3-*O*-glucoside (y = 97,787x − 743,469; R^2 = 0.999); [B]- pelargonidin-3-*O*-glucoside (y = 43,781x − 275,315; R^2 = 0.999).

Figure 2 shows a summary of the diverse stages used for optimization procedure, in order to recover the anthocyanin compounds from the red rubin basil leaves. The experimental values of the 28 experimental runs of the circumscribed central composite design (*CCCD*) design are presented in Table 2.

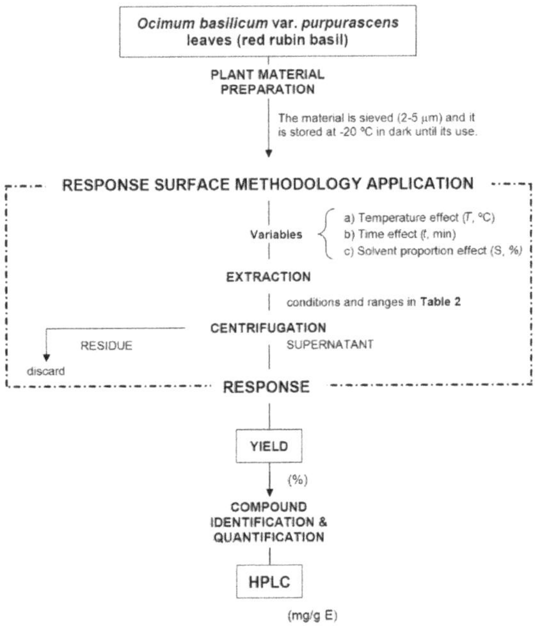

Figure 2. Diagram of the different steps carried out for optimizing the conditions that maximize the extraction responses of the anthocyanin compounds and the total extracted residue (Yield, %).

Table 2. The first part describes the experimental design that was applied in this work. The independent variables are presented in coded and natural values. The second part shows the response values for the detected anthocyanin compounds (mg/g E) and extraction yield (%) achieved for all the 28 experimental conditions performed for the HAE by the RSM design.

	Five-Level *CCCD* Experimental Design																												
	Runs	**1**	**2**	**3**	**4**	**5**	**6**	**7**	**8**	**9**	**10**	**11**	**12**	**13**	**14**	**15**	**16**	**17**	**18**	**19**	**20**	**21**	**22**	**23**	**24**	**25**	**26**	**27**	**28**
Coded values	X_1: Time (t)	−1	−1	−1	−1	1	1	1	1	1.68	−1.68	0	0	0	0	−1.68	−1.68	−1.68	−1.68	1.68	1.68	1.68	1.68	0	0	0	0	0	0
	X_2: Temp. (T)	−1	−1	1	1	−1	−1	1	1	0	0	−1.68	1.68	0	0	−1.68	−1.68	1.68	1.68	−1.68	−1.68	1.68	1.68	0	0	0	0	0	0
	X_3: Solvent (S)	−1	1	−1	1	−1	1	−1	1	0	0	0	0	−1.68	1.68	−1.68	1.68	−1.68	1.68	−1.68	1.68	−1.68	1.68	0	0	0	0	0	0
Natural values	X_1: t (min)	40.3	40.3	40.3	40.3	99.7	99.7	99.7	99.7	120.0	20.0	70.0	70.0	70.0	70.0	20.0	20.0	20.0	20.0	120.0	120.0	120.0	120.0	70.0	70.0	70.0	70.0	70.0	70.0
	X_2: T (°C)	37.2	37.2	72.8	72.8	37.2	37.2	72.8	72.8	55.0	55.0	25.0	85.0	55.0	55.0	25.0	25.0	85.0	85.0	25.0	25.0	85.0	85.0	55.0	55.0	55.0	55.0	55.0	55.0
	X_3: S (%)	20.3	79.7	20.3	79.7	20.3	79.7	20.3	79.7	50.0	50.0	50.0	50.0	0.0	100.0	0.0	100.0	0.0	100.0	0.0	100.0	0.0	100.0	50.0	50.0	50.0	50.0	50.0	50.0
	Response Variables for RSM Application																												
	P1	3.34	5.61	4.50	5.75	3.41	5.80	3.55	4.96	3.93	5.08	5.05	5.52	3.17	5.78	1.96	6.36	4.56	5.83	2.27	5.86	1.42	5.11	4.93	4.91	5.34	5.35	5.24	4.84
	P2	2.47	4.24	2.64	4.15	2.38	4.27	2.26	3.72	2.55	3.32	3.15	2.92	2.31	5.31	1.78	4.85	2.50	4.11	1.83	4.22	1.45	3.83	3.43	3.54	3.47	3.48	3.44	3.29
	P3	3.94	5.53	4.71	5.37	3.60	6.52	3.90	5.78	3.93	4.89	5.29	5.74	2.81	5.13	2.10	4.25	2.98	3.16	1.59	4.63	1.48	5.42	6.33	7.04	6.70	6.71	6.94	6.67
	P4	2.95	5.68	2.82	5.05	2.93	6.60	2.71	5.01	3.70	4.26	4.98	4.97	1.66	4.60	1.79	5.27	1.47	4.67	1.59	4.61	1.39	5.12	4.59	4.57	4.78	4.79	4.40	4.42
	P5	7.61	13.13	8.39	13.66	7.30	13.00	8.15	13.66	10.54	12.68	15.27	17.09	2.64	9.81	1.87	8.32	1.47	10.19	1.59	8.67	1.39	11.07	16.99	16.40	17.62	17.66	15.93	16.40
	P6	3.47	5.69	3.78	5.34	3.48	4.99	3.52	5.02	3.91	5.09	5.68	5.74	1.70	4.01	1.74	4.70	1.47	3.76	1.59	3.99	1.37	3.25	6.35	6.10	6.62	6.64	6.34	6.31
	P7	2.52	4.25	2.62	3.98	2.43	3.59	2.41	3.62	3.12	3.52	3.80	3.64	1.75	4.07	1.74	4.83	1.91	3.98	1.59	4.07	1.37	3.28	3.73	3.75	3.85	3.86	3.69	3.69
	P8	3.53	5.93	3.67	5.39	3.40	6.22	3.36	5.03	3.95	5.09	5.42	5.28	1.96	3.84	1.85	4.65	1.47	3.82	1.59	3.96	1.37	3.49	5.86	5.64	5.88	5.89	6.01	5.57
	P9	5.10	8.37	5.33	7.82	4.78	8.55	4.94	7.82	5.46	7.15	8.25	8.59	2.07	5.44	1.94	5.27	1.47	4.98	1.59	4.70	1.40	5.41	9.99	9.52	10.25	10.27	9.49	9.71
	P10	14.92	21.10	16.36	24.14	14.84	18.68	16.01	24.14	21.50	23.40	26.73	30.48	6.96	14.17	2.29	9.46	2.35	13.48	1.59	10.19	1.43	17.38	32.57	32.82	33.99	34.06	32.48	33.27
	P11	10.18	14.08	13.19	15.07	9.34	12.26	10.90	13.27	12.83	17.10	16.97	19.23	5.98	5.71	2.31	5.13	5.67	5.26	1.27	4.82	1.16	5.32	20.04	19.79	20.91	20.95	21.22	19.94
	P12	2.91	4.60	3.27	4.31	2.66	4.10	2.76	3.81	3.22	4.13	4.19	4.26	2.20	4.09	1.92	4.80	2.74	4.17	1.59	4.29	1.41	3.54	3.93	3.80	3.83	3.84	3.94	3.83
	P13	4.84	7.26	5.77	6.99	4.30	7.53	4.89	6.72	5.63	7.31	7.30	7.85	2.73	4.57	2.08	4.75	3.32	4.12	1.59	4.08	1.40	4.15	8.06	8.08	8.34	8.36	8.75	8.38
	TAC	67.78	105.46	77.06	107.01	64.86	102.12	69.37	102.56	84.25	103.00	112.07	121.30	37.96	76.52	25.38	72.65	33.37	71.53	21.22	68.10	18.04	76.37	116.79	115.97	111.59	111.85	117.87	116.32
	Yield	36.35	28.58	38.26	31.41	36.08	29.95	39.41	32.14	38.12	37.53	34.75	37.90	38.84	18.08	35.62	13.22	38.27	17.80	35.42	16.52	41.00	20.24	35.68	34.54	35.68	35.61	35.54	35.40

P: anthocyanin compound; TAC: Total anthocyanin content.

The content in individual (**P1** to **P13**) and grouped (TAC – total anthocyanin compounds) anthocyanin compounds were used as criteria to maximize their content and to optimize the extraction conditions of HAE from red rubin basil leaves under RSM assessment. The values of the extraction yield were also considered, and ranged from 13.22 to 41.00%, with the experimental runs no. 16 and 21, respectively (Table 2). In total, 15 response variables are taking into account for the optimization processes.

2.2. Theoretical Response Surface Models

Evaluating the precision of theoretical models to predict and comprehended the effects of independent variables in some response variable is necessary. This, as in many research fields, is achieved by fitting these models to the experimental values. In this study, a non-linear algorithm (least-squares estimates) has been used to adjust the response values (Table 2) to a second order polynomial model. The estimated coefficient values obtained from the polynomial model of Equation (1) and the coefficient of correlation (R^2) for each parametric response of the extraction process are shown in Table 3.

$$y = b_0 + \sum_{i=1}^{n} b_i X_i + \sum_{\substack{i=1 \\ j>i}}^{n-1} \sum_{j=2}^{n} b_{ij} X_i X_j + \sum_{i=1}^{n} b_{ii} X_i^2 \quad (1)$$

The parametric values obtained, not only it allows to translate response patterns, it also helps to undestand the complexity of the possible interactions between variables. However, some of the parameters of Equation (1) whose coefficients were non-significant (*ns*) at a 95% confidence level ($\alpha = 0.05$) were not used for building the model. By means of the statistic lack of fit it is possible to prove the adequacy of the obtained models and in this way it was demonstrated that a considerable improvement was not achieved by means of the inclusion of the statistically *ns* parametric values. Each of the 15 assessed responses can be seen in models in Table 4 getting in all cases R^2 coefficients higher than 0.92 (Table 3). According to this value, it can be said that the percentage of variability of each response can be explained by the model.These workable models were applied in the subsequent prediction and optimization steps, with a good agreement between the experimental and predicted values, which indicates that the variation is explained by the independent variables.

Although the obtained model coefficients (Table 3) cannot be associated with physical or chemical significance and are empirical, they can however be used to predict the results of untested extraction conditions [17]. As the effect sign marks the performance of the response, if a factor has a positive effect, the response is higher at the high level. On the other hand, the response is lower at the high level when a factor has a negative effect. Therefore, the weight of the corresponding variable will be more important the higher the absolute value of a coefficient. Certain characteristics relating to the general effects of the variables based on mathematical expressions can be observed in Table 4. The relevance of the significant parametric values can be order as a function of the variables involved in a decreasing form as $S > t >> T$. Previous authors that work with similar matrices [14], have concluded that the most relevant variable on the HAE extraction of bioactive compounds is S. As for the study of the linear, quadratic, and interactive parametric effects of the developed equations, it allowed to conclude that all these parameters play an important and significant role in all evaluated responses. For the linear effect, the variables S and t had strong values, while the effect of T was less important in almost all cases. All independent variables had moderate quadratic or nonlinear effects. As for the interactions of the variable (tT, TS and tS), these were of minor importance. The results obtained were represented in the response surface plots that can be seen below so that in this way one can see in a more obvious way the combined effects as well as to be able to visually describe the tendencies of extraction. The optimal HAE conditions, that maximize their retrieval from red rubin basil leaves, are presented in Table 3.

Table 3. Estimated coefficients and R^2 determined for the models obtained for individual and grouped anthocyanin compounds and extraction yield (Table 3), and optimal HAE conditions and response values.

Response variables	Fitting Coefficients Obtained after Applying the Second-Order Polynomial Equation with Interactive Terms											Optimal Processing Conditions and Response Values			
	Intercept	Linear Effect			Quadratic Effect			Interactive Effect			R^2	t (min)	T (°C)	S (%)	*Optimum*
	b_0	b_1 (t)	b_2 (T)	b_3 (S)	b_{11} (t^2)	b_{22} (T^2)	b_{33} (S^2)	b_{12} (tT)	b_{13} (tS)	b_{23} (TS)					
P1	5.06 ± 0.15	−0.28 ± 0.09	ns	0.92 ± 0.09	−0.16 ± 0.11	ns	−0.17 ± 0.11	−0.17 ± 0.06	0.07 ± 0.06	−0.15 ± 0.06	0.9441	81.06 ± 2.08	20.00 ± 1.73	100.00 ± 1.58	6.56 ± 0.31
P2	3.37 ± 0.10	−0.15 ± 0.06	ns	0.76 ± 0.06	−0.20 ± 0.07	ns	0.11 ± 0.07	−0.04 ± 0.00	ns	−0.07 ± 0.04	0.9556	64.04 ± 5.07	20.00 ± 0.43	100.00 ± 9.11	5.15 ± 0.36
P3	31.23 ± 1.68	ns	0.96 ± 0.95	ns	−2.14 ± 1.15	ns	−5.40 ± 1.15	ns	0.98 ± 0.68	ns	0.9359	70.00 ± 3.94	90.00 ± 6.07	50.00 ± 3.80	32.85 ± 2.47
P4	4.65 ± 0.19	ns	−0.20 ± 0.11	1.06 ± 0.11	−0.11 ± 0.01	ns	−0.41 ± 0.13	ns	ns	ns	0.9225	70.00 ± 5.49	20.00 ± 1.94	88.44 ± 5.39	5.67 ± 1.05
P5	15.59 ± 0.62	ns	0.33 ± 0.22	2.42 ± 0.35	−0.95 ± 0.42	ns	−2.85 ± 0.42	ns	ns	ns	0.9449	70.00 ± 1.26	90.00 ± 5.25	62.60 ± 5.86	16.66 ± 1.76
P6	5.90 ± 0.27	−0.16 ± 0.15	−0.10 ± 0.01	0.74 ± 0.15	−0.31 ± 0.18	ns	−0.89 ± 0.18	ns	ns	ns	0.9336	62.47 ± 1.24	20.00 ± 0.31	62.27 ± 3.93	6.24 ± 0.56
P7	3.63 ± 0.08	−0.16 ± 0.05	−0.09 ± 0.05	0.70 ± 0.05	−0.08 ± 0.06	ns	−0.23 ± 0.06	ns	ns	−0.07 ± 0.03	0.9701	40.93 ± 1.60	20.00 ± 1.49	100.00 ± 6.12	4.59 ± 0.34
P8	5.59 ± 0.18	−0.13 ± 0.10	−0.14 ± 0.10	0.77 ± 0.10	−0.23 ± 0.12	ns	−0.80 ± 0.12	ns	ns	ns	0.9456	61.55 ± 4.64	20.00 ± 1.03	64.30 ± 1.89	6.03 ± 0.55
P9	9.06 ± 0.43	ns	−0.12 ± 0.02	1.15 ± 0.24	−0.62 ± 0.30	ns	−1.52 ± 0.30	ns	ns	ns	0.9286	70.00 ± 1.16	20.00 ± 1.24	61.18 ± 2.85	9.47 ± 1.32
P10	29.87 ± 1.32	ns	0.85 ± 0.74	2.89 ± 0.74	−2.07 ± 0.91	ns	−6.27 ± 0.91	ns	ns	ns	0.9377	70.00 ± 1.42	90.00 ± 2.46	56.85 ± 0.94	31.63 ± 2.42
P11	8.93 ± 0.25	−0.32 ± 0.14	−0.12 ± 0.14	0.63 ± 0.14	−0.47 ± 0.17	ns	−1.64 ± 0.17	ns	ns	−0.10 ± 0.01	0.9359	59.99 ± 5.19	20.00 ± 0.59	57.18 ± 0.16	9.28 ± 0.38
P12	3.91 ± 0.06	−0.22 ± 0.03	−0.03 ± 0.03	0.65 ± 0.03	−0.06 ± 0.04	ns	−0.25 ± 0.04	−0.05 ± 0.02	ns	−0.09 ± 0.02	0.9757	36.49 ± 0.30	20.00 ± 1.21	98.54 ± 9.29	4.71 ± 0.20
P13	7.93 ± 0.20	−0.26 ± 0.11	−0.16 ± 0.11	0.73 ± 0.11	−0.37 ± 0.03	ns	−1.37 ± 0.13	ns	0.09 ± 0.08	−0.10 ± 0.01	0.9451	60.74 ± 0.00	20.00 ± 1.30	59.46 ± 0.04	8.38 ± 0.29
TAC	109.78 ± 2.73	−1.93 ± 1.54	1.07 ± 0.32	14.30 ± 1.54	−6.20 ± 1.87	ns	−17.00 ± 1.87	ns	ns	ns	0.9577	65.37 ± 3.62	90.00 ± 1.17	62.50 ± 4.24	114.74 ± 0.58
Yield	36.43 ± 1.46	0.49 ± 0.88	1.19 ± 0.87	−5.56 ± 0.87	ns	ns	−3.09 ± 0.84	ns	ns	ns	0.9592	120.00 ± 2.62	90.00 ± 7.72	23.23 ± 0.91	41.77 ± 1.59

ns: non-significant coefficient; R^2: Correlation coefficient; **P**: anthocyanin compound; TAC: total anthocyanin content.

Table 4. Mathematical models produced after fitting Equation (1) to the data set (individual and grouped values).

Anthocyanin Compounds	Equations	Equation Numbers
P1	$Y_{P1} = 5.06 - 0.28t + 0.92S - 0.16t^2 - 0.17S^2 - 0.17tT + 0.07tS - 0.15TS$	Equation (2)
P2	$Y_{P2} = 3.37 - 0.15t + 0.76S - 0.20t^2 - 0.11S^2 - 0.04tT - 0.07TS$	Equation (3)
P3	$Y_{P3} = 31.23 + 0.96T - 2.14t^2 - 5.40S^2 + 0.98tS$	Equation (4)
P4	$Y_{P4} = 4.65 - 0.20T + 1.06S - 0.11t^2 - 0.41S^2$	Equation (5)
P5	$Y_{P5} = 15.59 + 0.33T + 2.42S - 0.95t^2 - 2.85S^2$	Equation (6)
P6	$Y_{P6} = 5.90 - 0.16t - 0.10T + 0.74S - 0.31t^2 - 0.89S^2$	Equation (7)
P7	$Y_{P7} = 3.63 - 0.16t - 0.09T + 0.70S - 0.08t^2 - 0.23S^2 - 0,07TS$	Equation (8)
P8	$Y_{P8} = 5.59 - 0.13t - 0.14 + 0.77S - 0.23t^2 - 0.80S^2$	Equation (9)
P9	$Y_{P9} = 9.06 - 0.12T + 1.15S - 0.62t^2 - 1.52S^2$	Equation (10)
P10	$Y_{P10} = 29.87 + 0.85T + 2.89S - 2.07t^2 - 6.27S^2$	Equation (11)
P11	$Y_{P11} = 8.93 - 0.32t - 0.12 + 0.63S - 0.47t^2 - 1.64S^2 - 0,10TS$	Equation (12)
P12	$Y_{P12} = 3.91 - 0.22t - 0.03T + 0.65S - 0.06t^2 - 0.25S^2 - 0,05tT - 0,09TS$	Equation (13)
P13	$Y_{P13} = 7.93 - 0.26t - 0.16T + 0.73S - 0.37t^2 - 1.37S^2 + 0.09tS - 0.10TS$	Equation (14)
TAC	$Y_{TAC} = 109.78 - 1.93t + 1.07T + 14.30S - 6.20t^2 - 17.00S^2$	Equation (15)
Yield	$Y_{Yield} = 36.43 + 0.49t + 1.19T - 5.56S - 3.09S^2$	Equation (16)

2.3. Final Effects of the Studied Conditions of HAE on the Target Responses and Optimal Values that Maximize the Responses

Figure 3 shows the response surface plots of extraction yield, *TAC* and two other representative anthocyanins extracted (**P1** and **P10**), as well as their statistical analysis. Inspecting the given surface plots of the extraction yield (Figure 3), it is conceivable to confirm that the measure of removed material increments to an ideal point and afterward, by and large, it diminishes as a component of the included variables. Subsequently, the ideal values can be found similar to a solitary point, which permits figuring the extraction conditions that lead to the most extreme flat out. This behaviour is common to almost all responses, allowing us to determine the conditions that maximize the responses. In consequence, the ideal extraction values for the reactions shown in Figure 3 were determined for the HAE conditions (Table 3), as summarized below:

For yield, the optimal HAE conditions were: t = 120.00 ± 2.62 min, T = 85.00 ± 7.72 °C and 23.23 ± 0.91% of ethanol (v/v), and produced 41.77 ± 1.59%.

For TAC, the optimal HAE conditions were: t = 65.37 ± 3.62 min, T = 85.00 ± 1.17 °C and 62.50 ± 4.24% of ethanol (v/v), and produced 114.74 ± 0.58 mg/g of E.

For **P1**, the optimal HAE conditions were: t = 81.06 ± 2.08 min, T = 25.00 ± 1.73 °C and 100.00 ± 1.58% of ethanol (v/v), and produced 6.56 ± 0.31 mg/g of E.

For **P10**, the optimal HAE conditions were: t = 70.00 ± 1.42 min, T = 85.00 ± 2.46 °C and 56.85 ± 0.94% of ethanol (v/v), and produced 31.63 ± 2.42 mg/g of E.

It is well-known that the utilization of high values of ethanol in the solvent, increases the extraction of bioactive compounds from plant materials [13]. The effects of the independent variables on the extraction of individual anthocyanin compounds from red rubin basil leaves are represented in 2D in Figure 4. The processing conditions that generated optimal response values (⊙) are numerically described in Table 3. The identified anthocyanin compounds were organized as a function of the maximum amount achieved (mg/g of extract) in a decreasing order as follows: **P3** (32.85) > **P10** (31.63) >> **P5** (16.66) >> **P9** (9.47) > **P11** (9.28) > **P13** (8.38) > **P1** (6.56) > **P6** (6.24) > **P8** (6.03) > **P4** (5.67) > **P2** (5.15) > **P7** (4.59) > **P2** (4.71).

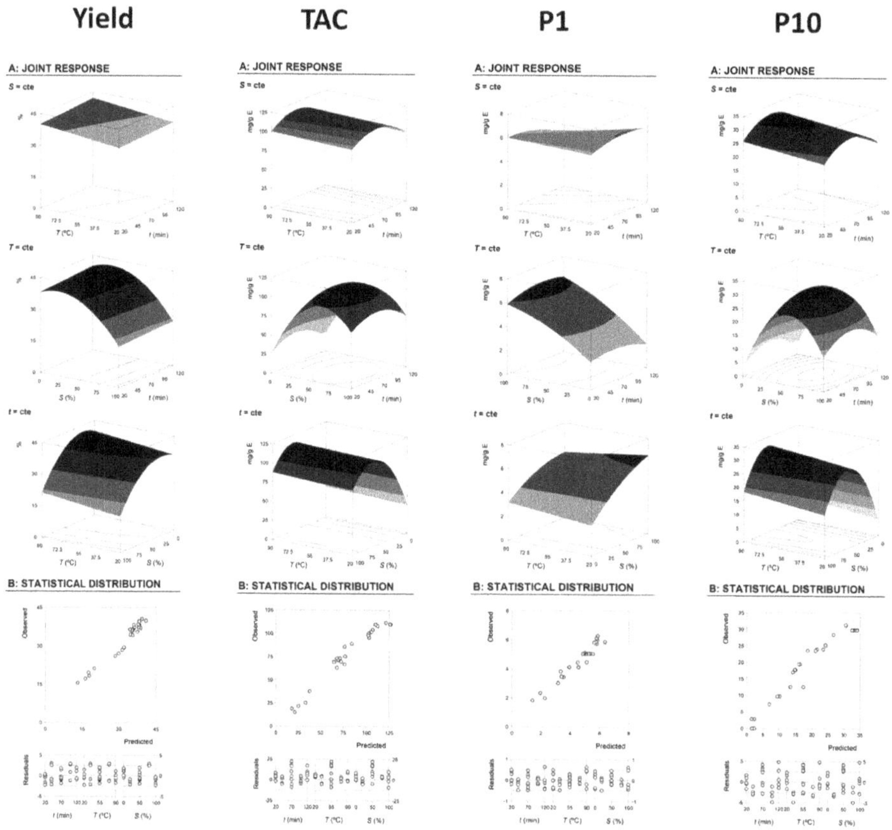

Figure 3. Illustrative representation of the extraction yield and grouped anthocyanin compounds (total anthocyanin acids, total flavonoids and total anthocyanin compounds) responses. The part A shows the 3D description as a function of each independent variable. The surfaces were constructed using the values presented in Table 3 and described by Equation (1). In each graph, the excluded variable was positioned at the optimum of their experimental domain (Table 3). Part B shows a summary of the goodness of fit using the observed/predicted and the residual distribution plots as a function of each variable.

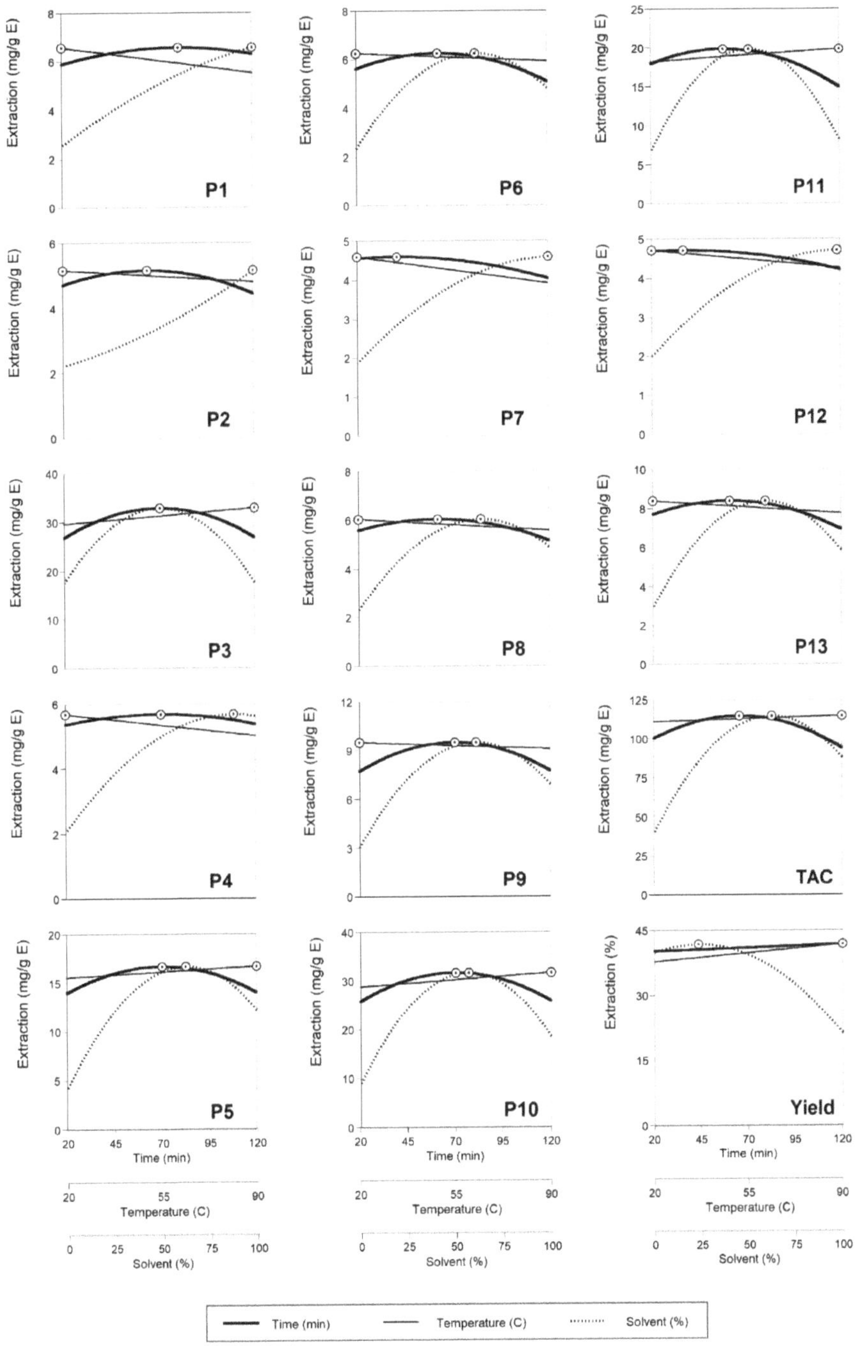

Figure 4. 2D graphical response of the effects of the independent variables on the extraction of anthocyanin compounds from red rubin basil leaves (see Figure 1 for peak identification). Dots (⊙) represent the optimal values. In each plot, each independent variable was positioned at the optimal value of the other two variables (Table 3).

The greater extraction values achieved under these optimized conditions highlight the suitability of HAE with RSM as an innovative process to recover a greater amount of anthocyanin compounds from red rubin basil leaves using shorter processing times and greener solvents.

2.4. Clustering of Anthocyanin Compounds According to the HAE Conditions that Maximize their Extraction

The maximum values for the response values of the different anthocyanin compounds and their concentrations if extracted under the optimal HAE conditions of the other compounds (Table 3) are presented in Table 5. The values of subparagraph (B) is the ratio of the optimum value of each compound between the maximum of the other compounds. When two compounds show values of 100%, i.e., the coefficient is 1, under the same conditions of HAE means that the optimal response value for both is in the same conditions. As example, the compounds **P1**, **P2**, **P4**, **P7** and **P12** were clustered in C1 under the same HAE conditions (Figure 5). By cons, if the coefficient is different from 1, it means that the conditions that are optimal for the extraction of a compound are not for the other (compounds **1** and **13**).

Table 5. Maximum response values of each anthocyanin compound and their values at the optimal processing conditions of the other compounds presented in Table 3.

(A) Maximum Response Values (mg/g of Extract) of the Individual Anthocyanin Compounds													
Peak:	**P1**	**P2**	**P3**	**P4**	**P5**	**P6**	**P7**	**P8**	**P9**	**P10**	**P11**	**P12**	**P13**
Optimum:	6.56	5.15	32.85	5.67	16.66	6.24	4.59	6.03	9.47	31.63	9.28	4.71	8.38
(B) Values of each Anthocyanin Compound at the Optimal Conditions of the other Compounds													
	P1	**P2**	**P3**	**P4**	**P5**	**P6**	**P7**	**P8**	**P9**	**P10**	**P11**	**P12**	**P13**
P1	1	0.99	0.77	0.96	0.81	0.84	0.95	0.85	0.83	0.79	0.81	0.94	0.82
P2	0.99	1	0.65	0.91	0.71	0.73	0.98	0.74	0.72	0.68	0.70	0.95	0.71
P3	0.45	0.42	1	0.63	0.97	0.87	0.33	0.85	0.88	0.99	0.88	0.33	0.88
P4	0.99	0.99	0.76	1	0.83	0.94	0.97	0.95	0.94	0.80	0.92	0.97	0.93
P5	0.65	0.66	0.97	0.80	1	0.93	0.61	0.93	0.93	0.99	0.92	0.61	0.93
P6	0.75	0.77	0.92	0.89	0.94	1	0.74	1.00	1.00	0.94	1.00	0.75	1.00
P7	0.97	0.99	0.76	0.97	0.80	0.90	1	0.91	0.88	0.79	0.87	1.00	0.88
P8	0.79	0.81	0.89	0.91	0.92	1.00	0.79	1	1.00	0.91	0.99	0.80	1.00
P9	0.72	0.72	0.93	0.86	0.96	1.00	0.66	0.99	1	0.95	0.99	0.66	0.99
P10	0.48	0.49	0.99	0.69	0.99	0.90	0.43	0.89	0.91	1	0.90	0.44	0.90
P11	0.61	0.63	0.94	0.80	0.93	0.99	0.61	0.99	0.99	0.94	1	0.63	1.00
P12	0.97	0.99	0.82	0.98	0.85	0.91	1.00	0.92	0.90	0.84	0.89	1	0.90
P13	0.69	0.70	0.91	0.84	0.91	1.00	0.67	1.00	1.00	0.92	1.00	0.68	1

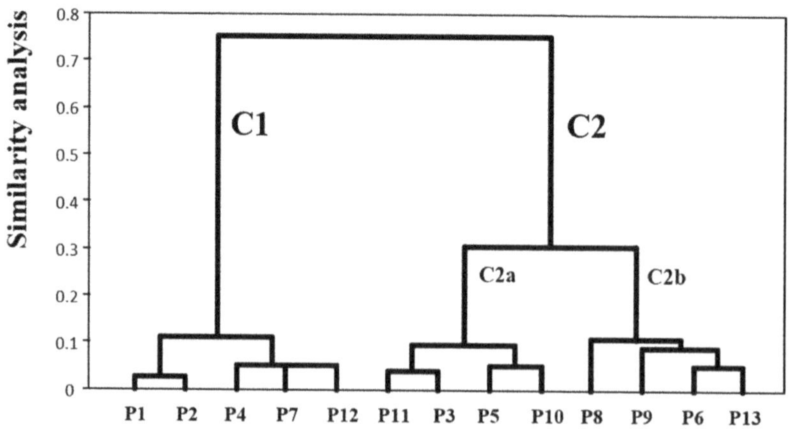

Figure 5. HCA dendrogram of anthocyanin compounds according to the HAE conditions that maximize their extraction from red rubin basil leaves.

In Table 5 it can be observed the formation of different groups of compounds of anthocyanin with maximum response values in conditions of HAE extraction similar. The division in these groups was made possible by the complete data set of Table 5 and by performing a multi objective optimization problem using an appropriate clustering algorithm. The results of Hierarchical Cluster Analysis (HCA) are presented in Figure 5. In the HCA dendrogram, the shorter distance between compounds, the higher similarity in terms of conditions that favour their extraction. Moreover, compounds belonging to the same group are better extracted under similar HAE conditions. Two significant clusters (C1 and C2), being the C2 divided in turn into 2 subgroups (a and b). Other less important subgroups were created, but they can be considered as a residual noise produced by the algorithm.

Cluster C1 included the compounds **P1**, **P2**, **P4**, **P7** and **P12**. The extraction of these compounds for maximize by medium t, high S and low/high T (Table 3 and Figure 3). The subgroups were mainly differentiated by the T values.

Cluster C2 included all other compounds **P11**, **P3**, **P5**, **P10**, **P8**, **P9**, **P6** and **P13**, which were subdivided in C2a and C2b. For maximizing the extraction of the compounds in C2a low T and medium S was used. On the other hand, the compounds in C2b was maximized when using high T and medium S.

Although it was expected that if the compounds have similar chemical characteristics also would have similar HAE conditions, the HCA analysis was an interesting and innovative approach in the field of extraction of high added-value compound from natural sources since this analysis highlighted suitable HAE conditions for maximize the simultaneous recovery of specific groups of compounds from red rubin basil leaves.

2.5. Dose-Response Analysis of the Solid-to-Liquid Effect at the Optimum Conditions

Thanks to the precise results obtained by HPLC, the S/L effect was tested under the optimal conditions provided for each extractive technique by the polynomial models, using the amount of anthocyanin as response. As confirmed by the preliminary results (data not shown), the maximum experimental value is close to 30 g/L, since at higher values of S/L it is observed experimental stirring, so an experiment was designed for each extractive process in which to check the S/L behaviour at values between 1 and 30 g/L. The obtained results are consistent with previous responses. It was observed that the effect caused by the S/L ratio follows a simple linear model with an intercept, and that this model follows a slightly decreasing pattern proportional to the increase of S/L in all the assays. However, that pattern, explained by the parametric coefficient of the slope, was non-significant with a confidence interval level of 95 % ($\alpha = 0.05$) and the decreasing effect was not taken into account for further analysis. In conclusion, it can be affirmed that the increase in the S/L ratio has very little effect on the *TAC* extraction, besides that saturation effects were not observed at any value below 30 g/L.

2.6. Evaluation of the Colorant Potential of the Extract Rich in Anthocyanin Compounds Obtained under Optimum Conditions from Leaves of O. basilicum var. purpurascens

The results of the chromatic analysis in the CIE $L^*a^*b^*$ colour space of the extract rich in anthocyanins present in the leaves of *O. basilicum* var. *purpurascens* are shown in Table 6. The colour of the pigmented extract showed an L^* value, lightness (0 to 100), of 20.5 ± 0.5; and in parameters a^* (colour intensity from green to red (−120 to 120)) and b^* (colour is evaluated at the intensity level from blue to yellow (−120 to 120)), the values were 33.0 ± 0.1 and 8.2 ± 0.4, respectively.

For a better understanding of the colour values, these were converted to RGB values and the colour obtained from the extract, red-berry, can be visualized. These results can be justified by the presence of anthocyanin compounds in the extract, which, in addition to having darker shades, are also characterized by blue, red and purple tones. The concentration of total anthocyanin compounds, obtained in the optimized extract, was similar to that predicted by the model.

Table 6. Amount of anthocyanins (cyanidin and pelargonidin derivatives) and color parameters under optimal conditions (mean ± SD).

Quantification (mg/g E)	L^*	a^*	b^*	Conversion Color to RGB Values
115.4 ± 0.4	20.5 ± 0.5	33.0 ± 0.1	8.2 ± 0.4	

L^* lightness; a^* chromatic axis from green (−) to red (+); b^* chromatic axis from blue (−) to yellow (+).

2.7. Evaluation of the Bioactive Properties of the Extract Rich in Anthocyanin Compounds Obtained under Optimal Conditions from Leaves of *O. basilicum* var. *purpurascens*

2.7.1. Antimicrobial Activity

Table 7 shows the results of the antimicrobial activity obtained from the extract rich in anthocyanins present in the leaves of *O. basilicum* var. *purpurascens*. The results demonstrate antibacterial activity of the pigmented extract for all microorganisms' strains. In this way, the best results are obtained against *Bacillus cereus* (*B.c.*) (MIC = 0.037 mg/mL; MBC = 0.075 mg/mL) and *Escherichia coli* (*E.c.*) (MIC = 0.037 mg/mL; MBC = 0.075 mg/mL) strains. However, the pigmented extract also showed a high activity against *Listeria monocytogenes* (*L.m.*) (MIC = 0.05 mg/mL; MBC = 0.075 mg/mL), *Staphylococcus aureus* (*S.a.*), *Enterobacter cloacae* (*En.cl.*) (MIC = 0.075 mg/mL, MBC = 0.15 mg/mL), and *Salmonella typhimirium* (*S.t.*) (MIC = 0.15 mg/mL; MBC= 0.30 mg/mL).

Table 7. Antibacterial activity (MIC and MBC, mg/mL) and antifungal activity (MIC and MFC, mg/mL) of the anthocyanins rich extract obtained under optimal extraction conditions.

		Antibacterial Activity					
		B.c.	***S.a.***	***L.m.***	***E.c.***	***En.cl.***	***S.t.***
Anthocyanins rich extract	MIC	0.037	0.075	0.05	0.037	0.075	0.15
	MBC	0.075	0.15	0.075	0.075	0.15	0.30
Streptomycin [(1)]	MIC	0.10	0.04	0.20	0.20	0.20	0.20
	MBC	0.20	0.10	0.30	0.30	0.30	0.30
Ampicillin [(1)]	MIC	0.25	0.25	0.40	0.40	0.25	0.75
	MBC	0.40	0.45	0.50	0.50	0.50	1.20
		Antifungal Activity					
		A.fun.	***A.o.***	***A.n.***	***P.f.***	***P.o***	***P.v.c.***
Anthocyanins rich extract	MIC	0.037	0.002	0.075	0.075	0.30	0.30
	MFC	0.075	0.075	0.15	0.15	0.45	0.45
Ketoconazole [(1)]	MIC	0.25	0.20	0.20	0.20	2.50	0.20
	MFC	0.50	0.50	0.50	0.50	3.50	0.30
Bifonazole [(1)]	MIC	0.15	0.10	0.15	0.20	0.20	0.10
	MFC	0.20	0.20	0.20	0.25	0.25	0.20

[(1)] Positive controls. *B.c.*: *Bacillus cereus*; *S.a.*: *Staphylococcus aureus*; *L.m.*: *Listeria monocytogenes*; *E.c.*: *Escherichia coli*; *En.cl.*: *Enterobacter cloacae*; *S.t.*: *Salmonella typhimurium*; *A.fum.*: *Aspergillus fumigatus*; *A.o.*: *Aspergillus ochraceus*; *A.n.*: *Aspergillus niger*; *P.f.*: *Penicillium funiculosum*; *P.o.*: *Penicillium ochrochloron*; *P.v.c.*: *Penicillium verrucosum* var. *cyclopium*. MIC—minimum inhibitory concentration; MBC—minimum bactericidal concentration; MFC—minimum fungicidal concentration.

Regarding antifungal activity, the extract showed a high potential against most of the tested fungi. *Aspergillus ochraceus* (*A.o.*) was the most susceptible species to the extract (MIC = 0.002 mg/mL; MFC = 0.075 mg/mL); however, no antifungal activity was observed against *Penicillium verrucosum* var. *cyclopium* (*P.v.c.*) (MIC = 0.30 mg/mL; MFC = 0.45 mg/mL). These results indicated a promising antimicrobial activity, and this can be explained due to the high concentration of anthocyanin compounds that have a high antimicrobial potential [18].

2.7.2. Cytotoxic Activity

Table 8 shows the results obtained in the cytotoxicity evaluation assays in extracts rich in anthocyanin compounds, obtained through optimal extraction conditions. The extract exhibited anti-proliferative capacity in HeLa (GI_{50} = 213 ± 9 μg/mL) and HepG2 (GI_{50} = 198 ± 9 μg/mL) tumour cell lines.

Table 8. Cytotoxic activity of the anthocyanins rich extract obtained under optimal extraction conditions (mean ± SD).

Tumor Cell Lines	Concentrations (GI_{50} Values, μg/mL)
MCF-7 (breast carcinoma)	>400
NCI-H460 (lung carcinoma)	>400
HeLa (cervical carcinoma)	213 ± 9
HepG2 (hepatocellular carcinoma)	198 ± 9
Non-Tumour Cells	
PLP2 (non-tumor porcine liver primary cells)	>400

GI_{50} values - concentration that inhibited 50% of cell growth. Ellipticin GI_{50} (positive control): 1.21 μg/mL (MCF-7), 1.03 μg/mL (NCI-H460), 0.91 μg/mL (HeLa), 1.10 μg/mL HepG2) and 2.29 μg/mL (PLP2).

These results may also be explained by the high levels of anthocyanin compounds present in the extract, since these molecules have been described, by several authors, as a potential anti-proliferative agent in tumor cell lines [19]. Regarding the assay performed on primary non-tumor cell culture (PLP2), the extract evidenced the absence of toxicity up to the maximal tested concentration (GI_{50} > 400 μg/mL).

3. Materials and Methods

3.1. Samples

Ocimum basilicum var. *purpurascens* (Lamiaceae) variety was obtained in Cantinho das Aromáticas, Vila Nova de Gaia, Portugal. The samples acquired were planted to grow in greenhouse at the Polytechnic Institute of Bragança and then collected (September 2017). The fresh leaves were separated through a mechanical procedure, posteriorly lyophilized (FreeZone 4.5, Labconco, Kansas City, MO, USA), reduced to a fine and homogeneous dried powder (~20 mesh) and stored protected from light and heat.

3.2. Heat-Assisted Extraction

Heat-Assisted Extraction (HAE) was performed in a water reactor agitated internally with a Cimarec™ Magnetic Stirrer at a constant speed (~500 *rpm*, Thermo Scientific, San Jose, CA, USA), following a procedure previously performed by Roriz et al. [20]. The powdered samples (300 mg) were extracted with solvent (20 mL of ethanol/water) under diverse conditions, as previously defined by the established RSM plan (Table 2). The ranges of the experimental design were: time (t or X_1, 20 to 120 min), temperature (T or X_2, 25 to 85 °C) and ethanol content (S or X_3, 0 to 100%). The solid-to-liquid ratio (S/L) was kept at 15 g/L for all conditions.

When all the individual extraction conditions were carried out, the samples were immediately centrifuged (4750× g during 20 min at 10 °C) and filtered (paper filter Whatman n° 4) to eliminate the non-dissolved material. The supernatant was collected and divided in two portions for HPLC and extraction yield analysis. The portion separated for HPLC analysis (2 mL) was filtered through a LC filter disk (0.22 μm), whereas the portion for the extraction yield determination (5 mL) was dried at 105 °C during 48 h and thereafter weighted.

3.3. Calculation of the Extraction Yield

The extraction yields (%) were calculated based on the dry weight (crude extract) obtained after evaporation of the solvent. In all cases, the filtrates were concentrated at 35 °C in a rotary evaporator (Büchi R-210, Flawil, Switzerland) under reduced pressure and the aqueous phase was then lyophilised to obtain a dried extract.

3.4. Chromatographic Analysis of Anthocyanin Compounds

The samples were analysed using Dionex Ultimate 3000 UPLC (Thermo Scientific, San Jose, CA, USA) coupled to a diode de array detector (chromatograms recorded at 520 nm) and to a Linear Ion Trap LTQ XL mass spectrometer (Thermo Finnigan, San Jose, CA, USA) equipped with an ESI source working in positive mode, following a procedure previously reported [21]. Quantitative analysis was performed using a calibration curve obtained using cyanidin-3-glucoside (y = 97,787x − 743,469; R^2 = 0.9993) and pelargonidin-3-glucoside (y = 43,781x − 275,315; R^2 = 0.9989) and results were expressed in mg per g of extract (mg/g E).

3.5. Experimental Design, Modelling and Optimization

3.5.1. Experimental Design

A RSM of five-level *CCCD* of 28 runs with 6 replicated values at centre points was applied to optimize the HAE conditions for the extraction of anthocyanin compounds. Coded and natural values of the independent variables X_1 (processing time (t), min), X_2 (temperature (T), °C) and X_3 (solvent (S), % of ethanol, v/v) are presented in Table 1.

3.5.2. Mathematical Modelling

The response surface models were fitted by means of least-squares calculation using the following second-order polynomial equation with interactive terms (Equation (1)). In this equation, Y represents the dependent variable (response variable) to be modelled, X_i and X_j are the independent variables, b_0 is the constant coefficient, bi is the coefficient of linear effect, b_{ij} is the coefficient of interaction effect, b_{ii} is the coefficient of quadratic effect, and n is the number of variables. The extraction yield and the individual and grouped anthocyanin compounds, 13 individual compounds plus the total anthocyanin content (*TAC*), were used as dependent variables.

3.5.3. Maximization of the Responses

For the extraction yield and the recovery of phenolic compounds responses, a *simplex* method was used for maximize the models developed of Equation (1) [22]. In all cases, restrictions were added to limit the values of the conditions assessed.

3.6. Gropping the Responses by Cluster Analyses

A cluster analysis was performed to group the anthocyanin compounds according to the extraction conditions that maximize their response values using the Excel add-in "XLSTAT 2016" (Addinsoft, Barcelana, Spain). A comparative agglomerative hierarchical clustering analysis (HCA) with automatic truncation based on entropy and Pearson correlation coefficient were used for clustering (similarity analysis).

3.7. Fitting Procedures and Statistical Analysis

Fitting procedures, coefficient estimates and statistical calculations were performed as previously described by Prieto and Vázquez [23]. In brief: (a) fitting procedure by nonlinear least-square (quasi-Newton) as provided by the Excel add-in "Solver"; (b) coefficient intervals determination by the Excel add-in "SolverAid"; and (c) the model consistency by common statistical tests for each model developed: (i) the Fisher F-test (α = 0.05); (ii) parametric assessment by the Excel add-in ";SolverStat"; (iii) the determination of R^2.

3.8. Preparation of the Extract Rich in Anthocyanin Compounds Obtained under Optimum Conditions from the Leaves of O. basilicum var. purpurascens

For the preparation of an extract rich in anthocyanin compounds, extraction from the leaves of *O. basilicum* var. *purpurascens* was performed, following the previously optimized procedure (Table 1). The samples (300 mg) were placed together ethanol/water (20 mL, 55:45, *v/v*) acidified with 0.25% citric acid (pH = 3) in a glass vial with a stopper. The extraction followed established conditions of temperature (T = 72 °C) and time (60 min). After the procedure described, the sample was centrifuged (Centurion K24OR, West Sussex, UK) at 5000 rpm for 5 min at 10 °C. They were then filtered through filter paper (Whatman n° 4) to remove suspended solids. The ethanol fraction was removed at a temperature of 35 °C and the aqueous fraction obtained was frozen and lyophilized (FreeZone 4.5), affording an extract rich in anthocyanin compounds. The lyophilized extract was stored away from the light for further analysis.

3.9. Evaluation of the Colorant Potential of the Extract Rich in Anthocyanin Compounds Obtained under Optimum Conditions from the Leaves of O. basilicum var. purpurascens

The evaluation of the colorant potential of the extract was carried out by measuring the colour and the measurement of the colouring compounds by chromatography, in order to corroborate the data provided by the MRS. The colour was measured using a colorimeter (model CR-400, Konica Minolta Sensing, Inc., Osaka, Japan) with an adapter for granular materials (model CR-A50), according to a procedure described by Pereira et al. [24]. The measurements were made in the CIE $L^*a^*b^*$ colour space, using the illuminant C and a diaphragm aperture of 8 mm. Data were processed with the "Spectra Magic Nx" (version CM-S100W 2.03.0006 software, Konica Minolta). Quantitation of anthocyanin compounds was accomplished by chromatography using an HPLC-DAD-ESI/MS system as described in Section 3.4.

3.10. Evaluation of the Bioactive Properties of the Extract Rich in Anthocyanin Compounds Obtained under Optimal Conditions from the Leaves of O. basilicum var. purpurascens.

3.10.1. Antimicrobial Activity

The antimicrobial activity was evaluated using the methodology described by Carocho et al. [25]. Gram-negative (*Enterobacter cloacae* (American Type Culture Collection (ATCC) 35030), *Escherichia coli* (ATCC 35210) and *Salmonella typhimurium* (ATCC 13311)) and Gram-positive (*Bacillus cereus* (clinical isolate), *Listeria monocytogenes* (NCTC (National collection of type cultures) 7973) and *Staphylococcus aureus* (ATCC 6538)) bacteria strains were used. For the calculation of the minimum inhibitory (MIC) and minimum bactericidal (MBC) concentrations, the microdilution method was applied and the results were expressed in mg/mL.

For the antifungal activity, a procedure previously described by Carocho et al. [25] was followed. *Aspergillus fumigatus* (ATCC 1022), *Aspergillus niger* (ATCC 6275), *Aspergillus ochraceus* (ATCC 12066), *Penicillium funiculosum* (ATCC 36839), *Penicillium ochrochloron* (ATCC 9112) and *Penicillium verrucosum* var. *cyclopium* (food isolate) were used. Minimum inhibitory concentration (MIC) and minimum fungicidal concentration (MFC) were also determined by using the microdilution method and the results were also expressed in mg/mL.

3.10.2. Cytotoxic Activity

The evaluation of the cytotoxic potential of the extract rich in anthocyanin compounds was performed by the Sulfarodamine B (SRB) assay previously described by Barros et al. [26] MCF-7 (breast carcinoma), NCI-H460 (lung carcinoma), HeLa (cervical carcinoma) and HepG2 (hepatocellular carcinoma) were used as human tumor cell lines. For the hepatotoxicity assay, the extract rich in anthocyanin compounds was tested in a primary non-tumor cell culture obtained from porcine liver (PLP2).

Ellipticine (Sigma-Aldrich, St. Louis, MO, USA) was used as the positive control and the results were expressed as GI_{50} values (sample concentration that inhibits the growth of cells by 50%), and expressed in μg/mL.

4. Conclusions

Colorants are one of the most important additives in terms of marketing, because their presence in food products is considered the principal factor influencing customer choice. To the authors' best knowledge, the potential industrial use of the anthocyanin compounds from red rubin basil leaves have not been explored previously. In such a context, the present work presents a new rapid method to extract anthocyanin compounds from red rubin basil leaves. RSM and other mathematical strategies were successfully employed to optimize extraction conditions that maximize the anthocyanin recovery to produce a rich extract with potential for industrial application as a natural colouring additive.

The scientific literature shows clear evidence that extraction procedures of target compounds from plant-based products, must be assessed individually. Therefore, a nonstop effort needs to be performed, because agro-industrial and food sectors are looking for byproduct valorisation into added-value products. However, in order to take full advantage of the technological advances, the extraction conditions need to be optimized. Mathematical solutions, such as RSM tools, could increase the efficiency and profitability of the process and help to change conventional extraction approaches.

In this study, the suitability of HAE for extracting anthocyanin compounds from red rubin basil leaves was demonstrated and the variables of t, T and S were combined in a five-level *CCCD* design coupled to RSM for optimization. According to the results, a good agreement between experimental and theoretical results was observed. In general, the recovery of anthocyanin compounds was maximized when high temperatures, high ethanol concentrations and medium extraction times were applied, validating this Heat-Assisted Extraction.

The colour analysis in the pigmented extract revealed interesting values, showing dark tones, more directed to a red tonality. It was also evident the antimicrobial and anti-proliferative potential against several strains and tumour cell lines, respectively, without presenting toxicity for non-tumor cells.

These results should promote interest in conducting further studies on *O. basilicum* varieties, highlighting the potential of ruby red basil as a potential source of natural and bioactive ingredients with application in several industrial factors, namely in the food and pharmaceutical areas.

Author Contributions: Conceptualization, L.B. and I.C.F.R.F.; Methodology, F.F., E.P., M.A.P., R.C.C., A.Ć., M.S., L.B. and I.C.F.R.F.; Writing—original draft, F.F., E.P., M.A P., M.S., J.S.-G., L.B. and I.C.F.R.F.

Funding: The authors are grateful to the Foundation for Science and Technology (FCT, Portugal) and FEDER under Program PT2020 for financial support to CIMO (UID/AGR/00690/2013), Lillian Barros and Ricardo C. Calhelha contracts. The authors are also grateful to the Interreg España-Portugal for financial support through the project 0377_Iberphenol_6_E). This work is funded by the European Regional Development Fund (ERDF) through the Regional Operational Program North 2020, within the scope of Project Mobilizador Norte-01-0247-FEDER-024479: ValorNatural®. Authors are also grateful to Ministry of Education, Science and Technological Development, Republic of Serbia, grant No. 173032. The authors thank the GAIN (Xunta de Galicia) for financial support (P.P. 0000 421S 140.08) to Miguel A. Prieto by a post-doctoral (modality B) grant.

Conflicts of Interest: The authors declare they have no conflict of interest.

References

1. Hoefkens, C.; Verbeke, W. Consumers' health-related motive orientations and reactions to claims about dietary calcium. *Nutrients* **2013**, *5*, 82–96. [CrossRef] [PubMed]
2. Martins, N.; Roriz, C.L.; Morales, P.; Barros, L.; Ferreira, I.C.F.R. Food colorants: Challenges, opportunities and current desires of agro-industries to ensure consumer expectations and regulatory practices. *Trends Food Sci. Technol.* **2016**, *52*, 1–15. [CrossRef]

3. Rodriguez-Amaya, D.B. Natural food pigments and colorants. *Curr. Opin. Food Sci.* **2016**, *7*, 20–26. [CrossRef]
4. Neri-Numa, I.A.; Pessoa, M.G.; Paulino, B.N.; Pastore, G.M. Genipin: A natural blue pigment for food and health purposes. *Trends Food Sci. Technol.* **2017**, *67*, 271–279. [CrossRef]
5. Almeida, H.H.S.; Barros, L.; Barreira, J.C.M.; Calhelha, R.C.; Heleno, S.A.; Sayer, C.; Miranda, C.G.; Leimann, F.V.; Barreiro, M.F.; Ferreira, I.C.F.R. Bioactive evaluation and application of different formulations of the natural colorant curcumin (E100) in a hydrophilic matrix (yogurt). *Food Chem.* **2018**, *261*, 224–232. [CrossRef] [PubMed]
6. Sigurdson, G.T.; Tang, P.; Giusti, M.M. Natural Colorants: Food Colorants from Natural Sources. *Annu. Rev. Food Sci. Technol* **2017**, *8*, 261–280. [CrossRef] [PubMed]
7. Ananga, A.; Georgiev, V.; Ochieng, J.; Phills, B.; Tsolov, V. Production of Anthocyanins in Grape Cell Cultures: A Potential Source of Raw Material for Pharmaceutical, Food, and Cosmetic Industries. In *The Mediterranean Genetic Code—Grapevine and Olive*; IntechOpen: London, UK, 2013; pp. 247–287.
8. Gerardi, C.; Tommasi, N.; Albano, C.; Blando, F.; Rescio, L.; Pinthus, E.; Mita, G. *Prunus mahaleb* L. fruit extracts: a novel source for natural food pigments. *Eur Food Res. Technol.* **2015**, *241*, 683–695. [CrossRef]
9. Jornal Oficial da União Europeia. Regulamento (UE) N°1129/2011. *Eur. Food Res. Technol.* **2011**, 25.
10. Flanigan, P.M.; Niemeyer, E.D. Effect of cultivar on phenolic levels, anthocyanin composition, and antioxidant properties in purple basil (*Ocimum basilicum* L.). *Food Chem.* **2014**, *164*, 518–526. [CrossRef]
11. Da Silva, F.J.; Nascimento, A.B.; Barbosa, L.N.; Magalhães, H.M. In vitro cultivation of purple basil *Ocimum basilicum* L. 'red rubin' at different levels of salts, charcoal, sucrose and potassium iodine. *Aust. J. Crop Sci.* **2017**, *11*, 1137–1145. [CrossRef]
12. El-Ziat, R.A.; Swaefy, H.M.; Esmail, S.E.A. The Response of Red Rubin Basil Plant to Organic Fertilizer and Humic Acid versus Chemical Fertilizers. *Middle East. J. Agric. Res.* **2018**, *7*, 740–751.
13. Alexandre, E.M.C.; Araújo, P.; Duarte, M.F.; de Freitas, V.; Pintado, M.; Saraiva, J.A. High-pressure assisted extraction of bioactive compounds from industrial fermented fig by-product. *Int. J. Food Sci. Technol.* **2017**, *54*, 2519–2531. [CrossRef]
14. Alexandre, E.M.C.; Araújo, P.; Duarte, M.F.; de Freitas, V.; Pintado, M.; Saraiva, J.A. Experimental design, modeling, and optimization of high-pressure-assisted extraction of bioactive compounds from pomegranate peel. *Food Bioprocess Tech.* **2017**, *10*, 886–900. [CrossRef]
15. Luna, M.C.; Bekhradi, F.; Ferreres, F.; Jordán, M.J.; Delshad, M.; Gil, M.I. Effect of Water Stress and Storage Time on Anthocyanins and Other Phenolics of Different Genotypes of Fresh Sweet Basil. *J. Agric. Food Chem* **2015**, *63*, 9223–9231. [CrossRef] [PubMed]
16. Phippen, W.B.; Simon, J.E. Anthocyanins in Basil (*Ocimum basilicum* L.). *J. Agric. Food Chem.* **1998**, *46*, 1734–1738. [CrossRef]
17. Ranic, M.; Nikolic, M.; Pavlovic, M.; Buntic, A.; Siler-Marinkovic, S.; Dimitrijevic-Brankovic, S. Optimization of microwave-assisted extraction of natural antioxidants from spent espresso coffee grounds by response surface methodology. *J. Clean. Prod.* **2014**, *80*, 69–79. [CrossRef]
18. Sun, X.-H.; Zhou, T.-T.; Wei, C.-H.; Lan, W.-Q.; Zhao, Y.; Pan, Y.-J.; Wu, V.C.H. Antibacterial effect and mechanism of anthocyanin rich Chinese wild blueberry extract on various foodborne pathogens. *Food Control* **2018**, *94*, 155–161. [CrossRef]
19. Zhou, L.; Wang, H.; Yi, J.; Yang, B.; Li, M.; He, D.; Yang, W.; Zhang, Y.; Ni, H. Anti-tumor properties of anthocyanins from *Lonicera caerulea* 'Beilei' fruit on human hepatocellular carcinoma: In vitro and in vivo study. *Biomed. Pharmacother.* **2018**, *104*, 520–529. [CrossRef]
20. Roriz, C.L.; Barros, L.; Prieto, M.A.; Morales, P.; Ferreira, I.C.F.R. Floral parts of *Gomphrena globosa* L. as a novel alternative source of betacyanins: Optimization of the extraction using response surface methodology. *Food Chem.* **2017**, *229*, 223–234. [CrossRef]
21. Gonçalves, G.A.; Soares, A.A.; Correa, R.C.G.; Barros, L.; Haminiuk, C.W.I.; Peralta, R.M.; Ferreira, I.C.F.R.; Bracht, A. Merlot grape pomace hydroalcoholic extract improves the oxidative and inflammatory states of rats with adjuvant-induced arthritis. *J. Funct. Foods* **2017**, *33*, 408–418. [CrossRef]
22. Vieira, V.; Prieto, M.A.; Barros, L.; Coutinho, J.A.P.; Ferreira, O.; Ferreira, I.C.F.R. Optimization and comparison of maceration and microwave extraction systems for the production of phenolic compounds from *Juglans regia* L. for the valorization of walnut leaves. *Ind. Crops Prod.* **2017**, *107*, 341–352. [CrossRef]

23. Prieto, M.A.; Vázquez, J.A. In vitro determination of the lipophilic and hydrophilic antioxidant capacity of unroasted coffee bean extracts and their synergistic and antagonistic effects. *Food Res. Int.* **2014**, *62*, 1183–1196. [CrossRef]
24. Pereira, E.; Antonio, A.L.; Barreira, J.C.M.; Barros, L.; Bento, A.; Ferreira, I.C.F.R. Gamma irradiation as a practical alternative to preserve the chemical and bioactive wholesomeness of widely used aromatic plants. *Food Res. Int.* **2015**, *67*, 338–348. [CrossRef]
25. Carocho, M.; Barros, L.; Calhelha, R.C.; Ćirić, A.; Soković, M.; Santos-Buelga, C.; Morales, P.; Ferreira, I.C.F.R. *Melissa officinalis* L. decoctions as functional beverages: a bioactive approach and chemical characterization. *Food Funct.* **2015**, *6*, 2240–2248. [CrossRef] [PubMed]
26. Barros, L.; Pereira, E.; Calhelha, R.C.; Dueñas, M.; Carvalho, A.M.; Santos-Buelga, C.; Ferreira, I.C.F.R. Bioactivity and chemical characterization in hydrophilic and lipophilic compounds of *Chenopodium ambrosioides* L. *J. Funct. Foods* **2013**, *5*, 1732–1740. [CrossRef]

Sample Availability: Samples are available from the authors.

molecules

MDPI

Article

Stability Analysis of Anthocyanins Using Alcoholic Extracts from Black Carrot (*Daucus Carota* ssp. *Sativus* Var. *Atrorubens* Alef.)

Guillermo Espinosa-Acosta [1], Ana L. Ramos-Jacques [2], Gustavo A. Molina [1], Jose Maya-Cornejo [3], Rodrigo Esparza [3], Angel R. Hernandez-Martinez [3], Itari Sánchez-González [4] and Miriam Estevez [3,*]

1 Posgrado en Ciencia e Ingeniería de Materiales, Centro de Física Aplicada y Tecnología Avanzada (CFATA), Universidad Nacional Autónoma de México (UNAM), Blvd. Juriquilla 3000, Querétaro, Mexico; senz_ad@hotmail.com (G.E.-A.); gustavomolina21@gmail.com (G.A.M.)

2 Independent Researcher, El Canto, Zibatá 76269, Querétaro, Mexico; al.ramos.jacques@gmail.com

3 Centro de Física Aplicada y Tecnología Avanzada (CFATA), Universidad Nacional Autónoma de México (UNAM), Blvd. Juriquilla 3000, Querétaro, México; iqm_jamc@yahoo.com.mx (J.M.-C.); resparza@fata.unam.mx (R.E.); arhm@fata.unam.mx (A.R.H.-M.)

4 Licenciatura en Tecnología, Centro de Física Aplicada y Tecnología Avanzada (CFATA), Universidad Nacional Autónoma de México (UNAM), Blvd. Juriquilla 3000, Querétaro, México; itari_13@hotmail.com

* Correspondence: miries@fata.unam.mx; Tel.: +52-442-238-1157

Received: 29 August 2018; Accepted: 22 October 2018; Published: 24 October 2018

Abstract: Anthocyanins are used for food coloring due their low toxicity and health benefits. They are extracted from different sources, but black carrot has higher anthocyanin content compared with common fruits and vegetables. Here, we study alcoholic anthocyanin extracts from black carrot to enhance their stability. The objective of our research is to determine if microencapsulation with tetraethyl orthosilicate (TEOS) is a feasible option for preventing black carrot anthocyanin degradation. Extraction solvents were solutions of (1) ethanol/acetic acid and (2) ethanol/citric acid. Samples were purified through a resin column and microencapsulated using TEOS. Fourier Transformed Infrared Spectroscopy (FTIR) spectra of samples were obtained, and degradation studies were performed under different conditions of UV radiation, pH and temperature. Antioxidant activity was evaluated with radical 2,2-diphenyl-1-picrylhydrazyl (DPPH) scavenging and electrochemical cupric reducing antioxidant capacity (CUPRAC). Color evaluation on food models were performed with CIE Lab at the beginning of experiments and after 25 days of storage. Results indicate that the more stable extracts against pH media changes are samples obtained with ethanol/acetic acid solution as extraction solvent. Extract purification through resin and TEOS microencapsulation had no significant effect on extract stability. In conclusion, although TEOS microencapsulation has proven to be effective for some dried materials from natural extracts in our previous research, we do not recommend its use for black carrot extracts considering our results in this particular case.

Keywords: anthocyanin; natural extract; tetraethyl orthosilicate; black carrot; antioxidant activity

1. Introduction

The use of extracts from natural sources as food coloring is an ongoing trend because, in general, they are Generally Recognized as Safe (GRAS) substances and bring health benefits for consumers [1]. A variety of natural colorants are used in the food industry, but there is still a concern about their production costs and stability and their performance had been studied for the past decade [2–6].

Substances like carotenoid, chlorophyll, turmeric, anthocyanin, and betalain extracts from natural sources impart a variety of colors [7,8].

Our hypothesis relies on the possibility of using microencapsulation (with TEOS) to increase stability of anthocyanins. Previously, we studied betalain extracts from *Beta vulgaris* and *Myrtillocactus geometrizans* and obtained dried materials that were microencapsulated using TEOS, obtaining an improvement in the materials' stability against UV light, pH and temperature [1].

Anthocyanins (polyhydroxy and polymethoxy derivatives of 2-phenylbenzopyrylium of flavylium salts) are a group of phenolic compounds that are responsible for the colors of flowers, fruits, and vegetables [9]. It has been reported that this group of compounds has antioxidant [10], antimutagenic, anticancer and antiobesity properties, and they reduce the risk of coronary heart disease [9,11,12]. The colors imparted by anthocyanins are bright and could be used in the food industry as a replacement for colorants like FD&C Red 40 [13]. The colors obtained are commonly red, orange, blue and purple, depending on the chemical structure of anthocyanins; but, structural transformations are induced by changes in the pH of the medium, that affect both color quality and intensity [14].

Common sources of anthocyanins are purple corn, red cabbage, purple sweet potato, apples, grapes, kiwi, red onions and several berries [6,15]. Some studies have used local plants like black carrot [6]. Black or purple carrot (*Daucus carota* ssp. *sativus* var. *atrorubens* Alef.) was originally from Turkey and the Middle and Far East [16], but recently, new varieties with high anthocyanin content had been cultivated in other parts of the world [12,13,16]. For those reasons, this source was selected for our study.

The stability of anthocyanins in plant extracts depends on the temperature and solids content; and by increasing these conditions, the degradation rates of anthocyanins increase too. To minimize this degradation, it is recommended to cool the concentrates as soon as produced. Other factors that affect color and stability of anthocyanins are concentration, light, presence of co-pigments, metallic ions, enzymes, sugars, proteins, and antiradical activity (which quantifies the ability of complex chemical structures to scavenge free radicals) [2,4,17–21].

Considering those factors, many efforts have been made to enhance anthocyanins' stability. Acylation of anthocyanins is the most commonly used method, as it has been reported that acylated cyanidin derivatives are more stable during prolonged storage compared to the corresponding non-acylated ones [12,13,16] Stintzing et al. [10] also confirmed that there is an increase in color strength through acylation. On the other hand, microencapsulation has been used to increase stability of natural colorants (anthocyanin and betalain derivatives) [2,4,17–19]. Then, considering these conditions, some efforts have been made to preserve natural extracts and colorants using different techniques, such as removing compounds through resin columns and microencapsulation with TEOS.

Here, we study the stability of black carrot extracts while modifying the experimental conditions, including extraction solvents, and we study the feasibility of microencapsulation with TEOS and the repercussion of this procedure on the stability of the extracts.

2. Results and Discussion

2.1. Anthocyanin Content

Major groups of substances were quantified with an UV-Vis spectroscopy analytic method. Figure 1 shows the Fourier Transformed Infrared (FTIR) spectra of samples extracted with ethanol/citric acid (BCS) and ethanol/acetic acid (BCA) and the IR spectra obtained by simulation. In Figure 1a, vibrations obtained for both samples showed a band at 980 cm^{-1} of an C-H in plane deformation, at 1070 cm^{-1}, corresponding to an aromatic ring C-H deformation, bands at 1620 and 1447 cm^{-1}, that correspond to vibration (C=O) of the benzopyran aromatic ring and 1590 cm^{-1} from the stretching vibration (C=C) of an aromatic ring, a band at 1235 cm^{-1}, that corresponds to stretching of pyran rings, typical of flavonoid compounds, and a band at 1335 cm^{-1}, that corresponds to C-O angular deformations of phenols, at 2830 and 2921 cm^{-1} due to symmetric and asymmetric C-H vibration respectively and 3269 cm^{-1} from O-H stretching vibration. In BCS samples there are two additional peaks, at 1710 cm^{-1} from a C=O stretching vibration and a 1180 cm^{-1} that correspond to C-O symmetric vibration, this is indicative that there are other acyl compounds in the extract that are account for a major proportion on the surface.

Becke 3-Lee-Yang-Parr (B3LYP) model simulation results are shown in Figure 1b, considering the cyanidin 3-O-glucoside molecule. The bands obtained are 3370 and 3290 cm^{-1} (O-H symmetric stretching vibration), 3220 cm^{-1} (C-H symmetric stretching), 1719 and 1689 cm^{-1} (C=C scissoring on pyran and phenolic group respectively), 1535 cm^{-1} (C-H scissoring), 1419 cm^{-1} (asymmetric ring vibration on plane), 1380 cm^{-1} (C-H deformation), 1213 cm^{-1} (C-O stretching), 1154 cm^{-1} (a scissoring plane vibration of phenol ring), 959 cm^{-1} (phenol ring C-H asymmetric stretching), 861 cm^{-1} (C-H phenol ring symmetric stretching) and 738 cm^{-1} (phenol ring C-H deformation).

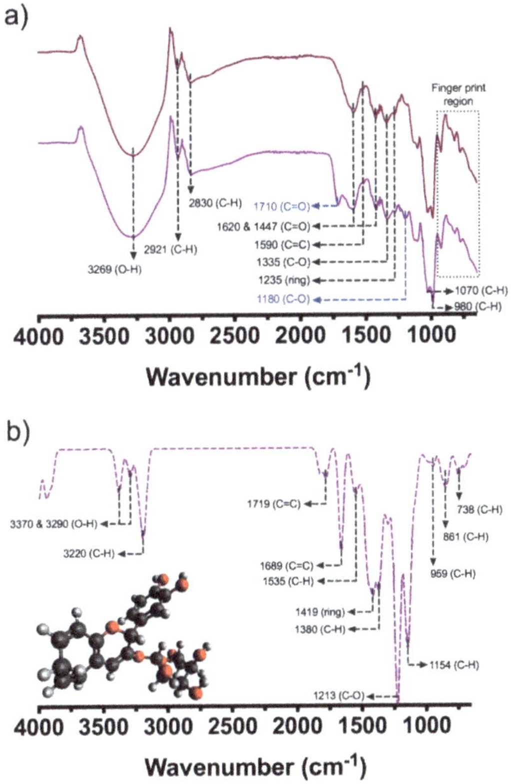

Figure 1. (**a**) Experimental FTIR spectra of black carrot extracts; samples extracted with ethanol/citric acid (BCS) and ethanol/acetic acid (BCA), (**b**) Simulated FTIR spectra using B3LYP calculation and cyanidin 3-O-glucoside molecule (used for simulation).

These results suggest that cyanidin 3-O-glucoside is the major anthocyanin component in black carrot, since they share similar vibrational frequencies and the same functional groups, like previously reported values for black carrot [20–23]. To confirm this notion, high performance liquid chromatography (HPLC) was performed for samples BCS and BCA. Figure 2a shows the HPLC chromatogram from sample BCS and Figure 2b shows the HPLC chromatogram for BCA. Peak identification was made using previous reports [22–24]. Table 1 shows the retention times of the characteristic peaks and compounds identified.

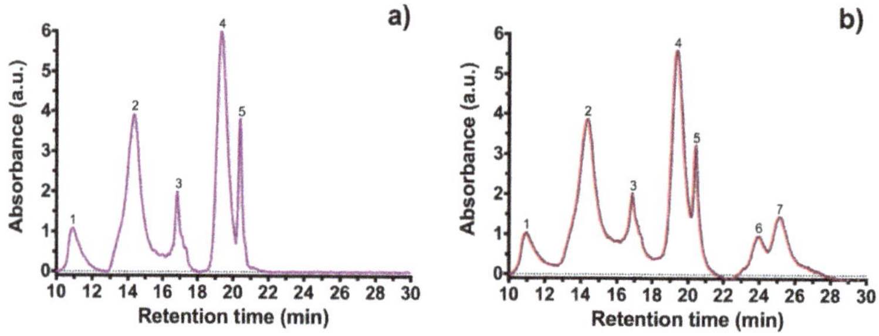

Figure 2. High performance liquid chromatography (HPLC) of black carrot extracts: (**a**) sample BCS (extracted with ethanol/citric acid). (**b**) sample BCA (extracted with ethanol/acetic acid). Peaks were identified and are shown in Table 1.

The chromatogram of sample BCS shows two additional peaks (anthocyanins reported as derived from feluric acid) compared with sample BCA. HPLC analysis confirmed the presence of cyanidin 3-O-glucoside molecule and its derivatives in the black carrot extracts obtained.

Table 1. Retention time and anthocyanin identification from black carrot extracts BCS (extracted with ethanol/citric acid) and BCA (extracted with ethanol/acetic acid).

Peak	Retention Time (min)		Anthocyanin
	BCS	BCA	
1	10.94	10.93	Cyanidin-3-xylosyl-glucosyl-galactoside
2	14.46	14.39	Cyanidin-3-xylosyl-galactoside
3	16.91	16.88	Sinapic acid derivative of cyanidin 3-xylosyl-glucosyl-galactoside
4	19.43	19.42	Ferulic acid derivative of cyanidin 3-xylosyl-glucosyl-galactoside
5	20.45	20.44	Coumaric acid derivative of cyanidin 3-xylosyl-glucosyl-galactoside
6	-	23.98	Feruic acid derivative of pelargonidin 3-xylosyl-glucosyl-galactoside
7	-	25.17	Ferulic acid derivative of peonidin 3-xylosyl-glucosyl-galactoside

Figure 3 shows the total anthocyanin content obtained by a differential pH analytical method. Using ethanol with acetic acid as extraction solvent (sample BCA) leads to more anthocyanin content compared with using ethanol and citric acid, which is congruent with results published before on anthocyanin quantification of extracts from other plants [24,25]. Anthocyanin content is reduced significantly after passing through an Amberlite XAD7 resin column (Figure 3a); this could be explained as a natural degradation process under the experimental conditions. Further chromatographic studies should investigate if 3-O-glucoside anthocyanin derivatives could be trapped in the resin.

The four extracts (BCA, BCS, BCAR, and BCSR) were microencapsulated using TEOS and their anthocyanin content is shown in Figure 3b; there was an average loss of 7 $\pm$ 0.1% and 2.5 $\pm$ 0.005% in the anthocyanin content for samples without passing through the resin column (TBCS, TBCA) and samples after passing through the resin column (TBCSR, TBCAR), respectively. That suggests that TEOS incorporation has a negligible effect in preventing degradation. This could be explained by hydrolysis and condensation processes of the alkoxide that could be favored by functional groups of the acids used. Also, bonds between Si-O and 3-O-glucoside structure, as proposed by other authors [1,3], could lead to less total anthocyanin content in microencapsulated samples than in samples without microencapsulation.

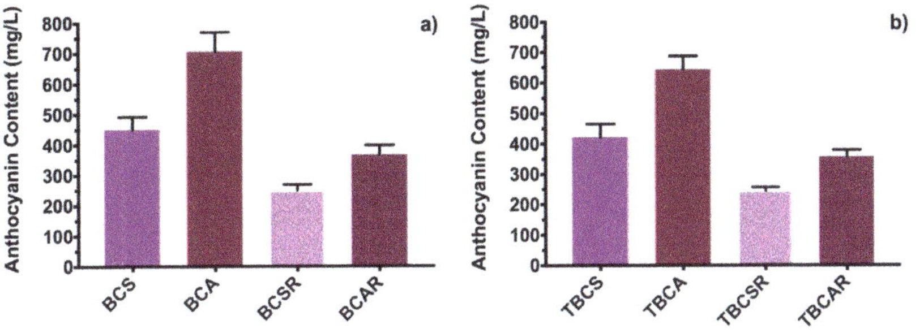

Figure 3. Anthocyanin content in black carrot extracts. (**a**) Samples extracted with ethanol/citric acid (BCS) and ethanol/acetic acid (BCA), and samples after resin column (BCSR and BCAR). (**b**) Microencapsulated samples using TEOS (TBCS, TBCA, TBCSR, TBCAR).

The anthocyanin contents for BCS, BCSR and BCAR and the same microencapsulated samples are in accordance with other microencapsulated powder samples reported for black carrot extracts [8]. BCA and TBCA extracts had the highest anthocyanin content compared with previously mentioned extracts, but it was not as high as the extraction reported using enzymes for other subspecies of black carrot [12,18], nevertheless the extraction method reported here, is inexpensive compared with others and it could be competitive for several industrial applications.

2.2. UV Radiation Study

In the UV radiation stability test (Figure 4a), there was a reduction of 19.81% and 17.99% of the total anthocyanin content for BCS and BCA, respectively. On the other hand, the extracts under resin purification BCSR and BCAR had a reduction of 7.06% and 12.55% from its total anthocyanin content. Considering statistical variations, it cannot be ensured that anthocyanin acylation results in protection against UV radiation.

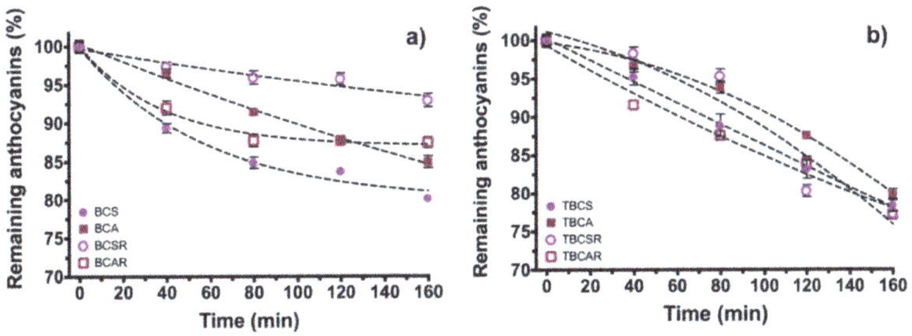

Figure 4. Anthocyanin content in black carrot extracts as function of time under UV radiation (315–400 nm); (**a**) Samples extracted with ethanol/citric acid (BCS) and ethanol/acetic acid (BCA), and samples after resin column (BCSR and BCAR). (**b**) Microencapsulated samples using TEOS (TBCS, TBCA, TBCSR, TBCAR).

Figure 4b shows samples microencapsulated with TEOS, and they follow a quasi-linear decay behavior instead an exponential decay (non-microencapsulated extracts). This can be seen in the reduction of total anthocyanin content which was 21.63%, 20.17%, 22.68%, and 22.82% for TBCS, TBCA, TBCSR and TBCAR, respectively, that are higher losses compared with samples without microencapsulation.

2.3. Thermal Stability

Sample BCA presents same trend compared with BCAR in thermal stability at different temperatures (Figure 5a–c). The same behavior was obtained for BCS and BCSR, therefore purifying extracts through resin column does not have a significant improvement in thermal stability. There is a difference in the behavior of the decay between extracts obtained with ethanol/acetic acid and ethanol/citric acid as extraction solvent; the first ones have a greater decay rate because their graphs (Figure 5) have a greater slope. This slope difference is more visible at high temperatures. On the other hand, considering anthocyanin content loss percentages of 9.28% and 9.31% at 40 °C, 20.08% and 23.96% at 60 °C, and 29.53% and 47.07% at 80 °C for BCS and BCA, respectively. The thermal stability behavior of anthocyanins as a function of time at 40 and 60 °C for TBCA and TBCAR samples is similar (Figure 5d,e). TBCS and TBCSR also had the same trend in thermal stability curves at 40 and 60 °C, from this behavior we assume that the resin column did not influence the thermal stability. Nevertheless, at 80 °C ethanolic extraction with acetic acid and encapsulated samples showed a rapid decay in anthocyanin content. Comparing the total anthocyanins loss of the encapsulated and non-encapsulated samples (9.28% and 14.93% for BCS and TBCS; 9.31% and 8.2% for BCA and TBCA, respectively) at 40 °C, it is possible to conclude that microencapsulation does not prevent thermal degradation in this case. In higher temperatures the same conclusion was obtained (for example at 80 °C: 29.53% and 35.91% for BCS and TBCS, 45.07% and 57.15% for BCA and TBCA, respectively).

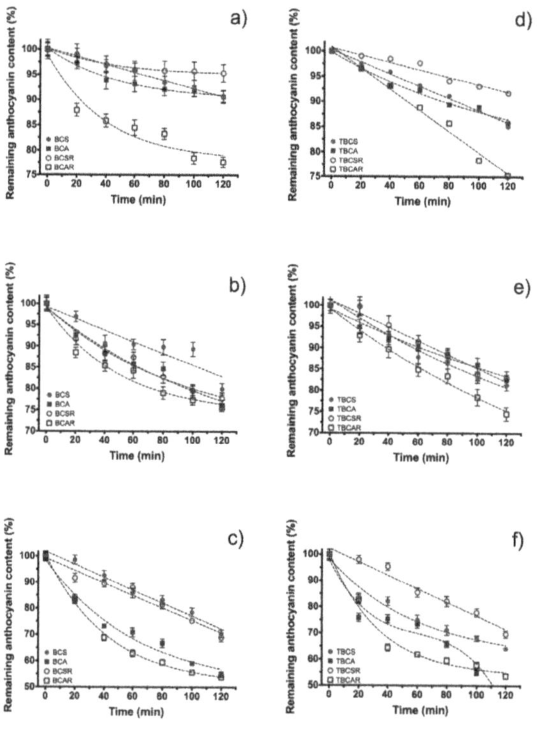

Figure 5. Anthocyanin content as function of time at different temperatures; (**a**,**d**) at 40 °C, (**b**,**e**) at 60 °C, and (**c**,**f**) at 80 °C. Samples extracted with ethanol/citric acid (BCS) and ethanol/acetic acid (BCA), samples after resin column (BCSR and BCAR), and microencapsulated samples using TEOS (TBCS, TBCA, TBCSR, TBCAR).

These results showed that the anthocyanins were more stable using ethanol with citric acid (as extraction solvent) and purified with resin at 40 °C (BCSR) and the stability can be slightly enhanced when TEOS microencapsulation is used at high temperatures. In this experiment, BCAR (extraction of ethanol acidified with acetic acid after purification) had the highest value of degradation and it is more clearly when microencapsulation with TEOS is used. For samples where resin is used, the results are consistent, even using TEOS for microencapsulation.

2.4. pH Storage Stability

The pH changes were evaluated using a short-term storage test during five days for analyzing monomeric anthocyanin content changes at acidic, neutral and alkaline pH and Figure 6 shows these changes.

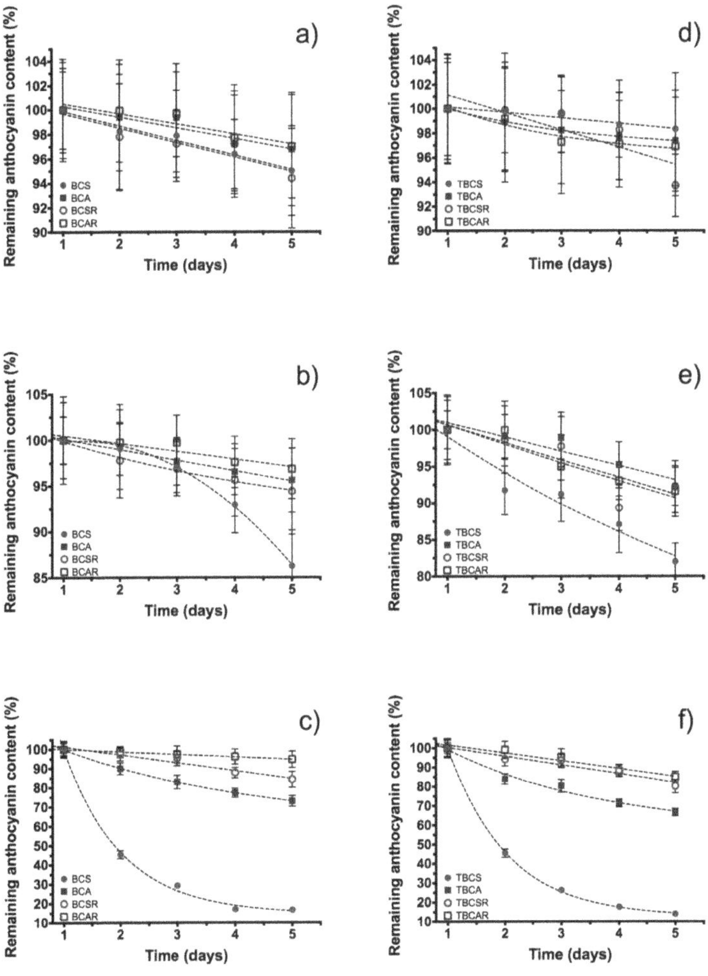

Figure 6. Anthocyanin content as function of storage time at different pH buffers; (**a**,**d**) pH = 4, (**b**,**e**) pH = 7 and (**c**,**f**) pH = 10. Samples extracted with ethanol/citric acid (BCS) and ethanol/acetic acid (BCA), samples after resin column (BCSR and BCAR), and microencapsulated samples using TEOS (TBCS, TBCA, TBCSR, TBCAR).

At acid pH (pH = 4, Figure 6a) there was an anthocyanin content loss of 4.98% and 3.22% for BCS and BCA, respectively, and for resin purified extracts (BCSR and BCAR) the losses were 5.58% and 3.01%. At neutral pH (pH = 7, Figure 6b), the content loss was BCS 13.76%, BCA 4.39%, BCSR 5.60% and BCAR 3.17%, and the major changes for the extracts were at alkaline pH (pH = 10, Figure 6c), since the samples with a higher loss from its initial value were 83.38% and 26.98% for BCS and BCA; at this pH, purified extracts had less degradation with a content loss of 15.57% for BCSR and 5.38% BCAR.

For microencapsulated samples at acid pH (Figure 6d) there was an anthocyanin content loss of 1.69% and 2.63% for TBCS and TBCA respectively and for TBCSR and TBCAR the loss was 5.36% and 3.10%. At neutral pH (Figure 6e) the content loss was 18.02% for TBCS, 7.8% for TBCAR, 7.7% for TBCSR and 8.39% for TBCAR and finally at alkaline pH (Figure 6f) major changes in content were found, such as 86.07% for TBCS, 33.48% for TBCA, 19.90% for TBCSR and 15.02% for TBCAR. Therefore, degradation of microencapsulated samples was reduced at acid pH. When the pH increases, degradation increases too due to the increase in alkalinity.

In both cases, microencapsulated and non-microencapsulated samples, the graphical tendency is the same and anthocyanins in BCAR samples were the most stable after the elapsed time under three-different conditions of pH, indicating that the anthocyanins extracted with ethanol/acetic acid are more stable to pH changes in comparison with the ethanol/citric acid extracts (higher anthocyanin content loss), which is highly evident at pH = 10.

Since the results from degradation studies of samples microencapsulated with TEOS showed no significant improvement in the stability of the extracts (except for the experiment at 60 °C), these samples were not analyzed for antioxidant activity and color in the food models. This decision was made considering also that the activity of nutraceutical compounds was reduced after treatment [3].

2.5. DPPH and Electrochemical CUPRAC Antioxidant Content Test

As seen from Figure 7a there is a direct relationship of anthocyanin content and antiradical activity; when the anthocyanin content is higher, the antioxidant effect increases. With more phenolic compounds, such as anthocyanins, a higher antiradical activity is expected.

The extract obtained with acetic acid leads to a higher yield of antiradical activity (614.52 μM TE g fw^{-1}) which is 15.5% higher compared with the extract obtained with citric acid. This was expected because the acylated nature of the extracted anthocyanin [25] confers higher antiradical activity than monomeric anthocyanins; also, the use of XAD7 resin reduced antiradical activity on 20.20% and 18.94% for BCS and BCA extracts, respectively.

In the case of CUPRAC test, Figure 7b shows cyclic voltammograms for the antioxidant agents obtained with different extraction methods. It can be observed that the initial potential for the BCAR and BCA was 0.433 and 0.428 V, respectively. Those values exhibited a shift to negative potentials compared with 0.454 and 0.488 V of BCSR and BCS, respectively. This shift to potential negative values is related with an increment in the amount of the complex $Cu(Nc)_2^+$ due to the capability of the antioxidant agent to donate an electron to the oxidized complex $Cu(Nc)_2^{2+}$ according to the following equation:

$$aCu(Nc)_2^{2+} + bAO_{red} \leftrightarrow cCu(Nc)_2^+ + dAO_{ox}^+ \quad (1)$$

where the AO_{red} is the reduced antioxidant agent and the AO_{oxi}^+ is the antioxidant agent when it was oxidized. This behavior is directly related with the antioxidant agent capability of the samples to promote the reduction reaction for the molecule that was previously oxidized ($Cu(Nc)_2^{2+}$). Also, the description above was based on the Nernst equation:

$$E = E^0 + \frac{RT}{F} \ln \frac{a_{Cu(Nc)_2^{2+}}}{a_{Cu(Nc)_2^+}} \quad (2)$$

where the potential of the reaction on the equilibrium were shifted to negative values owing an increase in the activity (concentration of $Cu(Nc)_2^{2+}$) of the products.

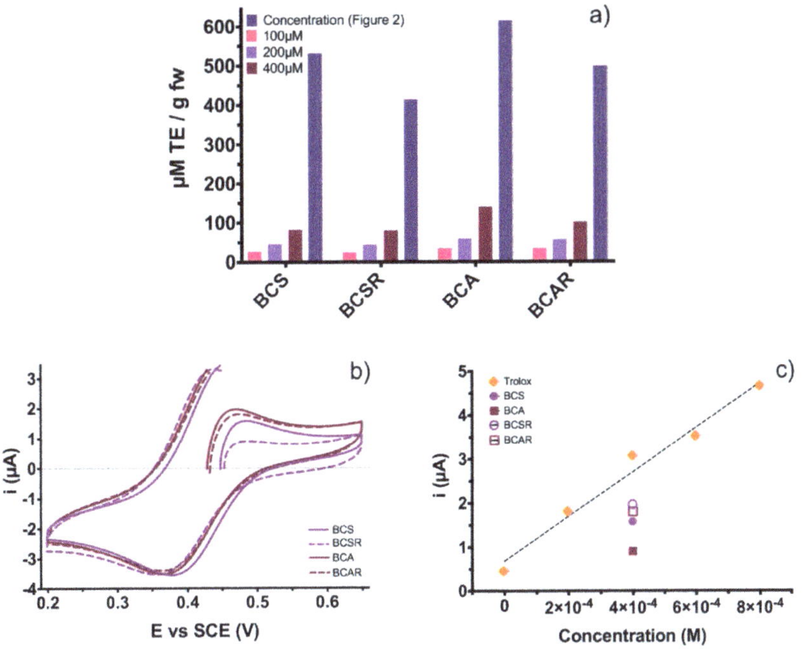

Figure 7. (**a**) Antiradical activity of black carrot extracts using DPPH assay at different concentrations. (**b**) Cyclic voltammograms for the four extracts (400 μM) in a CUPRAC solution. The scan began at the open circuit potential (OCP) with a sweep velocity of 100 mV·s^{-1}. (**c**) Electrochemical antioxidant activity and compared with Trolox activity at the same concentration. Samples extracted with ethanol/citric acid (BCS) and ethanol/acetic acid (BCA), samples after resin column (BCSR and BCAR).

Furthermore, the peak current for the oxidation process in the voltammograms has a direct relation with the concentration of $Cu(Nc)_2^+$ because if the electrolyte has a higher amount of $Cu(Nc)_2^+$ we obtained a major amount of oxidizer molecules ($Cu(Nc)_2^{2+}$) resulting in an increment of the current value. This is possible because the current is directly proportional for the concentration of the species in the reaction ($i \propto$ C) [26,27].

The Trolox calibration curve was obtained from the oxidation peak current and was used to analyze the antioxidant activity for the black carrot extracts obtained by the different extraction methods (r^2 = 0.9768). It is proposed that ratio between current peak of the black carrot and the current peak of Trolox (at same concentration, 400 μM) allows determining the antioxidant capability of each black carrot extract. This method is similar to calculating the Trolox Equivalent Antioxidant Capacity (TEAC), where calculations were made by the ratio of molar absorptivity of problem species and Trolox under the corresponding conditions [28]. In order to have cyanidin 3-O-glucoside at a 400 μM concentration, samples were diluted (Figure 3a) and the molecular weight of the anthocyanin was used for the analytical pH differential method.

The calculated Electrochemical Antioxidant Capacity (EAC) is shown in Figure 7c, giving the following results: 0.684, 0.346, 0.748 and 0.591 for BCS, BCA, BCSR and BCAR, respectively. These EAC results exhibited a behavior in accordance with the DPPH colorimetric method. BCA extract had the highest antiradical activity of all samples, followed by BCS, then BCAR and finally BCSR. Also, analyzing the antioxidant activity for black carrot extracts, values of current were under the values of the Trolox calibration plot. The decrease of the antiradical activity is not as higher than the anthocyanin content in Figure 3a; this is because electrochemical CUPRAC methods measure compounds related with the antiradical activity at an electron level. These CUPRAC test shows that our samples could have several 3,7-diglucoside derivatives and other phenolic compounds. The antiradical activity of black carrots extracts was higher than other reported values of several extracts [29].

2.6. Color of Black Carrot Extracts on Food Models

Figure 8a–d show the average results from the image analysis for food models using the black carrot extracts and Red FD&C analysis for comparison. All samples (in yogurt and jelly), had a light brown color tendency (hue angle below 2°) at the beginning of the experiment. The jelly has more saturated colors than yogurt and this is due to the base color of the food models (white vs. pale yellow). Red FD&C had the highest luminosity in yogurt and is more saturated than BCS and BCA samples (chroma value). FD&C in jelly is darker, but has the same color saturation than BCS, i.e. in the food model; FD&C and BCS have the same color saturation for the human eye. BCA sample is less saturated in jelly. In the case of BCSR and BCAR, in yogurt, the color is less saturated but darker; and for jelly, they have almost the same saturation (slightly less saturated) but it has a much brighter color.

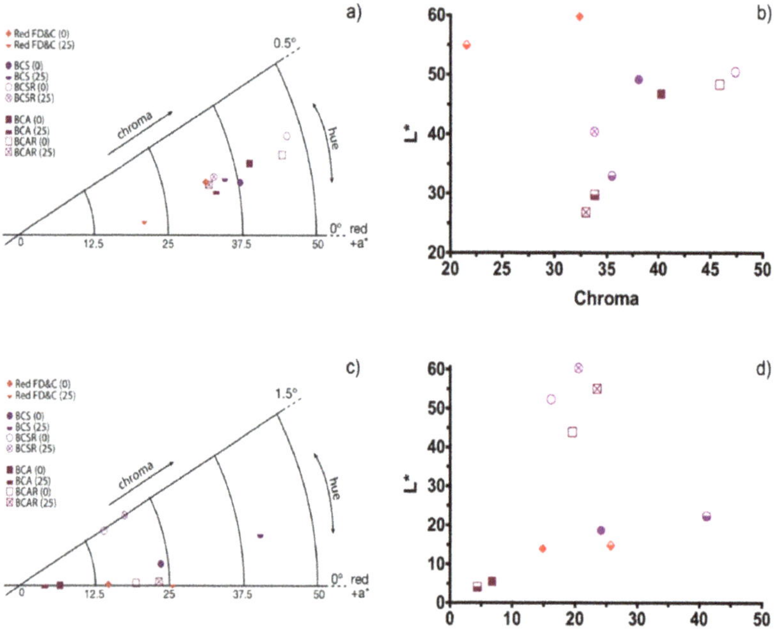

Figure 8. (**a**,**c**) Sectional polar and (**b**,**d**) cartesian diagram of color for the black carrots extracts and the red FD&C on food models (yogurt and jelly) at 0 and 25 days of storage.

As seen in Table 2, after the 25-day storage time the color differences (ΔE) in yogurt, of all samples, have values higher than five, which indicates that the color difference at the beginning and after the elapsed time is visually evident; also, samples have higher saturation (lowest chroma value) after storage and specifically the black carrots extract samples get darker since L* value is lower. BCS samples showed similar color difference after storage time compared with red FD&C.

Table 2. Color comparison between food models (yogurt and jelly) using black carrot extracts after 25 days in storage.

Sample	ΔE Yogurt	ΔE Jelly
Red FD&C	12.62	10.92
BCS	16.45	17.61
BCSR	24.27	9.21
BCA	18.31	2.82
BCAR	25.12	12.71

For jelly samples, only sample BCA has a value lower than five, which indicates that the color difference could be distinguished but was not as evident as the rest of the samples; also, BCA samples are the only ones with different saturation and luminosity trends, they had lower chroma values and are darker. The rest of the samples are less saturated (higher chroma value) and brighter. The samples with resin purification (BCSR and BCAR) had a similar color difference compared to red FD&C, except BCA.

For yogurt, BCSR and BCAR samples have the highest color change indicating that cyanidin-3-glucoside derivative anthocyanins are not suitable to be used in this food model, because the acetate group caused important appearance differences under these conditions; for jelly, BCS and BCAR had the highest color changes but BCSR and BCA the lowest compared with Red FD&C. Pigment concentration of the samples must change for the specific commercial use, for instance, strawberry yogurt samples have different L*, a* and b* values [3].

3. Materials and Methods

Fresh black carrot (*Daucus Carota* var. *L.* ssp. *sativus* var. *atrorubens* Alef.) were cultivated in Tlaxcala, Mexico and donated by a local cultivator, then were washed with tap water and stored in 3 kg perforated plastics bags and kept at −20 °C until further use. All reactants used were analytical grade. Ethanol 99.5%, citric acid 99.5%, Amberlite XAD7 resin, tetraethyl orthosilicate (TEOS) 98%, potassium chloride, 6-hydroxy-2,5,7,8-tetramethylchromane-2-carboxylic acid (Trolox), and 2,2-diphenyl-1-picrylhydrazyl (DPPH), $CuCl_2 \cdot 2H_2O$ (99%), neocuproine, and ammonium acetate (98.0%) were supplied by Sigma Aldrich (Toluca, Mexico). Acetic acid, sodium acetate, hydrochloric acid, buffer solutions (pH: 4, 7 & 10) and potassium hydroxide were supplied by J.T. Baker (Mexico City, Mexico). Methanol was supplied by Macron (Hamilton, PA, USA). All experiments were carried out using distilled water obtained from an Elix Advantage Water Purification System (Queretaro, Mexico).

3.1. *Anthocyanins Extraction and TEOS Microencapsulation*

Acidified ethanol was used as extraction solvent for all samples. Anthocyanins were obtained by blending 150 mL of acidified ethanol (citric/acetic acid solution 85:15 *v*/*v*) and 150 g of sliced frozen carrots (without thawing) with a Grinder 6807 blender Oster (Mexico) for 20 min. Solids were removed by filtration using a 100-mesh sieve filter. The liquid phase (extract) was labeled as BCS (acidified with citric acid) and BCA (acidified with acetic acid). Extracts BCS and BCA were introduced, separately, to a resin column containing Amberlite XAD7 resin for removing non-aromatic compounds from the extract. Flow rate was 32 mL/min (20BV/h) and a solution of 95 mL of ethanol and 5 mL of acidic water (pH = 1.4) was used as eluent. Finally, solvent excess was evaporated at 40 °C using vacuum (R-100, Büchi, Mexico). Samples obtained after this procedure were labeled BCSR and BCAR. All samples were stored in sealed amber glass vials at 4 °C until further use. Then samples were submitted to microencapsulation using TEOS with a procedure reported elsewhere [1,3]. The samples after microencapsulation were labeled as TBCS, TBCA, TBCSR and TBCAR. A flow diagram is shown in Figure 9.

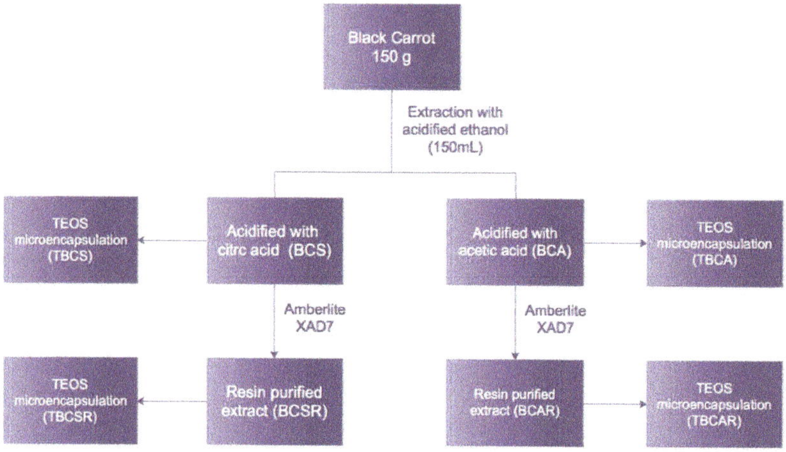

Figure 9. Flow diagram of experimental processes for obtaining samples. Samples extracted with ethanol/citric acid (BCS) and ethanol/acetic acid (BCA), samples after resin column (BCSR and BCAR), and microencapsulated samples using TEOS (TBCS, TBCA, TBCSR, TBCAR).

3.2. Anthocyanin Content

It has been reported that black carrot contains different monomeric anthocyanins with different sugar moieties such as peonidin, pelargonidin and cyanidin, being cyanidin 3-O-R the major compound in the total anthocyanin content [20–23]. In order to confirm this and to justify the use of cyanidin 3-O-glucoside for analytical measurements, Fourier Transform Infrared Spectroscopy (FTIR) spectra under attenuated total reflectance (ATR) was measured by a Frontier MIR/NIR Spectrometer (Perkin Elmer, Waltham, MA, United States), for dried samples of BCA and BCS and then compared with a frequency quantum chemical calculation of the same molecule (Figure 1c).

For the frequency calculation of cyanidin 3-O-glucoside molecule a personal computer running Gaussian 98W [30] was used. The geometry was fully optimized assuming Cs point group symmetry using the Becke 3-Lee-Yang-Parr (B3LYP), supplemented with the standard 6-31 + G basis sets. The simulated IR spectra were plotted using Avogadro molecular viewer [31] and the vibrational modes were analyzed and compared to the experimental data as mentioned.

Total anthocyanins content (monomeric anthocyanins) was determined using the pH-differential method reported by Giusti and Wrolstad [32], using a molar extinction coefficient of 26900 $M^{-1}cm^{-1}$ that corresponds to cyanidin 3-O-glucoside. The average molecular weight used was 756.87 g mol^{-1} of the anthocyanins according to previous studies from black carrot [32–34]. A VWR 1600-PC spectrophotometer and 1 cm path length glass cells were used; measurements were performed scanning from 700 to 400 nm at room temperature (~24 °C). Finally, High pressure liquid chromatography (HPLC) was performed on a Flexar LC (Perkin Elmer) system using a 250 mm × 4.6 mm C18 reverse phase column and BCS and BCA samples were measured according to conditions reported elsewhere [34,35]; using a flow rate of 1.0 mL/min and the chromatographs were recorded at 520 nm using PDA Plus Detector coupled with the equipment.

3.3. Degradation Studies

Black carrot extracts were transferred into vials with screw caps to perform degradation tests. All experiments were performed in triplicate and the referred anthocyanin content was normalized to 200 mg/L in all the samples in order to make direct comparison between samples so the results are presented as remaining percentage of cyanidin 3-O-glycoside.

The ASTM D 4320 method was used to determine anthocyanin stability against UV radiation. Black carrot extracts were exposed to UV lamp irradiation (315–400 nm) for 160 min; samples were placed within 15 cm from the source. Measurements were taken every 40 minand the temperature was kept constant at 25 °C under a working area of 3.5 m^2 isolated from other light sources [1,3].

For thermal stability studies, samples were placed in a preheated water bath at 80 °C, 60 °C and 40 °C. Samples were removed from water bath every 20 min, up to 120 minand rapidly cooled to room temperature. Anthocyanin content was analyzed immediately [35].

The anthocyanin stability in storage was also studied at three different pH conditions (4, 7 and 10) at room temperature (~24 °C). For this purpose, 3 mL of phosphate buffer solution were prepared at the required pH conditions and then colored with 500 μL of black carrot extract concentrate; these colored solutions were used without further treatment. Finally, anthocyanin storage stability was determined every 24 h during 5 days for each of the black carrot extracts [12].

3.4. Antioxidant Activity

The DPPH radical scavenging activity assay was performed according to several methods described previously [22,36–38] in which radical scavenging activities were determined by testing the extracts with the free radical DPPH and monitoring their absorbance decrease at 515 nm using a 1600-PC spectrophotometer (VWR, Graumanngasse, Vienna) with a 1 cm path length glass cells. Control assays using the black carrot extracts were performed in order to obtain their absorbance contributions. A solution of 50 μM DPPH was prepared using buffered methanol, which was prepared by mixing methanol with acetic acid buffer solution (0.1 M, pH 5.5) [36]. Then, 2.85 mL of DPPH solution were mixed with 150 μL of each extract and were left reacting 30 minat room temperature (~24 °C). Antiradical activity was expressed as Trolox equivalents per gram of fresh weight (μM TE $g \cdot fw^{-1}$), which was calculated from the equation obtained using a linear regression after plotting the known absorbance with different Trolox concentrations, from 1 to 800 μM and r^2 = 0.9657.

A cupric reducing antioxidant capacity (CUPRAC) solution [24,26] was prepared to determine antioxidant capacity via electrochemical tests using $CuCl_2$ with a concentration of 3 mM in distilled water. Also, a solution of neocuproin at 6 mM in ethanol was prepared. In order to control pH of the main solution, a 1.2 M ammonium acetate buffer solution was prepared (pH = 7), then pH was adjusted adding 1.2 M HCl and 1.2 M NaOH as required. Concentrations of Trolox were varied from 1 to 800 μM in ethanol for obtaining a Trolox standard curve. 2 mL of each $CuCl_2$, neocuproin, ammonium acetate buffer, Trolox and distilled water solutions were prepared and mixed. The 10 mL solution was stirred for 15 minand N_2 was bubbled into it for 5 min. The same procedure was followed to evaluate antioxidant capacity adding anthocyanins extracts instead of Trolox solution and the values are reported as a comparison between the analytical response vs. concentration plot during the antioxidant quantification because the slope is dependent on the stochiometric relationship between the antioxidant-oxidant species involved, which is related to the electron transfer per molecule pair.

The electrochemical tests were performed in a three-electrode electrochemical cell. A calomel Hg/Hg_2Cl_2 (saturated with KCl) was used as reference electrode, a graphite bar was used as counter electrode and a glassy carbon (3 mm) electrode was used as working electrode. Before each electrochemical measurement, the working electrode was polished with aluminum oxide powder followed by ultrasonic stirring during 10 min; this process was repeated 3 times. The voltammograms were obtained in a Bio-LogicVP-50 potentiostat (Bio-Logic Science Instruments, Seyssinet-Pariset, France) with a sweep velocity of 100 mV s^{-1}, starting the voltammograms from the open circuit potential (OCP) that was determined when the potential did not show a variation higher than 1 mV per second.

3.5. Extract and Food Models Color Determination

A custom MATLAB script (Mathworks Inc., Natick, MA, USA) was used to measure lightness and chromaticity coordinates in the L* a* b* color space (CIELAB) according to CIE standard illuminant A (typical, domestic, tungsten-filament lighting with correlated color temperature of 2856 K). L* indicates lightness, a* and b* are chromaticity coordinates, h (hue), c (chroma) and ΔE (color change) were calculated from a* and b* values. Digital images from samples were taken using a Sony digital camera α99II coupled with a Vario Sonnar T* 24–70 mm lens (Sony Corporation, Tokyo, Japan) under the same light conditions; the images were cropped to 1024 × 1024 pixels and then processed with the afore mentioned script. Additionally, samples were measured in two different colored food models: yogurt and jelly. For yogurt food model, 10 g of commercial yogurt (Yoplait natural yogurt, Sigma Alimentos Lácteos México, Queretaro, Mexico) were colored using 10 mg of calculated anthocyanins from each extract and jelly was prepared using jelly powder (Coloidales Duche, Ciudad de México, Mexico) dissolved in boiling water (1:3 ratio) and 10 g of the mixture were colored using the same calculated amount of anthocyanins from each extract. Samples were measured at the beginning of the experiment and after 25 days, and compared with a colored yogurt/jelly with 10 mg of Red FD&C (Red Currant 12.5%, Colores Duche, Ciudad de México, Mexico) under the same conditions.

4. Conclusions

Anthocyanins were extracted from black carrot with ethanol/citric acid and ethanol/acetic acid for comparison between total anthocyanin content, and stability against media changes and antioxidant capacity was obtained to analyze samples in food models. Microencapsulation with TEOS was performed with the objective of enhancing anthocyanin stability. Extracts had the highest degradation in alkaline pH, and BCAR was the most stable sample to pH media changes. The antiradical activity of black carrots extracts was higher than other reported values, and when anthocyanin content is higher, the antioxidant effect increases. Results of UV radiation and thermal stability tests indicate that TEOS microencapsulation provides a negligible improvement in anthocyanins' stability. In conclusion, extraction with ethanol/acetic acid is the most convenient and stable treatment against pH media changes. Purification with resin and TEOS microencapsulation did not increase stability of the black carrot extracts. TEOS microencapsulation has proven to be effective (enhancing stability) for some dried materials from natural extracts in our previous research, but we do not recommend its use for materials obtained from black carrot extracts. Even though anthocyanins are already used in the food industry in beverages, our samples were not suitable for the yogurt or jelly model selected except for BCA sample in jelly that has the highest antioxidant activity, this gives it potential for being a functional natural colorant in this specific food model.

Author Contributions: Data and experimentation, G.A.M. and A.R.H.-M.; Formal analysis, R.E., A.R.H.-M. and M.E.; Funding acquisition, M.E.; Investigation, G.E.-A., I.S.-G. and J.M.-C.; Methodology, J.M.-C. and A.R H.-M.; Supervision, A.R.H.-M. and M.E.; Validation, A.L.R.-J.; Writing—original draft, G.A.M., A.R.H.-M. and M.E.; Writing—review & editing, A.L.R.-J.

Funding: This research received no external funding.

Acknowledgments: Authors would like to acknowledge the grant obtained from the Dirección General de Asuntos del Personal Académico (DGAPA) of Universidad Nacional Autónoma de México (UNAM) through "Programa de Apoyo a Proyectos de Investigación e Innovación Tecnológica" (PAPIIT No. IN209517). One of the authors, MSc Guillermo Espinosa-Acosta—fellow number 388907-; acknowledges the financial support received from CONACYT at the first period of the 2015 National Scholarships Call. Also, the authors are especially grateful to Perkin Elmer Mexico for obtaining the chromatograms shown in Figure 2 of this manuscript. Finally, authors are grateful to Bernardino Rodriguez-Morales and Adrian Hendrik Oskam Voorduin for their technical support; and also, to Gerardo Fonseca, and José Antonio Pérez Guzman, for their valuable comments about the manuscript, and to Angel Luis Rodríguez for his support on the development of the MATLAB script used for color analysis.

Conflicts of Interest: The authors declare no conflict of interest

References

1. Hernández-Martínez, A.R.; Torres, D.; Molina, G.A.; Esparza, R.; Quintanilla, F.; Martínez-Bustos, F.; Estevez, M. Stability comparison between microencapsulated red-glycosidic pigments and commercial fd&c red 40 dye for food coloring. *J. Mater. Sci.* **2017**, *52*, 5014–5026. [CrossRef]

2. García-Tejeda, Y.V.; Salinas-Moreno, Y.; Hernández-Martínez, Á.R.; Martínez-Bustos, F. Encapsulation of purple maize anthocyanins in phosphorylated starch by spray drying. *Cereal Chem. J.* **2015**, *93*, 130–137. [CrossRef]
3. Molina, A.G.; Hernández-Martínez, R.A.; Cortez-Valadez, M.; García-Hernández, F.; Estevez, M. Effects of tetraethyl orthosilicate (teos) on the light and temperature stability of a pigment from *Beta vulgaris* and its potential food industry applications. *Molecules* **2014**, *19*. [CrossRef] [PubMed]
4. Mahdavi, S.A.; Jafari, S.M.; Ghorbani, M.; Assadpoor, E. Spray-drying microencapsulation of anthocyanins by natural biopolymers: A review. *Dry. Technol.* **2014**, *32*, 509–518. [CrossRef]
5. Shahid, M.; Mohammad, F. Recent advancements in natural dye applications: A review. *J. Clean. Prod.* **2013**, *53*, 310–331. [CrossRef]
6. He, J.; Giusti, M.M. Anthocyanins: Natural colorants with health-promoting properties. *Annu. Rev. Food Sci. Technol.* **2010**, *1*, 163–187. [CrossRef] [PubMed]
7. Rohring, B. Eating with Your Eyes: The Chemistry of Food Colorings. 2015. Available online: https://www.acs.org/content/acs/en/education/resources/highschool/chemmatters/past-issues/2015-2016/october-2015/food-colorings.html (accessed on 25 August 2018).
8. Ersus, S.; Yurdagel, U. Microencapsulation of anthocyanin pigments of black carrot (*Daucus carota* L.) by spray drier. *J. Food Eng.* **2007**, *80*, 805–812. [CrossRef]
9. Moldovan, B.; David, L.; Chişbora, C.; Cimpoiu, C. Degradation kinetics of anthocyanins from european cranberrybush (*Viburnum opulus L.*) fruit extracts. Effects of temperature, ph and storage solvent. *Molecules* **2012**, *17*. [CrossRef] [PubMed]
10. Stintzing, F.C.; Stintzing, A.S.; Carle, R.; Frei, B.; Wrolstad, R.E. Color and antioxidant properties of cyanidin-based anthocyanin pigments. *J. Agric. Food Chem.* **2002**, *50*, 6172–6181. [CrossRef] [PubMed]
11. Blando, F.; Calabriso, N.; Berland, H.; Maiorano, G.; Gerardi, C.; Carluccio, M.; Andersen, Ø. Radical scavenging and anti-inflammatory activities of representative anthocyanin groupings from pigment-rich fruits and vegetables. *Int. J. Mol. Sci.* **2018**, *19*, 169. [CrossRef] [PubMed]
12. Lee, J.; Koo, N.; Min, D.B. Reactive oxygen species, aging, and antioxidative nutraceuticals. *Compr. Rev. Food Sci. Food Saf.* **2004**, *3*, 21–33. [CrossRef]
13. Valko, M.; Leibfritz, D.; Moncol, J.; Cronin, M.T.D.; Mazur, M.; Telser, J. Free radicals and antioxidants in normal physiological functions and human disease. *Int. J. Biochem. Cell Biol.* **2007**, *39*, 44–84. [CrossRef] [PubMed]
14. Braga, A.R.C.; Murador, D.C.; de Souza Mesquita, L.M.; de Rosso, V.V. Bioavailability of anthocyanins: Gaps in knowledge, challenges and future research. *J. Food Compos. Anal.* **2018**. [CrossRef]
15. Akhavan Mahdavi, S.; Jafari, S.M.; Assadpoor, E.; Dehnad, D. Microencapsulation optimization of natural anthocyanins with maltodextrin, gum arabic and gelatin. *Int. J. Biol. Macromol.* **2016**, *85*, 379–385. [CrossRef] [PubMed]
16. Gandomi, H.; Abbaszadeh, S.; Misaghi, A.; Bokaie, S.; Noori, N. Effect of chitosan-alginate encapsulation with inulin on survival of *Lactobacillus rhamnosus GG* during apple juice storage and under simulated gastrointestinal conditions. *LWT Food Sci. Technol.* **2016**, *69*, 365–371. [CrossRef]
17. Kırca, A.; Özkan, M.; Cemeroğlu, B. Effects of temperature, solid content and ph on the stability of black carrot anthocyanins. *Food Chem.* **2007**, *101*, 212–218. [CrossRef]
18. Khandare, V.; Walia, S.; Singh, M.; Kaur, C. Black carrot (*Daucus carota* ssp. *Sativus*) juice: Processing effects on antioxidant composition and color. *Food Bioprod. Process.* **2011**, *89*, 482–486. [CrossRef]
19. Kırca, A.; Özkan, M.; Cemeroğlu, B. Stability of black carrot anthocyanins in various fruit juices and nectars. *Food Chem.* **2006**, *97*, 598–605. [CrossRef]
20. Türkyılmaz, M.; Yemiş, O.; Özkan, M. Clarification and pasteurisation effects on monomeric anthocyanins and percent polymeric colour of black carrot (*Daucus carota* L.) juice. *Food Chem.* **2012**, *134*, 1052–1058. [CrossRef] [PubMed]
21. Montilla, E.C.; Arzaba, M.R.; Hillebrand, S.; Winterhalter, P. Anthocyanin composition of black carrot (*Daucus carota* ssp. *sativus* var. *atrorubens* Alef.) cultivars antonina, beta sweet, deep purple, and purple haze. *J. Agric. Food Chem.* **2011**, *59*, 3385–3390. [CrossRef] [PubMed]
22. Algarra, M.; Fernandes, A.; Mateus, N.; de Freitas, V.; Esteves da Silva, J.C.G.; Casado, J. Anthocyanin profile and antioxidant capacity of black carrots (*Daucus carota* L. ssp. *sativus* var. *atrorubens* Alef.) from cuevas bajas, spain. *J. Food Compos. Anal.* **2014**, *33*, 71–76. [CrossRef]

23. Kammerer, D.; Carle, R.; Schieber, A. Detection of peonidin and pelargonidin glycosides in black carrots (*Daucus carota* ssp. *sativus* var. *atrorubens* Alef.) by high-performance liquid chromatography/electrospray ionization mass spectrometry. *Rapid Commun. Mass Spectrom.* **2003**, *17*, 2407–2412. [CrossRef] [PubMed]
24. Castañeda-Ovando, A.; Pacheco-Hernández, M.d.L.; Páez-Hernández, M.E.; Rodríguez, J.A.; Galán-Vidal, C.A. Chemical studies of anthocyanins: A review. *Food Chem.* **2009**, *113*, 859–871. [CrossRef]
25. Gerardi, C.; Tommasi, N.; Albano, C.; Blando, F.; Rescio, L.; Pinthus, E.; Mita, G. *Prunus mahaleb* L. Fruit extracts: A novel source for natural food pigments. *Eur. Food Res. Technol.* **2015**, *241*, 683–695. [CrossRef]
26. Cárdenas, A.; Gómez, M.; Frontana, C. Relationship between the chemical structures of antioxidants and the differences in their cupric ion reducing antioxidant capacity (CUPRAC) by electrochemical methods. *J. Electroanal. Chem.* **2014**, *729*, 116–120. [CrossRef]
27. Pletcher, D.; Greff, R.; Peat, R.; Peter, L.M.; Robinson, J. *Instrumental Methods in Electrochemistry*; Woodhead Publishing: Cambridge, UK, 2001.
28. Arts, M.J.T.J.; Haenen, G.R.M.M.; Voss, H.-P.; Bast, A. Antioxidant capacity of reaction products limits the applicability of the Trolox Equivalent Antioxidant Capacity (TEAC) assay. *Food Chem. Toxicol.*. **2004**, *42*, 45–49. [CrossRef] [PubMed]
29. Seeram, N.P.; Aviram, M.; Zhang, Y.; Henning, S.M.; Feng, L.; Dreher, M.; Heber, D. Comparison of antioxidant potency of commonly consumed polyphenol-rich beverages in the united states. *J. Agric. Food Chem.* **2008**, *56*, 1415–1422. [CrossRef] [PubMed]
30. *Gaussian 09*, revision A. 02; Gaussian, Inc.: Wallingford, CT, USA, 2009.
31. Hanwell, M.D.; Curtis, D.E.; Lonie, D.C.; Vandermeersch, T.; Zurek, E.; Hutchison, G.R. Avogadro: An advanced semantic chemical editor, visualization, and analysis platform. *J. Cheminform.* **2012**, *4*, 17. [CrossRef] [PubMed]
32. Giusti, M.M.; Wrolstad, R.E. Characterization and measurement of anthocyanins by uv-visible spectroscopy. In *Current Protocols in Food Analytical Chemistry*; John Wiley & Sons, Inc.: Hoboken, NJ, USA, 2001.
33. Wrolstad, R.E.; Durst, R.W.; Lee, J. Tracking color and pigment changes in anthocyanin products. *Trends Food Sci. Technol.* **2005**, *16*, 423–428. [CrossRef]
34. Alasalvar, C.; Al-Farsi, M.; Quantick, P.C.; Shahidi, F.; Wiktorowicz, R. Effect of chill storage and modified atmosphere packaging (MAP) on antioxidant activity, anthocyanins, carotenoids, phenolics and sensory quality of ready-to-eat shredded orange and purple carrots. *Food Chem.* **2005**, *89*, 69–76. [CrossRef]
35. Dai, Y.; Rozema, E.; Verpoorte, R.; Choi, Y.H. Application of natural deep eutectic solvents to the extraction of anthocyanins from *Catharanthus roseus* with high extractability and stability replacing conventional organic solvents. *J. Chromatogr. A* **2016**, *1434*, 50–56. [CrossRef] [PubMed]
36. Sharma, O.P.; Bhat, T.K. DPPH antioxidant assay revisited. *Food Chem.* **2009**, *113*, 1202–1205. [CrossRef]
37. Thaipong, K.; Boonprakob, U.; Crosby, K.; Cisneros-Zevallos, L.; Hawkins Byrne, D. Comparison of abts, dpph, frap, and orac assays for estimating antioxidant activity from guava fruit extracts. *J. Food Compos. Anal.* **2006**, *19*, 669–675. [CrossRef]
38. Molyneux, P. The use of the stable free radical diphenylpicrylhydrazyl (DPPH) for estimating antioxidant activity. *Songklanakarin J. Sci. Technol.* **2004**, *26*, 211–219.

Sample Availability: Samples of the compounds are available from the authors.

molecules

MDPI

Article

Betanin, a Natural Food Additive: Stability, Bioavailability, Antioxidant and Preservative Ability Assessments

Davi Vieira Teixeira da Silva, Diego dos Santos Baião, Fabrício de Oliveira Silva, Genilton Alves, Daniel Perrone, Eduardo Mere Del Aguila and Vania M. Flosi Paschoalin *

Instituto de Química, Universidade Federal do Rio de Janeiro, Av. Athos da Silveira Ramos 149, 21941-909 Rio de Janeiro, Brazil; daviufrj@outlook.com (D.V.T.d.S.); diegobaiao20@hotmail.com (D.d.S.B.); silvafo@live.com (F.d.O.S.); geniltonalves@gmail.com (G.A.); perrone@iq.ufrj.br (D.P.); emda@iq.ufrj.br (E.M.D.A.)

* Correspondence: paschv@iq.ufrj.br; Tel.: +55-21-3938-7362; Fax: +55-21-3938-7266

Academic Editors: Lillian Barros and Isabel C.F.R. Ferreira
Received: 11 January 2019; Accepted: 27 January 2019; Published: 28 January 2019

Abstract: Betanin is the only betalain approved for use in food and pharmaceutical products as a natural red colorant. However, the antioxidant power and health-promoting properties of this pigment have been disregarded, perhaps due to the difficulty in obtaining a stable chemical compound, which impairs its absorption and metabolism evaluation. Herein, betanin was purified by semi-preparative HPLC-LC/MS and identified by LC-ESI(+)-MS/MS as the pseudomolecular ion m/z 551.16. Betanin showed significant stability up to −30 °C and mild stability at chilling temperature. The stability and antioxidant ability of this compound were assessed during a human digestion simulation and ex vivo colon fermentation. Half of the betanin amount was recovered in the small intestine digestive fluid and no traces were found after colon fermentation. Betanin high antioxidant ability was retained even after simulated small intestine digestion. Betanin, besides displaying an inherent colorant capacity, was equally effective as a natural antioxidant displaying peroxy-radical scavenger ability in pork meat. Betanin should be considered a multi-functional molecule able to confer an attractive color to frozen or refrigerated foods, but with the capacity to avoid lipid oxidation, thereby preserving food quality. Long-term supplementation by beetroot, a rich source of betanin, should be stimulated to protect organisms against oxidative stress.

Keywords: beetroot; betalains; semi-preparative RP-HPLC; in vitro human gastrointestinal digestion; ex vivo colon fermentation; antioxidant ability; malonildialdehyde

1. Introduction

Beetroot (*Beta vulgaris L.*) is a vegetable presenting significant scientific interest, mainly because it is a rich source of nitrate (NO_3^-), a compound with beneficial cardiovascular health effects, through the endogen production of nitric oxide (NO) [1]. Moreover, beetroots are the main source of betalains, a heterocyclic compound and water-soluble nitrogen pigment, which can be subdivided into two classes according to their chemical structure: betacyanins, such as betanin, prebetanin, isobetanin and neobetanin, responsible for red-violet coloring, and betaxanthins, responsible for orange-yellow coloring, comprising vulgaxanthin I and II and indicaxanthin [2,3]. Betalains are present in the tuberous part of beetroots, conferring its red-purple coloration.

Betanin (betanidin 5-*O*-β-D-glucoside) is the most abundant betacyanin and the only one approved for use as a natural colorant in food products, cosmetics and pharmaceuticals, under code EEC No. E 162 by the European Union and under Section 73.40 in Title 21 of the Code of Federal

Regulations (CFR) stipulated by the Food and Drug Administration (FDA) in the United States [4–6] (Supplementary File—Figure S1A).

In the food industry, synthetic antioxidants are added to foods containing fat, especially meats, with the purpose of delaying oxidative processes that result in undesirable sensorial changes, decreased shelf life and nutritional value and the formation of secondary compounds potentially harmful to health [7,8]. However, data in the literature have associated the synthetic antioxidants BHT (butylated hydroxytoluene) and BHA (butylated hydroxyanisole) with possible deleterious health effects, as they have been reported as potential tumor promoters, following the chronic administration of these compounds to animals [9,10]. This has motivated the replacement of synthetic antioxidants with natural antioxidants extracted from foodstuffs [11,12].

Betanin can be used as a powerful antioxidant in the food industry in extract or powder form, and is also applied as a natural pigment. Its antioxidant activity in biological lipid environments has been demonstrated in human macromolecules, such as low density lipoproteins, membranes and whole cells [13]. Furthermore, betanin has attracted attention due to its anti-inflammatory and hepatic protective functions in human cells. This compound is able to modulate redox-mediated signal transduction pathways involved in inflammation responses in cultured endothelia cells, and has also displayed antiproliferative effects on human tumor cell lines [14,15]. In both healthy and tumoral human hepatic cell lines, betanin can induce the translocation of the erythroid 2-related factor 2 (Nrf2) antioxidant response element (ARE) from the cytosol to the nuclear compartment, which controls the mRNA and protein levels of detoxifying/antioxidant enzymes, including GSTP, GSTM, GSTT, GSTA (glutathione *S*-transferases), NQO1 (NAD(P)H quinone dehydrogenase 1) and HO– (heme oxygenase-1), in these cells, exerting hepatoprotective and anticarcinogenic effects [16].

Several betanin purification techniques have been reported, involving distinct steps and methodologies in order to purify this compound from vegetal sources, including complex food matrices such as beetroot. Among the methods employed for betanin purification, high-performance liquid chromatography (HPLC) and other chromatographic methods using reverse phase columns seem to provide the best balance between speed and efficiency [17]. However, no other studies have evaluated the stability of this molecule during storage conditions and its antioxidant capability after purification and during storage with the aim of use as a food additive.

In this context, the aim of the present study was to optimize a methodology applied for betanin purification in large amounts, using fresh juice obtained from red beetroot (*Beta vulgaris* L. species). In addition, the chemical stability and bioactivity of the purified molecule were also assessed, through two different viewpoints: (i) as a food preservative and (ii) as a *in natura* or processed food matrix component after consumption and simulated gastrointestinal and ex vivo colon fermentation processes.

2. Results and Discussion

2.1. Betanin Purification

The chromatographic profile of betanin and isobetanin in fresh beetroot juice was compared to a commercial standard (Figure 1A,B). In the fresh juice chromatogram, betanin corresponds to the major peak (RT = 18.17 min), followed by its isomer, isobetanin (RT = 19.27 min). Betanin is found in large amounts relative to its isobetanin isomer in fresh beetroot juice, which is similar to the ratio between betanin and isobetanin previously reported by Gonçalves et al. [17]. The betanin concentrations in fresh juice and in purified samples were calculated in comparison to a standard curve, described by the equation $y = 5077078x - 370531$ ($R^2 = 0.9993$). Linearity was obtained at betanin concentration ranging from 0 to 500 $mg{\cdot}mL^{-1}$. After purification an HPLC-diode array detector (DAD) analysis indicated a single peak at the maximum absorption λ (535–540 nm), characteristic for betacyanins (Figure 1C). Purified betanin can be preserved at 4 °C for at least 20 days and for 275 days if maintained at −30 °C (Figure 1D). On the contrary of what has been described previously [17], betanin preparations turned brown during shelf life, due to the possible action of polyphenol oxidases (PPO enzymes) in the present study, whereas no sign of degradation was observed in the purified betanin. Temperature control, pH and non-exposure to light throughout the purification process reduced the chances of sample decomposition. Such precautions may have had a positive influence on the stability of the molecule, as demonstrated later in the stability assessment.

Purification technique yield is one of the most economically important aspects when obtaining natural food additives. The betanin extraction and purification yield from fresh beetroot juice and purified betanin recovered after chromatographic separation and mobile phase evaporation was calculated from an initial 500 g beetroot mass. The betanin concentration in juice was of 1.19 g mL^{-1} and the amount of purified betanin recovered after purification by chromatographic separation and mobile phase evaporation was of 48 mg·mL^{-1} corresponding to 4% of the initial betanin concentration.

As noted, in Figure 1C the beginning of betanin isomerization to isobetanin occurred, culminating with co-elution and a small base widening. The analytical column used for the separation and the elution conditions were not able to clearly separate the peaks in the initial isomerization stage, but the presence of the already formed isobetanin was well-evidenced and the peaks were clearly separated during storage (Figure 1D). It is noteworthy that betanin isomerization can be considered a structural modification rather than a degradation reaction. Isobetanin (2*S*/15*R*) differs from betanin (2*S*/15*S*) by the spatial conformation of the carbonyl group at carbon 15, but exhibits similar chromatic properties to betanin with no observable color change [18].

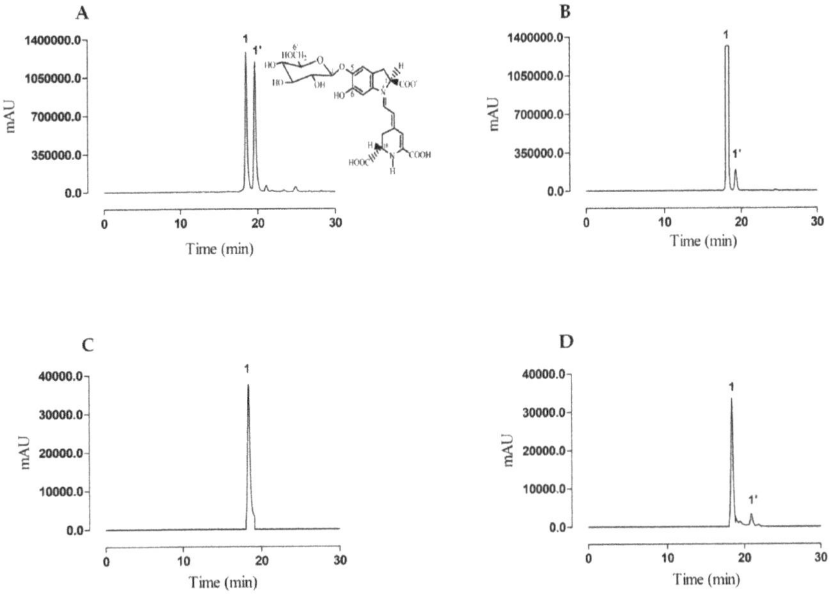

Figure 1. Betanin separation by high-performance liquid chromatography diode array detector (HPLC-DAD) monitored at 536 nm. (**A**) Betanin standard chromatographed in the analytical HPLC column, (**B**) fresh beetroot juice sample chromatographed in semi-preparative HPLC, (**C**) betanin purified by semi-preparative HPLC and separated using an analytical HPLC column and (**D**) betanin evaluated after 275 days of freezing and chromatographed using an analytical HPLC column. Betanin (peak 1) and isobetanin (peak 1′). The betanin chemical structure from red beet was reproduced from Cai et al. [19].

2.2. HPLC-ESI(+)-MS/MS Analysis

HPLC-purified betanin identification was performed by mass spectrometry, as displayed in Figure 2A,B. The pseudomolecular ion *m/z* 551.16 corresponding to betanin (Figure 2A) and its characteristic fragmentation (Figure 2B), the ion m/z 389.11 corresponding to its aglycone form, betanidin, a precursor structure of betacyanins (Supplementary File—Figure S1B), were observed, corroborating previous findings reported by Gonçalves et al. [17] and Netzel et al. [20].

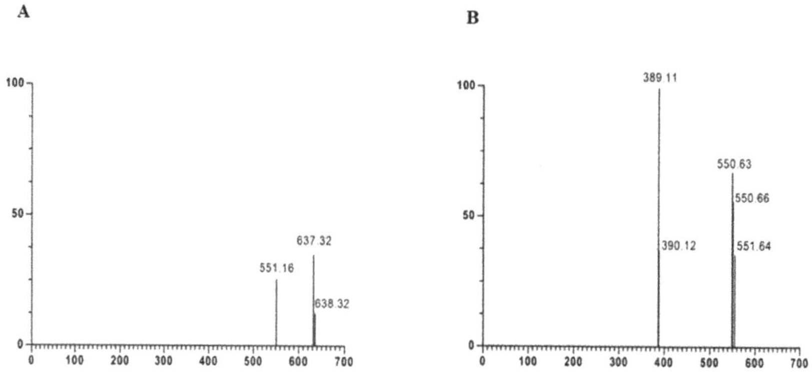

Figure 2. Identification of purified betanin by HPLC-ESI(+)-MS/MS. (**A**) betanin m/z 551 $[M + H]^+$, (**B**) fragmentation of purified betanin m/z from the MS/MS of 551 $[M + H]^+$.

It can be suggested that the m/z 637 adduct may be the result of the condensation of the formic acid (molecular weight 46) used in betanin purification and a decarboxylated betanin derivative, forming 6′-*O*-malonyl-2-decarboxyl-betanin (molecular weight 592), a decarboxylated betanin derivative [21]. The mass spectrometry analysis confirmed that the purification procedure was successful in isolating betanin in its purified form.

2.3. Storage Stability

Betanin was stable for over 275 days (9 months) when stored at −30 °C at pH 7 (Figure 1D) and at least for 20 days when stored at 4 °C. No significant loss of the betanin samples was observed before and after frozen storage (21.30 ± 1.98 mg·mL^{-1} versus 17.23 ± 4.82 mg·mL^{-1}, $p < 0.001$, respectively). The initial concentration of purified betanin samples during storage at refrigerator temperature was unaltered during the first 20 days (12.13 ± 0.70 mg·mL^{-1} versus 13.51 ± 1.01 mg·mL^{-1}, $p < 0.001$) but on the 27th day, a loss of 25% of the initial betanin concentration was observed (9.72 ± 0.60 mg·mL^{-1}).

Most betalains, including betanin, are under-utilized as colorants in processed foods due to reports concerning poor stability compared with the shelf-life of foods. In standard storage conditions betacyanin stability in spray dried beetroot powder was reported by Kaimainen et al. [22], who assessed the product stability by HPLC at 535 nm and pH 5 during storage for 25 weeks at different temperatures, including frozen (−20 °C) and chilled (4 °C). Betacyanins in beetroot powder remained unchanged during 4 months under freezing. However, stability under refrigeration was not well established. It is noteworthy that in that study, beside the addition of sweeteners, betanin was not in its purified form, but rather protected by the food matrix that naturally contains antioxidants and chelating agents, which may exert protective effects on the chemical structure of betanin maintained under freezing conditions [18]. Factors such as temperature, pH, type of buffer solution and the presence or absence of oxygen can affect betanin stability during storage. Betanine degradation results in the formation of betalamic acid and cyclo-dopa-5-*O*-glycoside. However, betanin can also display the ability to degrade and regenerate continuously during storage, as the reaction is reversible, thus maintaining betanin concentrations [23]. This continuous betanin regeneration capacity during the storage process is still not well-elucidated in the literature [2,23,24]. Therefore, the stability results found in the present study indicate that betanin, if used as an additive in refrigerated or frozen foods, would remain stable during the time and temperature recommended by the legislation as ideal for preserving meats and meat-derivatives from 6 to 12 months at temperatures below −18 °C and for 5 days at 4 °C [25].

2.4. Lipid Peroxidation Inhibition in Meat Matrices

Betanin's ability to inhibit lipid peroxidation process in meat was assessed by thiobarbituric acid reactive substance (TBARS) determination. Control samples (without antioxidants) showed the highest malondialdehyde (MDA) concentrations when compared to samples treated with betanin, BHA or BHT during 9 days of storage, determined on the 3rd, 6th and 9th days. Betanin 2% was equally effective in inhibiting lipid peroxidation when compared to the synthetic antioxidants BHA and BHT on the 3th and 6th day of storage ($p < 0.05$). Although the amounts of MDA in meat samples treated with betanin 2% (*w*/*w*) (5.07 ± 0.03 mg·kg^{-1}) were 22% and 16% higher than those found in samples treated with BHA (4.47 ± 0.10 mg·kg^{-1}) and BHT (4.28 ± 0.38 mg·kg^{-1}) on the 9th day of storage, the meat samples treated with betanin 2% (*w*/*w*) still presented lower MDA concentrations than the samples with no preservative addition (7.15 ± 0.07 mg·kg^{-1} vs 5.07 ± 0.03 mg·kg^{-1}) (Figure 3).

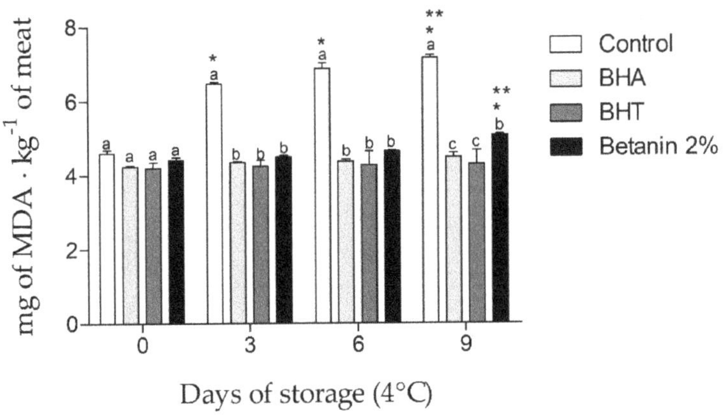

Figure 3. Lipid oxidation in ground pork loin evaluated by the production of malondialdehyde (MDA) during 9 days of storage at 4 °C. Control H_2O-DD, BHA (buthylated hydroxyanisole), BHT (butylated hydroxytoluene), Betanin 2% (*w*/*w*). Data are expressed as the means ± SD of three independent determinations. Different letters indicate differences between days at a significance level of $p < 0.01$. The symbol * ($p < 0.05$) indicates differences compared to day 0. The symbol ** ($p < 0.05$) indicates differences compared to day 3.

Few studies assessing the effect of the addition of beetroot or beetroot-extracted compounds on the oxidative stability of foods susceptible to lipid oxidation are available. The effect of beetroot inclusion as a mayonnaise ingredient promoted a higher inhibitory effect on lipid oxidation compared to the commercial antioxidant [26]. Contradictory results, however, were observed in fermented meat sausages, since no beetroot effect was observed [27].

To the best of our knowledge, the present study is the first to evaluate the use of purified betanin as a natural antioxidant in food matrices. Lipid oxidation is one of the main factors affecting food quality and is directly related to nutritional value and sensorial characteristics. The present study indicates that betanin used as an additive at the concentration of 2% (*w*/*w*) is a potential substitute for synthetic antioxidants in the preservation of refrigerated meat. Furthermore, betanin can exert its maximum protective effect against lipid oxidation for 6 days exceeding the 5-day shelf-life recommended for refrigerated meats [25].

2.5. Betanin Chemical Stability during In Vitro Simulated Gastrointestinal Digestion

A 23 mg·mL^{-1} dose was used to assess betanin bioavailability during in vitro human simulated gastrointestinal digestion by continuous multistage steps. A small loss was observed in betanin content after the oral phase digestion. However, more important decreases were observed after the gastric simulated digestion, reaching 65% of the initial sample content, and lowering to 46%, after small intestine simulated digestion (Table 1). No betanin was detected after the ex vivo colon fermentation assay, where the remaining betanin recovered at the end of the in vitro simulated gastrointestinal digestion, corresponding to 54% of the original sample, was assayed by ex vivo colon fermentation (Supplementary File—Figure S2).

Table 1. Betanin concentrations during in vitro simulated gastrointestinal digestion.

	Pre-Digestion	Oral Fluid	Gastric Fluid	Small Intestine Fluid	Colon Fermentation Fluid
Betanin content (mg·mL^{-1})	23.05 ± 0.61 [a]	21.44 ± 2.03 [a]	14.84 ± 0.11 [b]	12.42 ± 0.01 [c]	0.0
Loss (mg·mL^{-1}) and loss percentage after pre-digestion	-	1.6 (≈7%)	8.2 (≈35%)	10.6 (≈46%)	-

Betanin availability was determined by reverse phase high-performance liquid chromatography diode array detector (RP-HPLC-DAD), assessed through changes in the peak area determined at 536 nm. In vitro human gastrointestinal digestion was sequentially simulated and samples were harvested at each phase. The ex vivo colon assay was performed incubating the digested material obtained after the entire in vitro gastrointestinal digestion with fresh feces donated by seven healthy volunteers. Data are expressed as the means ± SD of three independent experiments. Different letters in the same line indicate significant differences between samples ($p < 0.01$).

Only a few studies presenting limitations are available on the chemical stability of the purified betanin in in vitro simulated digestion through the gastrointestinal tract. In a previous report, betanin was degraded by 75% and 35% after the gastric and intestinal phases, respectively [28]. However, the sample was added directly to each fluid—gastric or intestinal, generating no information about physiological digestion in the digestive tract. In the present study, an in vitro simulated digestion was conducted sequentially, where betanin was added to simulated oral fluid and digestion was observed by a sequential in vitro system, transferring aliquots to the simulated gastric fluid, resulting in a decrease of 35% after gastric digestion. This 35% decrease in betanin contents observed after gastric digestion is due to its impaired stability at acidic pH 2. It is known that betalains exhibit stability at pH ranging from 3 and 7 [18]. A significant decrease in betacyanin stability in a solution containing hydrochloric acid at pH 2.0 at 37 °C was observed, whereas betacyanins maintained at pH 4.7 were less susceptible to degradation [29]. In acid pH, the betanin structure can be degraded in C-17 decarboxylation, dehydrogenation and cleavage of betalamic acid and cyclo-Dopa-5-*O*-β-glycoside [2,30]. Herein, in addition to exposure to acidic pH, the absence of the food matrix may have exacerbated the betanin susceptibility to gastric fluid degradation, since it has been previously demonstrated that betanin and its isobetanin isomer (unpurified) can be protected from stomach digestion by the food matrix [31].

In addition, an overall decrease in betanin content to approximately 46% was noted when the simulated gastric fluid digestion was performed, followed by an 11% decrease during intestinal digestion. Besides the absence of the protective effect of the food matrix, the influence of pH on betanin stability is reinforced by the data reported herein, comparing the percentage of loss in the gastric phase to the intestinal phase (35% vs 11%). Small intestine pH is around 6.5, matching the reported pH-range of betanin stability and corroborating the lower betanin degradation at the small intestine [31].

Several polyphenols are described as reaching the large intestine, where they are absorbed following metabolization by colon bacteria consortia [32]. In the present study, betanin was detectable by HPLC at the 536–540 nm range, while no other metabolite derived from its biosynthetic conversion was observed. Although betanin was shown to be stable at 30 °C and 4 °C, the 24 h exposure to

37 °C and environmental conditions in the colon lumen may still promote its degradation. In a murine model study, only 3% of betanin administered to animals were recovered in feces after 24 h, indicating that colon absorption is not likely [28]. Permeability and solubility are important barriers concerning colon absorption in humans that, alongside the metabolic activity of bacterial consortia and the mild temperatures within the large intestine lumen point to improbable colon absorption and/or maintenance of the chemical stability of the betanin molecule.

2.6. Betanin Antioxidant Activity throughout Simulated Human Gastrointestinal Digestion

Betanin in its purified form was able to inhibit the OH-radical in the total antioxidant potential (TAP) assay. The OH-radical is considered the most reactive oxidant in living organisms, generated by the Fenton reaction [33] (Supplementary File—Figure S3). In the ferric reducing ability of plasma (FRAP) assay, betanin was effective in reducing the ferric ion of the tripyridyltriazine complex (Fe^{3+}-TPTZ) to the ferrous ion (Fe^{2+}-TPTZ), reflecting its ability to donate electrons and reduce reactive species. In addition, betanin was effective in reducing the $ABTS^{+}$ radical, as observed in the trolox equivalent antioxidant capacity (TEAC) and oxygen radical antioxidant capacity (ORAC) assays (Table 2). In addition, betanin showed a high TAP value after both oral and intestinal digestion.

Table 2. Total betanin antioxidant potential and antioxidant activity pre and post in vitro simulated human gastrointestinal digestion.

		TAP (%)	FRAP µmoL ($Fe^{2+}\cdot L^{-1}$)	TEAC µmoL (Trolox·L^{-1})	ORAC µmoL (Trolox·L^{-1})
Pre-digestion	Betanin	75.42 ± 5.91 ^{b}	518.31 ± 3.31 ^{c}	3932.02 ± 94.42 ^{a}	1992.44 ± 214.31 ^{ab}
Post-digestion	Oral fluid	80.71 ± 0.92 ^{b}	585.82 ± 13.23 ^{b}	4964.03 ± 5.31 ^{a}	2217.53 ± 10.31 ^{a}
	Gastric fluid	55.11 ± 9.23 ^{c}	400.02 ± 12.43 ^{d}	1382.94 ± 4.91 ^{b}	1475.41 ± 18.73 ^{c}
	Small intestine fluid	96.63 ± 0.61 ^{a}	1053.81 ± 164.64 ^{a}	4312.71 ± 651.81 ^{a}	2199.71 ± 19.75 ^{a}

Betanin antioxidant potential and antioxidant activity were evaluated before and after the simulated human gastrointestinal digestion using different assays, namely FRAP, TEAC and ORAC. Data are expressed as the means ± SD from three independent experiments. Different letters in the same column indicate difference at a significance level of $p < 0.001$.

The high antioxidant activity of betanin is well-documented [34]. However, to promote health beneficial in human beings, the chemical structure of the ingested betanin and its antioxidant properties should be maintained in the gastrointestinal absorption site. The antioxidant activities of purified betanin following the final sequential digestive process through the gastrointestinal apparatus increased when compared to pre-digested samples, as demonstrated in the FRAP and TAP assays (Table 2).

When assessing each digestion fluid, the antioxidant activity of betanin evaluated by the FRAP assay was increased in the post-oral digestion assays. The FRAP assay increments can be attributed to natural antioxidants originally present in human saliva [35]. After small intestine digestion, purified betanin increased or maintained the antioxidant activities evaluated by the TAP, FRAP, TEAC and ORAC assays when compared to the oral digestion processes (Table 2).

Additionally, the pH of the different fluids in the human body undergoes variations, influencing the stability and bioactivity of betanin in the different digestive tract compartments. Salivary fluid present a pH of about 7.4, whereas stomach is maintained between 1.5 and 3.5, while the abdominal cavities, including the small and large intestine, display a pH of 7.4.

The antioxidant ability of betanin was reduced in the acidic pH of the gastric fluid, evidencing the pH-dependence of the free radical-scavenging activity of betanin, but, countering its antioxidant activity in the alkaline environment of the small intestine lumen, where the antioxidant activity assessed by all assays was either increased or maintained at the same level of the oral fluid, as mentioned previously (Table 2). The alternate antioxidant activity of betanin between low and high levels in the simulated gastric compartment and in the small intestine indicates that the chemical structure of betanin was maintained unaltered following acidic pH exposure during the simulated gastric digestion, but should be attributed to the protonation of the betanin molecule, favoring the maintenance of free radical-scavenging activity until absorption in the gastrointestinal tract.

Changes in pH fluids induce changes in the chemical structure of betanin along the gastroenteric apparatus. Betanin presents its cationic form at pH < 2, zwitterionic form at pH = 2, anionic mono at 2 < pH < 3 with deprotonated C2-COOH and C15-COOH groups, anionic bis at 3.5 < pH < 7 with C2-COOH, C15-COOH and C17-COOH groups deprotonated and anionic tris at 7.5 < pH < 9 with all carboxyl deprotonated groups, in addition to the C6-OH group on the phenolic ring (Supplementary File—Figure S4) [36]. The increase in betanin bioactivity, when in an alkaline pH environment, as found in the simulated small intestine fluid, can be ascribed to its ability to donate H^+ and electrons when altering from the cationic to the mono, bis and tri deprotonated states. The free radical scavenging activity of betanin at different pH (from 2 to 9) was previously determined through the TEAC assay, phenolic O–H homolytic bond dissociation energy (OH BDE), ionization potential (IP) and deprotonation energy (DE) [37]. The TEAC assay indicated that the antioxidant activity of betanin is dependent on pH, and very high above pH > 4. Moreover, with the gradual increase of the deprotonation of the betanin molecule (mono, bi- and tri-deprotonated) according to increasing pH, BDE and PI values decreased. This implies that, at slightly alkaline pH, betanin becomes a better hydrogen and electron donor, increasing its radical-scavenger ability as observed in the antioxidant assays in the simulated small intestine fluid when compared to the gastric fluid.

Although almost 46% (11 mg) of betanin content were chemically modified in the gastric tract, the antioxidant power of the remaining betanin, 54% of the original amount, corresponding to 12 mg found in the small intestine fluid (21 μmol) seems to be enough to promote lipid oxidation inhibition, since betanin in the range of 0.3–1.9 μmol has been found to inhibit lipid peroxidation in biological membranes, in a linoleate emulsion catalyzed by the "free iron" redox cycle, in H_2O_2-activated metmyoglobin and in lipoxygenase activity [38].

3. Material and Methods

3.1. Standards and Reagents

The betanin standard ($C_{24}H_{26}N2O_{13}$), sulfuric acid (H_2SO_4), boric acid (H_3BO_3), formic acid (CH_2O_2), hydrochloric acid (HCL), terephthalic acid (TPA, $C_8H_6O_4$), ethylenediaminetetraacetic acid (EDTA), 6-hydroxy-2-5-7-8-tetramethylchromo-2-carboxylic acid (Trolox), ascorbic acid ($C_6H_8O_6$), sodium hydroxide (NaOH), potassium permanganate ($KMnO_4$), hydrogen peroxide (H_2O_2), potassium sulphate (K_2SO_4), ferrous sulphate ($FeSO_4$), methyl red, bromocresol green, petroleum ether, anhydrous sodium acetate (CH_3COONa), tetrabutylammonium perchlorate ($C_{16}H_{36}N.H_2PO_4$), vanillin, tripyridyltriazine (TPTZ, $C_{18}H_{12}N_6$), iron chloride ($FeCl_3$), dibasic sodium phosphate ($NaH_2PO_4.H_2O$), monobasic sodium phosphate ($NaH_2PO_4.H_2O$), sodium chloride (NaCl), anhydrous monobasic sodium phosphate (Na_2HPO_4), sodium bicarbonate ($NaHCO_3$) C-211,2,2′-Azobis (2-methylpropionamidine), dihydrochloride (AAPH), 2,2′-azinobis [3-ethylbenzothiazoline-6-sulfonic acid]-diammonium salt (ABTS, $C_{18}H_{24}N_6O_6S_4$), sodium fluorescein ($C_{20}H_{12}O_5$), potassium hydroxide (KOH), ammonium thiocyanate (NH_4SCN), trichloroacetic acid (Cl_3CCOOH) were purchased from Sigma-Aldrich Chemical Co. (São Paulo, SP, Brazil). Methanol (MeOH), ethanol, acetone, and acetonitrile were purchased from Tedia Company Inc. (Rio de Janeiro, RJ, Brazil). Buthylated hydroxyanisole (BHA), butylated hydroxytoluene (BHT), 1,1,3,3-tetramethoxypropane ($(CH_3O)_2CHCH_2CH(OCH_3)_2$) and 2-thiobarbituric acid ($C_4H_4N_2O_2S$) were purchased from Sigma-Aldrich Co. HPLC grade Milli-Q water (Merck Millipore, Burlington, MA, USA) was used throughout the experiments.

3.2. Betanin Purification

3.2.1. Sample Preparation

Red beetroot was peeled, sliced and homogenized using a centrifuge food processor EC 700 (Black and Decker, São Paulo, Brazil). The homogenates were centrifuged at 15,000× *g* for 30 min at 25 °C and filtered through a PTFE filter membrane 25 mm, pore size 0.45 μm (Merck-Millipore). The supernatants (4 mL) were concentrated under reduced pressure (18 mbar, 25 °C) and resuspended in 2 mL deionized water.

3.2.2. HPLC Betanin Purification

Concentrated beetroot juice was purified by RP-HPLC. The HPLC apparatus consisted in an LC-20A Prominence, (Shimadzu®, Kyoto, Japan) equipped with a quaternary pump and a DAD model SPD-M20A (Shimadzu®, Kyoto, Japan). A 15 μm Phenomenex C18 column (250 × 21.2 mm I.D., Torrance, California, USA) connected to an FRC-10A fraction collector (Shimadzu®) was used in the semi-preparative HPLC. The elution conditions were performed according to Cai et al. [39] with modifications. Solvent A was 1% formic acid, and solvent B was 80% methanol at a linear gradient (0–25 min, 11–55%). The injection volume was 100 μL and a flow rate of 5.5 mLmin^{-1} was used. Separations were monitored at 536 nm and, after purification, magenta fractions, containing betanin, were concentrated by a rotary evaporator (Rotavapor® R-215, Buchi, São Paulo, Brazil) at 24 °C, 150 rpm and a water bath at 40 °C. The extracts were then suspended in 1 mL deionized water and stored at −30 °C under an N_2 atmosphere for further analysis. The purified betanin was analyzed using a Nucleosil 100-C18 column (250 × 4.6 mm I.D., 5 μm) with 30 μL injection volume and a flow rate of 1.0 mL min^{-1}. The mobile phase and gradient conditions were similar to the purification step and betanin concentrations were quantified in comparison to a betanin standard solution (Sigma-Aldrich Co.).

3.3. Betanin Identification by Liquid Chromatography Positive Ion Electrospray Ionization Tandem Mass Spectrometry (LC-ESI(+)-MS/MS)

Mass spectrometry was performed as described by Gonçalves et al. [17]. The RP-HPLC purified fraction was ionized in the positive mode and ions were monitored in the full scan mode (range of m/z 50–1500). The ESI(+)-MS/MS analysis was carried out on a Bruker Esquire 3000 Plus Ion Trap Mass Spectrometer (Bruker Co., Billerica, MA, USA) equipped with an electrospray source in the positive ion mode. Nitrogen was used as the nebulizing (45 psi) and drying gas (6 L·min^{-1}, 300 °C) and helium as the buffer gas (4 × 10^{-6} mbar). The high capillary voltage was set to 3500 V. To avoid space–charge effects, smart ion charge control (ICC) was set to an arbitrary value of 50.000. Betanin identification was based on its mass (550 g·mol^{-1}) and by similarity with the commercial standard and literature-available spectra [39].

3.4. Storage Stability

The stability of purified betanin during refrigeration (4 °C) and freezing (−30 °C) was evaluated by RP-HPLC-DAD (Shimadzu®, Kyoto, Japan), monitoring changes in the area under the chromatogram peak obtained at 536 nm, in similar conditions as those described for the betanin analysis.

3.5. Betanin Ability to Inhibit Lipid Peroxidation in Meat

Betanin ability to inhibit lipid peroxidation was evaluated by MDA determination in meat TBARS assay, as described previously [40] with modifications. A sample of ground pork loin (500 g) from a local butcher shop in Rio de Janeiro, Brazil and divided into 4 portions and treated as follow: (i) ground pork loin non-treated by antioxidants; (ii) ground pork loin treated with betanin (2%; *w*/*w*); (iii) ground pork loin treated with BHT (0.01%); (iv) ground pork loin treated with BHA (0.01%). MDA extraction was performed in 3.0 g of each meat sample homogenized with 9 mL of 7.5% TCA. The homogenate was centrifuged at 3000× *g* for 15 min at 25 °C and filtered through Whatman n° 4 paper (Merck Millipore Co). TMP (the MDA standard) at 3.2 mM in 0.1 M HCl (stock solution) was kept for 2 h at room temperature in the dark. After hydrolysis, the TMP solution was diluted with 7.5% TCA to the concentrations of 1, 2, 4, 8, 16 and 32 μM. After, 1 mL of MDA at different concentrations or 7.5% TCA solution (blank) was transferred into a screw-cap tube and 1 mL of 20 mM TBA solution was added. The tubes were heated in a boiling water bath at 90 °C for 30 min and cooled in tap water for 10 min. Absorbance of the MDA-TBA adducts were measured at 532 nm on a spectrophotometer DU®530 (Beckman Coulter Inc., Brea, CA, USA). Because betanin absorbs light in the range of 530–540 nm, additional blanks containing betanin (1 or 2%), TCA or TBA (no meat) were used to correct the overestimation of the TBA-MDA adduct absorbance. The concentration of MDA was expressed in mg of MDA per kg of meat (mg of MDA·kg^{-1} meat) at each treatment along the 9 days storage at 4 °C.

3.6. TAP Determination

Betanin samples were analyzed as described previously [41]. Samples were diluted (1:10) and centrifuged at 4500× g for 10 min, and the supernatants were then filtered through 0.45 μm cellulose membranes (Merck Millipore Co). The resulting samples were transferred to amber vials and incubated at 37 °C for 10 min with a solution containing 1 mM Fe^{2+}, 10 mM H_2O_2 and 1 mM terephthalic acid (TPA) in 50 mM phosphate buffer pH 7.4. Hydroxyterephthalic acids (HTPA) were detected by HPLC. TAP measurements were obtained by the difference between the chromatogram surface area generated in the Fenton reaction with and without the sample.

3.7. Antioxidant Activity Determination by Different Assays

3.7.1. FRAP Determination

FRAP assays were performed using a modification of the method described by Benzie and Strain [42]. Betanin samples were diluted (1:10) and then mixed thoroughly with the FRAP reagent. Standard $FeSO_4$ solutions were used and absorbances at 593 nm were determined on a V–530 UV/VIS spectrophotometer (Jasco®, Easton, PA, USA). The FRAP results for each sample were evaluated in triplicate and expressed as μmol of Fe^{2+}·L^{-1}.

3.7.2. TEAC Determination

TEAC assays were performed using a modification of the method described by Re et al. [43]. The ABTS radical cation ($ABTS^{\bullet+}$) was generated by chemical reaction of ABTS with $K_2S_2O_8$ in the dark at room temperature for 12–16 h. Each betanin sample was mixed with the $ABTS^{\bullet+}$ reagent and absorbances at 720 nm were determined using a V–530 UV/VIS spectrophotometer (Jasco®). TEAC results were determined in triplicate and were associated to the $ABTS^{\bullet+}$ inhibition percentage by antioxidants present in the samples. The TEAC results for each beetroot sample were evaluated in triplicate and expressed as μmol of Trolox·L^{-1}.

3.7.3. ORAC Determination

The ORAC assay was performed according to Zuleta et al. [44], with modifications. Sample absorbances were determined on a Wallac 1420 VICTOR multilabel counter (Perkin–Elmer Inc, Waltham, MA, USA) with fluorescence filters at an excitation wavelength of 485 nm and emission wavelength of 535 nm. A fluorescein stock solution was prepared by weighing 3 mg of fluorescein followed by dissolution in 100 mL of phosphate buffer (75 mM, pH 7.4). The fluorescein stock solution was stored in complete darkness under refrigeration. The fluorescein working solution (78 nM) was prepared daily by dilution of 0.100 mL of the fluorescein stock solution in 100 mL of phosphate buffer. The AAPH radical (221 mM) was prepared daily by mixing 600 mg of AAPH in 10 mL phosphate buffer. A 25 μM Trolox solution was used as reference standard, prepared daily in phosphate buffer from a 4 mM stock standard solution kept in a freezer at 20 °C. A total of 100 μL of fluorescein (78 nM) and 100 μL of the samples, blanks (phosphate buffer), or standards (25 μM of Trolox) were added to each well, followed by 50 μL of AAPH (221 mM). ORAC values, expressed as μM Trolox equivalents were calculated by applying the following formula:

$$\text{ORAC } (\mu\text{M Trolox equivalents}) = \frac{C_{Trolox} \cdot (AUC_{Sample} - AUC_{Blank}) \cdot k}{(AUC_{Trolox} - AUC_{Blank})}$$

where C_{Trolox} is the Trolox concentration (μM), k is the sample dilution factor, and AUC is the area below the fluorescence decay curve of the samples, blanks and Trolox, respectively, calculated using the GraphPad Prism v.5 software package (GraphPad Software Inc., San Diego, CA, USA). ORAC determinations were performed in triplicate and expressed as mmol Trolox equivalents·100 g^{-1}.

3.8. Simulated Betanin In Vitro Human Gastrointestinal Digestion and Ex Vivo Colon Fermentation (Supplementary File—Figure S5)

Betanin concentrations after in vitro oral, gastric and small intestine digestion and ex vivo colon fermentation were evaluated by RP-HPLC while antioxidant activity was evaluated by TAP, FRAP, TEAC and ORAC assays, as described previously. Samples were analyzed in triplicate.

The in vitro human simulated gastrointestinal digestion, including the oral, gastric and small intestine phases, was performed according to Oomen et al., [45] and Sagratini et al. [46], with modifications. For the OD simulation, 1 mL betanin aliquots at 23 mg·mL^{-1} were placed in a glass jar followed by the addition of 3 mL of human saliva, and incubated at 37 °C for 1 min under orbital agitation at 260 rpm in a Sorvall ST 16R centrifuge (Thermo Scientific™, Waltham, MA, USA) to complete the OD.

A 2.5 mL aliquot of artificial gastric fluid containing 2.75 g of NaCl, 0.27 g of NaH_2PO_4, 0.82 g of KCl, 0.42 g of $CaCl_2$, 0.31 g of NH_4Cl, 0.65 g of glucose, 0.085 g of urea, 3.0 g of mucine, 2.64 g of swine gastric pepsin, 1.0 g of bovine albumin, 8.3 mL of HCl was added to the oral fluid sample to a final volume of 500 mL and the pH was adjusted to 2.0 with 5 M HCl. The glass jars were then resealed with a rubber septum and the atmosphere was saturated with N_2 and incubated at 37 °C for 2 h under orbital shaking at 260 rpm to complete the GD.

The gastric fluid had its pH adjusted to 6.0 with $NaHCO_3$ and 2.0 mL of artificial small intestine fluid containing 6.75 g of NaCl, 0.517 g of KCl, 0.205 g of $CaCl_2$, 3.99 g of $NaHCO_3$, 0.06 g of KH_2PO_4, 0.0375 g of $MgCl_2$, 0.1375 g of urea, 25.0 g of swine bile, 4.0 g of swine pancreatin, 1.2 g of albumin bovine and 0.185 mL HCl were added to a final volume of 500 mL. The glass jars were then resealed and the atmosphere was saturated with N_2 and incubated at 37 °C for 2 h under orbital shaking at 260 rpm to complete the ID. At the end of each simulated gastrointestinal step (oral, gastric and small intestine), aliquots were collected and centrifuged (3000× g, 15 min, 25 °C). The supernatants were then filtered through 0.45 and 0.22 μm membranes, followed by Amicon ultra filtration using 10 kDa cut-off membranes.

Ex vivo Colon Fermentation

The ex vivo colon fermentation assay was performed according to Hu et al. [47] with modifications, in accordance to the ethical standards of the declaration of Helsinki after approval by the Hospital Universitário Clementino Fraga Filho/Universidade Federal do Rio de Janeiro Education and Research Committee, under No. 512.84.

The ex vivo assay was performed using fresh feces donated by seven healthy volunteers (4 men and 3 women), recruited according to the following criteria: age between 18 and 50, eutrophic (BMI between 18.5 and 24.9 kg m^2), absence of gastrointestinal diseases, displaying one bowel movement every two days and up two bowel movements per day, with no medication and/or food supplements used 90 days prior to the feces collection.

The feces were homogenized in a nutrient-rich medium (0.5 g·10 mL^{-1}), prepared according to McDonald et al. [48], where the medium was autoclaved and saturated with CO_2 in an anaerobic chamber for 48 h. A 5 mL aliquot of this mixture was added to the digested material after the in vitro gastrointestinal digestion. The mixture was then incubated at 37 °C, under orbital shaking at 50 rpm for 48 h. The ex vivo colon fermentation assay was independently repeated three times.

3.9. Statistical Analyses

A one-way analysis of variance (ANOVA) with repeated measurements were performed to identify differences in TAP and antioxidant activities (FRAP, TEAC and ORAC) before and after the simulated gastrointestinal digestion (pre-digestion, oral, gastric, small intestine and colon phases). In addition, a two-way analysis of variance (ANOVA) with repeated measurements was performed to identify differences in MDA concentrations in the lipid peroxidation assay between each type of antioxidant and between each experiment day, evaluated by TBARS assay. When a significant F was found, an additional *post hoc* analysis was performed by a Bonferroni correction. Data were expressed as means ± standard deviation (SD). The statistical analyses were performed using Graphpad Prism software version 5 for Windows® (GraphPad Software, San Diego, CA, USA).

4. Conclusions

The decomposition of food components and bioactive or additive compounds along the gastrointestinal trait is a helpful tool to assess their potential positive and negative effects on health, as well as to evaluate possible toxicity and/or safety usage. The prediction of the in vivo absorption of such compounds can aid in compound usage regulation, establishing safety dosages in food.

Betanin in its purified form can be very stable during storage at low temperature and alkaline pH, so it may be useful as food colorant and antioxidant additive in meat and meat derivatives as a substitute for synthetic antioxidants. According to the chemical stability at both refrigerator and freezer temperatures, betanin can be used as a food colorant or preservative in almost all foodstuffs, including those stored at −30 °C, such as bacon, sausages, vegetables, ham, corned beef, ice cream and sherbet (www.fda.gov/food/guidance/regulation) and those preserved at 4–8 °C, including minced meat and fresh meat, yogurts and desserts, since their shelf life is lower than the 20 days in which betanin is stable.

In addition, betanin maintained its bioactivity during the simulated digestive process, presenting high TAP, FRAP, TEAC and ORAC values in the intestinal phase. Although the exact parts of the gastrointestinal tract in where betanin is absorbed still require elucidation, it can be suggested that absorption may occur in the small intestine. The chemical integrity of betanin and its antioxidant activity can be considered potential aid against diseases caused by oxidative stress.

These novel findings reinforce the importance of the regular uptake of red beetroot and its derivative products. The formulation of new dietary supplements or processed foods can include purified betanin, not only as a natural food colorant or preservative, but also as a bioactive compound that may act as an adjuvant in the treatment and prevention of chronic diseases related to oxidative stress in humans.

Supplementary Materials: Figure S1: Chemical structure of betanin (A) and betanid in (B); Figure S2: Betanin chromatograms before and after each in vitro digestion phase assessed by RP-HPLC equipped with DAD detector (536 nm). Purified betanin (A), after oral digestion (B), after gastric digestion (C), after small intestine digestion (D); Figure S3: Betanin total antioxidant potential (TAP) assessed by RP-HPLC equipped with a fluorescence detector (312/428 nm). Hydroxyterephthalic acid (HTPA) chromatograms of generated in the Fenton reaction without any sample (A), after betanin addition (B), after oral digestion (C), after gastric digestion (D), after small intestine digestion (E); Figure S4: Influence of pH on betanin chemical structure charge changes in an aqueous solution according to Frank et al. [35]. Figure S5: Simulated digestion scheme.

Author Contributions: D.V.T.d.S., E.M.D.A. and V.M.F.P. conceptualized and designed the research; D.V.T.d.S., D.d.S.B., F.d.O.S., G.A. and D.P. compiled, quantified and performed all experiments; D.V.T.d.S., D.d.S.B., E.M.D.A. and V.M.F.P. evaluated and interpreted all the results; D.V.T.d.S. and D.d.S.B. prepared the figures; D.V.T.d.S and V.M.F.P. wrote the manuscript; D.V.T.S., D.d.S.B. and V.M.F.P. edited and revised the manuscript.

All authors critically revised the manuscript concerning important intellectual content, and read and approved the final manuscript.

Funding: The authors acknowledge financial support from CNPq (Conselho Nacional de Desenvolvimento Científico e Tecnológico), CAPES (Coordenação de Aperfeiçoamento de Pessoal de Nível Superior) and FAPERJ (Fundação Carlos Chagas Filho de Amparo à Pesquisa do Estado do Rio de Janeiro).

Conflicts of Interest: This article does not report any studies with human or animal subjects. The authors declare no conflicts of interest.

References

1. Baião, D.S.; da Silva, D.V.; Del Aguila, E.M.; Paschoalin, V.M.F. Nutritional, Bioactive and Physicochemical Characteristics of Different Beetroot Formulations, Food Additives. 6 September 2017. Available online: https://www.intechopen.com/books/food-additives/nutritional-bioactive-and-physicochemical-characteristics-of-different-beetroot-formulations (accessed on 3 December 2018).
2. Azeredo, M.C. Betalains: Properties, sources, applications, and stability—A review. *Int. J. Food Sci. Technol.* **2009**, *44*, 2365–2376. [CrossRef]
3. Saponjac, V.T.; Canadanovic-Brunet, J.; Cetkovic, G.; Jakisic, M.; Djilas, S.; Vulic, J.; Stajcic, S. Encapisulation of beetroot pomace extract: Rsm optimization, storage and gastrointestinal stability. *Molecules* **2016**, *21*, 584. [CrossRef] [PubMed]
4. European Food Safety Authority. Scientific opinion on the re-evaluation of beetroot red (E 162) as a food additive. *EFSA J.* **2015**, *13*, 4318. [CrossRef]
5. US Food and Drug Administration (FDA). Code of Federal Regulations. 2009. Available online: https://www.accessdata.fda.gov/scripts/cdrh/cfdocs/cfcfr/CFRSearch.cfm?fr=73.260 (accessed on 28 January 2019).
6. Nemzer, B.; Pietrzkowski, Z.; Sporna, A.; Stalica, P.; Thresher, W.; Michałowski, T.; Wybraniec, S. Betalainic and nutritional profiles of pigment-enriched red beet root (*Beta vulgaris* L.) dried extracts. *Food Chem.* **2011**, *127*, 42–53. [CrossRef]
7. Gómez, M.; Lorenzo, J.M. Effect of packaging conditions on shelf-life of foal fresh meat. *Meat Sci.* **2012**, *91*, 513–520. [CrossRef] [PubMed]
8. Falowo, A.B.; Fayemi, P.O.; Muchenje, V. Natural antioxidants against lipid-protein oxidative deterioration in meat and meat products: A review. *Food Res. Int.* **2014**, *64*, 171–181. [CrossRef]
9. Ito, N.; Fukushima, S.; Tsuda, H.; Shirai, T.; Tatematsu, M.; Imaida, K. Modification of chemical carcinogenesis by antioxidants. *Princess Takamatsu Symp.* **1983**, *14*, 381–389. [CrossRef]
10. Williams, G.M.; Wang, C.X.; Iatropoulos, M.J. Toxicity studies of butylated hydroxanisole and butylated hydroxytoluene. II. Chronic feeding studies. *Food Chem. Toxicol.* **1990**, *28*, 799–806. [CrossRef]
11. Cunha, L.C.M.; Monteiro, M.L.G.; Costa-Lima, B.R.C.; Guedes-Oliveira, J.M.; Alves, V.H.M.; Almeida, A.L.; Tonon, R.V.; Rosenthal, A.; Conte-Junior, C.A. Effect of microencapsulated extract of pitaya (*Hylocereuscostaricensis*) peel on color, texture and oxidative stability of refrigerated ground pork patties submitted to high pressure processing. *Innov. Food Sci. Emerg. Technol.* **2018**, *49*, 136–145. [CrossRef]
12. Muíño, I.; Fuente, J.; Pérez, C.; Apeleo, E.; Pérez-Santaescolástica, C.; Cañeque, V.; Lauzurica, S.; Bermejo-Poza, R.; Díaz, A.M.T. Use of red wine polyphenols as a natural preservative in health-promoting omega-3 fatty acids-enriched lamb patties. *Molecules* **2018**, *23*, 3080. [CrossRef] [PubMed]
13. Khan, M.I. Plant betalain: Safety, antioxidant activity, clinical efficacy, and bioavailability. *Compr. Rev. Food Sci. Food Saf.* **2015**, *15*, 316–330. [CrossRef]
14. Gentile, C.; Tesoriere, L.; Allegra, M.; Livrea, M.A.; D'Alessio, P. Antioxidant betalains from cactus pear (*Opuntiaficus-indica*) inhibit endothelial ICAM-1 expression. *Ann. N. Y. Acad. Sci.* **2004**, *1028*, 481–486. [CrossRef] [PubMed]
15. Kapadia, G.J.; Azuine, M.A.; Rao, G.S.; Arai, T.; Lida, A.; Tokuda, H. Cytotoxic effect of the red beetroot (*Beta vulgaris* L.) extract compared to doxorubicin (*Adriamycin*) in the human prostate (PC-3) and breast (MCF-7) cancer cell lines. *Anticancer Agents Med. Chem.* **2011**, *11*, 280–284. [CrossRef] [PubMed]
16. Krajka-Kuźniak, V.; Paluszczak, J.; Szaefer, H.; Baer-Dubowska, W. Betanin, a beetroot component, induces nuclear factor erythroid-2-related factor 2-mediated expression of detoxifying/antioxidant enzymes in human liver cell lines. *Br. J. Nutr.* **2013**, *110*, 2138–2149. [CrossRef] [PubMed]

17. Gonçalves, L.C.P.; Trassi, M.A.S.; Lopes, N.B.; Dör, F.A.; dos Santos, M.T.; Baader, J.W.; Oliveira Jr, V.X.; Bastos, E.L. A comparative study of the purification of betanin. *Food Chem.* **2012**, *131*, 231–238. [CrossRef]
18. Herbach, K.M.; Stintzing, F.C.; Carle, R. Betalain stability and degradation—Structural and chromatic aspects. *J. Food Sci.* **2006**, *71*, R41–R50. [CrossRef]
19. Cai, Y.; Sun, M.; Corke, H. Antioxidant activity of betalains from plants of the *Amaranthaceae*. *J. Agric. Food Chem.* **2003**, *51*, 2288–2294. [CrossRef]
20. Netzel, M.; Stintzing, F.C.; Quaas, D.; Stra, B.G.; Carle, R.; Bitsch, R.; Bitsch, I.; Frank, T. Renal excretion of antioxidative constituents from red beet in humans. *Food Res. Int.* **2005**, *38*, 1051–1058. [CrossRef]
21. Belhadj, S.I.; Najar, T.; Abderrabba, M. Chemical and antioxidant properties of betalains. *J. Agric. Food Chem.* **2017**, *65*, 675–689. [CrossRef]
22. Kaimainen, M.; Laaksonen, O.; Järvenpää, E.; Sandell, M.; Huopalahti, R. Consumer acceptance and stability of spray dried betanin in model juices. *Food Chem.* **2015**, *187*, 398–406. [CrossRef] [PubMed]
23. Han, D.; Kim, S.J.; Kim, S.H.; Kim, D.M. Repeated regeneration of degraded red beet juice pigments in the presence of antioxidants. *J. Food Sci.* **1998**, *63*, 69–72. [CrossRef]
24. Kujala, T.S.; Loponen, J.M.; Klika, K.D.; Pihlaja, K. Phenolics and betacyanins in red beetroot (*Beta vulgaris*) root: Distribution and effect of cold storage on the content of total phenolics and three individual compounds. *J. Agric. Food Chem.* **2000**, *48*, 5338–5342. [CrossRef] [PubMed]
25. FDA (US Food and Drug Administration)—Refrigerator & Freezer Storage Chart. Available online: Chttps://www.fda.gov/downloads/food/resourcesforyou/healtheducators/ucm109315.pdf. (accessed on 2 December 2018).
26. Raikos, V.; McDonagh, A.; Ranawana, V.; Duthie, G. Processed beetroot (*Beta vulgaris* L.) as a natural antioxidant in mayonnaise: Effects on physical stability, texture and sensory attributes. *Food Sci. Hum. Wellness* **2016**, *5*, 191–198. [CrossRef]
27. Sucu, C.; Yildiz, G.T. The investigation of the use of beetroot powder in Turkish fermented beef sausage (sucuk) as nitrite alternative. *Meat Sci.* **2018**, *140*, 158–166. [CrossRef] [PubMed]
28. Krantz, C.; Monier, M.; Wahlström, B. Absorption, excretion, metabolism and cardiovascular effects of beetroot extract in the rat. *Food Cosmet. Toxicol.* **1980**, *18*, 363–366. [CrossRef]
29. Watts, A.R.; Lennard, M.S.; Mason, S.L.; Tucker, G.T.; Woods, H.F. Beeturia and the biological fate of beetroot pigments. *Pharmacogenetics* **1993**, *3*, 302–311. [CrossRef]
30. Herbach, K.M.; Stintzing, R.C. Impact of thermal treatment on color and pigment pattern of Red beet (*Beta vulgaris* L.) preparations. *J. Food Sci.* **2004**, *69*, 491–498. [CrossRef]
31. Tesoriere, L.; Fazzari, M.; Angileri, F.; Gentile, C.; Livrea, M.A. In vitro digestion of betalainic foods. Stability and bioaccessibility of betaxanthins and betacyanins and antioxidative potential of food digesta. *J. Agric. Food Chem.* **2008**, *56*, 10487–11092. [CrossRef]
32. Williamson, G.; Clifford, M.N. Role of the small intestine, colon and microbiota in determining the metabolic fate of polyphenols. *Biochem. Pharmacol.* **2017**, *139*, 24–39. [CrossRef]
33. Birben, E.; Sahiner, U.M.; Sackesen, C.; Erzurum, S.; Kalayci, O. Oxidative stress and antioxidant defense. *World Allergy Organ. J.* **2012**, *5*, 9–19. [CrossRef] [PubMed]
34. Esatbeyoglu, T.; Wagner, A.E.; Schini-Kerth, V.B.; Rimbach, G. Betanin—A food colorant with biological activity. *Mol. Nutr. Food Res.* **2015**, *59*, 36–47. [CrossRef] [PubMed]
35. Nagler, R.M.; Klein, I.; Zarzhevsky, N.; Drigues, N.; Reznick, A.Z. Characterization of the differentiated antioxidant profile of human saliva. *Free Rad. Biol. Med.* **2002**, *32*, 268–277. [CrossRef]
36. Frank, T.; Stintzing, F.C.; Carle, R.; Bitsch, I.; Quaas, D.; Strass, G.; Bitsch, R.; Netzel, M. Urinary pharmacokinetics of betalains following consumption of red beet juice in healthy humans. *Pharmacol. Res.* **2005**, *52*, 290–297. [CrossRef] [PubMed]
37. Gliszczyńska-Swigło, A.; Szymusiak, H.; Malinowska, P. Betanin, the main pigment of red beet: Molecular origin of its exceptionally high free radical-scavenging activity. *Food Addit. Contam.* **2006**, *23*, 1079–1087. [CrossRef] [PubMed]
38. Kanner, J.; Harel, S.; Granit, R. Betalains–A new class of dietary cationized antioxidants. *J. Agric. Food Chem.* **2001**, *49*, 5178–5185. [CrossRef] [PubMed]
39. Cai, Y.; Sun, H.; Corke, H. HPLC Characterization of betalains from plants in the *Amaranthaceae*. *J. Chromatogr. Sci.* **2005**, *43*, 454–460. [CrossRef]

40. Jung, S.; Nam, K.C.; Jo, C. Detection of malondialdehyde in processed meat products without interference from the ingredients. *Food Chem.* **2016**, *209*, 90–94. [CrossRef]
41. Silva, D.V.; Silva, F.O.; Perrone, D.; Pierucci, A.P.T.R.; Conte-Junior, C.A.; Alvares, T.S.; Del Aguila, E.M.; Paschoalin, V.M.F. Physicochemical, nutritional, and sensory analyses of a nitrate-enriched beetroot gel and its effects on plasmatic nitric oxide and blood pressure. *Food Nut. Res.* **2016**, *60*, 29909. [CrossRef]
42. Benzie, I.F.; Strain, J.J. The ferric reducing ability of plasma (FRAP) as a measure of "antioxidant power": The FRAP assay. *Ana. Biochem.* **1996**, *239*, 70–76. [CrossRef]
43. Re, R.; Pellegrini, N.; Proteggente, A.; Pannala, A.; Yang, M.; Rice-Evans, C. Antioxidant activity applying an improved ABTS radical cation decolorization assay. *Free Rad. Biol. Med.* **1999**, *26*, 1231–1237. [CrossRef]
44. Zuleta, A.; Esteve, M.J.; Frígola, A. ORAC and TEAC assays comparison to measure the antioxidant capacity of food products. *Food Chem.* **2009**, *114*, 310–316. [CrossRef]
45. Oomen, A.G.; Rompelberg, C.J.; Bruil, M.A.; Dobbe, C.J.; Pereboom, D.P.; Sips, A.J. Development of an in vitro digestion model for estimating the bioaccessibility of soil contaminants. *Arch. Environ. Contam. Toxicol.* **2003**, *44*, 281–287. [CrossRef] [PubMed]
46. Sagratini, G.; Caprioli, G.; Maggi, F.; Font, G.; Giardinà, D.; Mañes, J.; Meca, G.; Ricciutelli, M.; Sirocchi, V.; Torregiani, E.; Vittori, S. Determination of soya saponins I and βg in raw and cooked legumes by solid phase extraction (SPE) coupled to liquid chromatography (LC)-mass spectrometry (MS) and assessment of their bioaccessibility by an in vitro digestion model. *J. Agric. Food Chem.* **2013**, *61*, 1702–1709. [CrossRef] [PubMed]
47. Hu, J.; Zheng, Y.L.; Hyde, W.; Hendrich, S.; Murphy, P.A. Human fecal metabolism of soya saponin I. *J. Agric. Food Chem.* **2004**, *52*, 2689–2696. [CrossRef] [PubMed]
48. McDonald, J.A.; Schroeter, K.; Fuentes, S.; Heikamp-Dejong, I.; Khursigara, C.M.; de Vos, W.M.; Allen-Vercoe, E. Evaluation of microbial community reproducibility, stability and composition in a human distal gut chemostat model. *J. Microbiol. Methods* **2013**, *95*, 167–174. [CrossRef]

Sample Availability: Samples of the compounds are available from the authors.

molecules

MDPI

Article

Influence of Oxygen-Containing Sulfur Flavor Molecules on the Stability of β-Carotene under UVA Irradiation

Gong-Liang Zhang [1], Hong-Yan Wu [2], Ying Liang [2], Jie Song [1], Wei-Qi Gan [1] and Hong-Man Hou [1,*]

1 School of Food Science and Technology, Dalian Polytechnic University, Dalian 116034, China; zgl_mp@163.com (G.-L.Z.); sj101521@163.com (J.S.); 17317138996@163.com (W.-Q.G.)

2 Graduate School of Environmental and Life Science, Okayama University, Okayama 700-8530, Japan; wuhongyan1908@hotmail.com (H.-Y.W.); dlliangying@163.com (Y.L.)

* Correspondence: houhongman2011@hotmail.com; Tel.: +86-411-86322020

Academic Editor: Lillian Barros

Received: 28 December 2018; Accepted: 13 January 2019; Published: 16 January 2019

Abstract: The influence of 11 kinds of oxygen-containing sulfur flavor molecules was examined on β-carotene stability under UVA irradiation in ethanol system. Both the effects of sulfides on dynamic degradation of β-carotene and the relation between structure and effect were investigated. The oxidation products of β-carotene accelerated by sulfides under UVA irradiation were also identified. The results indicated that the disulfides had more obvious accelerative effects on the photodegradation of β-carotene than mono sulfides. The degradation of β-carotene after methyl (2-methyl-3-furyl) disulfide (MMFDS), methyl furfuryl disulfide (MFDS) and bis(2-methyl-3-furyl) disulfide (BMFDS) exposure followed first-order kinetics. Furan-containing sulfides such as MMFDS and BMFDS showed more pronounced accelerative effects than their corresponding isomers. The oxidation products were identified as 13-*cis*-β-carotene, 9,13-di-*cis*-β-carotene and all-*trans*-5,6-epoxy-β-carotene. These results suggest that both the sulfur atom numbers and the furan group in oxygen-containing sulfides play a critical role in the photooxidation of β-carotene.

Keywords: oxygen-containing sulfur flavor molecules; β-carotene; bis(2-methyl-3-furyl) disulfide (BMFDS); oxidation products

1. Introduction

Carotenoids are natural pigments of the isoprenoid family, commonly biosynthesized in fruits and vegetables [1], presenting potential physiological benefits, such as antioxidants in food and pro-vitamin A activity [2]. As one of the most commonly used carotenoids, β-carotene is expected to be conducive to health because of its valuable nutritional properties and antioxidant capacities, which confer on this compound an important role in lowering the risk of cataracts [3], inhibiting age-related macular degeneration [4], and enhancing the prevention of cardiovascular diseases [5].

However, due to its poor water solubility [6] and low bioavailability [7,8] during food processing and storage, widespread applications of β-carotene in food matrices normally suffer considerable challenges. Moreover, the restriction of β-carotene utilization as a nutritional ingredient in the food industry is currently also attributed to the existence of numerous unsaturated groups, resulting in high vulnerability to degradation reaction when exposed to light, heat and other external factors [9,10]. It has been reported that strong illumination can influence the stability of β-carotene extracted from palm oil, unveiling the formation of cis isomers [11]. The in-depth study carried out by Ayu et al. [12], who investigated interactive influence of tocopherols, tocotrienols, and β-carotene in the process of photooxidation of red palm oil, suggesting that the degradation of β-carotene easily occurs under the irradiation of light. Apart from that, a comparison of β-carotene degradation under different UV stresses was conducted by Chen et al. [13], which showed that the longer the wavelength applied, the faster the degradation rate.

Several reports have also focused on the effects of chemical substances and their stability. The presence of 1,4-dimethylnaphthalene-1,4-endoperoxide and lycopene had the potential to induce the generation of (9Z)-, (13Z)- and (15Z)-β-carotene, which was associated with the formation of singlet oxygen [14]. Lewis acids, including titanium tetrachloride and ferric chloride, can catalyze the degradation of β-carotene to form an intermediate radical carbocation [15].

Currently, more than 300 sulfides have been registered as Generally Recognized as Safe (GRAS) substances with various threshold limits, making them the critical food flavors [16]. Biological functions including antithrombotic [17], antimicrobial [18], anticancer [19], and anti-inflammatory activities [20] in combination with their attractive odor characteristics such as garlic, onion, meat and nut flavors, have increased their feasibility of acting as food additives. In our previous studies, we discovered that dimethyl sulfides exerted apoptosis-inducing effects in leukemia cell lines via the generation of reactive oxygen species, especially for dimethyl trisulfide(Me_2S_3) and dimethyl tetrasulfide(Me_2S_4) [21]. Furthermore, β-carotene combined with Me_2S_4 under UVA irradiation presented a synergistic action in inhibiting the viability of HL-60 cells viability, and elevating caspase-3 levels [22], mostly like probably raising the possibility of the reaction between sulfides and β-carotene assisted by UVA.

In this study, we selected 11 kinds of oxygen-containing sulfur flavor molecules, commonly used in the food industry, as experimental materials to examine their influence on β-carotene stability under UVA irradiation. Moreover, both the dynamic analysis of β-carotene degradation and the structural effects of sulfides that accelerated the degradation of β-carotene were investigated to provide a clearer and better comprehension of their acceleration effects. Furthermore, the oxidation products of β-carotene under UVA irradiation were also analyzed in order to elucidate its degradation mechanism.

2. Results

2.1. The Effects of Oxygen-Containing Sulfur Flavor Molecules on β-Carotene Degradation under UVA Irradiation

The structures of 11 kinds of oxygen-containing sulfur flavor molecules are shown in Table 1. Most of the sulfides contain furan or furfuryl group, such as 2-methyl-3-(methylthio) furan (MMTF) and methyl furfuryl disulfide (MFDS). Some sulfides contain different numbers of sulfur atoms but have the same side chain groups, such as MMTF and methyl (2-methyl-3-furyl) disulfide (MMFDS), difurfuryl sulfide (DFS) and difurfuryl disulfide (DFDS). Furthermore, it is worth noting the existence of isomers, such as bis(2-methyl-3-furyl) disulfide (BMFDS) and DFDS, MMFDS and MFDS.

The effects of 11 kinds of oxygen-containing sulfur flavor molecules assisted by UVA irradiation on β-carotene stability are shown in Figure 1A. After irradiating under UVA for 60 min in which the light intensity was 2.5 mW/cm^2, the contents of β-carotene in all groups showed a reducing trend. To gain more knowledge about the correlation between the structure of coexistent sulfides and the degradation ratios of β-carotene, the remaining amounts of β-carotene were compared according to the structural characteristics of coexistent sulfides. The amounts of β-carotene treated with MMFDS and BMFDS were dramatically decreased by approximately 96.05% and 99.70%, respectively ($p < 0.05$). Likewise, the remaining amount of β-carotene in the presence of MFDS declined approximately 43.64%. These findings proved the fact that natural sulfur substance may affect β-carotene stability.

2.2. The Effects of Furan-Containing Sulfides on β-Carotene Degradation under UVA Irradiation

The order of the reaction with respect to the photodegradation of β-carotene was acquired according to Equation (2) to examine the changes of β-carotene concentration with time after UVA irradiation.

Table 1. Structure of eleven oxygen-containing sulfur flavor molecules.

FEMA Number	Name	Abbreviation	Structure
3573	Methyl (2-Methyl-3-furyl) disulfide	MMFDS	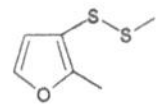
3362	Methyl furfuryl disulfide	MFDS	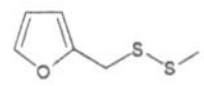
2720	Methyl 3-methylthiopropionate	MMTP	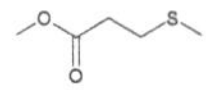
3949	2-Methyl-3-(methylthio) furan	MMTF	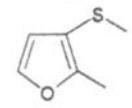
2747	3-(Methylthio) propionaldehyde	MTPD	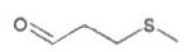
3343	Ethyl 3-methylthiopropionate	EMTP	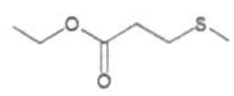
3674	Ethyl 3-(furfurylthio) propionate	EFTP	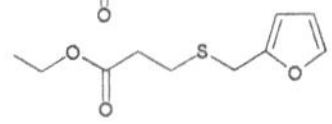
3161	Furfuryl isopropyl sulfide	FIS	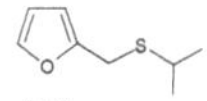
3238	Difurfuryl sulfide	DFS	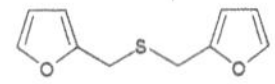
3146	Difurfuryl disulfide	DFDS	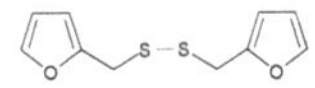
3259	Bis (2-methyl-3-furyl) disulfide	BMFDS	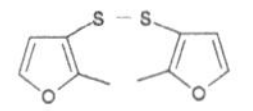

FEMA: Flavor and Extract Manufacturers Association of the United States.

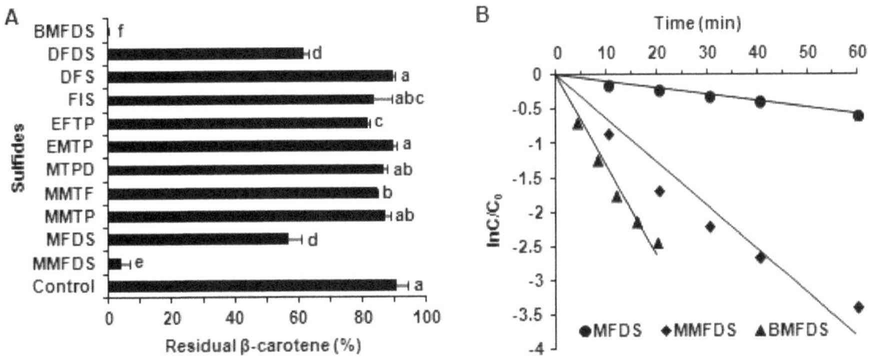

Figure 1. The effect of oxygen-containing sulfides on the degradation of β-carotene in ethanol under UVA irradiation. β-Carotene was treated with various oxygen-containing sulfur flavor molecules under UVA irradiation for 60 min (**A**) or within 60 min (**B**) in ethanol system. The residual β-carotene (**A**) and the first-order kinetics curve of β-carotene degradation (**B**) were determined. The bar results are expressed as means ± SD from three independent replicates. Different small letters show significant differences ($p < 0.05$).

As presented in Figure 1B, β-carotene degradation upon exposure to MMFDS, BMFDS and MFDS followed first-order kinetics, consistent with the kinetic model in dichloromethane system [23]. The corresponding kinetic parameters are listed in Table 2. It can be seen that the presence of BMFDS in ethanol significantly improved the k and shortened the $t_{1/2}$ of β-carotene degradation, compared with the other two sulfides. The k and $t_{1/2}$ were 0.131 min^{-1} and 5.29 min for BMFDS, while they were 0.0633 min^{-1} and 10.95 min for MMFDS and 0.0095 min^{-1} and 72.96 min for MFDS, respectively.

Furthermore, it should also be noticed that these three kinds of sulfides (MMFDS, BMFDS and MFDS) having obviously promoting effects on the degradation of β-carotene contain at least two sulfur atoms, which is consistent with the previous report [24]. These results suggest that sulfides with more sulfur atoms might trigger a stronger chemical effect on β-carotene stability in ethanol model system.

Table 2. Degradation kinetics parameters of β-carotene in the presence of BMFDS, MMFDS, MFDS under UVA irradiation in ethanol system.

Sulfides	K (min^{-1})	R^2	$t_{1/2}$ (min)
MFDS	0.0095	0.9959	72.96
MMFDS	0.0633	0.9356	10.95
BMFDS	0.131	0.9720	5.29

BMFDS: Bis (2-methyl-3-furyl) disulfide, MMFDS: Methyl (2-Methyl-3-furyl) disulfide, MFDS: Methyl furfuryl disulfide.

2.3. Kinetics of β-Carotene Degradation Treated with BMFDS under UVA Irradiation

To further clarify the acceleration effect of BMFDS, the degradation kinetics parameters of β-carotene under UVA irradiation were determined. As shown in Table 3, the photodegradation of β-carotene treated by BMFDS in an ethanol system followed first-order kinetics. In terms of the rate constant k, the degradation degree of β-carotene treated with BMFDS was much higher, approximately 156 times than that in the control group. Therefore, our findings demonstrated that β-carotene degradation was followed first-order kinetics after treatment with BMFDS.

Table 3. Degradation kinetics parametersβ-carotene in the presence of BMFDS under UVA irradiation.

	R Zero-Order (c)	R First-Order (ln c)	R Second-Order (1/c)	k (min^{-1})	R^2
Control	0.9468	0.9512	0.9554	0.0012	0.9785
BMFDS	0.6376	0.9766	0.9141	0.1879	0.9537

c is the concentration of reactant.

2.4. The Analysis of Photooxidation Products of β-Carotene Treated with BMFDS under UVA Irradiation

According to the remarkable acceleration effect of BMFDS, HPLC-DAD-APCI-MS combined with Raman spectroscopy was applied to make a preliminary identification about the oxidation products of β-carotene, given their low contents and rather complex process of products collection. The chromatographic and spectral data of the oxidation products of β-carotene by HPLC-DAD-ACPI-MS are shown in Figure 2 and Table 4. The chromatographic peaks of leading β-carotene oxidation products had the same peak time in both the experimental group and the control group. Therefore, it was supposed that β-carotene treated with or without BMFDS under UVA irradiation had the same oxidation products (Figure 2A). In comparison to the retention time of a standard product (24.322 min) (Table 4), peak 3 was confirmed as all-*trans*-β-carotene (24.319 min) (Figure 3A). The UV spectra data are depicted in Figure 2B, and among the four obvious peaks, peak 2 and peak 4 were identified tentatively as 13-*cis*-β-carotene [25] and 9,13-di-*cis*-β-carotene [26], respectively (Figure 3C,D) according to the spectral characteristics and Q-ratios (Table 4), which were stipulated as the absorbance ratio of the middle main absorption peak to the cis peak.

Table 4. Product identification of β-carotene treated with BMFDS under UVA irradiation.

Peak	Retention Time (min)	Molecular Ion, Formula	Isomer	λ (nm)		Q-Ratio	
				Found	Reported	Found	Reported
1	19.467	553, $C_{18}H_{57}O$	All-*trans*-5,6-expoxy-β-carotene	446	444		
2	21.953	537, $C_{18}H_{57}$	13-*cis*-β-carotene	340 446	340 445	0.41	0.41
3	24.319	537, $C_{18}H_{57}$	All-*trans*-β-carotene	451	455		
4	25.399	537, $C_{18}H_{57}$	9,13-di-*cis*-β-carotene	350 446	335 439	0.16	0.16

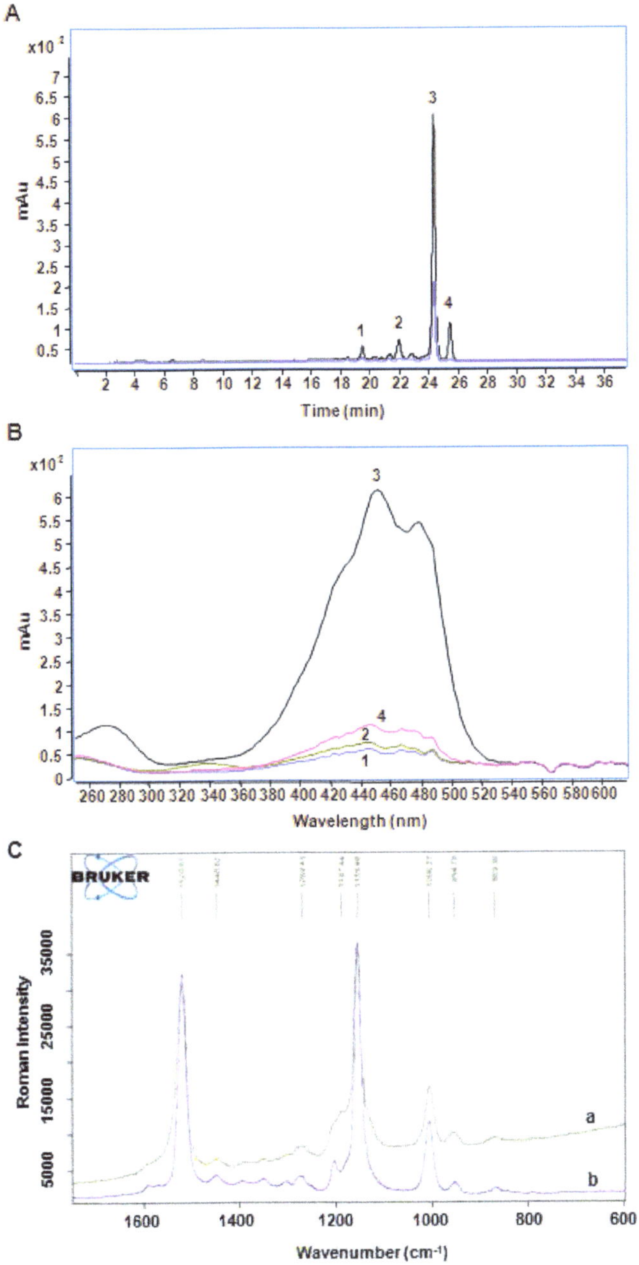

Figure 2. The products of β-carotene treated with BMFDS under UVA irradiation. HPLC chromatograms (**A**) and UV-Vis spectra (**B**) of β-carotene oxidation products induced by BMFDS treatment for 5 min in ethanol under UVA irradiation. The blue line in HPLC chromatograms is referred to standard β-carotene. Peak identification for (**A**) and (**B**): Peak 1, all-*trans*-5,6-expoxy-β-carotene; Peak 2, 13-*cis*-β-carotene; Peak 3, all-*trans*-β-carotene; Peak 4, 9,13-di-*cis*-β-carotene. Raman spectra (**C**) at different wave number corresponded to β-carotene under UVA irradiation for 5 min with BMFDS (**a**) and without BMFDS (**b**), respectively.

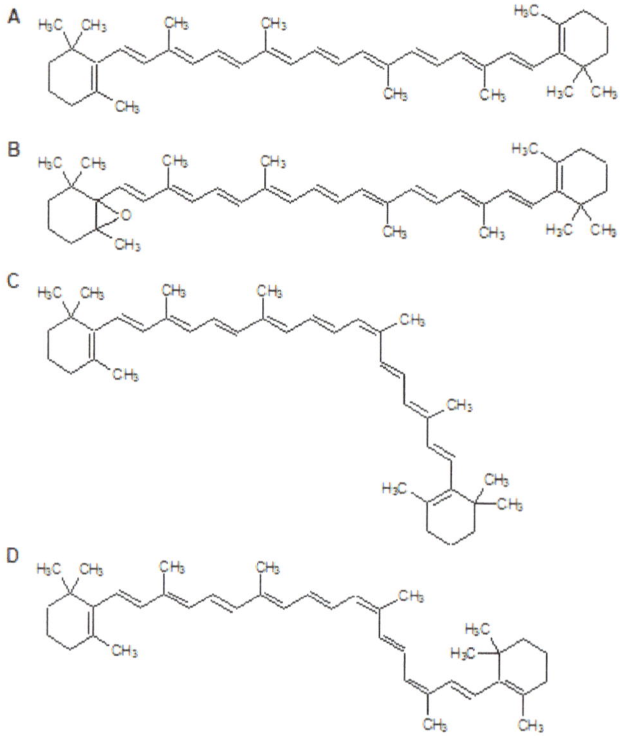

Figure 3. The structure of β-carotene and its oxidation products induced by BMFDS under UVA irradiation. (**A**) all-*trans*-β-carotene; (**B**) all-*trans*-5,6-epoxy-β-carotene; (**C**) 13-*cis*-β-carotene; (**D**) 9,13-di-*cis*-β-carotene.

The Raman spectra at various wavenumber (cm^{-1}) of β-carotene in ethanol system treated with or without BMFDS under UVA irradiation are displayed in Figure 2C. By comparison of Figure 2C (a) and Figure 2C (b), both υ1 (–C=C–) at 1520.67 cm^{-1} and υ2 (–C=C–C=C–) at 1155.48 cm^{-1} can be observed. However, the distinction existed in the appearance of a new polar function group (C–O) presented in Figure 2C (a). Combined with its m/z of 553 (Table 4), this new vibration peak was identified tentatively as all-*trans*-5,6-epoxy-β-carotene (Figure 3B). As would have been expected, mono-epoxide is susceptible to generate from different sorts of carotenoids.

3. Discussion

β-Carotene was susceptible to be affected when exposed to external factors, consistent with previous reports, which considered that light can exert influence on β-carotene degradation [27–29]. In this study, sulfides have also showed to be involved in interfering the stability of β-carotene, which agreed with the report of Wei et al., implying that the stability of β-carotene can be significantly influenced when chitosan-(−)-epigallocatechin-3-gallate conjugates on β-carotene emulsions covered by sodium caseinate [30]. Moreover, in agreement with previous studies, both the number of sulfur atoms and the type of side group can affect the accelerated degradation of β-carotene under UVA irradiation [24,31]. It has been reported that the coexistence of disulfides can remarkably decrease the residual ratios of β-carotene to approximately 51.8–69.1%, while the presence of mono sulfides did not show obvious accelerating effects compared to the absence of mono sulfides [24].

β-Carotene degradation upon exposure to MMFDS, BMFDS and MFDS followed first-order kinetics, consistent with the kinetic model in dichloromethane system [23]. On the basis of the previous report which focused on the phenomenon of the existence of isomers [32], we also studied the accelerated effects of side groups among these sulfides on the degradation of β-carotene. Although MMFDS and MFDS both possess the same molecular formula ($C_6H_8OS_2$), MMFDS showed a stronger accelerated degradation effect than MFDS. It is presumably because there is a methyl group and a furan group on the side of the disulfide bond in MMFDS, while a methyl group and a furfuryl group exist on the side of the disulfide bond in MFDS. For DFDS and its corresponding isomer BMFDS. Similarly, there is a furan group on both ends of the disulfide bond in BMFDS, while there is a furfuryl group on each side of the disulfide bond in DFDS. Their different abilities to promote the degradation of β-carotene can be related to the existence of various side groups. These results may account for the fact that furan-containing sulfur flavor molecules (MMFDS and BMFDS) showed a much more remarkable acceleration effect on the degradation of β-carotene than furfuryl-containing sulfur flavor molecules (MFDS and DFDS, respectively). Therefore, the number of sulfur atoms and the furan group in oxygen-containing sulfur flavor molecules may play a critical role in the accelerated degradation of β-carotene under UVA irradiation in ethanol system.

Several studies have investigated the order of kinetics on the degradation of β-carotene in different model systems under various conditions. It also followed a first order reaction under ambient storage, ultraviolet radiation and even heat treatments [13]. The photodegradation of β-carotene treated by BMFDS in ethanol system followed first-order kinetics, which agreed with previous studies carried out in food model systems such as carrots [33], oil/carrot emulsion system [34], oil model systems [10,35] and pulp or juices [36,37]. Ferreira et al. observed a first-order reaction for β-carotene degradation in a low-moisture and aqueous model system, as well as in lyophilized guava under different processing and storage conditions [38].

In addition, our results were in good agreement with Li et al. who summarized that the *trans-cis* isomerization of carotenoids can be generated via contacting with acids, thermal treatment or light [39]. It has been a long time since 13-*cis*-β-carotene was recognized as one of the main *cis* forms of β-carotene in food [40]. Chen et al. even analyzed it by different processing means, including over-heating and (non)-iodine-catalyzed photodegradation [41]. In addition, 9,13-di-*cis*-β-carotene was also confirmed as a common β-carotene degradation product according to Glaser et al. [42]. Moreover, our founding was consistent with Handelman et al., who had detected 5,6-epoxide of β-carotene through utilizing HPLC with mass analysis [43]. Similarly, Zeb identified all-*trans*-5,6-epoxy-β-carotene by an HPLC system and single ion monitoring mass spectrometry as well [44].

4. Materials and Methods

4.1. Materials and Chemicals

Eleven kinds of oxygen-containing sulfur flavors and β-carotene were obtained from Sigma-Aldrich (St. Louis, MO, USA). The structures of these sulfides are presented in Table 1. Methanol and methyl *tert*-butyl ether (MTBE)were procured from Damao Chemical factory (Tianjin, China) and Fisher Scientific (Pittsburgh, PA, USA), respectively. The other chemicals and reagents were of analytical grade.

4.2. Preparation of the Model Systems

For the preparation of the ethanol model system, 1 mg β-carotene was dissolved in 15 mL ethanol according to Onsekizoglu et al. [27] with minor modifications. The β-carotene solution was prepared daily and kept in the dark at 4 °C before use. Stock solutions of 11 kinds of sulfides were prepared in ethanol at a concentration of 10 mM and kept at 4 °C prior to use.

4.3. Kinetic Analysis of β-Carotene Degradation

The working solutions of β-carotene were transferred into quartz cuvettes, followed by the addition of 10 μL sulfur flavors. The control was performed with 10 μL ethanol. Then, the mixture was treated by UVA light (2.5 mW/cm^2) with the aim of assessing the degradation kinetics of β-carotene treated with sulfide. The degradation of β-carotene was measured immediately in a UV-1750 spectrophotometer (Shimadzu, Tokyo, Japan) at the wavelength of 450 nm for 60 min, which was monitored every 10 min. All measurements were performed in triplicate and data are expressed as mean of three independent experiments.

4.4. Degradation Kinetics Modeling

The trial-and-error procedure was carried out in accordance with the integral method outlined by Sánchezet al. [45] to determine the reaction order of theβ-carotene degradation. Different order models can be represented as follows:

$$c - c_0 = -kt \tag{1}$$

$$\ln c/c_0 = -kt \tag{2}$$

$$1/c - 1/c_0 = kt \tag{3}$$

In these formulas, c (μM) is thereactantconcentrationat a given time, c_0 (μM) is the initial reactant concentration, k (min^{-1}) is the degradation rate constant, and t (min) is the treatment time.

4.5. Analysis of β-Carotene Treated with UVA Irradiation and BMFDS

The working solutions of β-carotene were transferred into quartz cuvettes, followed by the addition of 10 μL BMFDS. The control was performed with 10 μL ethanol. Then, the mixture containing β-carotene and BMFDS was placed under a UV lamp (Shimadzu, Japan) with an intensity of 2.5 mW/cm^2 for 5 min, followed by drying completely under a nitrogen stream. The residue was redissolved in 0.1 mL MTBE before use.

The further analysis was carried out and relative parameters were applied according to Santos et al. [46]. Briefly, once redissolved, the solution was passed through a 0.22 μm filter, followed by the injection into an HPLC-DAD-APCI-MS system (Agilent, Santa Clara, CA, USA) for closer analysis. A YMC C_{30} column (250 × 4.6 mm, 5 μm) and gradient mobile phase of methanol-MTBE-water (85:15:5, $v/v/v$) and MTBE (100%) were used for β-carotene detection.

4.6. Determination of Degradation Products by Raman Spectroscopy

The Raman spectra of the degradation products were recorded on a Raman spectrometer (Bruker Instruments Inc., Bill-erica, MA, USA). The wave number was in the range of 400–4000 cm^{-1} using the 785 nm as the excitation line. The power was 10 Mw while the integration time was 20 s.

4.7. Statistical Analysis

All the data were expressed as the mean ± SD or mean and subjected to the Student's *t*-test for statistical analysis. Statistical significance was considered at a $p < 0.05$.

5. Conclusions

Overall, through applying the results we obtained and drawing upon the information provided by other studies, it is supposed that the oxidation products of β-carotene treated with BMFDS under UVA irradiation in ethanol system might include 13-*cis*-β-carotene, 9,13-di-*cis*-β-carotene and all-*trans*-5,6-epoxy-β-carotene. Our results might shed new light on the accelerative effect of BMFDS on the photodegradation of β-carotene. More insights into the mechanism involved in degradation of oxidation products of BMFDS-treated β-carotene should be further studied further.

Author Contributions: Methodology, Y.L. and W.-Q.G. formal analysis, J.S. data curation, Y.L. and J.S.; writing—original draft preparation, G.-L.Z. and H.-Y.W.; writing—review and editing, G.-L.Z. and H.-M.H.; supervision, H.-M.H.; funding acquisition, G.-L.Z. All authors discussed the results and approved the final manuscript.

Funding: This research was funded by The National Natural Science Foundation of China, grant numbers [31571888 and 31201419].

Conflicts of Interest: The authors declare no conflict of interest.

References

1. Durante, M.; Lenucci, M.S.; D'Amico, L.; Piro, G.; Mita, G. Effect of drying and co-matrix addition on the yield and quality of supercritical CO_2 extracted pumpkin (*Cucurbita moschata* Duch.) oil. *Food Chem.* **2014**, *148*, 314–320. [CrossRef] [PubMed]
2. Lemmens, L.; Colle, I.; Buggenhout, S.V.; Palmero, P.; Loey, A.V.; Hendrickx, M. Carotenoid bioaccessibility in fruit-and vegetable-based food products as affected by product (micro) structural characteristics and the presence of lipids: A review. *Trends Food Sci. Technol.* **2014**, *38*, 125–135. [CrossRef]
3. Hou, Z.Q.; Liu, Y.W.; Lei, F.; Gao, Y.X. Investigation into the in vitro release properties of β-carotene in emulsions stabilized by different emulsifiers. *LWT Food Sci. Technol.* **2014**, *59*, 867–873. [CrossRef]
4. Gul, K.; Tak, A.; Singh, A.K.; Singh, P.; Yousuf, B.; Wani, A.A. Chemistry, encapsulation, and health benefits of β-carotene—A review. *Cogent. Food Agric.* **2015**, *1*, 1018696. [CrossRef]
5. Tanaka, T.; Shnimizu, M.; Moriwaki, H. Cancer chemoprevention by carotenoids. *Molecules* **2012**, *17*, 3202–3242. [CrossRef]
6. Huang, Q.R.; Yu, H.L.; Ru, Q.M. Bioavailability and delivery of nutraceuticals using nanotechnology. *J. Food Sci.* **2010**, *75*, 50–57. [CrossRef] [PubMed]
7. Salvia-Trujillo, L.; Qian, C.; Martín-Belloso, O.; McClements, D.J. Influence of particle size on lipid digestion and beta-carotene bioaccessibility in emulsions and nanoemulsions. *Food Chem.* **2013**, *141*, 1472–1480. [CrossRef]
8. Salvia-Trujillo, L.; Qian, C.; Martin-Belloso, O.; McClements, D.J. Modulating β-carotene bioaccessibility by controlling oil composition and concentration in edible nanoemulsions. *Food Chem.* **2013**, *139*, 878–884. [CrossRef]
9. Boon, C.S.; Mcclements, D.J.; Weiss, J.; Decker, E.A. Factors influencing the chemical stability of carotenoids in foods. *Crit. Rev. Food Sci.* **2010**, *50*, 515–532. [CrossRef]
10. Achir, N.; Randrianatoandro, V.A.; Bohuon, P.; Laffargue, A.; Avallone, S. Kinetic study of β-carotene and lutein degradation in oils during heat treatment. *Eur. J. Lipid Sci. Technol.* **2010**, *112*, 349–361. [CrossRef]
11. Guan, Y.; Wu, J.; Zhong, Q. Eugenol improves physical and chemical stabilities of nanoemulsions loaded with β-carotene. *Food Chem.* **2016**, *194*, 787–796. [CrossRef] [PubMed]
12. Ayu, D.F.; Andarwulan, N.; Hariyadi, P.; Purnomo, E.H. Effectof tocopherols, tocotrienols, β-carotene and chlorophyll on the photo-oxidative stability of red palm oil. *Food Sci. Biotechnol.* **2016**, *25*, 401–407. [CrossRef] [PubMed]
13. Chen, H.Q.; Zhong, Q.X. Thermal and UV stability of β-carotene dissolved in peppermint oil microemulsified by sunflower lecithin and Tween 20 blend. *Food Chem.* **2015**, *174*, 630–636. [CrossRef]
14. Heymann, T.; Heinz, P.; Glomb, M.A. Lycopene inhibits the isomerization of β-carotene during quenching of singlet oxygen and free radicals. *J. Agric. Food Chem.* **2015**, *63*, 3279–3287. [CrossRef]
15. Rajendran, V.; Chen, B.H. Isomerization of β-carotene by titanium tetrachloride catalyst. *J. Chem. Sci.* **2007**, *119*, 253–258. [CrossRef]
16. Liu, Y.P.; Chen, H.T.; Yin, D.C.; Sun, B.G. Synthesis and odor evaluation of five new sulfur-containing ester flavor compounds from 4-ethyloctanoic acid. *Molecules* **2010**, *15*, 5104–5111. [CrossRef] [PubMed]
17. Apitz-Castro, R.; Badimon, J.J.; Badimon, L. Effect of ajoene, the major antiplatelet compound from garlic, on platelet thrombus formation. *Thromb. Res.* **1992**, *68*, 145–155. [CrossRef]
18. Naganawa, R.; Iwata, N.; Ishikawa, K.; Fukuda, H.; Fujino, T.; Suzuki, A. Inhibition of microbial growth by ajoene, a sulfur-containing compound derived from garlic. *Appl. Environ. Microb.* **1996**, *62*, 4238–4242.
19. Taylor, P.; Noriega, R.; Farah, C.; Abad, M.J.; Arsenak, M.; Apitz, R. Ajoene inhibits both primary tumor growth and metastasis of B16/BL6 melanoma cells in C57BL/6 mice. *Cancer Lett.* **2006**, *239*, 298–304. [CrossRef]
20. Dirsch, V.M.; Vollmar, A.M. Ajoene, a natural product with nonsteroidal anti-inflammatory drug (NSAID)-like properties. *Biochem. Pharmacol.* **2001**, *61*, 587–593. [CrossRef]

21. Zhang, G.L.; Wu, H.T.; Zhu, B.W.; Shimoishi, Y.; Nakamura, Y.; Murata, Y. Effect of dimethyl sulfides on the induction of apoptosis in human leukemia Jurkat cells and HL-60 cells. *Biosci. Biotechnol. Biochem.* **2008**, *72*, 2966–2972. [CrossRef] [PubMed]
22. Zhang, G.L.; Wu, H.T.; Zhu, B.W.; Shimoishi, Y.; Nakamura, Y.; Murata, Y. Induction of apoptosis by beta-carotene and dimethyl tetrasulfide assisted by UVA irradiation in HL-60 cells. *Biosci. Biotechnol. Biochem.* **2009**, *73*, 1014–1020. [CrossRef] [PubMed]
23. Sun, Y.; Ma, G.; Ye, X.; Kakuda, Y.; Meng, R. Stability of all-trans-beta-carotene under ultrasound treatment in a model system: Effects of different factors, kinetics and newly formed compounds. *Ultrason. Sonochem.* **2010**, *17*, 654–661. [CrossRef] [PubMed]
24. Arita, S.; Ando, S.; Hosoda, H.; Sakaue, K.; Nagata, T.; Murata, Y.; Shimoishi, Y.; Tada, M. Acceleration effect of sulfides on photodegradation of carotenoids by UVA irradiation. *Biosci. Biotechnol. Biochem.* **2005**, *69*, 1786–1789. [CrossRef] [PubMed]
25. Lowe, G.M.; Vlismas, K.; Graham, D.L.; Carail, M.; Caris-Veyrat, C.; Young, A.J. The degradation of (all-*E*)-β-carotene by cigarette smoke. *Free Radic. Res.* **2009**, *43*, 280–286. [CrossRef] [PubMed]
26. Dan, Q.; Chen, Z.R.; Li, H.R. Effect of heating on solidβ-carotene. *Food Chem.* **2009**, *112*, 344–349.
27. Onsekizoglu, P.; Gökmen, V.; Acar, J. Degradation of β-carotene with the effects of light and sulfur dioxide may be responsible for the formation of white spot in dried apricots. *Eur. Food. Res. Technol.* **2005**, *221*, 357–360. [CrossRef]
28. Spada, J.C.; Noreña, C.P.Z.; Marczak, L.D.F.; Tessaro, I.C. Study on the stability of β-carotene microencapsulated with pinhão (*Araucaria angustifolia* seeds) starch. *Carbohydr. Polym.* **2012**, *89*, 1166–1173. [CrossRef]
29. Chen, B.H.; Chen, T.M.; Chien, J.T. Kinetic model for studying the isomerization of alpha- and beta-carotene during heating and illumination. *J. Agric. Food Chem.* **1994**, *42*, 2391–2397. [CrossRef]
30. Wei, Z.H.; Gao, Y.X. Physicochemical properties of β-carotene bilayer emulsions coated by milk proteins and chitosan–EGCG conjugates. *Food Hydrocoll.* **2016**, *52*, 590–599. [CrossRef]
31. Zhang, G.L.; Zhu, B.W.; Nakamura, Y.; Shimoishi, Y.; Murata, Y. Structure-dependent photodegradation of carotenoids accelerated by dimethyl tetrasulfide under UVA irradiation. *Biosci. Biotechnol. Biochem.* **2008**, *72*, 2176–2183. [CrossRef]
32. Zhang, G.L.; Liang, Y.; Zhu, J.Y.; Jia, Q.; Gan, W.Q.; Sun, L.M.; Hou, H.M. Oxidative stress-mediated antiproliferative effects of furan-containing sulfur flavors in human leukemia Jurkat cells. *Food Chem.* **2015**, *180*, 1–8. [CrossRef] [PubMed]
33. Lavelli, V.; Zanoni, B.; Zaniboni, A. Effect of water activity on carotenoid degradation in dehydrated carrots. *Food Chem.* **2007**, *104*, 1705–1711. [CrossRef]
34. Knockaert, G.; Pulissery, S.K.; Lemmens, L.; Van, B.S.; Hendrickx, M.; Van, L.A. Carrot beta-carotene degradation and isomerization kinetics during thermal processing in the presence of oil. *J. Agric. Food Chem.* **2012**, *60*, 10312–10319. [CrossRef]
35. Aparicio-Ruiz, R.; Mínguez-Mosquera, M.I.; Gandul-Rojas, B. Thermal degradation kinetics of lutein, β-carotene and β-cryptoxanthin in virgin olive oils. *J. Food Compos. Anal.* **2011**, *24*, 811–820. [CrossRef]
36. Ahmed, J.; Shivhare, U.S.; Sandhu, K.S. Thermal degradation kinetics of carotenoids and visual color of papaya puree. *J. Food Sci.* **2002**, *67*, 2692–2695. [CrossRef]
37. Saxena, A.; Maity, T.; Raju, P.S.; Bawa, A.S. Degradation kinetics of colour and total carotenoids in jackfruit (*Artocarpus heterophyllus*) bulb slices during hot air drying. *Food Bioprocess Technol.* **2012**, *5*, 672–679. [CrossRef]
38. Ferreira, J.E.M.; Rodriguez-Amaya, D.B. Degradation of lycopene and β-carotene in model systems and in lyophilized guava during ambient storage: Kinetics, structure, and matrix effects. *J. Food Sci.* **2008**, *73*, 589–594. [CrossRef]
39. Li, D.J.; Xiao, Y.D.; Zhang, Z.Y.; Liu, C. Light-induced oxidation and isomerization of all-*trans*-β-cryptoxanthin in a model system. *J. Photochem. Photobiol. B* **2015**, *142*, 51–58. [CrossRef]
40. Chen, B.H.; Chen, Y.Y. Stability of chlorophylls and carotenoids in sweet potato leaves during microwave cooking. *J. Agric. Food Chem.* **1993**, *41*, 1315–1320. [CrossRef]
41. Chen, B.H.; Huang, J.H. Degradation and isomerization of chlorophyll a and β-carotene as affected by various heating and illumination treatments. *Food Chem.* **1998**, *62*, 299–307. [CrossRef]

42. Glaser, T.; Lienau, A.; Zeeb, D.; Krucker, M.; Dachtler, M.; Albert, K. Qualitative and quantitative determination of carotenoid stereoisomers in a variety of spinach samples by use of MSPD before HPLC-UV, HPLC-APCI-MS, and HPLC-NMR online coupling. *Chromatographia* **2003**, *57*, S19–S25. [CrossRef]
43. Handelman, G.J.; Kuijk, F.J.G.M.V.; Chatterjee, A.; Krinsky, N.I. Characterization of products formed during the autoxidation of β-carotene. *Free Radic. Biol. Med.* **1991**, *10*, 427–437. [CrossRef]
44. Zeb, A. Oxidation and formation of oxidation products of β-carotene at boiling temperature. *Chem. Phys. Lipids* **2012**, *165*, 277–281. [CrossRef] [PubMed]
45. Sánchez, A.M.; Carmona, M.; Ordoudi, S.A.; Tsimidou, M.Z.; Alonso, G.L. Kinetics of individual crocetin ester degradation in aqueous extracts of saffron (*Crocus sativus* L.) upon thermal treatment in the dark. *J. Agric. Food Chem.* **2008**, *56*, 1627–1637. [CrossRef]
46. Santos, J.; Herrero, M.; Mendiola, J.A.; Oliva-Teles, M.T.; Ibáñez, E.; Delerue-Matos, C.; Oliveira, M.B.P.P. Assessment of nutritional and metabolic profiles of pea shoots: The new ready-to-eat baby-leaf vegetable. *Food Res. Int.* **2014**, *58*, 105–111. [CrossRef]

Sample Availability: Samples of the compounds are not available from the authors.

molecules

MDPI

Article

Influence of Pickling Process on *Allium cepa* and *Citrus limon* Metabolome as Determined *via* Mass Spectrometry-Based Metabolomics

Mohamed A. Farag [1,2,*], Ahmed F. Tawfike [3,4], Marwa S. Donia [5], Anja Ehrlich [6] and Ludger A. Wessjohann [6,*]

1 Pharmacognosy Department, College of Pharmacy, Cairo University, Kasr el Aini St., Cairo 11562, Egypt
2 Department of Chemistry, School of Sciences & Engineering, The American University in Cairo, New Cairo 11835, Egypt
3 Department of Computational and Analytical Science, Rothamsted Research, Harpenden AL5 2JQ, UK; ahmed.tawfike@rothamsted.ac.uk
4 Department of Pharmacognosy, Faculty of Pharmacy, Helwan University, Cairo 11795, Egypt
5 Pharmacognosy Department, College of Pharmacy, Suez Canal University, Ismailia 41522, Egypt; mdonia00@gmail.com
6 Leibniz Institute of Plant Biochemistry, Department of Bioorganic Chemistry, Weinberg 3, D-06120 Halle (Saale), Germany; Anja.Ehrlich@ipb-halle.de
* Correspondence: Mohamed.farag@pharma.cu.edu.eg (M.A.F.); Ludger.wessjohann@ipb-halle.de (L.A.W.); Tel.: +20-100-4142567 (M.A.F.); Fax: +0202-25320005 (M.A.F.)

Academic Editors: Lillian Barros and Isabel C.F.R. Ferreira
Received: 4 January 2019; Accepted: 17 February 2019; Published: 7 March 2019

Abstract: Brine, the historically known food additive salt solution, has been widely used as a pickling media to preserve flavor or enhance food aroma, appearance, or other qualities. The influence of pickling, using brine, on the aroma compounds and the primary and secondary metabolite profile in onion bulb *Allium cepa* red cv. and lemon fruit *Citrus limon* was evaluated using multiplex metabolomics technologies. In lemon, pickling negatively affected its key odor compound "citral", whereas monoterpene hydrocarbons limonene and γ-terpinene increased in the pickled product. Meanwhile, in onion sulphur rearrangement products appeared upon storage, i.e., 3,5-diethyl-1,2,4-trithiolane. Profiling of the polar secondary metabolites in lemon fruit via ultra-performance liquid chromatography coupled to MS annotated 37 metabolites including 18 flavonoids, nine coumarins, five limonoids, and two organic acids. With regard to pickling impact, notable and clear separation among specimens was observed with an orthogonal projections to least squares-discriminant analysis (OPLS-DA) score plot for the lemon fruit model showing an enrichment of limonoids and organic acids and that for fresh onion bulb showing an abundance of flavonols and saponins. In general, the pickling process appeared to negatively impact the abundance of secondary metabolites in both onion and lemon, suggesting a decrease in their food health benefits.

Keywords: *Allium cepa* red cv.; *Citrus limon*; pickling; volatiles; SPME; chemometrics

1. Introduction

Pickling—using food additives, i.e., a highly concentrated salt solution (brine)—is a well-known method for preserving foods including vegetables that has been used for thousands of years and remains in use today. Undoubtedly, the most unambiguous role of pickling for human nutrition is to make the nutrients naturally present in the original food materials more palatable while also preserving their quality. Nevertheless, pickling processes can have several effects on food nutritional qualities [1]. The effect of the pickling preservation method and storage time on vitamins and amino acid composition in garlic was reported by Montano et al., 2004 [2]. In addition, the influence of pickling with and without fermentation was revealed, with fermented products showing higher α-tocopherol, riboflavin, and total amino acid levels except for thiamine and arginine. Another study on Chinese cabbage reported that the nitrogenous content, viz. nitrate and nitrite, in vegetables was dramatically affected by fermentation [3]. Other studies [4,5] demonstrated a decrease in the level of nitrates during the natural fermentation of cabbage and radish. The risk of the reduction of nitrates to nitrite lies in the possible induction of methemoglobinemia [6]. The metabolic changes associated with pickling may thus develop some distinctive properties related to its organoleptic characters, shelf life, and/or safety [7,8]. In terms of human health benefits, pickled foods produced by fermentation have been recognized as protective for the gut [9] via the microbial and enzymatic processing of food ingredients. Fermentation can improve protein and fiber digestibility by improving micronutrient bioavailability and decaying anti-nutritional factors [10]. Fermentation processes can also reduce toxic compounds such as aflatoxins concurrent with the production of antimicrobial factors such lactic acid, bacteriocins, carbon dioxide, and ethanol, which prevent the growth of food-borne pathogens [11,12].

Citrus fruits are enriched in polyphenols which mainly include flavanones and, to a lesser extent, flavones, flavonols, coumarins, and phenolic acids, viz. ferulic, *p*-coumaric, sinapic, and caffeic acids [13]. Phenolic compounds are well known for their myriad health benefits, as they exhibit antioxidant, anticancer, and anti-inflammatory properties, mostly attributed to their remarkable free radical scavenging activities [14]. A study on four Tunisian *Citrus* varieties revealed the abundance of phenolic acids (73.13–86.40%) [15] among their secondary metabolite profiles, whereas camphene was found to be a major component in *C. limon*. *Citrus* fruit is also abundant in coumarins, limonoids, sterols, and flavonoids, which all exhibit various bioactivities [16].

Onion bulbs (*Allium cepa*) encompass a myriad of bioactive natural product classes which are linked to its health benefits, i.e., anticancer or chemo-preventive agents [17]; antimicrobial [18], antiviral [19], and antioxidant activities [20]; protective activities against cardiovascular diseases [21], diabetes [22], neurodegenerative disorders, and cataract formation [23,24]. Among the secondary metabolites found in onion, flavonoids, fructans, and organosulphur compounds were mainly found to mediate its health benefits [25–27]. We have previously reported on the flavor makeup of *A. cepa* red cv. grown in the Siwa Oasis, Egypt, revealing its enrichment in sulphur compounds [28]. Solid-phase microextraction (SPME) is regarded as a powerful analytical tool for volatiles extraction without solvent consumption and with no heat application, which aids in maintaining the original food aroma without the formation of artifacts as compared to steam distillation [29]. Further, confined direct MS analysis with real-time ion source and its applications in the analysis of volatile organic compounds of *Citrus limon* (lemon) and *Allium cepa* (onion) was reported [30]. In terms the secondary non-volatile metabolite composition of *A. cepa*, flavonols and acylated amino acids were the major forms as analyzed via UPLC-QTOFMS [28]. Metabolomics provides a holistic overview for the chemical profile of an organism (set of metabolites) exposed to different conditions. It can provide a shorter and more accessible route for visualizing metabolic difference in food matrices [31]. Significant strides have been made in the analysis of plant foods, mostly using different metabolomics platforms data [32]. The application of mass spectrometry (MS) coupled to an array of separation techniques [33,34] helps to simultaneously analyze a relatively large number of metabolites in a high-throughput, reproducible, and sensitive manner [35]. This study aims to explore the impact of pickling as a preservation method on the metabolite profile. Onion and lemon were employed as two different models considering their different compositional profiles, as these represent major food crops preserved using pickling. The effect of pickling, with and without fermentation, in terms of processing and storage time on organosulphur compounds of garlic has been previously reported [36]. This study reported the drawbacks of fermentation on the bioactive sulphur compounds, namely γ-glutamyl peptides and *S*-alk(en)yl-L-cysteine sulfoxides. For the comprehensive analysis of metabolites, solid-phase microextraction (SPME) coupled with GC-MS was employed for volatiles profiling, whereas UPLC-QTOFMS was used to capture the polar non-volatile secondary metabolites. Results from the current study reveal for the apparent negative effect of pickling on the bioactive and flavor chemical composition in both lemon and onion.

2. Results and Discussion

2.1. Profiling of Onion and Lemon Volatiles Using SPME/GC-MS and Multivariate Data Analysis

Three independent biological replicates representing each specimen including fresh and pickled samples were extracted and analyzed under identical conditions. The volatile analysis of lemon samples (Figure 1) revealed for the presence of 31 volatile components, with monoterpene hydrocarbons representing the major class at ca. 96.6% and 97.5% in the fresh and pickled lemon, respectively (Table 1). The fresh lemon monoterepenes pool was characterized by a relatively high β-pinene level (33%) versus an enrichment of limonene and δ-terpinene in pickled lemon. With regard to oxygenated monoterpenes, α/β-citral (odor key marker) was remarkably reduced in pickled lemon and likely accounted for the lower aroma strength of pickled lemon. Furthermore, total sesquiterpene hydrocarbons were found to be slightly higher in fresh lemon with (*E*)-α-bergamotene as major component compared with a relative increase of (*E*/*Z*)-α-farnesene and β-bisabolene in pickled lemon. SPME/GC-MS volatiles analysis of fresh and pickled *A. cepa* onion bulbs detected the presence of 20 volatile components belonging to both sulphur and non-sulphur compounds (Figure 1). Sulphur volatiles were dominant with 96.6% and 98.1% in pickled and fresh samples, respectively. Several of the volatile components either disappeared or were reduced in pickled onion, as shown in Table 2. Moreover, significant increases in the levels of allyl methyl trisulfide, 3,5-diethyl-1,2,4-trithiolane, and 2-hexyl-5-methyl-3(2*H*)-furanone were detected in pickled onion. Such chemical change in volatiles makeup due to pickling is likely to affect its organoleptic properties such as taste and aroma.

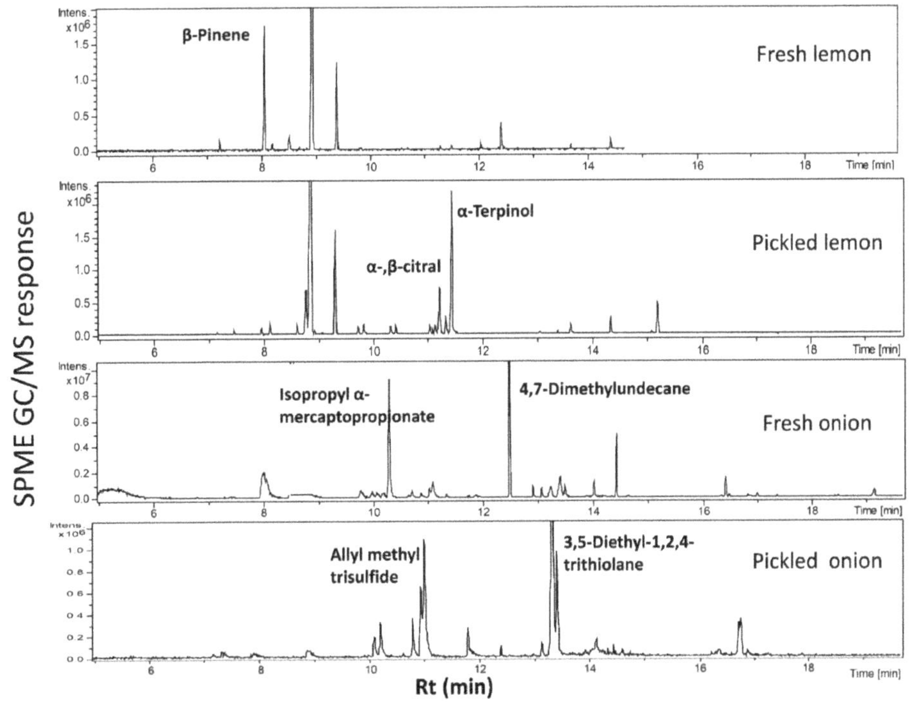

Figure 1. Solid-phase microextraction (SPME)/GC-MS chromatogram of headspace volatiles collected from fresh and pickled onion bulb and lemon fruit.

Table 1. Relative percentiles of volatile components to total peak areas detected in fresh and pickled lemon fruit using SPME/GC-MS measurements (n = 3).

Peak	rt (min)	KI	ID	Average ± Std. dev. Fresh	Average ± Std. dev. Pickled
1	7.03	903	α-Thujene	0.15 ± 0.11	0.04 ± 0.03
2	7.18	912	α-Pinene [a]	0.37 ± 0.48	0.35 ± 0.27
3	7.48	929	Fenchene	0.04 ± 0.04	0.02 ± 0.02
4	8.03	961	β-Pinene [a]	33.24 ± 5.89	23.44 ± 1.26
5	8.16	967	β-Myrcene [a]	0.15 ± 0.20	0.36 ± 0.28
6	8.64	995	1,4-Cineole [a]	0.02 ± 0.02	0.07 ± 0.06
7	8.66	996	2-Carene	2.12 ± 2.69	0.17 ± 0.13
8	8.80	1005	Melilotal	0.05 ± 0.07	0.39 ± 0.31
9	8.94	1014	Limonene [a]	32.10 ± 10.45	43.49 ± 5.16
10	9.38	1042	γ-Terpinene	12.39 ± 8.90	20.97 ± 3.21
11	9.76	1066	Terpinolene	0.10 ± 0.13	0.52 ± 0.41
12	9.86	1073	p-Cymenene	0.02 ± 0.03	0.19 ± 0.15
13	10.43	1110	α-Linalool [a]	0.01 ± 0.02	0.37 ± 0.29
14	11.06	1154	Bicyclo[3.2.1]oct-3-en-2-one, 4-methyl-	0.01 ± 0.02	0.36 ± 0.29
15	11.10	1158	α-Phellandren-8-ol	0.02 ± 0.03	0.03 ± 0.02
16	11.22	1166	1-Terpinen-4-ol	0.99 ± 0.81	0.89 ± 0.71
17	11.33	1173	p-Cymen-8-ol	0.02 ± 0.03	0.05 ± 0.04
18	11.43	1180	α-Terpineol	0.39 ± 0.49	3.33 ± 2.66
19	11.56	1189	Unknown	0.15 ± 0.18	2.50 ± 2.00
20	11.97	1220	β-Citral	3.46 ± 4.47	0.00 ± 0.00
21	12.36	1249	α-Citral [a]	10.86 ± 11.79	0.00 ± 0.00
22	13.16	1311	δ-Elemene	0.02 ± 0.03	0.08 ± 0.07
23	13.37	1328	6-Dimethyl-2,6-octadien-8-yl acetate	0.70 ± 0.90	0.24 ± 0.19
24	14.28	1402	(*Z*)-β-Farnesene	0.33 ± 0.46	0.23 ± 0.18
25	14.34	1407	(*E*)-α-Bergamotene	2.29 ± 2.56	0.25 ± 0.20
26	14.70	1437	Unknown sesquiterpene	0.00 ± 0.00	0.01 ± 0.01
27	15.04	1464	Unknown sesquiterpene	0.00 ± 0.00	0.05 ± 0.04
28	15.08	1468	(*Z,E*)-α-Farnesene	0.02 ± 0.03	0.80 ± 0.63
29	15.12	1478	β-Bisabolene	0.01 ± 0.01	0.69 ± 0.54
30	15.80	1523	epi-α-Eudesmol	-	0.01 ± 0.01
31	16.06	1541	Unknown sesquiterpene	-	0.08 ± 0.07
			Total monoterpene hydrocarbons %	80.66	89.55
			Total oxygenated monoterpenes %	15.83	5.49
			Total sesquiterpene hydrocarbons %	3.37	2.44
			Total oxygenated sesquiterpenes %	0.7	0.25

[a] denotes volatiles confirmed using the authentic standard in addition to RI and MS spectral matching.

Table 2. Relative percentiles of volatile components to total peak areas detected in fresh and pickled onion bulb using SPME/GC-MS measurements (n = 3).

Peak	rt (min)	KI	Name	Average ± Std. dev. Fresh	Average ± Std. dev. Pickled
1	7.44	924	1-propenyl methyl disulphide	2.91 ± 2.19	-
2	7.99	958	Dimethyl trisulfide [a]	3.24 ± 3.00	2.31 ± 3.18
3	9.79	1068	2-Acetylpyrrole	0.26 ± 0.15	0.00 ± 0.00
4	10.12	1089	Unknown sulphur	3.45 ± 0.36	6.71 ± 1.27
5	10.31	1101	Isopropyl α-mercaptopropionate	10.42 ± 1.94	8.35 ± 0.70
6	10.89	1143	Diethanol disulphide	12.20 ± 5.82	10.90 ± 2.89
7	10.99	1149	Methyl pentyl disulfide	0.24 ± 0.10	-
8	11.11	1157	Allyl methyl trisulfide	11.48 ± 5.08	25.25 ± 7.56
9	11.83	1209	β-Hydroxyethyl phenyl ether	0.38 ± 0.20	-
10	11.88	1212	Dimethyl tetrasulfide	3.35 ± 3.67	-
11	12.16	1234	3-Isopropylbenzaldehyde	0.26 ± 0.05	-
12	12.32	1246	4,7-Dimethylundecane	0.20 ± 0.10	-
13	12.92	1292	(Allylsulfanyl)acetonitrile	2.72 ± 3.67	-
14	13.23	1316	Dipropyl trisulfide	4.44 ± 1.26	1.28 ± 2.21
15	13.37	1328	3,5-Diethyl-1,2,4-trithiolane	18.95 ± 7.15	32.08 ± 4.53
16	13.42	1332	Diallyl trisulfide isomer	0.75 ± 0.42	-
17	14.43	1415	2-Hexyl-5-methyl-3(2*H*)-furanone	0.42 ± 0.43	3.44 ± 1.06
18	15.41	1495	Unknown hydrocarbon	0.34 ± 0.29	-
19	16.76	1589	4,6-Diethyl-1,2,3,5-tetrathiolane	10.03 ± 9.17	9.69 ± 1.13
20	17.00	1605	2,4-Dimethyl-5,6-dithia-2,7-Nonadienal	13.96 ± 12.73	-
			Total sulphur compounds %	98.14	96.57

[a] denotes volatiles confirmed using the authentic standard in addition to RI and MS spectral matching.

Multivariate data analysis was further applied to the volatiles abundance GC-MS dataset to visually demonstrate similarities and differences among specimens. Principal component analysis (PCA) was used as an unsupervised clustering model to highlight the variation based on volatiles abundance peak data. The PCA score plot (Figure S1) modeling fresh and pickled lemon showed no tight clustering of replicates, especially for pickled lemon. Consequently, supervised multivariate data analysis, i.e., OPLS, was employed for samples classification. OPLS-DA outscores PCA in the separation of the predictive variation from orthogonal variation and enhances data interpretation [37]. The OPLS-DA score plot (Figure 2A) showed better discrimination between the sample groups with the pickled group clustering separately from the fresh group. The S-loading plot derived from the OPLS-DA model is represented in Figure 2C. The figure highlights limonene and δ-terpinene as the most discriminatory aroma compounds for the pickled lemon, whereas β-pinene and α-citral were found distinctive for fresh lemon. Citral is a key flavor compound that accounts for lemon aroma [38]. Similarly, the OPLS-DA score plot of the pickled onion model (Figure 2C) discriminated between both sample groups, with 2-hexyl-5-methyl-3-(2*H*)-furanone and 3,5-diethyl-1,2,4-trithiolane (identified as markers for pickled onion) being highly correlated to the variation of the pickled onion. 2-Hexyl-5-methyl-3(2*H*)-furanone has previously been reported in "shallot" onion [39] as a decomposition fatty acid product derived from the onion lachrymatory factor [40], suggesting that sulphur compounds also undergo such degradation during pickling. Similarly, 3,5-diethyl-1,2,4-trithiolane may occur due to the thermal decomposition of cysteine [41], as reported in heat-prepared *Allium* essential oil [39].

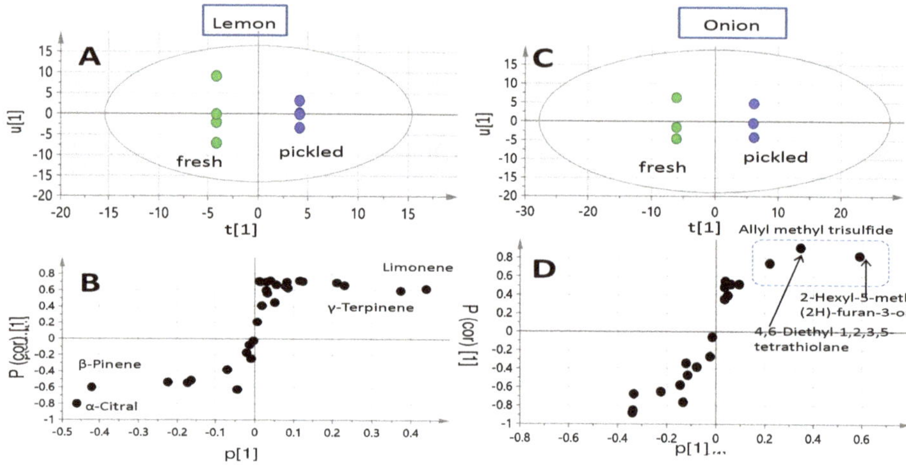

Figure 2. OPLS-DA score plot derived from modeling volatile metabolites. The metabolites were analyzed via SPME/GC-MS, and were derived fresh and pickled lemon fruit (**A**) and onion bulb (**C**), each modeled one at a time ($n = 3$). The respective *S*-plots, (**B**) and (**D**), show the covariance p[1] against the correlation p(cor)[1] of the variables of the discriminating component of the OPLS-DA model. Cut-off values of $p < 0.05$ were used; selected variables are highlighted in the *S*-plot and identifications are discussed in the text.

2.2. Profiling of Onion and Lemon Secondary Metabolites via UPLC-QTOFMS and Multivariate Data Analysis

To assess changes in secondary metabolite compositions of brine-preserved lemon and onion, non-targeted profiling was conducted using ultra-performance liquid chromatography (UPLC) coupled to a photodiode array and high resolution QTOF-MS operated in the negative- and positive-ionization modes (Figure S2). It should be noted that extracts were analyzed in both positive and negative-ion electrospray ionization (ESI) MS modes (Supplementary Figure S2) as changes in ESI polarity can often circumvent or significantly alter competitive ionization and suppression effects, revealing otherwise suppressed metabolite signals. We have previously reported on the UPLC-MS characterization of onion bulb, as published elsewhere [28], and we now present an annotation of metabolites for lemon fruit. Compared to the positive-ion ESI mode, the negative-ion MS spectra revealed better sensitivity and more observable peaks in the case of lemon. Nevertheless, it did not yield as much fragmentation information as in the positive-ion mode. Thus, the positive-ion mode provided more structural context and the negative-ion mode greater sensitivity. In total, 14 peaks from lemon fruit were annotated based on their negative-ionization mass spectral data versus 32 in the positive-ion mode, see Table 3. This is the first profile of *C. limon* fruit methanol extract from Egypt obtained via a UPLC-QTOFMS platform and it is presented herein as part of this study towards the composition of fresh *C. limon*. LC-MS and NMR were reported for the metabolite profiling of citrus oil derived from Italy [42]. A representative UPLC-MS base peak chromatogram (BPC) of *C. limon* is shown in (Figure S2A). The identities, retention times (rt), UV characteristics, and observed molecular and fragment ions for individual secondary metabolites are presented in Table 3. A total of 49 metabolites were detected, of which 36 were annotated. Metabolites belonged to several natural product classes including flavonoids, limonoids, coumarins, and phenolic/organic acids.

Table 3. Peak annotations of metabolites in fresh *C. limon* fruit methanol extract using UPLC-PDA-QTOFMS in negative- and positive-ionization modes.

Peak	rt (min)	M + H/M − H	Error (ppm)	Formula	MSMS	Name	Class
1	0.51	339.1088	2.909	$C_{12}H_{21}O_{11}^-$	-	Methyl-*O*-hexosyl pentoside	Sugar
2	0.66	381.078958	−2.67	$C_{17}H_{17}O_{10}^+$	-	Hydroxybergaptol hexoside	Coumarin
3	0.67	191.01953	0.9	$C_6H_7O_7^-$	111	(Iso)citric acid	Organic acid
4	0.96	205.0351	0.82	$C_7H_9O_7^-$	173, 111	Methylated (iso)citric acid	Organic acid
5	8.74	375.09094	−1.25	$C_{15}H_{19}O_{11}^+$	357, 339, 231, 137	Unknown	
6	8.95	357.11874	0.73	$C_{16}H_{21}O_9^-$	195	*O*-Feruloyl-hexoside	Phenolic acid
7	9.03	595.16473	−1.02	$C_{27}H_{31}O_{15}^+$	449, 287	Luteolin *O*-rhamnosyl-hexoside	Flavone-*O*-glycoside
8	9.31	625.17462	−1.69	$C_{28}H_{33}O_{16}^+$	607, 589, 487	Diosmetin-*C*,*C*-di-hexoside	Flavone-*C*-glycoside
9	9.72	565.15454	−0.64	$C_{26}H_{29}O_{14}^+$	433, 415, 313	Apigenin-*C*-hexosyl-*O*-pentoside	Flavone-*C*-glycoside
10	9.85	433.11185	−1.07	$C_{21}H_{21}O_{10}^+$	415, 367, 313	Apigenin-*C*-hexoside	Flavone-*C*-glycoside
11	10.1	463.12177	−1.72	$C_{22}H_{23}O_{11}^+$	445, 301	Diosmetin-*C*-hexoside	Flavone-*C*-glycoside
12	10.23	625.17487	−1.44	$C_{28}H_{33}O_{16}^+$	-	Diosmetin-*C*,*C*-di-hexoside isomer	Flavone-*C*-glycoside
13	10.23	579.16962	−1.21	$C_{27}H_{31}O_{14}^+$	433, 271	Apigenin-*O*-rhamnosyl-hexoside	Flavone-*O*-glycoside
14	10.33	609.1799	−1.41	$C_{28}H_{33}O_{15}^+$	463. 301	Diosmetin-*O*-rhamnosyl-hexoside	Flavone-*O*-glycoside
15	10.43	611.19482	−2.23	$C_{28}H_{35}O_{15}^+$	465, 449, 303	Hesperetin-*O*-rhanmosyl hexoside	Flavanone-*O*-glycoside
16	10.63	395.09641	−0.86	$C_{18}H_{19}O_{10}^+$	377, 291, 147	Hydroxybergaptol-*O*-methyl ether hexoside	Coumarin
17	10.71	509.1282	−0.77	$C_{23}H_{25}O_{13}^+$	347	Limocitrin-*O*-hexoside	Flavonol-*O*-glycoside
18	10.8	797.21155	−1.93	$C_{35}H_{41}O_{21}^+$	433, 347	Limocitrin-*O*-glycosyl malonate	Flavonol
19	10.82	767.20105	−1.87	$C_{34}H_{39}O_{20}^+$	317	Gossypetin rhamnosyl dideoxyhexosyl hexoside	Flavonol-*O*-glycoside
20	10.84	717.22156	−2.1	$C_{31}H_{41}O_{19}^+$	391, 255	Xyloccensin Q/R/S	Limonoid
21	10.97	687.21094	−2.15	$C_{30}H_{39}O_{18}^+$	669, 303	Unknown hesperitin glycoside	Flavanone-*O*-glycoside
22	11.06	653.1687	−2.53	$C_{29}H_{33}O_{17}^+$	347	Unknown	Flavonol-*O*-glycoside
23	11.13	479.11685	−1.55	$C_{22}H_{23}O_{12}^+$	461, 287	Luteolin dimethyl ether-*O*-hexoside	Flavone-*O*-glycoside
24	11.20	371.1485	−2.087	$C_{17}H_{23}O_9^-$	-	Citrusin E	Phenolic acid
25	11.26	305.10092	−1.04	$C_{16}H_{17}O_6^+$	203	Unknown	
26	12.00	345.06082	0.33	$C_{17}H_{13}O_8^-$	330	Limocitrin	Flavonol
27	12.39	247.05951	−0.59	$C_{13}H_{11}O_5^+$	232	Unknown	
28	12.75	515.19141	0.24	$C_{27}H_{31}O_{10}^-$	469	Methyllimonexic acid	Limonoid
29	12.78	471.19974	−1.6	$C_{26}H_{31}O_8^+$	425,409, 367	Limonine	Limonoid
30	12.97	287.0907	−0.7	$C_{16}H_{15}O_5^+$	203	Trichoclin	Coumarin
31	13.00	327.14435	0.52	$C_{16}H_{23}O_7^-$	173, 111	Unknown	
32	13.12	293.17548	0.74	$C_{17}H_{25}O_4^-$	236, 221	Unknown terpene	Terpene
33	13.13	515.22675	−0.81	$C_{28}H_{35}O_9^+$	455, 411, 393, 369	Nomilinic acid	Limonoid
34	13.54	455.20485	−1.6	$C_{26}H_{31}O_7^+$	437, 409, 393	Deoxylimonin/obacunone	Limonoid
35	13.55	269.08	0.95	$C_{16}H_{13}O_4^-$	214, 201	Unknown terpene	Terpene

Table 3. *Cont.*

Peak	rt (min)	M + H/M − H	Error (ppm)	Formula	MSMS	Name	Class
36	13.59	309.21	−0.05	$C_{18}H_{29}O_4^-$	291, 251	Unknown	
37	13.67	299.0924	1.04	$C_{17}H_{15}O_5^-$	284	Trihydroxy-dimethylflavanone	Flavanone
38	14.21	233.04378	−0.67	$C_{12}H_9O_5^+$	218, 205, 173	Hydroxybergapten	Coumarin
39	14.27	301.10632	−0.73	$C_{17}H_{17}O_5^+$	233	Hydroxy-dimethoxyflavanone	Flavanone
40	14.27	339.0621	−3.09	$C_{22}H_{11}O_4^+$	-	Unknown	
41	15.00	313.14386	0.42	$C_{19}H_{21}O_4^-$	176	Geranyloxy hydroxy coumarin.	Fatty acyl coumarin
42	15.45	481.2717	−5.06	$C_{29}H_{37}O_6^-$	-	Dihydrobergamotin octanal acetal	Coumarin
43	15.65	377.11435	−2.87	$C_{26}H_{17}O_3^+$	241	Unknown	
44	16.07	369.16898	−0.67	$C_{22}H_{25}O_5^+$	233	Hydroxybergaptol-*O*-dimethyl-octadienyl-Me ether	Fatty acyl coumarin
45	16.17	339.15781	−1.28	$C_{21}H_{23}O_4^+$	203	Bergaptin	Fatty acyl coumarin
46	16.19	337.14368	0.24	$C_{21}H_{21}O_4^-$	267, 254	Geranyloxypsoralen (bergaptin)	Coumarin
47	16.22	329.17377	−0.97	$C_{20}H_{25}O_4^+$	193, 184	Geranyloxy-methoxycoumarin	Fatty acyl coumarin
48	16.30	271.2276	0.71	$C_{16}H_{31}O_3^-$	225	Hydroxypalmitic acid	Fatty acid
49	16.52	339.19971	−1.63	$C_{15}H_{31}O_8^-$	183	Unknown	Fatty acid

2.2.1. Phenolic Acids

As precursors for most phenolic metabolites, phenolic acids are commonly reported in metabolite profiling studies of functional foods, often conjugated with sugars as in the case of *C. limon*. Phenolic acids are also known to contribute, among others, to the flavor/taste of foods. In this study, two phenolic acid glycosides were found to be predominant in lemon fruit extracts, namely 6-*O*-Feruloyl-hexoside (peak 6, *m*/*z* 341.087 $[M-H]^-$ and citrusin E (peak 24, *m*/*z* 353.0872 $[M-H]^-$) preferentially ionized in negative-ionization mode.

2.2.2. Flavonoids

Flavonoids were detected as the most abundant class, represented by 18 peaks belonging to flavanone, flavone, and flavonols subclasses, principally as *O*-glycosides. In MS analysis, the nature of the sugars could be revealed by the elimination of the sugar residue, which would have a mass of 162 amu (hexose; glucose or galactose), 146 amu (rhamnose), or 132 amu (pentose; xylose or arabinose). The MS spectrum interpretation allowed for the annotation of hesperitin signals (*m*/*z* 301.27, $C_{16}H_{13}O_6^-$) in peaks 37 and 39, luteolin (*m*/*z* 285.21, $C_{15}H_9O_6^-$) in peaks 7 and 23, apigenin (*m*/*z* 269.21, $C_{15}H_9O^-$) in peak 13, diosmetin (*m*/*z* 299.26, $C_{16}H_{11}O_6^-$) in peak 14, and limocitirin (*m*/*z* 345.21, $C_{17}H13O_8^-$) in peaks 17, 18, and 26. The readily cleaved sugar moieties from aglycone yielded a fragment mass in these respective peaks. In contrast, several flavonoid peaks (8–12) showed intense molecular ion peaks, with $[M-90-H]^-$ and $[M-120-H]^-$ fragments exclusively in flavone peaks of the MSn spectra indicative of sugar cleavage in *C*-glycoside peaks 8–12. The fragmentation pathways most characteristic of *C*-glycosyl flavonoids include dehydration (−18 amu) and cross-ring cleavage (0,2 and 0,3) of the sugar moiety, i.e., −120 amu and −90 amu for the *C*-hexosides; −90 amu and −60 amu for the *C*-pentosides. Citrus fruits are rich dietary sources of flavonoids [43] and thus exhibit a wide range of pharmaceutical properties including anti-atherogenic, anti-inflammatory, antitumor, and antioxidant activities [44].

2.2.3. Limonoids/Coumarins

Limonoids (abundant in lemon) belonging to tetranortriterpenes were annotated in peaks 48, 49, and 54 as glycosidic conjugates and in peak 65 as the aglycone "limonin", eluting much later at a relatively less polar eluent composition. This the first report of methyllimonexic acid in *C. limon*. Aside from their bitterness, imparting their taste to lemon, limonoids are regarded as an important constituent in food diets including *Citrus* fruit owing to their anti-obesity and anti-hyperglycemic effects [45]. Several coumarins (furano-type) were also detected, associated with peaks 56 (*m*/*z* 201.0186 $[M-H]^-$), 62 (*m*/*z* 371.1488 $[M-H]^-$), and 75 (*m*/*z* 297.1519 $[M-H]^-$) and annotated as bergaptol, 6′,7′-dihydroxybergamottin, and geranyloxycoumarin, respectively. Lipophilic coumarins acylated with a fatty acid group and eluting in the fatty acid region were also assigned in peaks 44, 45, and 47. Except for bergapten and dihydrobergamotin octanal acetal, these coumarins are annotated for the first time in *C. limon* and might also account for the food properties or health effects observed in grapefruit.

The UPLC-QTOFMS dataset was further subjected to multivariate data analysis to help determine which analytical platform provides a better prediction of pickling impact on *Allium* and *Citrus* metabolism. PCA was initially constructed to reveal for the metabolites variation among fresh and pickled specimens based on the abundance of peaks. The PCA score plots (Figure S3a,b) demonstrated a clear segregation between the fresh and pickled product in both lemon and onion with a total covered variance of 86% and 94%, respectively. The OPLS-DA analysis was further employed to identify metabolites that were highly correlated with the sample type. Compared to GC-MS, the UPLC-QTOFMS-derived OPLS model was found to be more predictive of the assessment of pickling, as revealed from the prediction power of their respective models. However, it should be noted that in contrast to the GC-MS-based model, the identified variant masses in case of UPLC-QTOFMS were based on MS spectral interpretation and not confirmed with authentic standards. A clear separation was detected from the OPLS-DA score plot (Figure 3A,C) for lemon and onion.

The S-loading plot of lemon (Figure 3B) revealed the presence of xyloccensin, citric acid, and its methyl ester as the most distinctive markers for fresh lemon. In contrast, onion fresh bulb was discriminated from the pickled one by the abundance of steroidal saponins in addition to quercetin glycosidic conjugates (Figure 3D). Xyloccensin exhibits an anti-hyperglycemic activity [46] and a decrease in its levels could ultimately affect lemon benefits. Moreover, the benefits of *Allium's* polyphenols, i.e., quercetin glycoside and diglycoside, are well recognized [47] and a decrease in their levels was also observed in the case of *A. cepa* red cv. examined herein. Whether these metabolites get solubilized in the brine solution surrounding the bulb or get degraded has yet to be determined.

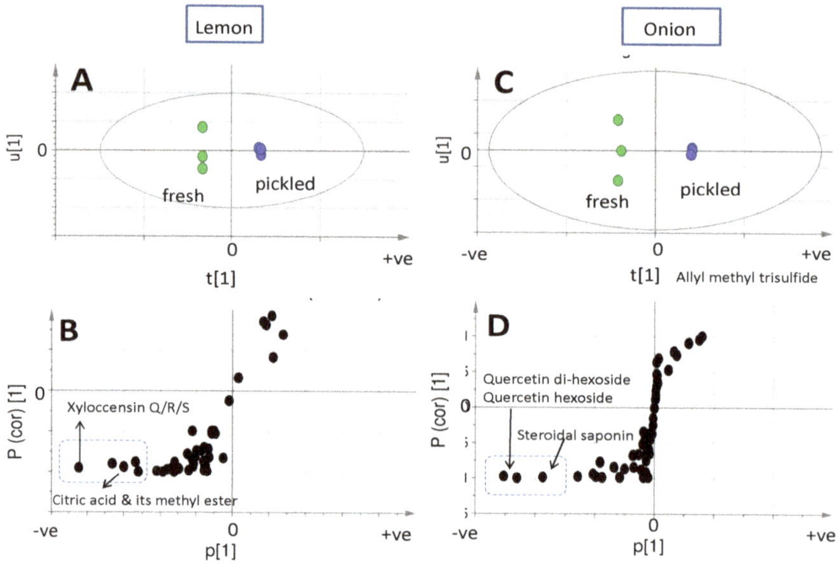

Figure 3. OPLS-DA score plot derived from modeling primary and secondary metabolites analyzed using UPLC-QTOFMS. Metabolites were derived fresh and pickled lemon fruit (**A**) and onion bulb (**C**), each modeled one at a time. The respective S-plots, (**B**) and (**D**), show the covariance p[1] against the correlation p(cor)[1] of the variables of the discriminating component of the OPLS-DA model. Cut-off values of $p < 0.05$ were used; selected variables are highlighted in the S-plot and identifications are discussed in the text.

To assess the validity of the UPLC-MS based models, Q2 and R2 values of all calculated models were found to be larger than 0.5 and close to 1. The permutation diagnostic analysis of 20 iterations provided a reference distribution of R2/Q2 values and hence indicated the statistical significance of these parameters, with most models showing a regression line crossing zero, with negative Q2 and R2 values close to 1, which signifies the model validation. Also, the *p*-value for each OPLS-DA model was calculated using CV-ANOVA and all *p*-value were below 0.005 in the pickled models of lemon and onion (Supplementary Figures S4 and S5). Nevertheless, the models derived from the volatiles GC-MS dataset exhibited lower model validation, showing smaller R2 and Q2 values and with *p*-values higher than 0.05, suggesting that the UPLC-MS-derived models provided a better assessment of the pickling impact on both foods

2.3. Onion and Lemon Primary Metabolite Profiling Using GC-MS

Although UPLC-QTOFMS revealed the variation in organic acids, viz. citric acid and its methyl derivative, as a major contributor to the segregation of pickled lemon from fresh ones, such a platform is not optimized for primary metabolite profiling [48]. To provide a more sensitive analysis of primary metabolites, viz. sugars and amino acids, compared to reversed-phase UPLC-QTOFMS, GC-MS was adopted for primary metabolite profiling in extracts prepared from fresh and pickled tissues. GC-MS analysis detected 52 primary metabolites belonging to organic acids, amino acids, fatty acids, sugars, and inorganic and nitrogenous compounds (Figure 4, Table S1). Primary metabolite analysis revealed a remarkable decrease in the total amount of organic acids, amino acids, and inorganic and nitrogenous compounds in the pickled tissue of both onion and lemon. In contrast, sugars and fatty acids were the major primary metabolites in pickled samples. Particularly in onion, fresh tissue was rich in malic acid (12.6%), proline (7%), fructofuranose (27%), and mannopyranose (11.6%). On the other hand, pickled onion showed dominancy in sugars, viz. maltose (16.3%) and sucrose (38%). In lemon, fresh tissue was also predominated by organic acids, viz. citric acid (37%), and sugar, viz. sucrose (28%), accounting for the sour taste of lemon concurrent with an abundance of sugar alcohol, viz. myo-inositol (21.9%) and fructofuranose (47%). A decrease in citric acid and its methyl ester levels in pickling can explain the less sour taste observed in pickled lemon and is in agreement with the UPLC-QTOFMS results. However, citric acid is considered a natural preservative that aids in smooth digestion and helps dissolve kidney stones [49]. Also of interest is the different sugar profile in the pickled product being more dominated by sugar alcohols, viz. myo-inositol, and monosaccharides, viz. fructose, compared to disaccharides in the fresh product, viz. sucrose. A decrease in disaccharide levels upon pickling in lemon could be attributed to the chemical hydrolysis of sucrose upon storage and especially due to the acidic nature of lemon, as an such effect was not observed in case of pickling *Allium* bulb. Sugar alcohols exhibit higher thermal stability and do not undergo any Milliard reaction compared to sugars [50].

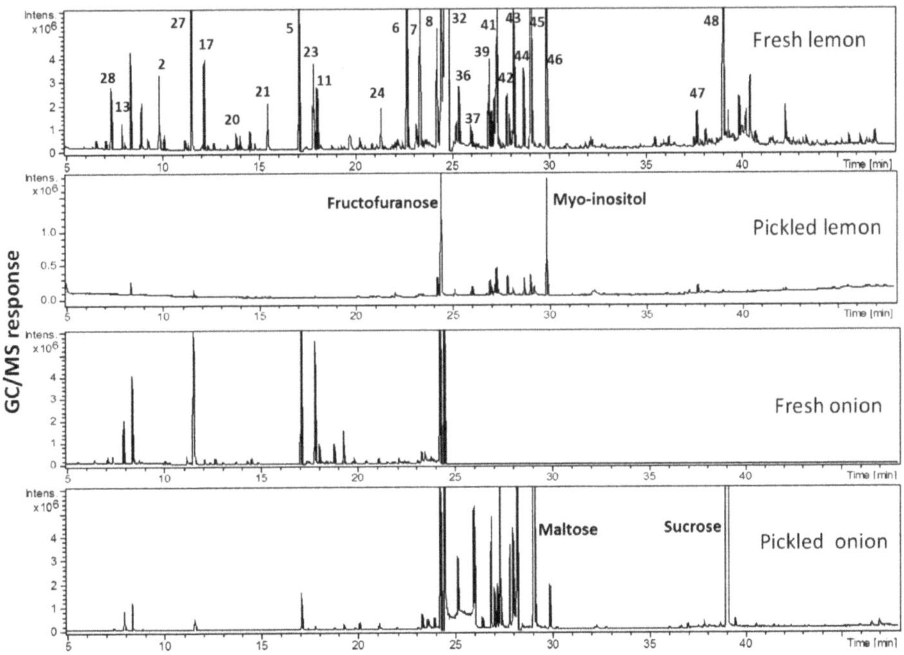

Figure 4. GC-MS chromatogram of silylated primary metabolites analyzed from fresh and pickled onion bulb and lemon fruit.

3. Materials and Methods

3.1. *Plant Material*

Allium sativum and *A. cepa* red cv. bulbs were collected fresh from the field at Siwa Oasis, Egypt during the month of May 2016. Lemon fruit was collected at the ripe stage from trees grown in Behira governorate, Egypt during the month of May 2016.

3.2. *Pickling Protocol*

Preparing pickled lemon and onion was made by packing directly with acidified brine after blanching. Before processing, defective fruits or bulbs were discarded. Lemon fruit and pealed onion bulbs were blanched separately in water at 90 °C for 5 min and immediately packed in dark glass jars that were closed tight, containing acidified brine solution using 5% acetic acid. Brine was prepared by dissolving NaCl in water at a concentration of 15%. Both pickled products were stored in a dark place away of light at room temperature for four weeks.

3.3. *Chemicals and Fibers*

SPME fiber of stableflex coated with divinylbenzene/carboxen/polydimethylsiloxane (DVB/CAR/PDMS, 50/30 μm) was purchased from Supelco (Oakville, ON, Canada). All chemicals and standards were purchased from Sigma Aldrich (St. Louis, MO, USA). Acetonitrile and formic acid (LC–MS grade) were obtained from J.T. Baker (Avantor, Netherlands). MilliQ water was used for UPLC-PDA-ESI-TOF-MS analysis.

3.4. *Headspace Volatiles Analysis of C. limon and A. cepa Bulbs*

The headspace-solid phase microextraction (HS-SPME) volatile analysis was carried out as mentioned previously [28]. (Z)-3-hexneyl acetate, dissolved in water to make a final concentration of 2 μg per 1.5-mL SPME screw cap vial, was used to spike 100 mg of dried, finely ground peeled bulbs of onion and lemon fruit placed in 1.5-mL glass vials. The SPME fiber was soaked in a vial containing plant material and placed in an oven of 50 °C for 30 min. The fiber was successively withdrawn into the needle and then injected into the injection port of the gas chromatography-mass spectrometer (GC-MS). A system blank containing no plant material was run as a control. SPME fibers were desorbed at 210 °C for 1 min in the injection port of a Shimadzu Model GC-17A gas chromatograph interfaced with a Shimadzu model QP-5000 mass spectrometer (Kyoto, Japan). Volatiles were separated on a DB5-MS column (30 m length, 0.25 mm inner diameter, and 0.25 μm film) (J & W Scientific, Santa Clara, CA, USA). Injections were made in the splitless mode for 30 s. The gas chromatograph was operated under the following conditions: injector 220 °C, column oven 38 °C for 3 min, then programmed at a rate of 12 °C/min to 180 °C, maintained at 180 °C for 5 min, and finally ramped at a rate of 40 °C min^{-1} to 220 °C and maintained for 2 min. He was used as the carrier gas at 1 mL/min^{-1}. The transfer line and ion-source temperatures were adjusted to be 230 and 180 °C, respectively. The HP quadrupole mass spectrometer was operated in the electron ionization mode at 70 eV. The scan range was set at m/z 40–500. Peaks were deconvoluted using AMDIS software (Gaithersburg, MD, USA www.amdis.net). Volatile components were identified by matching their retention indices (RI) relative to n-alkanes (C_6–C_{20}), as well as mass comparison to NIST and WILEY library databases and with standards.

3.5. *GC-MS Analysis of Silylated Primary Metabolites in Fruit Pulp and Skin*

Freeze-dried fruit or bulb powders (100 mg) were extracted by adding 5 mL 50% MeOH. Then, 100 μL of this 50% aqueous extract was evaporated under nitrogen until dryness was reached. Next, 150 μL of *N*-methyl-*N*-(trimethylsilyl)-trifluoroacetamide (MSTFA) was added and incubated at 60 °C for 45 min for derivatization. The primary metabolites were analyzed using GC-MS. Silylated products were purified on an Rtx-5MS (30 m length, 0.25 mm inner diameter, and 0.25 μm film) column.

Injections were made in a (1:15) split mode with the following conditions: injector 280 °C, column oven 80 °C for 2 min, rate 5 °C/min to 315 °C, maintained at 315 °C for 12 min. He was used as the carrier gas at 1 mL/min^{-1}. The transfer line and ion-source temperatures were set to 280 and 180 °C, respectively. Compounds were identified as described in [51].

3.6. Extraction Procedure and Sample Preparation for UPLC-PDA-MS Analysis

Dried lyophilized food material was grounded separately in a mortar with liquid nitrogen. A powdered aliquot (ca. 30 mg) was then homogenized with 2.5 mL 70% MeOH containing 5 mg/mL umbelliferone (internal standard) with sonication for 30 min. Extracts were vortexed, centrifuged at 3000× *g* for 30 min to remove debris, and stored. Chromatographic separation was performed on an Acquity UPLC system (Waters Corp., Milford, MA, USA) equipped with an HSS T3 column (100 × 1.0 mm, particle size 1.8 mm; Waters Corp.) applying the following binary gradient at a flow rate of 150 mL/min: 0 to 1 min, isocratic 95% A (water:formic acid, 99.9:0.1, *v*/*v*), 5% B (acetonitrile:formic acid, 99.9:0.1, *v*/*v*); 1 to 16 min, linear from 5 to 95% B; 16 to 18 min, isocratic 95% B; 18 to 20 min, isocratic 5% B.

3.7. GC-MS and UPLC-QTOFMS Data Processing for Multivariate Analysis

MS peak abundance of volatiles and secondary metabolites were extracted using MET-IDEA software with default parameter settings for GC-MS and LC-MS files [52]. The aligned peak abundance data table was normalized to a spiked internal standard and further exported to principal component analysis (PCA) and partial least squares-discriminant analysis (OPLS-DA) using SIMCA-P version 13.0 software package (Umetrics, Umeå, Sweden). All variables were mean-centered and scaled to Pareto variance.

4. Conclusions

Pickling using brine, an old traditional process used to preserve food products, was investigated for the first time using an MS-based metabolomics approach in two food materials—lemon and onion. Metabolomics was effectively implemented to highlight the similarities and differences between fresh and pickled products. Chemical analysis revealed a decline in lemon sensory volatiles, i.e., α,β-citral, as well as the degradation and rearrangement of sulphur compounds in onion to generate 2-hexyl-5-methyl-3(2*H*)-furanone and 3,5-diethyl-1,2,4-trithiolane, which are sulphur rearrangement products. A marked decrease in citric acid levels in lemon is likely to account for the less sour taste of its fruit. Nevertheless, it should be noted that the chemical changes observed in this study need to be further complemented with a sensory analysis to prove whether they are ultimately reflected in changes in aroma or taste. The impact of pickling on the non-volatile polar bioactive metabolites was revealed via UPLC-QTOFMS analysis and suggested for a lower health value of pickled product relative to the fresh specimens. The effect of blanching as part of the pickling was also not separately assessed, as samples were only analyzed after the pickling step.

Supplementary Materials: The followings are available online, Table S1: Relative percentiles of silylated primary metabolites detected in fresh and pickled onion bulb and lemon fruit using GC-MS measurements; Figure S1: SPME/GC-MS-based PCA score plot derived from modeling the pickling effect on lemon fruit to assess the effect of pickling on metabolite composition; Figure S2. UPLC-QTOFMS chromatogram of metabolites analyzed from fresh and pickled lemon fruit (A) and onion bulb (B) in negative- and positive-ionization modes, showing qualitative differences upon pickling in each examined food; Figure S3: UPLC-QTOFMS-based PCA score plot derived from modeling the pickling effect on lemon fruit (A) and onion bulb red cv. (B) separately to assess the effect of pickling on metabolite composition.

Author Contributions: M.A.F. and L.A.W. conceived and designed the experiments; M.A.F. and A.E. performed the experiments and the measurements; A.F.T. wrote the manuscript, M.S.D. performed the data analysis; M.A.F. revised the paper.

Funding: Mohamed A. Farag wishes to thank the Alexander von Humboldt foundation, Germany for the financial support and the American University of Cairo Internal Research Support Grant (RSG1-18).

Conflicts of Interest: The authors declare no conflict of interest. The founding sponsors had no role in the design of the study; in the collection, analyses, or interpretation of data; in the writing of the manuscript, and in the decision to publish the results.

References

1. McFeeters, R.F. Effects of Fermentation on the Nutritional Properties of Food. In *Nutritional Evaluation of Food Processing*; Karmas, E., Harris, R.S., Eds.; Springer Netherlands: Dordrecht, The Netherlands, 1988; pp. 423–446.
2. Montano, A.; Casado, F.J.; de Castro, A.; Sanchez, A.H.; Rejano, L. Vitamin content and amino acid composition of pickled garlic processed with and without fermentation. *J. Agric. Food Chem.* **2004**, *52*, 7324–7330. [CrossRef] [PubMed]
3. Ji, F.-D.; Ji, B.-P.; Li, B.O.; Lu, F.E.I. Effect of fermentation on nitrate, nitrite and organic acid contents in traditional pickled chinese cabbage. *J. Food Process. Preserv.* **2009**, *33*, 175–186. [CrossRef]
4. Guo, X.H.; Yang, J.S.; Zhang, J.J. The study of nitrite peakvalue change mechanism and inhibition methods during Chinese cabbage fermentation. *Chin. Food Ferment. Ind.* **1989**, *15*, 26–34.
5. He, S.L.; Li, B.; Ji, B.P. Study on nitrate reductase activityduring the fermentation of pickled vegetable. *Chin. Food Technol.* **2005**, *1*, 94–97.
6. Chan, T.Y.K. Vegetable-borne nitrate and nitrite and the risk of methaemoglobinaemia. *Toxicol. Lett.* **2011**, *200*, 107–108. [CrossRef] [PubMed]
7. Holzapfel, W. Use of starter cultures in fermentation on a household scale. *Food Control* **1997**, *8*, 241–258. [CrossRef]
8. Leroy, F.; De Vuyst, L. Lactic acid bacteria as functional starter cultures for the food fermentation industry. *Trends Food Sci. Technol.* **2004**, *15*, 67–78. [CrossRef]
9. Con, A.H.; Karasu, N. Determination of Antagonistic Starter Cultures for Pickle and Olive Fermentation Processes. *Czech J. Food Sci.* **2009**, *27*, 185–193.
10. Giraffa, G. Studying the dynamics of microbial populations during food fermentation. *FEMS Microbiol. Rev.* **2004**, *28*, 251–260. [CrossRef] [PubMed]
11. Holzapfel, W.H. Appropriate starter culture technologies for small-scale fermentation in developing countries. *Int. J. Food Microbiol.* **2002**, *75*, 197–212. [CrossRef]
12. Hansen, E.B. Commercial bacterial starter cultures for fermented foods of the future. *Int. J. Food Microbiol.* **2002**, *78*, 119–131. [CrossRef]
13. Abad-García, B.; Garmón-Lobato, S.; Sánchez-Ilárduya, M.B.; Berrueta, L.A.; Gallo, B.; Vicente, F.; Alonso-Salces, R.M. Polyphenolic contents in Citrus fruit juices: Authenticity assessment. *Eur. Food Res. Technol.* **2014**, *238*, 803–818. [CrossRef]
14. Spinola, V.; Pinto, J.; Castilho, P.C. Identification and quantification of phenolic compounds of selected fruits from Madeira Island by HPLC-DAD-ESI-MS(n) and screening for their antioxidant activity. *Food Chem.* **2015**, *173*, 14–30. [CrossRef] [PubMed]
15. Tounsi, M.S.; Wannes, W.A.; Ouerghemmi, I.; Jegham, S.; Ben Njima, Y.; Hamdaoui, G.; Zemni, H.; Marzouk, B. Juice components and antioxidant capacity of four Tunisian Citrus varieties. *J. Sci. Food Agric.* **2011**, *91*, 142–151. [CrossRef] [PubMed]
16. Hamdan, D.; El-Readi, M.Z.; Tahrani, A.; Herrmann, F.; Kaufmann, D.; Farrag, N.; El-Shazly, A.; Wink, M. Secondary metabolites of ponderosa lemon (*Citrus pyriformis*) and their antioxidant, anti-inflammatory, and cytotoxic activities. *Zeitschrift fur Naturforschung C J. Biosci.* **2011**, *66*, 385–393. [CrossRef]
17. Guyonnet, D.; Siess, M.H.; Le Bon, A.M.; Suschetet, M. Modulation of Phase II Enzymes by Organosulfur Compounds from Allium Vegetables in Rat Tissues. *Toxicol. Appl. Pharmacol.* **1999**, *154*, 50–58. [CrossRef] [PubMed]
18. Rose, P.; Whiteman, M.; Moore, P.K.; Zhu, Y.Z. Bioactive S-alk(en)yl cysteine sulfoxide metabolites in the genus Allium: The chemistry of potential therapeutic agents. *Nat. Prod. Rep.* **2005**, *22*, 351–368. [CrossRef] [PubMed]
19. Wu, C.P.; Calcagno, A.M.; Hladky, S.B.; Ambudkar, S.V.; Barrand, M.A. Modulatory effects of plant phenols on human multidrug-resistance proteins 1, 4 and 5 (ABCC1, 4 and 5). *FEBS J.* **2005**, *272*, 4725–4740. [CrossRef] [PubMed]

20. Gülşen, A.; Makris, D.P.; Kefalas, P. Biomimetic oxidation of quercetin: Isolation of a naturally occurring quercetin heterodimer and evaluation of its in vitro antioxidant properties. *Food Res. Int.* **2007**, *40*, 7–14. [CrossRef]
21. Liu, L.; Yeh, Y.Y. S-alk(en)yl cysteines of garlic inhibit cholesterol synthesis by deactivating HMG-CoA reductase in cultured rat hepatocytes. *J. Nutr.* **2002**, *132*, 1129–1134. [CrossRef] [PubMed]
22. Babu, P.S.; Srinivasan, K. Influence of dietary capsaicin and onion on the metabolic abnormalities associated with streptozotocin induced diabetes mellitus. *Mol. Cell. Biochem.* **1997**, *175*, 49–57. [CrossRef] [PubMed]
23. Kaur, C.; Joshi, S.; Kapoor, H.C. Antioxidants in onion (*Allium cepa* L.) cultivars grown in India. *J. Food Biochem.* **2009**, *33*, 184–200. [CrossRef]
24. Corzo-Martínez, M.; Corzo, N.; Villamiel, M. Biological properties of onions and garlic. *Trends Food Sci. Technol.* **2007**, *18*, 609–625. [CrossRef]
25. Ames, B.N.; Shigenaga, M.K.; Hagen, T.M. Oxidants, antioxidants, and the degenerative diseases of aging. *Proc. Natl. Acad. Sci. USA* **1993**, *90*, 7915–7922. [CrossRef] [PubMed]
26. Paganga, G.; Miller, N.; Rice-Evans, C.A. The polyphenolic content of fruit and vegetables and their antioxidant activities. What does a serving constitute? *Free Radic. Res.* **1999**, *30*, 153–162. [CrossRef] [PubMed]
27. Rice-Evans, C.; Miller, N.; Paganga, G. Antioxidant properties of phenolic compounds. *Trends Plant Sci.* **1997**, *2*, 152–159. [CrossRef]
28. Farag, M.A.; Ali, S.E.; Hodaya, R.H.; El-Seedi, H.R.; Sultani, H.N.; Laub, A.; Eissa, T.F.; Abou-Zaid, F.O.F.; Wessjohann, L.A. Phytochemical Profiles and Antimicrobial Activities of *Allium cepa* Red cv. and A. sativum Subjected to Different Drying Methods: A Comparative MS-Based Metabolomics. *Molecules* **2017**, *22*, 761. [CrossRef] [PubMed]
29. Marsili, R. Analysis of musty microbial metabolites by stir bar sorptive extraction. In *Flavor, Fragrance, and Odor Analysis*, 2nd ed.; CRC Press: Boca Raton, FL, USA, 2011; pp. 63–92.
30. Li, Y. Confined direct analysis in real time ion source and its applications in analysis of volatile organic compounds of *Citrus limon* (lemon) and *Allium cepa* (onion). *Rapid Commun. Mass Spectrom.* **2012**, *26*, 1194–1202. [CrossRef] [PubMed]
31. Tawfike, A.F.; Viegelmann, C.; Edrada-Ebel, R. Metabolomics and dereplication strategies in natural products. *Methods Mol. Biol.* **2013**, *1055*, 227–244. [PubMed]
32. Farag, M.A. Comparative mass spectrometry & nuclear magnetic resonance metabolomic approaches for nutraceuticals quality control analysis: A brief review. *Recent Pat. Biotechnol.* **2014**, *8*, 17–24. [PubMed]
33. Wolfender, J.L.; Marti, G.; Thomas, A.; Bertrand, S. Current approaches and challenges for the metabolite profiling of complex natural extracts. *J. Chromatogr. A* **2015**, *1382*, 136–164. [CrossRef] [PubMed]
34. Wolfender, J.L.; Rudaz, S.; Choi, Y.H.; Kim, H.K. Plant metabolomics: From holistic data to relevant biomarkers. *Curr. Med. Chem.* **2013**, *20*, 1056–1090.
35. Villas-Boas, S.G.; Mas, S.; Akesson, M.; Smedsgaard, J.; Nielsen, J. Mass spectrometry in metabolome analysis. *Mass Spectrom. Rev.* **2005**, *24*, 613–646. [CrossRef] [PubMed]
36. Beato, V.M.; Sanchez, A.H.; de Castro, A.; Montano, A. Effect of processing and storage time on the contents of organosulfur compounds in pickled blanched garlic. *J. Agric. Food Chem.* **2012**, *60*, 3485–3491. [CrossRef] [PubMed]
37. Song, H.H.; Kim, D.Y.; Woo, S.; Lee, H.K.; Oh, S.R. An approach for simultaneous determination for geographical origins of Korean Panax ginseng by UPLC-QTOF/MS coupled with OPLS-DA models. *J. Ginseng Res.* **2013**, *37*, 341–348. [CrossRef] [PubMed]
38. McGorrin, R. Character-impact flavor and off-flavor compounds in foods. In *Flavor, Fragrance, and Odor Analysis*, 2nd ed.; CRC Press: Boca Raton, FL, USA, 2011; pp. 207–262.
39. Tocmo, R.; Lin, Y.; Huang, D. Effect of processing conditions on the organosulfides of shallot (*Allium cepa* L. Aggregatum group). *J. Agric. Food Chem.* **2014**, *62*, 5296–5304. [CrossRef] [PubMed]
40. Mondy, N.; Duplat, D.; Christides, J.P.; Arnault, I.; Auger, J. Aroma analysis of fresh and preserved onions and leek by dual solid-phase microextraction-liquid extraction and gas chromatography-mass spectrometry. *J. Chromatogr. A* **2002**, *963*, 89–93. [CrossRef]
41. Yu, A.-N.; Tan, Z.-W.; Wang, F.-S. Mechanism of formation of sulphur aroma compounds from l-ascorbic acid and l-cysteine during the Maillard reaction. *Food Chem.* **2012**, *132*, 1316–1323. [CrossRef] [PubMed]

42. Marti, G.; Boccard, J.; Mehl, F.; Debrus, B.; Marcourt, L.; Merle, P.; Delort, E.; Baroux, L.; Sommer, H.; Rudaz, S.; et al. Comprehensive profiling and marker identification in non-volatile citrus oil residues by mass spectrometry and nuclear magnetic resonance. *Food Chem.* **2014**, *150*, 235–245. [CrossRef] [PubMed]
43. Wang, S.C.; Yang, C.K.; Tu, H.; Zhou, J.J.; Liu, X.Q.; Cheng, Y.J.; Luo, J.; Deng, X.X.; Zhang, H.Y.; Xu, J. Characterization and Metabolic Diversity of Flavonoids in Citrus Species. *Sci. Rep.* **2017**, *7*. [CrossRef] [PubMed]
44. Kumar, S.; Singh, A.K.; Verma, S.K.; Misra, R.; Seniya, C. Antibacterial and Phyto-chemical Analysis of Some Medicinal Plants and their Efficacy on Multidrug Resistant Bacteria. *J. Pure Appl. Microbiol.* **2013**, *7*, 2191–2204.
45. Gualdani, R.; Cavalluzzi, M.M.; Lentini, G.; Habtemariam, S. The Chemistry and Pharmacology of Citrus Limonoids. *Molecules* **2016**, *21*, 1530. [CrossRef] [PubMed]
46. Zhou, Z.-F.; Kong, L.-Y.; Kurtán, T.; Liu, H.-L.; Mándi, A.; Li, J.; Gu, Y.-C.; Guo, Y.-W. Four Phragmalin Orthoesters from the Chinese Mangrove Xylocarpus granatum. *Planta Med.* **2014**, *80*, 949–954. [CrossRef] [PubMed]
47. Quideau, S.; Deffieux, D.; Douat-Casassus, C.; Pouysegu, L. Plant polyphenols: Chemical properties, biological activities, and synthesis. *Angew. Chem.* **2011**, *50*, 586–621. [CrossRef] [PubMed]
48. Farag, M.A.; Porzel, A.; Wessjohann, L.A. Comparative metabolite profiling and fingerprinting of medicinal licorice roots using a multiplex approach of GC-MS, LC-MS and 1D NMR techniques. *Phytochemistry* **2012**, *76*, 60–72. [CrossRef] [PubMed]
49. Gonzalez-Molina, E.; Dominguez-Perles, R.; Moreno, D.A.; Garcia-Viguera, C. Natural bioactive compounds of *Citrus limon* for food and health. *J. Pharm. Biomed. Anal.* **2010**, *51*, 327–345. [CrossRef] [PubMed]
50. Neumann, H.; Niedermaier, S.; Gschwander, S.; Schossig, P. Cycling stability of D-mannitol when used as phase change material for thermal storage applications. *Thermochim. Acta* **2018**, *660*, 134–143. [CrossRef]
51. Farag, M.A.; Rasheed, D.M.; Kamal, I.M. Volatiles and primary metabolites profiling in two Hibiscus sabdariffa (roselle) cultivars via headspace SPME-GC-MS and chemometrics. *Food Res. Int.* **2015**, *78*, 327–335. [CrossRef] [PubMed]
52. Broeckling, C.D.; Reddy, I.R.; Duran, A.L.; Zhao, X.C.; Sumner, L.W. MET-IDEA: Data extraction tool for mass spectrometry-based metabolomics. *Analyt. Chem.* **2006**, *78*, 4334–4341. [CrossRef] [PubMed]

Sample Availability: Samples of the compounds are not available from the authors.

molecules

Article

Improvement of Quality Properties and Shelf Life Stability of New Formulated Muffins Based on Black Rice

Constantin Croitoru [1], Claudia Mureşan [2], Mihaela Turturică [3], Nicoleta Stănciuc [3], Doina Georgeta Andronoiu [3], Loredana Dumitraşcu [3], Vasilica Barbu [3], Elena Enachi (Ioniţă) [3], Georgiana Horincar (Parfene) [3] and Gabriela Râpeanu [3,*]

1 Academy of Agricultural and Forestry Sciences, 61 Marasti Blvd, 011464 Bucharest, Romania; c.croitoru@sodinal.com
2 Faculty of Food Engineering, Tourism and Environmental Protection, Aurel Vlaicu University of Arad, 2 Elena Dragoi Street, 310330 Arad, Romania; claudia.muresan@uav.ro
3 Integrated Center for Research, Expertise and Technological Transfer in Food Industry, Faculty of Food Science and Engineering, Dunarea de Jos University of Galati, 111 Domnească Street, 800201 Galati, Romania; mihaela.turturica@ugal.ro (M.T.); nsava@ugal.ro (N.S.); Georgeta.Andronoiu@ugal.ro (D.G.A.); ldumitrascu@ugal.ro (L.D.); vbarbu@ugal.ro (V.B.); elena.ionita@ugal.ro (E.E.); gparfene@ugal.ro (G.H.)
* Correspondence: Gabriela.Rapeanu@ugal.ro or grapeanu@ugal.ro; Tel.: +4-0336-130177; Fax: +4-0236-460165

Received: 25 October 2018; Accepted: 19 November 2018; Published: 21 November 2018

Abstract: Effects of partial (50%) and total replacement of wheat flour with black rice flour on the phytochemical, physico-chemical, sensorial, and textural properties of muffins were studied. Partial or total replacement of wheat flour with black rice flour in muffins improved their nutritional and antioxidative properties with a positive effect on microbiological and color stability during the storage period in accelerated conditions. The low gluten muffins had an anthocyanin content of 27.54 ± 2.22 mg cyanidin-3-glucoside (C3G)/100 g dry weight (DW), whereas the gluten free muffins had 46.11 ± 3.91 mg C3G/100 g DW, with significant antioxidant values. Retention of 60% and 64% for anthocyanins and 72% and 80% for antioxidant activity after baking was found. The fracturability and hardness scores increased with the addition of black rice flour, whereas firmness and chewiness increased for gluten free muffins. The confocal analysis revealed a tendency of glucidic components to aggregate, with gathers of small bunches of black rice starch granules comprising anthocyanin. The results allowed designing two new value added bakery products, low and free gluten muffins, with significant high amounts of bioactive compounds, suggesting the functional potential of black rice flour.

Keywords: black rice flour; anthocyanins; antioxidant activity; low gluten muffins; added value products

1. Introduction

Muffins are sweet baked products highly appreciated by consumers due to their good taste and soft texture, perfect for breakfast, brunch and snacks. Muffin composition is a fat in water emulsion obtained from an egg-sugar-water-fat mixture as a continuous phase, and air bubbles represent a discontinuous phase where the flour is dispersed. Muffins are generally associated with a high porous spongy texture [1,2]. Traditionally, a muffin recipe is composed of wheat flour, vegetable oil, eggs and milk [3]. For this reason, many people with celiac disease are unable to consume this type of product since they are made with wheat flour.

The demand for low gluten and gluten-free products is increasing because it is well known that celiac disease is a common lifelong disorder, affecting 1% of the world's population [4–6]. The reaction to gluten ingestion for those who sufferer from celiac disease is the inflammation of the small intestine leading to malabsorption of the nutrients [7,8]. However, a gluten free diet is characterized by low daily energy intake combined, with an unbalanced macronutrient content, compared to a balanced normal daily diet [9]. In avoiding the use of gluten in foods, significant technological and quality problems will have to be solved. Their sensorial properties are still different from similar products containing gluten. The main ingredients of gluten free cereal products are gluten-free flours, corn, rice and potato starches and different hydrocolloids that slow down the gluten viscoelastic properties. Recent studies were performed to improve the nutritional profile of gluten-free products by using pseudo-cereals as functional gluten-free ingredients [10–12]. In recent years, there have been many research projects for the development of gluten free sweet bakery products aimed to improve the organoleptic properties of the finished products [13,14].

Rice is a suitable cereal for developing gluten free products because it has a low level of prolamine and is hypoallergenic [15]. The main ingredient of the gluten free muffins, cake or cupcakes recipes is the rice flour [13,14,16], or different starch sources, such as corn, potato and wheat [17].

Rice flour has a big potential to be a wheat flour substitute in muffins because it has been used before to prepare gluten free bakery products, such as breads and cakes, which are traditionally made with wheat flour [8]. However, less information is available on the use of rice flour for gluten free products such as muffins. Several researchers have developed gluten free products using starches, dairy products, probiotics, gums and hydrocolloids to improve the structure and taste of the products [8,18].

Black rice was grown at a small scale in the early history of agriculture. In fact, black rice is considered to have the highest nutritional profile of all the cereals. The interest in black rice is growing because is gluten free, cholesterol-free, and low in sugar, salt and fat. Among these properties it also contains anthocyanins, antioxidants, B and E vitamins, iron, thiamine, magnesium, niacin and phosphorous and high fiber content. There are a lot of scientific studies showing that black rice powder is one of nature's most well-balanced foods [19]. Black rice anthocyanins are about 26.3% and the most effective constituents in a percentage of 90% are represented by the cyanidin-3-*O*-glucoside and peonidin-3-*O*-glucosid anthocyanin [20]. Anthocyanins represent the flavonoid pigments of the black rice and they are a source of antioxidants that have the ability to inhibit the formation or to reduce the concentrations of reactive cell damaging free radicals [21].

The aim of the present study was to obtain value added low gluten and gluten-free black rice based muffins. In order to demonstrate the added-value of the products, the muffins were tested for total phyto-chemical, physico-chemical and microbiological properties, textural and sensorial analysis. An accelerated storage test for phytochemicals and microbiological stability was performed over 21 days at a temperature of 25 °C.

2. Results and Discussion

2.1. Black Rice Flour, Batter and Muffin Characterization

The black rice flours were characterized in terms of anthocyanins content using the chromatographic technique. Figure 1 shows a typical HPLC chromatogram, where the major anthocyanin found was cyanidin-3-glucoside. Four compounds were found in the black rice flour extract among which only three of them were identified as follows: Peak 1, cyanidin-3,5-diglucoside (1.08 mg/100 g dry weight (DW)); peak 2, cyanidin-3-glucoside (176.83 mg/100 g DW); peak 3, peonidin-3-glucoside (7.08 mg/100 g DW); and peak 4, unidentified. The total anthocyanin content in black rice flour was 192.36 ± 1.14 mg (C3G)/100 g DW. The results are similar with ones reported by Bordiga et al. [22] and Melini et al. [23] who studied the same variety of black rice. Bolea et al. [24] reported significant lower quantities of 21.00 μg/g cyanidin-3-glucoside and 0.10 μg/g peonidin-3-glucoside in the whole black rice flour.

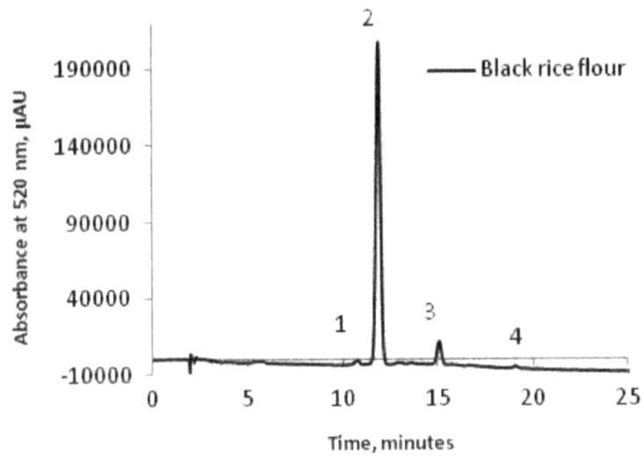

Figure 1. HPLC chromatogram of anthocyanins from black rice flour.

Cyanidin-3-glucoside and peonidin-3-glucoside have been previously identified as the main anthocyanins present in the black rice [25,26]. According to other studies reported by Zhang et al. [27,28] five anthocyanins have been separated and identified in waxy and non-waxy black rice. These anthocyanins were malvidin, pelargonidin-3,5-diglucoside, cyanidin-3-glucoside, cyanidin-3,5- diglucoside and peonidin-3-glucoside.

The presence of gluten in black rice flour has been checked with a gluten ELISA assay and found no gluten presence in black rice flour.

The phytochemical characteristics of batters are described in Table 1. The physico-chemical and phytochemical characteristics of muffins are presented in Table 2. The total anthocyanin content (TAC) in S2 batter was 69.93 ± 2.34 mg cyanidin-3-glucoside (C3G)/100 g DW and 125.4 ± 6.64 mg C3G/100 g DW for S3 batter. After baking, the TAC was 27.54 ± 2.22 mg C3G//100 g DW for S2 muffins and 46.11 ± 3.91 mg C3G//100 g DW for S3 muffins, respectively. Therefore, the retention of TAC in the S2 and S3 after baking was approximately 60% and 64%. The total polyphenolic content (TPC) in batters were 254.1 ± 5.52 mg gallic acid (GA)/100 g DW and 307.3 ± 1.02 mg GA/100 g DW in S2 and S3, respectively, whereas baking caused a decrease to 170.3 ± 4.55 mg GA/100 g DW and 226.5 ± 2.14 mg GA/100 g DW.

Table 1. Phytochemical characteristics of batters.

Phytochemical Properties	Samples		
	S1	S2	S3
Total anthocyanin content (TAC), mg cyanidin-3-glucoside (C3G)/100 g dry weight (DW)	n.d.	69.93 ± 2.34 [a]	125.4 ± 6.64 [b]
Total polyphenolic content (TPC), mg gallic acid (GA)/100 g DW	82.1 ± 1.06 [a]	254.1 ± 5.52 [b,c]	307.3 ± 1.02 [b]
Total flavonoid content (TFC), mg catechin equivalents (CE)/100 g DW	71.2 ± 1.44 [a]	149.4 ± 3.10 [b]	187.1 ± 5.04 [c]
Antioxidant activity, mM 6-Hydroxy-2,5,7,8-tetramethylchromane-2-carboxylic acid (Trolox)/100 g DW	152.8 ± 2.10 [a]	611.2 ± 8.32 [b,c]	552.71 ± 5.06 [c]

* Values with different letters in the same raw are significantly different ($p < 0.05$).

Therefore, the retention of TPC in the S2 and S3 after baking was approximately 67% and 74%, respectively. Total flavonoid content (TFC) in batter were 149.4 ± 3.10 mg catechin equivalents (CE)/100 g DW and 187.1 ± 5.04 mg CE/100 g DW, respectively. Baking caused a slight decrease in TFC to 133.4 ± 1.88 mg CE/100 g DW and to 158.6 ± 1.02 mg CE/100 gDW, respectively, with a retention of 89% in S2 and 85% in S3. The antioxidant activities in batter were 611.2 ± 8.32 mM 6-Hydroxy-2,5,7,8-tetramethylchromane-2-carboxylic acid (Trolox)/100g DW and 687.72 ± 4.11 mM Trolox/100g DW for S2 and S3 respectively, whereas in muffins the corresponding values were 445.89

± 2.22 mM Trolox/100g DW and 552.71 ± 5.06 mM Trolox/100g DW. Coefficients of 72% and 80% respectively were found for antioxidant retention in muffins after cooking.

Table 2. Physico-chemical and phytochemical characteristics of muffins.

Physico-Chemical and Phytochemical Properties		Samples S1	S2	S3
Proteins, g/100 g		11.69 ± 0.57 [b]	12.16 ± 1.16 [a]	12.71 ± 0.92 [a]
Fats, g/100 g		20.17 ± 1.37 [b]	20.22 ± 0.45 [c]	18.37 ± 1.91 [a]
Carbohydrates, g/100 g		45.44 ± 2.60 [a]	42.91 ± 1.68 [b,c]	42.38 ± 2.51 [c]
Moisture, g/100 g		20.60 ± 0.11 [b]	22.53 ± 0.23 [c]	24.13 ± 0.15 [d]
Ash, g/100 g		2.10 ± 0.01 [a]	2.18 ± 0.01 [a]	2.41 ± 0.01 [b]
Energy value, %:				
kcal		421.81	413.83	396.71 [a]
kJ		1763.18	1729.82	1658.24 [a]
TAC, mg C3G/100 g DW		n.d.	27.54 ± 2.22 [a]	46.11 ± 3.91 [b]
TPC, mg GA/100 g DW		64.4 ± 3.16 [a]	170.3 ± 4.55 [b]	226.5 ± 2.14 [v]
TFC, mg CE/100 g DW		57.2 ± 0.94 [a]	133.4 ± 1.88 [b]	158.6 ± 1.02 [c]
Antioxidant activity, mM Trolox/100 g DW		124.6 ± 3.20 [a]	445.89 ± 2.22 [b,c]	552.71 ± 5.06 [c]
Colorimetric parameters	L^*	80.41 ± 9.13 [a]	27.71 ± 0.15 [b]	19.6 ± 3.58 [b,c]
	a^*	0.06 ± 0.001 [a]	8.47 ± 1.08 [b]	6.53 ± 0.95 [c]
	b^*	51.83 ± 1.15 [a]	7.31 ± 0.41 [b]	1.49 ± 0.14 [c]

* Values with different letters in the same raw are significantly different ($p < 0.05$).

However, it is difficult to estimate the effect of food matrices on the different phytochemicals, due to the complexity and different processing parameters. For example, retention of malvidin in the bun and biscuit after baking were 95.9% and 98.6%, respectively as reported by Karakaya et al. [29]. Similarly, a significant decrease ranging from 37.5% to 70% in the TAC content was determined during snack production by Nemś et al. [30], whereas Barti et al. [31] found a decrease in the anthocyanin content of breads produced using purple and blue wheat flours during the baking process. As expected regarding the colorimetric parameters, a lower L^* value was observed for S3, whereas a^* and b^* values suggested a red dark (S2) to red brown (S3) color, with a pleasant taste due to the presence of black rice.

2.2. Sensory Analysis

Table 3 shows the average scores of sensorial attributes evaluated by the panelists.

Table 3. Sensory characteristics of muffins.

Sensorial Attribute	Samples S1	S2	S3
Color	1.82 ± 0.87 [a]	5.63 ± 1.2	6.27 ± 1.27
Surface humidity	3.27 ± 1.84 [a]	3.82 ± 0.98 [a]	4.72 ± 1.19
Cross section appereance	1.73 ± 1.10	1.64 ± 0.92	2.55 ± 1.7
Denseness	2.82 ± 1.47	2.82 ± 1.25	2.82 ± 1.94
Fracturability	2.46 ± 1.7	2.82 ± 1.33	3.64 ± 1.7
Hardness	2.64 ± 1.57 [a]	3.73 ± 1.35 [a]	4.36 ± 1.5
Cohesivity	5.46 ± 1.21	4.73 ± 1.00	4.55 ± 1.44
Moistness of mass	3.36 ± 1.75	3.46 ± 1.58	3.81 ± 2.27
Taste	6.00 ± 0.89	5.09 ± 1.22	4.90 ± 1.38
Sweetness	4.90 ± 1.51	4.27 ± 1.67	4.63 ± 1.7
Overall acceptability	5.90 ± 0.83	5.18 ± 0.98	5.18 ± 1.4

[a] Based on Dunnett multiple comparisons with a control, means on the same row that do not share a letter are significantly different ($p < 0.05$).

Control muffins (S1) as expected showed the lightest color, while S3, was the darkest ($p < 0.001$). Surface humidity was perceived as being higher ($p < 0.05$) for S3 compared with S1 and S2. The scores given for fracturability and hardness attribute increased with the addition of black rice flour, reaching a maximum for S3.

The taste of all samples was appreciated; however the control sample was evaluated by the panelists with the highest score. Some panelists perceived that samples with black rice flour contained some crispy particles as compared with the control sample. Overall acceptability indicates that the panelists liked the analyzed muffins, regardless of whether or not they contained black rice flour. These results indicate that gluten free muffins obtained with black rice flour could be an alternative for people suffering from gluten intolerance.

2.3. Texture Analysis

Texture parameters revealed by instrumental analysis are shown in Table 4. Firmness, defined as the maximum force required to compress the samples in the first cycle, varied between 4.75 ± 0.16 N for S1 and 6.67 ± 0.02 N for S3. Similar values for firmness were reported by Demirkesen et al. [32] and Wronkowska et al. [33] for bread formulated with rice, wheat, chestnut flour or buckwheat. The smaller value of control firmness could be explained by the presence of glutenin and prolamin (the major fractions of gluten) which are responsible for the porous network in muffins.

Table 4. Texture parameters of muffins.

Textural Parameters, Unit	Samples		
	S1	S2	S3
Firmness, N	4.75 ± 0.16 [a]	5.82 ± 0.26	6.67 ± 0.02
Cohesiveness, dimensionless	0.37 ± 0.01	0.35 ± 0.02	0.33 ± 0.02
Springiness, mm	6.95 ± 0.06	6.83 ± 0.08	6.57 ± 0.23
Chewiness, mJ	10.15 ± 0.23	12.21 ± 0.25	15.13 ± 0.17

[a] Mean of the five determinations ± standard deviation.

In the samples containing black rice flour, the reduced porosity led to a higher resistance during compression. Cohesiveness, determined as the ratio between the resistance of the samples during the second and the first compression, and springiness, defined as the deformation recovered between the two compression cycles, showed the highest values for S1 sample. These values may be due to the presence of glutenin, which is responsible for elastic and cohesive properties of dough [32]. Chewiness, described as the energy required to disintegrate the food during mastication, raised from 10.15 ± 0.23 mJ for S1 sample, to 12.21 ± 0.23 mJ for S2 sample and 15.13 ± 0.17 mJ for S3.

2.4. Confocal Microscopy Analysis

The wheat flour that was analyzed as the first control sample (control 1) contained starch granules that can be grouped into three categories. The major category (about 60%) was displayed as large, lenticular or disc shaped granules with a diameter > 10 μm. Approximately 30% of the wheat starch granules were spherical, medium-sized (3–10 μm), while 10% were really small grains (under 3 μm) with irregular forms (as it can be seen in Figure 2a). The heterogeneity of the wheat flour starch granules could be attributed to the wheat variety (soft or hard wheat), the amylose content, and especially the moment in which is formed during anthesis or their different times of formation during grain development [34–38]. There are also many studies that confirm the presence of amylose in the peripheral region of the starch granules as it was likewise assessed in our study. As such, in Figure 2a it can be observed an interaction between the lipophilic dye molecules and some granules that afterwards displayed a green border whereas in the central location (in the hilum) longer amylopectin chains were noticed to form several inclusion complexes with the ligands as it was also suggested by Manca et al. [39].

(a) (b1) (b2)

Figure 2. Confocal laser scanning microscopy images of control 1—wheat flour (**a**), and control 2—black rice flour (**b1** and **b2**).

The starch granule sizes of the black rice flower (control 2) were somewhere around 2–10 µm, similar to the results obtained by BeMiller & Whistler [40]. Rice starch granules were polygonal, irregular in shape [41] with sharp angles, and without any obvious concentric striations, hilum or cleft (Figure 2(b1)), most of them were grouped into large aggregates as can be seen in Figure 2(b2). The same characteristics were also reported by Leewatchararongjaroen & Anuntagool [42].

The confocal analysis of the muffins samples displayed a much greater complexity due to the different biochemical composition of the ingredients that were used for the recipe (butter, sugar, eggs and wheat flour and black rice flour in variable proportions). It was more difficult to distinguish the components in the cooked samples, possibly as a result of the complex interactions between the gelatinized or expanded starch, denatured proteins and lipids. When only the wheat flour was used (S1 sample), in the texture of the baked dough, large wheat starch intact granules were also observed, the granules being isolated or grouped into the complex protein matrix. The size of the isolated particles was variable, from 9.56 to 43.44 µm, and the largest conglomerates exceeded 100 µm (Figure 3(S1)). By increasing the proportion of black rice flour, the glucidic components displayed a more obvious tendency towards aggregation so that around the large granules of the wheat starch gathered small bunches of black rice starch granules that come with the intake of anthocyanins (in green) (Figure 3(S2)). Confocal images taken for the muffins prepared with simple black rice flour frequently showed huge clusters (over 200 µm in size) consisting of starch granules, most of them having expanded due to the cooking temperature (about 80 µm in diameter) and at the same time being strongly colored in green due to the presence of anthocyanins, as it can be seen in Figure 3(S3).

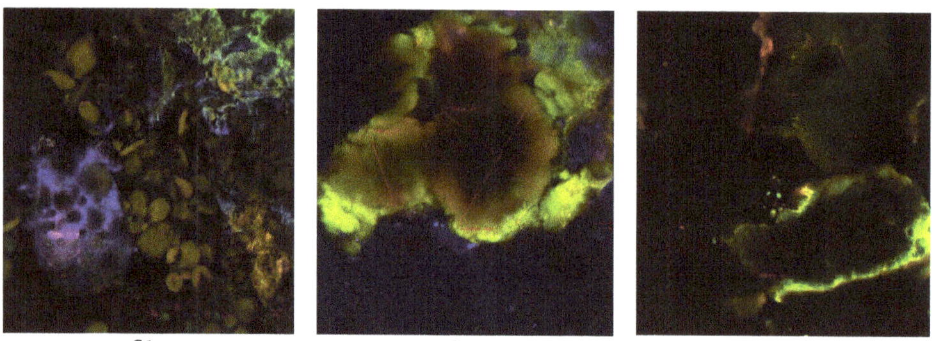

S1 **S2** **S3**

Figure 3. Confocal laser scanning microscopy images of muffin samples: (**S1**) (muffins with wheat flour), (**S2**) (muffins with 1:1 wheat and black rice flour) and (**S3**) (muffins with black rice flour).

Our results were similar to those obtained by Malik et al. [43]. It has been found that by replacing the wheat flour with rice flour in pastries or baked goods, the firmness and the sensorial attributes of freeze-thawed cake are improved due to a low amylose content of rice flour [44]. Furthermore, black rice flour also brings additional bioactive compounds such as anthocyanin pigments that are valuable in improving the food functionality.

2.5. Anthocyanin in Vitro Digestibility

To evaluate the anthocyanins in vitro digestibity of new formulated muffins, simulated digestions conditions were applied. The digestion pattern of formulated muffins is given in Figure 4. As can be seen from Figure 4a, the maximum release of anthocyanins registered for the S3 sample was 14.23 ± 1.02% after 120 min of reaction. The digestion of S3 samples was limited with a maximum release of 7.22 ± 0.69% after 120 min of reaction. The results presented in Figure 4b during duodenal digestion revealed that the anthocyanin release was faster in the case of S3 compared with S2. From our results, it seems that less than 26% of the anthocyanins in S2 and 18% in S3 were retained in the formulated muffins during in vitro digestion.

Therefore, it can be appreciated that anthocyanins were slowly released from the muffins under simulated digestion conditions. Our results are similar with those reported by Sari et al. [45], suggesting that curcumin is released slowly from the nanoemulsion under simulated digestion conditions. Our in vitro digestibility results support a slowly release of anthocyanins from the food matrices during simulated gastric digestion and a significant release of the bioactive compounds into the gut.

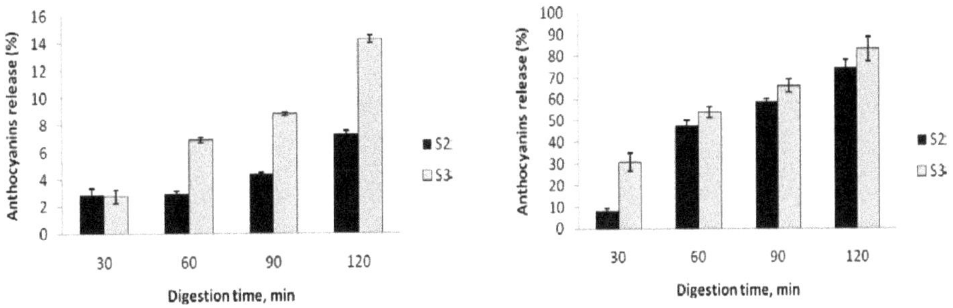

(a) (b)

Figure 4. The patterns of gastric (**a**) and duodenal (**b**) digestion of formulated muffins S2 (muffins with 1:1 wheat and black rice flour) and S3 (muffins with black rice flour).

2.6. Shelf-Life Assessment

To evaluate the phytochemicals and antioxidant activity and color stability in the newly formulated matrix, the samples were stored at a temperature of 25 °C for 21 days. At every seven days, the following parameters were measured: Total polyphenolic, flavonoids and anthocyanins content, antioxidant activity, color parameters and molds and yeasts. Data from Figure 5 showed the total anthocyanin content, total polyphenols, total flavonoids content and antioxidant activity changes during the storage period. The anthocyanin content significantly decreased, up to 50% in S2 and 33% in S3 in the first 14 days of storage, whereas degradation continued up to 68% and 39%, respectively after 21 days, probably due to degradation reactions (Figure 5a).

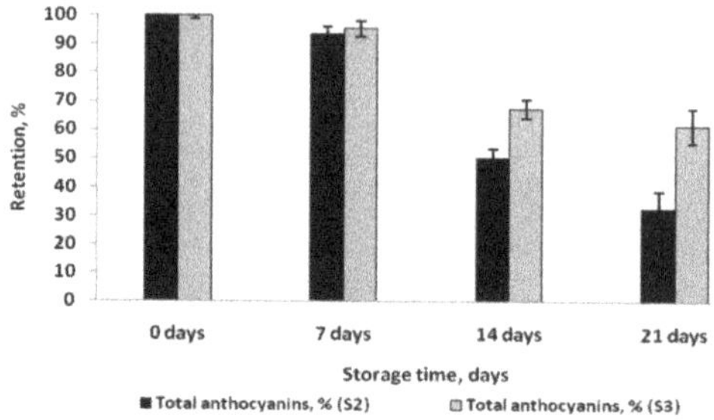

(a)

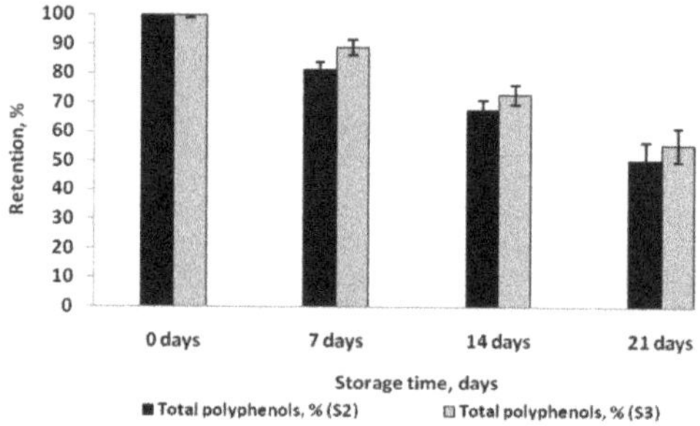

(b)

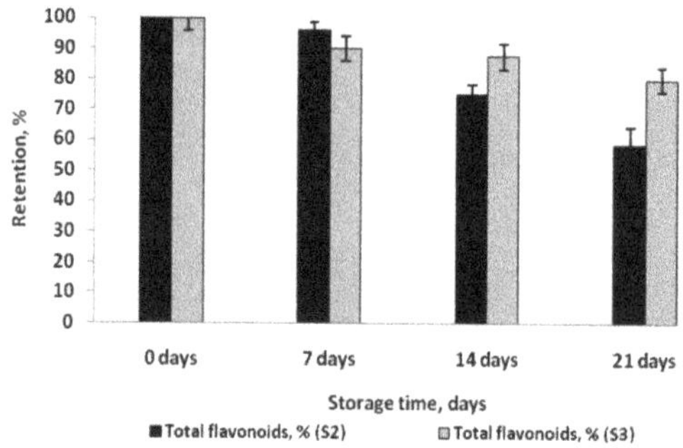

(c)

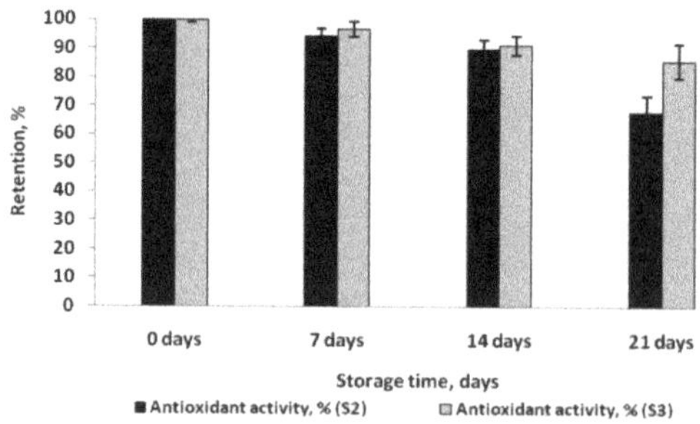

(d)

Figure 5. The retention in anthocyanis content (**a**), total polyphenols (**b**), total flavonoids (**c**) and antioxidant activity (**d**) of muffins during storage at a temperature of 25 °C for 21 days.

It can be noticed that the anthocyanin's degradation was more significant in S2 compared to S3, probably due to the higher concentration of the polyphenolic compounds, which exhibited a protective action. Regarding the total polyphenolic content in muffins during storage, the decreases were up to 50% (S2) and 45% (S3), and for flavonoids, up to 42% (S2) and 22% (S3) (Figure 5b,c). However, a slow decrease in antioxidant activity was found in all samples (Figure 5d). Therefore, after 14 days of storage, antioxidant activity decreases by 11% and 9% in S2 and S3, respectively, and by 33% and 15% after 21 days of storage. As expected, the decrease in antioxidant activity was lower in S3 when compared with S2, due to the higher concentration of polyphenolic compounds. However, it seems that the anthocyanin degradation does not significantly affect the antioxidant activity in all tested samples. Malvidin stability in the anthocyanin enriched bun after 21 days at room temperature was significantly lower than those of buns stored for 7 days. However, Karakaya et al. [29] reported that storage of 21 days at room temperature did not cause huge losses in anthocyanin contents of the bun and biscuits.

Table 5 shows the variation of color parameters. A slight increase in *L* * values can be observed with increasing storage time for all samples, likely due to anthocyanin degradation. Significant differences in brightness ($p < 0.05$) can also noticed, both in terms of samples and period of storage. The control sample (S1) had *a* * value close to 0 with no variation during storage time, while for S2 and S3 samples the *a* * value increased. Our results are in line with ones reported by Ursache et al. [46].

Table 5. Colorimetric analysis of muffins.

Storage Period, Days	S1			S2			S3		
	Colorimetric Parameters								
	L *	a *	b *	L *	a *	b *	L *	a *	b *
0	80.41 ± 9.13 ^{c,d}	0.06 ± 0.001 ^{b,c}	51.83 ± 1.15 ^{a,b}	27.71 ± 0.15 ^{b,c,d}	8.47 ± 1.08 ^{a,b,c,d}	7.31 ± 0.41 ^{a,b,c}	19.6 ± 3.58 ^{c}	6.53 ± 0.95 $^{a b}$	1.49 ± 0.14 ^{a,b,c}
7	88.02 ± 0.83 ^{a}	0.10 ± 0.001 ^{b}	60.11 ± 4.96 ^{b,c}	29.42 ± 0.30 ^{d}	9.10 ± 1.63 ^{a,b}	8.40 ± 0.83 ^{a}	23.9 ± 1.96 ^{b,c}	7.30 ± 1.30 ^{a,b,d}	1.71 ± 0.26 ^{a,b,c}
14	110.41 ± 5.25 ^{a,b,c,d}	0.18 ± 0.011 ^{b}	73.25 ± 2.49 ^{d}	35.80 ± 73.74 ^{a,b}	10.30 ± 1.84 ^{a}	9.20 ± 0.94 ^{c}	28.8 ± 2.49 ^{d}	11.6 ± 0.22 ^{c}	2.25 ± 0.66 ^{a}
21	119.63 ±4.60 ^{b,c,d}	0.21 ± 0.057 ^{b}	75.53 ± 4.31 ^{a,b}	38.51 ± 1.41 ^{a,b,c}	11.11 ± 1.10 ^{a,b,c,d}	9.62 ± 0.31 ^{b,c}	30.1 ± 3.63 ^{a,b}	13.2 ± 0.95 ^{a,b}	3.51 ± 0.51 ^{b,c,d}

Values with different letters in the same column are significantly different ($p < 0.05$) (L *—lightness, a *—redness, b *—yellowness).

The *b* * values which denote a yellow color of the samples, had higher values for S1 and lower S3 baked with only black rice flour.

From microbiological point of view the results suggested that value-added muffins are microbiologically satisfactory during the accelerated storage test compared to control (Table 6).

Table 6. Yeasts and molds during storage (colony forming unit CFU/g).

Samples	Storage Period, Days			
	0	7	14	21
S1	<10	$1.33 \cdot \times 10^2 \pm 0.13$	$2.59 \times 10^3 \pm 0.08$	$5.16 \times 10^5 \pm 1.10$
S2	<10	<10	<100	<100
S3	<10	<10	<10	<100

3. Materials and Methods

3.1. Materials and Chemicals

2,2-Diphenyl-1-picrylhydrazyl (DPPH), 6-Hydroxy-2,5,7,8-tetramethylchromane-2-carboxylic acid (Trolox), Folin-Ciocalteu reagent, sodium carbonate, sodium hydroxide, sodium acetate, sodium nitrite, potassium chloride, aluminum chloride, gallic acid, catehine, potassium persulfate, formic acid, ethanol and methanol (HPLC grade), cyanidin and peonidin standards were obtained from

Sigma Aldrich Steinheim, Germany. For muffin preparation, coconut butter with 80% fat content, brown sugar, hen eggs, wheat flour and black rice (*Oryza sativa L.* ssp. *Japonica*, Nerone variety, Italy) were purchased from the local supermarket, Galati, Romania.

3.2. Batters and Muffins Preparation

Preparation of the muffin batter was performed with the following steps: The coconut butter was mixed continuously with salt and brown sugar until the sugar is dissolved and a foam is formed; then the eggs were added, alternately with wheat flour (**S1 sample**, considered as control), wheat and black rice flour (1:1) (**S2 sample**), black rice flour (**S3 sample**), and finally the baking powder was added. The batter was mixed for 10 min at 300 rpm in order to get a uniform composition. Finally, the batter was filled in paper cups and baked at 185 °C for 25 min in a convection oven with forced air circulation. Products from each recipe were produced, baked and analyzed in two independent batches. The muffins were packed in vacuum bags at 800 mbar and stored at a temperature of 25 °C for 21 days.

3.3. HPLC Technique

The chromatographic analysis of the anthocyanins from black rice flour was performed as described earlier by Bolea et al. [24]. HPLC analysis was performed using a Surveyor HPLC system, controlled by an Xcalibur software system (Finnigan Surveyor LC, Thermo Scientific, Waltham, MA, USA). The anthocyanins detected in black rice flour were analyzed at a wavelength of 520 nm. The column used for this analysis was a C18 BDS Hypersil (150 mm × 4.6 mm, 5 mm). The gradient used for the elution of the anthocyanins was: 0–20 min: 9–35% (A), 20–30 min: 35% (A), 30–40 min: 35–50% (A), 40–45 min: 50–9% (A), with an injection volume of 10 μL, and the flow rate maintained at 1.000 mL/min.

3.4. Sensorial Analysis

A panel consisting of 11 different panelists aged between 29–50 years old performed the sensorial analysis of gluten free and added value muffins according to seven point hedonic scale. The panelists assessed the muffins samples for color (light to dark), surface humidity (none to very high), cross section appearance (non uniform to uniform), denseness (dense to airy), fracturability (low to very high), hardness (low to very hard), cohesiveness (none to tight mass), moistness of mass (low to very high), taste (dislike very much to like extremely), and overall likeability of the product (dislike to like extremely). Muffins samples were served in random order to panelists on white papers. Water was used for mouth rising before and between samples.

3.5. Physico-Chemical, Phytochemical and Microbiological Analysis of Muffins

Standardized and validated laboratory methods were used to determine the physico-chemical characteristics of muffins, in terms of moisture, fat, protein, carbohydrates, ash and energy value.

The phytochemical content of extract (total polyphenols, total flavonoids, total monomeric anthocyanins) and antioxidant activity were determined as described by Turturică et al. [47]. In brief, 1 g of muffin was crushed and then mixed with 8 mL of 70% ethanol and 1 mL HCl 1N. The mixture was stirred for 8 h at room temperature on an orbital shaker at 150 rpm. After centrifugation at 6000 rpm for 10 min, the supernatant was collected and concentrated at 40 °C to dryness under reduced pressure (AVC 2-18, Christ, UK). The extracts were redissolved in 2 mL of MiliQ water and used for phytochemical analysis.

3.6. Microbiological Assessment

The microbiological shelf life of muffins was evaluated by monitoring fungal growth over 1, 7, 14 and 21 days of storage. The standard pour plate method described by Ursache et al. [46] was used to count the number of yeasts and molds. The results were expressed as CFU/g of sample. Each sample was analysed in duplicate on each day of storage.

3.7. Textural Analysis of Muffins

The textural analysis was achieved by the Texture Profile Analysis (TPA) Method, using the Brookfield CT3-1000 analyzer. The sample preparation consisted of removing the rind of the muffins and cutting the core in cube shapes with the sight length of 15 mm. Then, a double compression was applied at a distance of 10 mm, at a speed of 1 mm/s, with no holding time between the two compression cycles. The trigger load was 0.02 N and the load cell was 1000 g. The compression was performed using an acrylic cylinder (diameter ~ 24.5 mm, height ~ 35 mm) (TA11/1000). The data were recorded and processed using the TexturePro CT V1.5 software. For each sample, five tests were performed. The textural parameters determined by TPA were firmness, cohesiveness, springiness and chewiness.

3.8. Confocal Microscopy Analysis

The comparative confocal analysis of the samples was performed in order to capture the structural, textural and compositional changes of the experimental variants, while for the control samples simple wheat flour and black rice flour were used. The confocal microscope that was used for the analysis is a Zeiss Axio Observer Z1 inverted microscope model (LSM 710) equipped with a laser scanning system: Diode laser (405 nm), Ar laser (458 nm, 488 nm and 514 nm), DPSS (561 nm pumped solid state diodes), and HeNe-laser (633 nm). The strong anthocyanin absorption in the visible range was registered between 465 nm and 550 nm [48] with an in vivo peak, between 537 nm and 542 nm [49]. The distribution of the pigments into the protein matrix was observed at the excitation wavelength of 488 nm and by applying the FS38 filter, whereas the emission was collected between 500–600 nm. The powder that was stained with two dyes, DAPI (1 μg/mL) and Red Congo (40 μM), in a ratio of 3:1:1, was observed using a 40x apochromatic objective (numerical aperture 1.4) and the FS49 and FS15 filters. The 3D images were rendered and analyzed with ZEN 2012 SP1 software (Black Edition).

3.9. In Vitro Digestibility

In vitro digestibility was performed by using a method described by Oancea et al. [50]. Briefly, 1 g of muffins (S2 and S3) was mixed with Tris-HCl buffer (10 mM, pH 7.7). The gastric digestion was performed by the addition of a simulated gastric fluid (SGF), which consisted of porcine pepsin (40 mg/mL in 0.1 M HCl) that was added to the initial mixtures in a ratio of 0.5 g of pepsin per 100 g of sample and the pH was adjusted to 2.0 with 6 M HCl. Regarding the enteric digestion step, the simulated intestinal fluid (SIF) consisted of a mixture containing pancreatin (2 mg/mL) and afterwards the resulting mixture was neutralized to pH 5.3 with 0.9 M sodium bicarbonate. The pH of the system was adjusted to 7.0 with 0.1M NaOH, prior to the incubation of the samples for 2 h. The incubation was performed in an SI—300R orbital shaking incubator (Medline Scientific, Oxfordshire, UK), at 100 rpm and 37 °C. The total anthocyanin's content of the samples was measured at every 30 min during the in vitro digestion.

3.10. Colorimetric Study

The color parameter values of the muffins were measured using the Minolta CR-410 Chroma Meter (Konica Minolta, Osaka, Japan) as described by Ursache et al. [46]. The results were expressed as L * (a lower value indicates a darker color, black: L * = 0 and white: L * = 100), a * (indicate the balance between red (>0), green (0) and blue (<0) color), and b * (the balance between yellow (>0) and blue (<0) color). All the measurements were performed in triplicates.

3.11. Storage Stability

No preservatives were used in the recipe formulation of the gluten free and added value muffins. Therefore, an accelerated storage stability test was performed during a period of 21 days at temperature of 25 °C. Duplicate samples were considered for determination of the molds and yeast, total polyphenolic content, total flavonoids and anthocyanins content, antioxidant activity and color at every 7 days.

3.12. Statistical Analysis of Data

Minitab 18 statistical processing software was employed to perform the statistical evaluation of the sensorial data. First, the data were checked for normality and homoscedasticity using the Ryan Joiner test and the Bartlett test. Then, one-way ANOVA was used to identify if panelists detect any differences between samples considering a significance level of 0.05. Post-hoc analysis via Dunnett multiple comparisons with a control were performed when appropriate. All data reported in this study represent the averages of duplicate analyses and is reported as mean ± standard error of the mean.

4. Conclusions

The muffins baked with black rice flour presented a high anthocyanin content and antioxidant activity compared with the control sample baked with wheat flour. The textural analysis suggested that the addition of black rice caused the increase of firmness, springiness and chewiness, while the cohesiveness was lower compared with the control sample and was related to a weaker binding between the constituents. Confocal images taken for the muffins baked with black rice flour showed huge clusters (over 200 μm in size) consisting of starch granules, most of them having expanded due to the cooking temperature (about 80 μm in diameter) and at the same time being strongly colored in green due to the presence of anthocyanins. Sensorial analysis showed that all samples were appreciated; some panelists even perceived that samples with black rice flour contain pleasant crispy particles compared with the control sample.

Storage stability of muffins revealed a decrease of anthocyanins, antioxidant activity and color parameters. The added value products showed a microbiological stability during the accelerated storage period, probably due to the presence of polyphenolic compounds. These results indicated that value added muffins obtained with black rice flour could be an alternative for people suffering from gluten intolerance, whereas proving a significant amount of polyphenolic content, with potentially beneficial effects on human health.

Author Contributions: G.R. and N.S. conceived and designed the experiments and reviewed the final manuscript; C.C.; C.M.; D.G.A.; L.D. and G.H. performed the experiments, analyzed the data and prepared the manuscript; M.T. performed the HPLC analysis; E.E. and V.B. performed the confocal microscopy analysis; G.R. reviewed the final manuscript.

Funding: This research received no external funding.

Acknowledgments: The Integrated Center for Research, Expertise and Technological Transfer in Food Industry is acknowledged for providing technical support (www.bioaliment.ugal.ro). The authors are grateful for the technical support offered by the Grant POSCCE ID 1815, cod SMIS 48745 (www.moras.ugal.ro).

Conflicts of Interest: The authors declare no conflict of interest.

References

1. Matos, M.E.; Sanz, T.; Rosell, C.M. Establishing the function of proteins on the rheological and quality properties of rice based gluten free muffins. *Food Hydrocoll.* **2014**, *35*, 150–158. [CrossRef]
2. Martínez-Cervera, S.; Sanz, T.; Salvador, A.; Fiszman, S.M. Rheological, textural and sensorial properties of low-sucrose muffins reformulated with sucralose/polydextrose. *LWT Food Sci. Technol.* **2012**, *45*, 213–220. [CrossRef]

3. Sanz, T.; Salvador, A.; Baixauli, R.; Fiszman, S.M. Evaluation of four types of resistant starch in muffins. II. Effects in texture, colour and consumer response. *Eur. Food Res. Technol.* **2009**, *229*, 197–204. [CrossRef]
4. Gujral, N.; Freeman, J.H.; Thomson, A. Celiac disease: Prevalence, diagnosis, pathogenesis and treatment. *World J. Gastroenterol.* **2012**, *18*, 6036–6059. [CrossRef] [PubMed]
5. Rai, S.; Kaur, A.; Singh, B. Quality characteristics of gluten free cookies prepared from different flour combinations. *J. Food Sci. Technol.* **2014**, *51*, 785–789. [CrossRef] [PubMed]
6. Shevkani, K.; Singh, N. Influence of kidney bean, field pea and amaranth protein isolates on the characteristics of starch-based gluten-free muffins. *Int. J. Food Sci. Technol.* **2014**, *49*, 2237–2244. [CrossRef]
7. Feighery, C.F. Coeliac disease. *Br. Med. J.* **1999**, *319*, 236–239. [CrossRef]
8. Man, S.; Păucean, A.; Muste, S.; Pop, A. Studies on the formulation and quality characteristics of gluten free muffins. *J. Agroaliment. Process. Technol.* **2014**, *20*, 122–127.
9. Bardella, M.T.; Fredella, C.; Prampolini, L.; Molteni, N.; Giunta, A.M.; Bianchi, P.A. Body composition and dietary intakes in adult celiac disease patients consuming a strict gluten-free diet. *Am. J. Clin. Nutr.* **2000**, *72*, 937–939. [CrossRef] [PubMed]
10. Islas-Rubio, A.R.; De la Barca, A.C.; Cabrera-Chávez, F.; Cota-Gastélum, A.G.; Beta, T. Effect of semolina replacement with a raw: Popped amaranth flour blend on cooking quality and texture of pasta. *LWT Food Sci. Technol.* **2014**, *57*, 217–222. [CrossRef]
11. Sakač, M.; Torbica, A.; Sedej, I.; Hadnađev, M. Influence of breadmaking on antioxidant capacity of gluten free breads based on rice and buckwheat flours. *Food Res. Int.* **2011**, *44*, 2806–2813. [CrossRef]
12. Sedej, I.; Sakač, M.; Mandić, A.; Mišan, A.; Pestorić, M.; Šimurina, O.; Čanadanović-Brunet, J. Quality assessment of gluten-free crackers based on buckwheat flour. *LWT Food Sci. Technol.* **2011**, *44*, 694–699. [CrossRef]
13. Gularte, M.A.; de la Hera, E.; Gómez, M.; Rosell, C.M. Effect of different fibers on the enrichment of gluten-free layer cake. *LWT Food Sci. Technol.* **2012**, *48*, 209–214. [CrossRef]
14. Gularte, M.A.; Gómez, M.; Rosell, C.M. Impact of legume flours on quality and in vitro digestibility of starch and protein from gluten-free cakes. *Food Bioprocess Technol. Int. J.* **2012**, *5*, 3142–3150. [CrossRef]
15. Omary, M.B.; Fong, C.; Rothschild, J.; Finney, P. Effects of germination on the nutritional profile of gluten-free cereals and pseudocereals: A review. *Cereal Chem.* **2012**, *89*, 1–14. [CrossRef]
16. Park, S.J.; Ha, K.-Y.; Shin, M. Properties and qualities of rice flours and gluten-free cupcakes made with higher-yield rice varieties in Korea. *Food Sci. Biotechnol.* **2012**, *21*, 365–372. [CrossRef]
17. Ronda, F.; Oliete, B.; Gómez, M.; Caballero, P.A.; Pando, V. Rheological study of layer cake batters made with soybean protein isolate and different starch sources. *J. Food Eng.* **2011**, *102*, 272–277. [CrossRef]
18. Schamne, C.; Dutcosky, S.D.; Demiate, I.M. Obtention and characterization of gluten free baked products. *Ciênc. Tecnol. Aliment.* **2010**, *30*, 741–750. [CrossRef]
19. Ujjawal, K.; Kushwaha, S. Black rice. In *Black Rice Research, History and Development*; Springer International Publishing AG: Basel, Switzerland, 2016; pp. 21–47, ISBN 978-3-319-30153-2.
20. Chang, K.K.; Kikuchi, S.; Kim, Y.K.; Park, S.H.; Yoon, U.; Lee, G.S.; Choi, J.W.; Kim, Y.H.; Park, S.C. Computational identification of seed specific transcription factors involved in anthocyanin production in black rice. *Biochip. J.* **2010**, *4*, 247–255. [CrossRef]
21. Adom, K.K.; Liu, R.H. Antioxidant activity of grains. *J. Agric. Food. Chem.* **2002**, *50*, 6170–6182. [CrossRef]
22. Bordiga, M.; Gomez-Alonso, S.; Locatelli, M.; Travaglia, F.; Coïsson, J.D.; Hermosin-Gutierrez, I.; Arlorio, M. Phenolics characterization and antioxidant activity of six different pigmented Oryza sativa L. cultivars grown in Piedmont (Italy). *Food Res. Int.* **2014**, *65*, 282–290. [CrossRef]
23. Melinia, V.; Panfilib, G.; Fratiannib, A.; Acquistuccia, R. Bioactive compounds in rice on Italian market: Pigmented varieties as a source of carotenoids, total phenolic compounds and anthocyanins, before and after cooking. *Food Chem.* **2019**, *277*, 119–127. [CrossRef]
24. Bolea, C.; Turturica, M.; Stanciuc, N.; Vizireanu, C. Thermal degradation kinetics of bioactive compounds from black rice flour (*Oryza sativa* L.) extracts. *J. Cereal Sci.* **2016**, *71*, 160–166. [CrossRef]
25. Yawadio, R.; Morita, N. Color enhancing effect of carboxylic acids on anthocyanins. *Food Chem.* **2007**, *105*, 421–427. [CrossRef]
26. Hou, Z.; Qin, P.; Zhang, Y.; Cui, S.; Ren, G. Identification of anthocyanins isolated from black rice (*Oryza sativa* L.) and their degradation kinetics. *Food Res. Int.* **2013**, *50*, 691–697. [CrossRef]

27. Zhang, M.W.; Guo, B.J.; Zhang, R.F.; Chi, J.W.; Wei, Z.C.; Xu, Z.H.; Zhang, Y.; Tang, X.J. Separation, purification and identification of antioxidant compositions in black rice. *Agr. Sci. China* **2006**, *39*, 153–160. [CrossRef]
28. Zhang, M.W.; Zhang, R.F.; Guo, B.J.; Chi, J.W.; Wei, Z.C.; Xu, Z.H. The hypolipidemic and antioxidative effects of black rice pericarp anthocyanin in rats. *Acta Nutrimenta Sin.* **2006**, *28*, 404–408.
29. Karakaya, S.; Simsek, S.; Tolga Eker, A.; Pineda-Vadillo, C.; Dupont, D.; Perez, B.; Viadel, B.; Sanz-Buenhombre, M.; Guadarrama Rodriguez, A.; Kertész, Z.; et al. Stability and bioaccessibility of anthocyanins in bakery products enriched with anthocyanins. *Food Funct.* **2016**, *7*, 3488–3496. [CrossRef] [PubMed]
30. Nemś, A.; Peksa, A.; Kucharska, A.Z.; Sokół-Łętowska, A.; Kita, A.; Drożdż, W.; Hamouz, K. Anthocyanin and antioxidant activity of snacks with coloured potato. *Food Chem.* **2015**, *172*, 175–182. [CrossRef] [PubMed]
31. Barti, P.; Albreht, A.; Skrt, M.; Tremlová, B.; Ošťádalová, M.; Šmejkal, K.; Vovk, I.; Ulrih Poklar, N. Anthocyanins in purple and blue wheat grains and in resulting bread: Quantity, composition, and thermal stability. *Int. J. Food Sci. Nutr.* **2015**, *66*, 514–519. [CrossRef] [PubMed]
32. Demirkesen, I.; Mert, B.; Sumnu, G.; Sahin, S. Rheological properties of gluten-free bread formulations. *J. Food Eng.* **2010**, *96*, 295–303. [CrossRef]
33. Wronkowska, M.; Troszynska, A.; Soral-Smietana, M.; Wolejszo, A. Effects of buckwheat flour (*Fagopyrum esculentum moench*) on the quality of gluten-free bread. *Pol. J. Food Nutr. Sci.* **2008**, *58*, 211–216.
34. Langeveld, S.M.J.; Van, W.R.; Stuurman, N.; Kijne, J.W.; de Pater, S. B-type granule containing protrusions and interconnections between amyloplasts in developing wheat endosperm revealed by transmission electron microscopy and GFP expression. *J. Exp. Bot.* **2000**, *51*, 1357–1361. [CrossRef] [PubMed]
35. Bechtel, D.B.; Wilson, J.D. Amyloplast formation and starch granule development in hard red winter wheat. *Cereal Chem.* **2003**, *80*, 175–183. [CrossRef]
36. Wilson, J.D.; Bechtel, D.B.; Todd, T.C.; Seib, P.A. Measurement of wheat starch granule size distribution using image analysis and laser diffraction technology. *Cereal Chem.* **2006**, *83*, 259–268. [CrossRef]
37. Xie, X.J.; Cui, S.W.; Li, W.; Tsao, R. Isolation and characterization of wheat bran starch. *Food Res. Int.* **2008**, *41*, 882–887. [CrossRef]
38. Kumar, R.; Kumar, A.; Sharma, N.K.; Kaur, N.; Chunduri, V.; Chawla, M.; Sharma, S.; Singh, K.; Garg, M. Soft and hard textured wheat differ in starch properties as indicated by trimodal distribution, morphology, thermal and crystalline properties. *PLoS ONE* **2016**, *11*, e0147622. [CrossRef] [PubMed]
39. Manca, M.; Woortman, A.J.J.; Mura, A.; Loos, K.; Loi, M.A. Localization and dynamics of amylose–lipophilic molecules inclusion complex formation in starch granules. *Phys. Chem. Chem. Phys.* **2015**, *17*, 7864–7871. [CrossRef] [PubMed]
40. BeMiller, J.N.; Whisler, R.L. *Starch: Chemistry and Technology*, 3rd ed.; Academic Press: New York, NY, USA, 2009; ISBN 9780080926551.
41. Li, W.H.; Bai, Y.F.; Mousaa-Saleh, A.S.; Zhang, Q.; Shen, Q. Effect of high hydrostatic pressure on physicochemical and structural properties of rice starch. *Food Bioproc. Technol.* **2011**, *5*, 2233–2241. [CrossRef]
42. Leewatchararongjaroen, J.; Anuntagool, J. Effects of Dry-Milling and Wet-Milling on Chemical, Physical and Gelatinization Properties of Rice Flour. *Rice Sci.* **2016**, *23*, 274–281. [CrossRef]
43. Malik, A.N.; Gaiani, C.; Fukai, S.; Bhandari, B. X-ray photoelectron spectroscopic analysis of rice kernels and flours: Measurement of surface chemical composition. *Food Chem.* **2016**, *212*, 349–357. [CrossRef]
44. Jongsutjarittam, N.; Charoenrein, S. Influence of waxy rice flour substitution for wheat flour on characteristics of batter and freeze-thawed cake. *Carbohydr. Polym.* **2013**, *97*, 306–314. [CrossRef] [PubMed]
45. Sari, T.P.; Mann, B.; Kumar, R.; Singh, R.R.B.; Sharma, R.; Bhardwaj, M.; Athira, S. Preparation and characterization of nanoemulsion encapsulating curcumin. *Food Hydrocoll.* **2015**, *43*, 540–546. [CrossRef]
46. Ursache, F.M.; Andronoiu, D.G.; Ghinea, I.O.; Barbu, V.; Ionită, E.; Cotârlet, M.; Dumitrascu, L.; Botez, E.; Râpeanu, G.; Stănciuc, N. Valorizations of carotenoids from sea buckthorn extract by microencapsulation and formulation of value-added food products. *J. Food Eng.* **2018**, *219*, 16–24. [CrossRef]
47. Turturică, M.; Stănciuc, N.; Bahrim, G.; Râpeanu, R. Effect of thermal treatment on phenolic compounds from plum (Prunus domestica) extracts—A kinetic study. *J. Food Eng.* **2016**, *171*, 200–207. [CrossRef]
48. Glover, B.J.; Martin, C. Anthocyanins. *Curr. Biol.* **2012**, *22*, 147–150. [CrossRef] [PubMed]

49. Merzlyak, M.N.; Chivkunova, O.B.; Solovchenko, A.E.; Naqvi, K.R. Light absorption by anthocyanins in juvenile, stressed, and senescing leaves. *J. Exp. Bot.* **2008**, *59*, 3903–3911. [CrossRef] [PubMed]
50. Oancea, A.M.; Aprodu, I.; Ghinea, I.O.; Barbu, V.; Ionită, E.; Bahrim, G.; Rapeanu, G.; Stanciuc, N. A bottom-up approach for encapsulation of sour cherries anthocyanins by using β-lactoglobulin as matrices. *J. Food Eng.* **2017**, *210*, 83–90. [CrossRef]

Sample Availability: Samples of the compounds are available from the authors.

molecules

MDPI

Article

The Use of Trisodium Citrate to Improve the Textural Properties of Acid-Induced, Transglutaminase-Treated Micellar Casein Gels

Hongliang Li [1], Chang Yang [2], Chong Chen [3], Fazheng Ren [1], Yuan Li [1], Zhishen Mu [2] and Pengjie Wang [1,4,*]

1 Beijing Advanced Innovation Center for Food Nutrition and Human Health, College of Food Science & Nutritional Engineering, China Agricultural University, Beijing 100083, China; l12wpj@cau.edu.cn (H.L.); renfazheng@cau.edu.cn (F.R.); 16828777@163.com (Y.L.)

2 Inner Mongolia Mengniu Dairy (Group) CO., Ltd., Hohhot 750306, China; fdl409@cau.edu.cn (C.Y.); muzhishencau@126.com (Z.M.)

3 Key Laboratory of Functional Dairy, Co-Constructed by Ministry of Education and Beijing Government, Beijing 100083, China; chenchong409@cau.edu.cn

4 Beijing Laboratory of Food Quality and Safety, Beijing Higher Institution Engineering Research Center of Animal Product, Beijing 100083, China

* Correspondence: wpj1019@cau.edu.cn; Tel.: +86-010-6273-7644

Academic Editor: Lillian Barros

Received: 3 June 2018; Accepted: 1 July 2018; Published: 4 July 2018

Abstract: In this study, the effect of trisodium citrate on the textural properties and microstructure of acid-induced, transglutaminase-treated micellar casein gels was investigated. Various concentrations of trisodium citrate (0 mmol/L, 10 mmol/L, 20 mmol/L, and 30 mmol/L) were added to micellar casein dispersions. After being treated with microbial transglutaminase (mTGase), all dispersions were acidified with 1.3% (*w*/*v*) gluconodelta-lactone (GDL) to pH 4.4–4.6. As the concentration of trisodium citrate increased from 0 mmol/L to 30 mmol/L, the firmness and water-holding capacity increased significantly. The final storage modulus (G′) of casein gels was positively related to the concentration of trisodium citrate prior to mTGase treatment of micellar casein dispersions. Cryo-scanning electron microscopy images indicated that more interconnected networks and smaller pores were present in the gels with higher concentrations of trisodium citrate. Overall, when micellar casein dispersions are treated with trisodium citrate prior to mTGase crosslinking, the resulted acid-induced gels are firmer and the syneresis is reduced.

Keywords: trisodium citrate; microbial transglutaminase; casein gels; textural properties; microstructures

1. Introduction

Yogurt is a very popular fermented milk-based product worldwide due to its nutritional and sensory properties [1]. However, the acid-induced milk gels often have textural defects such as low firmness (fragile structure) and syneresis (low water holding capacity) during storage or following mechanical damage [2], which greatly reduce their sensory properties. Therefore, the development of acid-induced milk gels with greater firmness and water holding capacity is an important issue.

The firmness and water-holding capacity of acid-induced milk gels are closely related to their network structures. The primary building blocks of acid-induced milk gel network structures are proteins. Microbial transglutaminase (mTGase) is an enzyme widely used to form covalent links between individual protein molecules [3–6]. The textural properties of acid-induced gels prepared from mTGase-treated milk proteins can be significantly enhanced [2,7]. Unfortunately, even though plenty of work has been carried out regarding the effect of mTGase on the textural properties of acid-induced milk gels, the gel defects including low firmness (fragile structure) and syneresis have not been completely prevented.

The related mechanism of the mTGase-induced changes in the texture of acidified milk protein gels is summarized below [8,9]. The primary building blocks of acid-induced milk gel network structures are caseins [10–12], which are cross-linked with calcium phosphate to form association colloids known as casein micelles [12–14]. Schorsch and others [15] investigated the effect of the mTGase treatment on the acid gelation of micellar caseins. The mTGase generated covalent bonds between the individual casein molecules on the micellar surface or within the micelles. This inhibited the division of casein micelles even though the solubilization of calcium phosphate upon acidification occurred [16]. In this case, the casein particles could not be adequately rearranged to form more interconnected gel networks.

The addition of calcium-chelating agents (for example, trisodium citrate, sodium phosphate) to micellar casein dispersions has been reported to dissociate the native casein micelles into smaller casein aggregates [12,17,18]. This may change the rearrangement of casein particles during gel formation and the contact between mTGase and caseins, which would alter the crosslink bonds between casein molecules before or during acidification. Therefore, the mTGase-treated smaller casein particles might create different gel network structures compared with the gels from acidified mTGase-treated native casein micelles.

In this study, the effect of the calcium-chelating agent, which is trisodium citrate (TSC), on the textural properties and microstructure of acid-induced, transglutaminase-treated micellar casein gels was investigated with special reference to the related mechanisms of action.

2. Results and Discussion

The building blocks of the acid-induced milk gel network structures are proteins, which include approximately 80% of caseins and 20% of whey proteins [10]. Whey proteins were susceptible to mTGase only after being denatured by heat treatment [19]. In addition, the denatured whey proteins play an important role in the textural properties of yogurt, but some denatured whey proteins form particles [20–22]. The particle diameters range from about 80 nm to 500 nm [23,24], which was in the particle size range of casein micelles. Therefore, in the complex system consisting of both casein micelles and denatured whey proteins, it would be difficult to distinguish the changes of casein micelles with different treatments. Due to the primary role of casein micelles in the formation of acid-induced milk gels, the whey proteins were not included in the present study, as reported in previous papers [6,15,25,26].

The preparation of micellar casein gels can be divided into three routes which are described in the Materials and Methods section.

In route 1, various amounts of trisodium citrate were added to the casein micelle dispersions before mTGase treatment, which was followed by acidified with GDL. For convenience, the dispersions were first treated with various amounts of trisodium citrate and then treated with mTGase, which are defined as TG, 10-TG, 20-TG, and 30-TG, respectively, where the numbers refer to 0 mmol/L trisodium citrate, 10 mmol/L trisodium citrate, 20 mmol/L trisodium citrate, and 30 mmol/L trisodium citrate, respectively.

In route 2, various amounts of trisodium citrate were added to the casein micelle dispersions after mTGase treatment, which was followed by acidification with GDL. For convenience, the dispersions are first treated with mTGase and then treated with various amounts of trisodium citrate, which are defined as TG, TG-10, TG-20, and TG-30, respectively, where the numbers refer to 10 mmol/L trisodium citrate, 20 mmol/L trisodium citrate, and 30 mmol/L trisodium citrate, respectively. TG refers to samples with no trisodium citrate added.

In route 3, micellar casein dispersions treated with 0 mmol/L trisodium citrate, 10 mmol/L trisodium citrate, 20 mmol/L trisodium citrate, and 30 mmol/L trisodium citrate (in the absence of mTGase) were directly acidified with GDL.

2.1. Particle Size of Casein Micelles and Soluble Calcium Contents

Table 1 shows the effects of trisodium citrate and routes on the average diameter of casein particles. Two-way ANOVA revealed that there was a significant effect of routes on the particle sizes. In route 1 or route 3, with increasing trisodium citrate concentrations, the average diameter of particles exhibited an obvious decreasing trend. This suggested that the particle size in casein micelle dispersions became smaller with increasing trisodium citrate concentrations. The soluble calcium concentrations in casein micelle dispersions containing 0 mM trisodium citrate, 10 mM trisodium citrate, 20 mM trisodium citrate, and 30 mM trisodium citrate were 8.6 ± 0.5 mM, 13.3 ± 0.8 mM, 23.7 ± 0.6 mM, and 23.6 ± 0.6 mM, respectively. Similar results can be also found in Reference [27]. It is well-established that calcium-chelating agents can disrupt the micellar framework by removing calcium from the micelles, which leads to the dissociation of casein micelles [17]. Therefore, it was expected that the presence of citrate ions would reduce the particle size of native casein micelles. However, in route 2, the diameter of particles in TG, TG-5, TG-10, TG-20, and TG-30 showed little variation. This may be due to cross-linking of caseins within the particles by mTGase-generated covalent casein networks. Although the colloidal calcium phosphate dissociated from the casein micelles in the presence of citrate ions, this did not disrupt the stability of the mTGase-treated micelles [4,28].

Table 1. The effects of trisodium citrate on the average diameter (nm) of casein particles [1].

Treatment	Trisodium Citrate Concentration (mM)			
	0	10	20	30
Route 1	144.0 ± 6.0 ^{a,A}	88.3 ± 2.9 ^{b,A}	24.7 ± 1.7 ^{c,A}	18.8 ± 1.8 ^{c,A}
Route 2	144.0 ± 6.0 ^{a,A}	129.7 ± 0.8 ^{a,B}	130.3 ± 1.2 ^{a,B}	137.4 ± 2.2 ^{a,B}
Route 3 (Without mTGase)	142.2 ± 2.2 ^{a,A}	79.4 ± 9.5 ^{b,A}	20.3 ± 2.9 ^{c,A}	13.6 ± 1.6 ^{c,A}

[1] Superscripts with different letters (a–c) and (A, B) for the same row and column, respectively, indicate significant differences ($p \leq 0.05$).

2.2. Gelation Kinetics

The effect of trisodium citrate on the storage modulus (G′) as a function of time after the addition of GDL is shown in Figure 1. As expected, the G′ increased after gelation, followed by a plateau during the gel stage. Similar results have been reported elsewhere [29,30]. It can be seen that trisodium citrate had a significant influence on the acid-induced gelation kinetics of mTGase-treated micellar caseins. In the three routes, gelation time (when $G' \geq 1$) was positively related to the trisodium citrate concentration. This may be because trisodium citrate could dissolve the colloid calcium phosphate in the micelles, which leads to the liberation of phosphate into the soluble phase. Due to the buffering capacity of phosphate, with the increasing concentrations of trisodium citrate, a slower acidification rate of the dispersions occurred [27]. It can also be observed that, in the same trisodium citrate concentration, the final G′ values of gels prepared in route 1 were significantly higher than route 3. This indicated the mTGase-crosslinking played a crucial role in the gel structure. In route 1, the final G′ of the gels was positively related to the trisodium citrate concentrations. As mentioned above, when the amount of trisodium citrate was increased, the native casein micelles dissociated into smaller casein particles. This increased the flexibility of the casein particles, which favored the rearrangement of caseins during gel formation. In addition, the surface area of the smaller casein particles was larger than the native casein micelles, which favored contact between mTGase and caseins. This enhanced the crosslink bonds between casein molecules before or during acidification [31]. Due to these two reasons, in route 1, the final G′ of gels increased with trisodium citrate concentration.

However, the final G′ of gels formed in route 2 decreased with increasing concentrations of trisodium citrate. The size of casein particles in route 2 showed little variation in different trisodium citrate concentrations. During acidification, calcium binding or the other non-covalent interactions may have favored the coagulation of individual micelles. Therefore, the presence of the calcium-chelating agent trisodium citrate in route 2 decreased the calcium binding or other non-covalent interactions (especially electrostatic repulsion) between casein particles. This could inhibit the interconnectivity between micelles, which would lead to lower final G′ values of the gels.

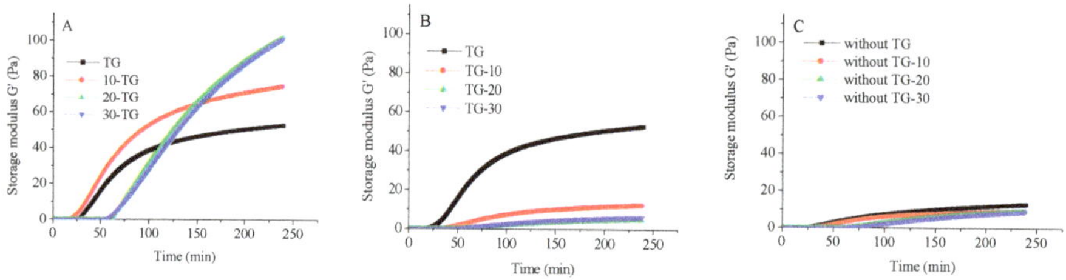

Figure 1. The evolution of the storage modulus as a function of time ((**A**) Route 1; (**B**) Route 2; (**C**) Route 3 (without mTGase)).

2.3. The Firmness of Acid-Induced Casein Gels

The results displayed in Table 2 illustrate the effect of trisodium citrate on the firmness of acid-induced micellar casein gels. Two-way ANOVA revealed that there was a significant route effect on the firmness of gels. In the samples without mTGase (route 3), it can be seen that the addition of trisodium citrate had little influence on the firmness of acid-induced casein gels. It was observed that the firmness of gels prepared in route 1 decreased in the following order: 30-TG or 20-TG > 10-TG > TG. These findings suggested that the firmness of GDL-induced micellar casein gels was positively related to the amount of trisodium citrate in route 1. The reasons for this was previously explained for the storage modulus (G′) of gels. It can also be observed that in the same trisodium citrate concentration, the firmness of gels prepared in route 1 was significantly higher than route 3. This indicated that the mTGase-crosslinking played a crucial role in the gel firmness.

The firmness of gels prepared in route 2 decreased with increasing trisodium citrate concentration. These trends are in agreement with the rheological results. It was also found that the firmness of gels prepared from 30-TG was significantly greater than the firmness of gels prepared from TG-30, which was the same as that in the other groups (20-TG versus TG-20 and 10-TG versus TG-10). This indicated that trisodium citrate treatment before the enzyme cross-linking of casein micelles resulted in firmer gels than those treated after cross-linking. However, trisodium citrate treatment after the enzyme cross-linking of casein micelles had no significant influence on gel firmness. The mTGase-treated casein micelles could not be dissociated into smaller particles even though colloidal calcium phosphate can be removed by trisodium citrate [32]. Therefore, the flexibility of the casein particles was not enhanced and the crosslink bonds between casein molecules were fewer in the gels prepared in route 2 than in route 1.

Table 2. The effects of trisodium citrate on the firmness (mN) of casein gels [1].

Treatment	Trisodium Citrate Concentration (mM)			
	0	10	20	30
Route 1	86.7 ± 6.1 ^{a,A}	114.8 ± 1.3 ^{b,B}	175.2 ± 1.3 ^{c,B}	179.4 ± 8.9 ^{c,C}
Route 2	86.7 ± 6.1 ^{a,A}	68.9 ± 1.6 ^{b,A}	65.0 ± 1.4 ^{b,A}	61.8 ± 1.9 ^{b,A}
Route 3 (Without mTGase)	67.8 ± 3.1 ^{a,A}	65.7 ± 5.7 ^{a,A}	71.5 ± 2.5 ^{a,A}	69.7 ± 1.2 ^{a,B}

[1] Superscripts with different letters (a–c) and (A–C) for the same row and column, respectively, indicate significant differences ($p \leq 0.05$).

2.4. Water Holding Capacity of Acid-Induced Casein Gels

The effect of trisodium citrate on the water-holding capacity of acid-induced micellar casein gels is shown in Table 3. Two-way ANOVA revealed that there was a significant effect of routes on the water-holding capacity of gels. It can be seen that, in the presence of trisodium citrate, little water in the gels prepared in route 1 was expelled by centrifugation. However, more than 30% of water was expelled in the gels prepared in route 2. This indicated that the gels prepared from the smaller particles in route 1 exhibited greater water retention than the gels prepared in route 2. This was consistent with the results of gel firmness since weak gels tend to flow and are more prone to shrinkage and subsequent expulsion of whey [33]. It can be also observed that, in the same trisodium citrate concentration, the water holding capacity of gels prepared in route 1 was significantly higher than in route 3. This indicated that the mTGase-crosslinking played a crucial role in improving the water-holding capacity.

Table 3. The effects of trisodium citrate on the water holding capacity (%) of casein gels [1].

Treatment	Trisodium Citrate Concentration (mmol L^{-1})			
	0	10	20	30
Route 1	65.4 ± 5.2 ^{a,A}	95.4 ± 0.8 ^{b,B}	98.7 ± 0.1 ^{c,B}	97.7 ± 0.2 ^{c,B}
Route 2	65.4 ± 5.2 ^{a,A}	59.4 ± 4.1 ^{a,A}	66.0 ± 2.3 ^{a,A}	64.8 ± 2.2 ^{a,A}
Route 3 (Without mTGase)	58.6 ± 1.3 ^{a,A}	63.5 ± 0.7 ^{b,A}	61.4 ± 1.6 ^{a,A}	60.1 ± 1.7 ^{a,A}

[1] Superscripts with different letters (a–c) and (A, B) for the same row and column, respectively, indicate significant differences ($p \leq 0.05$).

2.5. Microstructure and Degree of Cross-Linking of Acid-Induced Casein Gels

The microstructures of acid-induced casein gels prepared in route 1 and route 2 are shown in Figure 2. It can be clearly seen that gels prepared in route 1 had a more interconnected network than gels prepared in route 2 at the given concentration of trisodium citrate. In addition, the gels prepared in route 1 had smaller pores and more interconnected networks were present in the gels at 30 mmol/L trisodium citrate than at 0 mmol/L trisodium citrate. These trends align with the firmness and rheological results.

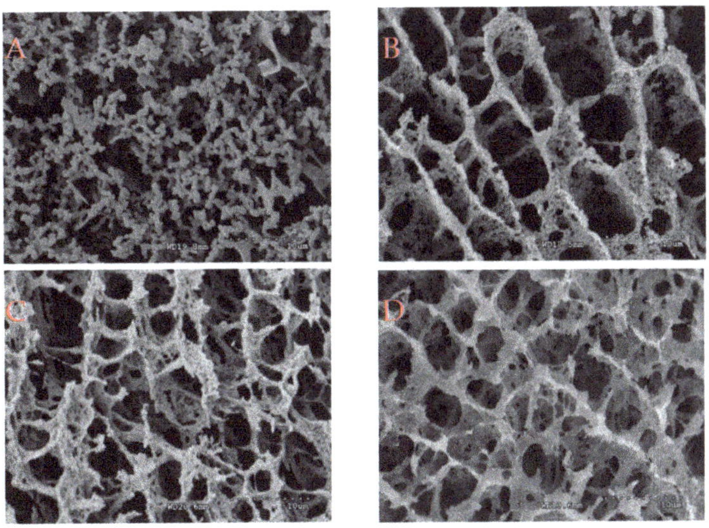

Figure 2. *Cont.*

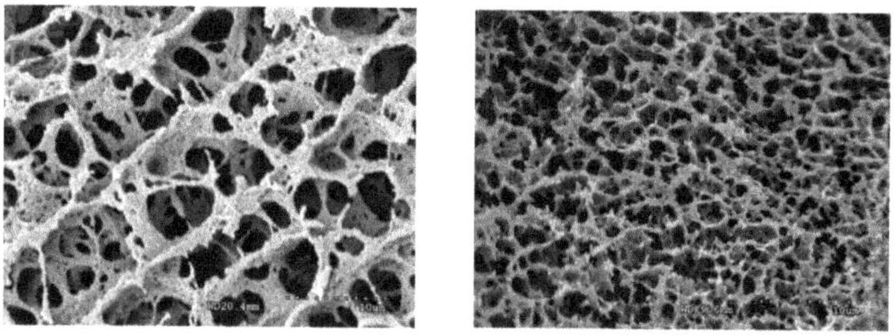

Figure 2. Cryo-scanning electron microscopy images of gels prepared from casein particle dispersions ((**A**) without mTGase and trisodium citrate, (**B**) With mTGase but without trisodium citrate, (**C**) TG-10 in route 2, (**D**) 10-TG in route 1, (**E**) TG-30 in route 2, and (**F**) 30-TG in route 2). Magnification = 3000×.

The cross-linking degrees of TG, TG-10, TG-20, and TG-30 were 19.9 ± 1.5%, 17.9 ± 0.5%, 21.2 ± 0.4%, and 21.7 ± 2.3%, respectively. This indicated that the trisodium citrate concentration in route 2 had little influence on the cross-linking degree of casein gels. The cross-linking degrees of TG, 10-TG, 20-TG, and 30-TG were 19.9 ± 1.5%, 29.8 ± 1.7%, 52.8 ± 1.6%, and 54.3 ± 1.9%, respectively. This indicated that, in route 1, the cross-linking degree of casein gels increased with trisodium citrate concentration (0 mmol/L to 20 mmol/L). As mentioned above, calcium-chelating agents can dissociate the native casein micelles into smaller particles [17]. This may have enhanced the contact between mTGase and caseins, which then increased the crosslink bonds between casein molecules before or during acidification.

2.6. Proposed Mechanism

The results presented above allowed us to explain the combined effect of trisodium citrate and mTGase on the gel properties of acid-induced micellar casein gels (Figure 3). In native casein micelles, the casein molecules are mainly cross-linked with colloidal calcium phosphate. In route 1, the citrate ion is a calcium chelating agent, which would disrupt the casein micelles by decreasing the [Ca^{2+}] and [Mg^{2+}] and colloidal calcium phosphate contents [27,34,35]. This favors the dissociation of native casein micelles into smaller casein particles [27]. Before acidification by GDL, the introduction of mTGase to the smaller casein particle dispersions resulted in partial cross-linkage of caseins within the particle matrix. In route 2, the introduction of mTGase to native casein micelles generated covalent protein networks. Trisodium citrate could bond to the calcium in the casein particles. However, the stability of the particle could not be disrupted due to the presence of covalent cross-linkage among casein molecules. This indicated that the flexibility of casein particles in route 2 was lower than in route 1. During acidification, the larger particles could not be adequately rearranged to form a more interconnected gel network. On acidification, due to the reduction of electrostatic repulsion, the casein particles began to aggregate. However, in route 1, due to their higher flexibility, the smaller casein particles were adequately rearranged to form more interconnected networks than the cross-linked native casein micelles. In this case, mTGase continued to crosslink two adjacent particles in the gel network even though the enzyme activity of mTGase gradually reduced with decreasing pH. Considering that the surface area of smaller casein particles was higher, the number of covalent bonds in gels in route 1 was larger than that in route 2, which led to denser and more stable gel networks. This may be the reason why the final G' of gels prepared using route 1 was significantly higher than the G' of gels prepared using route 2.

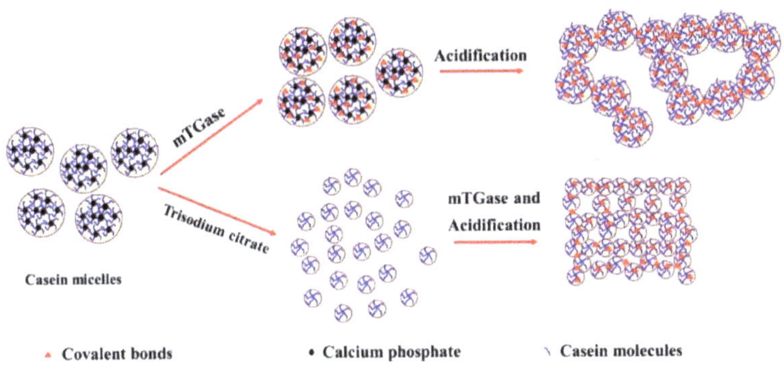

Figure 3. The schematic representation of the role of trisodium citrate in the formation of acid-induced, transglutaminase-treated casein gels.

3. Materials and Methods

3.1. Materials

Micellar casein powder was purchased from Develing International (Bunschoten, The Netherlands). The contents of total solids, ash, fat, lactose, and protein in the micellar powder were 96.2%, 7.6%, 1.3%, 2.7%, and 85.6%, respectively. Calcium-insensitive mTGase was purchased from C&P Group GmbH (Rosshaupten, Germany) and the enzyme activity was 200 U/g. GDL, 2,4,6-trinitro-benzensulfonic acid (TNBS), and trisodium citrate were purchased from Sigma-Aldrich (St. Louis, MO, USA). All other chemical reagents used were of an analytical grade.

3.2. Preparation of Acid-Induced Micellar Casein Gels

3.2.1. Preparation of the Micellar Casein or mTGase Dispersions

The micellar casein powder (2.5%, w/v) was hydrated in simulated milk ultrafiltrate (SMUF) at 55 °C for 40 min. During this process, the dispersion was stirred at 600 rpm. The SMUF was prepared based on a previously reported method [36]. The mTGase was hydrated in ultrapure water to a content of 4.0% (w/v) and stirred at 600 rpm for 1 h at 25 °C. The enzyme dispersion was further centrifuged (TDL-5C, Shanghai Anting Scientific Instrument Co. Ltd., Shanghai, China) at 3000× g at 25 °C for 15 min to remove the impurities at the bottom of the centrifuge tubes. The mTGase and casein micelle dispersions were stored at 4 °C prior to use.

3.2.2. Pretreatment of Casein Micelle Dispersions

The casein micelle dispersions were pre-warmed to 42 °C, which was followed by the slow addition (with continuous stirring) of trisodium citrate powder into the dispersions to 0 mmol/L, 10 mmol/L, 20 mmol/L, and 30 mmol/L, respectively. For convenience, the citrate ions in the original SMUF were not included in the calculation of trisodium citrate concentration in this study. The dispersions containing different concentrations of trisodium citrate were magnetically stirred at 600 rpm for 5 min. The pH of the dispersions was adjusted to 6.6 with 0.5 mol/L of HCl. Afterward, a certain volume of the mTGase dispersions was added to the casein micelle dispersions. The final ratio of the mTGase enzyme activity to the mass of micellar caseins was 10 U/g. The samples were incubated in a water bath at 42 °C for 60 min, which was followed by acidification with GDL (as described in the following section).

For comparison, after mTGase treatment at 42 °C for 60 min, trisodium citrate-free casein micelle dispersions were added to trisodium citrate at concentrations of 10 mmol/L, 20 mmol/L, and 30 mmol/L, which was followed by acidification with GDL (as described in the following section). The trisodium citrate-free casein micelle dispersions in this paper refer to the dispersions without the addition of citrate before enzyme cross-linking of caseins even though SMUF contained approximately 10 mmol/L citrate ions.

For another comparison, micellar casein dispersions treated with 0 mmol/L trisodium citrate, 10 mmol/L trisodium citrate, 20 mmol/L trisodium citrate, and 30 mmol/L trisodium citrate (in the absence of mTGase) were also investigated.

In summary, the preparation of casein micelle dispersions can be divided into three routes, which are shown schematically in Figure 4.

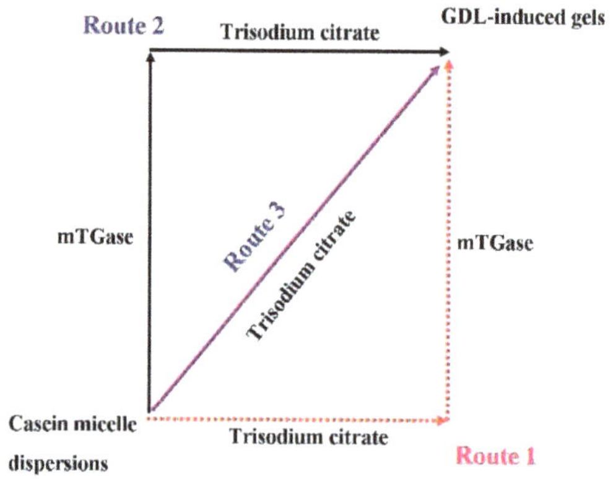

Figure 4. The schematic representation of the gel preparation route.

3.2.3. Preparation of Acid-Induced Casein Gels

Following incubation, GDL was directly added to the dispersions under magnetic stirring (600 rpm) for 3 min at 25 °C. Afterward, the samples were left to stand in a water bath at 43 °C for 4 h at a pH of 4.3–4.6. The concentrations of GDL in the different dispersions prepared in Section 3.2.2 were 1.3% (*w*/*v*). The casein gels in the beaker were cooled by ice water and stored at 4 °C for further analysis.

3.2.4. Characterization of Casein Micelle Dispersions or Gels

The sizes of the casein particles were determined using a dynamic light scattering (DLS) instrument (Nano ZS, Malvern, UK) at 20 °C. The treated samples were diluted 500 times with water [37]. Separation of calcium in the soluble and micellar phases was carried out by ultracentrifugation at 80,000× *g* for 3 h at 20 °C (CP 80MX, Hitachi, Tokyo, Japan) [34,38,39]. The calcium concentration in the soluble phase was analyzed by atomic absorption spectrometry (725-ES, Varian, Palo Alto, CA, USA).

The firmness of the casein gels was measured using a texture analyzer (TA-XT Plus, Stable Micro Systems, Godalming, Surrey, UK) in a single compression cycle test with a 5 kg load cell [40]. The set gels from cold storage (4 °C) were incubated at 25 °C for 60 min. The forces that pushed the cylindrical probe vertically into the gels (to a depth of 20 mm) at a speed of 60 mm/min were determined. Data were collected and analyzed to determine peak compression force.

The gel formation was determined using an MCR302 rheometer (Anton-Paar, Graz, Austria), based on the reported methods with some modifications [41]. An aliquot (15 mL) of the casein dispersions prepared in Section 3.2.2 (containing 1.3% GDL) was transferred to a concentric cylinder. A droplet of soybean oil was added onto the surface to prevent water evaporation. The gels were oscillated at a frequency of 0.1 Hz with an applied strain of 1% which was tested to be within the linear viscoelastic region. The test temperature was kept at 43 °C. Gelation time was defined as the point when the storage modulus (G′) of acid-induced gels $\geq$ 1 Pa [42].

The water holding capacity (WHC) of casein gels was determined based on a modified procedure [43]. A total of 8 mL of the pretreated casein micelle dispersions was poured into a 10 mL centrifuge tube and was acidified, which is described in Section 3.2.3. After acidification, the gels (Y) were centrifuged at 2800× g for 3 min at 25 °C. The mass of expelled water (W) was carefully collected and weighed. The WHC was calculated based on the formula below.

$$\text{WHC} = (\text{Y} - \text{W})/\text{Y} \times 100\% \quad (1)$$

The cross-linking degree of casein gels was measured based on a previously reported method with some modifications [44]. The final gels were mashed and washed with ultrapure water four times and then the samples were freeze-dried to a constant weight. One mL of 4 % (w/v) NaHCO3 solution and 1 mL of 0.1% (w/v) TNBS solution were added to the freeze-dried casein gels (5 mg). Following incubation at 40 °C for 3 h, 6 mol/L HCl (2 mL) was added to the samples and the temperature was raised to 60 °C. After reacting for 1.5 h, the samples were diluted with water (5 mL). The absorbance of the samples was determined at 345 nm using a U-2900 spectrometer (Hitachi, Ltd., Tokyo, Japan). The cross-linking degree (DE) was calculated based on the formula below.

$$\text{DE\%} = (1 - (A_S/m_s)/(A_{NS}/m_{Ns})) \times 100\% \quad (2)$$

where A_S and A_{NS} represent the absorbance of samples and non-cross-linked casein micelles, respectively. m_s and m_{Ns} represent the mass of samples and non-cross-linked casein micelles, respectively. To remove the ions, the non-cross-linked casein micelles were acidified at a pH of 4.6 and freeze-dried to a constant weight.

The microstructures of the gels were observed by using S-3000N cryo-scanning electron microscopy (Hitachi Co., Tokyo, Japan), which was described in previous reports [45] with some modifications. Gels were deposited in the cryo-specimen holder (Quorum PP3000T, East Grinstead, UK) and fixed in slush nitrogen (<180 °C). After being fractured, the gels were sublimated at −85 °C for 35 min and sputter-coated with gold.

3.2.5. Statistical Analysis

The experiments were carried out in triplicate. The means and standard deviations were calculated from these measurements. Data were tested using a two-way analysis of variance in order to determine the differences in particle size, firmness, and water-holding capacity induced by routes and trisodium citrate concentrations. SPSS 18 software for Windows (IBM, Ehningen, Germany) was used as the statistical analysis software. Duncan's test was applied to determine the statistical differences between the means.

4. Conclusions

In this study, the effect of trisodium citrate on the textural properties and microstructure of acid-induced, transglutaminase-treated micellar casein gels was investigated. It was concluded that the casein micelles that were first treated by trisodium citrate (10–30 mmol/L) and then cross-linked with mTGase (route 1) resulted in more stable acid gels than those directly cross-linked with mTGase (route 2) or without mTGase (route 3). The trisodium citrate (10–30 mmol/L) could dissociate the native casein micelles into smaller casein particles. In the presence of mTGase, the smaller particles formed acid-induced gels with greater firmness, higher water-holding capacity, more interconnected networks, and smaller pores than native casein micelles.

At the present time, mTGase is listed as a food ingredient in several commercial acid-induced milk gel products sold in many countries. It is used to improve the textural properties of acid-induced gel products. However, textural defects such as low firmness (fragile structure) and syneresis cannot be completely avoided. The method proposed in route 1 provides a novel strategy for improving the textural properties of acid-induced milk gel products.

Author Contributions: H.L. and P.W. conducted the conceptualization. F.R. managed the methodology. Z.M. handled the software. Y.L. validated the results. H.L. conducted the formal analysis. C.C. took part in data curation. H.L. wrote and prepared the original draft. P.W. reviewed and edited the manuscript. P.W. supervised the project. C.Y. conducted the ultracentrifugation and gelation kinetics experiments. P.W. took part in project administration and funding acquisition.

Funding: This study was supported by the Supported by the Open Project Program of Beijing Key Laboratory of Flavor Chemistry (No. SPFW-2018-YB01), Beijing Technology and Business University (BTBU), and 111 project from the Education Ministry of China (No. B18053).

Acknowledgments: We thank Yanbao Tian for help in cryo-SEM sample preparation and observation at the Institute of Genetics and Developmental Biology Chinese Academy of Sciences.

Conflicts of Interest: The authors declare no conflict of interest.

References

1. Nguyen, P.T.M.; Kravchuk, O.; Bhandari, B.; Prakash, S. Effect of different hydrocolloids on texture, rheology, tribology and sensory perception of texture and mouthfeel of low-fat pot-set yoghurt. *Food Hydrocoll.* **2017**, *72*, 90–104. [CrossRef]
2. Faergemand, M.; Qvist, K.B. Transglutaminase: Effect on rheological properties, microstructure and permeability of set style acid skim milk gel. *Food Hydrocoll.* **1997**, *11*, 287–292. [CrossRef]
3. Raak, N.; Rohm, H.; Jaros, D. Enzymatic cross-linking of casein facilitates gel structure weakening induced by overacidification. *Food Biophys.* **2017**, *12*, 261–268. [CrossRef]
4. Raak, N.; Rohm, H.; Jaros, D. Cross-linking with microbial transglutaminase: Isopeptide bonds and polymer size as drivers for acid casein gel stiffness. *Int. Dairy J.* **2016**, *66*, 49–55. [CrossRef]
5. Jaros, D.; Schwarzenbolz, U.; Raak, N.; Lobner, J.; Henle, T.; Rohm, H. Cross-linking with microbial transglutaminase: Relationship between polymerisation degree and stiffness of acid casein gels. *Int. Dairy J.* **2014**, *38*, 174–178. [CrossRef]
6. Schorsch, C.; Carrie, H.; Clark, A.H.; Norton, I.T. Cross-linking casein micelles by a microbial transglutaminase conditions for formation of transglutaminase-induced gels. *Int. Dairy J.* **2000**, *10*, 519–528. [CrossRef]
7. Vasbinder, A.J.; Rollema, H.S.; Bot, A.; de Kruif, C.G. Gelation mechanism of milk as influenced by temperature and pH; studied by the use of transglutaminase cross-linked casein micelles. *J. Dairy Sci.* **2003**, *86*, 1556–1563. [CrossRef]
8. Hinz, K.; Huppertz, T.; Kelly, A.L. Susceptibility of the individual caseins in reconstituted skim milk to cross-linking by transglutaminase: Influence of temperature, pH and mineral equilibria. *J. Dairy Res.* **2012**, *79*, 414–421. [CrossRef] [PubMed]
9. Smiddy, M.A.; Martin, J.; Kelly, A.L.; de Kruif, C.G.; Huppertz, T. Stability of casein micelles cross-linked by transglutaminase. *J. Dairy Sci.* **2006**, *89*, 1906–1914. [CrossRef]
10. Bijl, E.; van Valenberg, H.J.F.; Huppertz, T.; van Hooijdonk, A.C.M. Protein, casein, and micellar salts in milk: Current content and historical perspectives. *J. Dairy Sci.* **2013**, *96*, 5455–5464. [CrossRef] [PubMed]
11. Huppertz, T. Chemistry of the Caseins. In *Advanced Dairy Chemistry*; Springer: Boston, MA, USA, 2013.
12. De Kruif, C.G.; Huppertz, T.; Urban, V.S.; Petukhov, A.V. Casein micelles and their internal structure. *Adv. Colloid Interface Sci.* **2012**, *171*, 36–52. [CrossRef] [PubMed]
13. Gonzalez-Jordan, A.; Thomar, P.; Nicolai, T.; Dittmer, J. The effect of pH on the structure and phosphate mobility of casein micelles in aqueous solution. *Food Hydrocoll.* **2015**, *51*, 88–94. [CrossRef]
14. Thorn, D.C.; Ecroyd, H.; Carver, J.A.; Holt, C. Casein structures in the context of unfolded proteins. *Int. Dairy J.* **2015**, *46*, 2–11. [CrossRef]
15. Schorsch, C.; Carrie, H.; Norton, I.T. Cross-linking casein micelles by a microbial transglutaminase: Influence of cross-links in acid-induced gelation. *Int. Dairy J.* **2000**, *10*, 529–539. [CrossRef]
16. Boenisch, M.P.; Lauber, S.; Kulozik, U. Effect of ultra-high temperature treatment on the enzymatic cross-linking of micellar casein and sodium caseinate by transglutaminase. *J. Food Sci.* **2004**, *69*, E398–E404. [CrossRef]

17. Thomar, P.; Nicolai, T. Dissociation of native casein micelles induced by sodium caseinate. *Food Hydrocoll.* **2015**, *49*, 224–231. [CrossRef]
18. Silva, N.F.; Casanova, F.; Gaucheron, F.; de Carvalho Teixeira, A.V.; da Silva, G.M.; Minim, L.A.; de Carvalho, A.F. Combined effect of transglutaminase and sodium citrate on the microstructure and rheological properties of acid milk gel. *Food Hydrocoll.* **2018**, *82*, 304–311. [CrossRef]
19. Sharma, R.; Lorenzen, P.C.; Qvist, K.B. Influence of transglutaminase treatment of skim milk on the formation of ε-(γ-glutamyl)lysine and the susceptibility of individual proteins towards crosslinking. *Int. Dairy J.* **2001**, *11*, 785–793. [CrossRef]
20. Anema, S.G. Effect of whey protein addition and pH on the acid gelation of heated skim milk. *Int. Dairy J.* **2018**, *79*, 5–14. [CrossRef]
21. Morand, M.; Guyomarc'h, F.; Pezennec, S.; Famelart, M.H. On how kappa-casein affects the interactions between the heat-induced whey protein/kappa-casein complexes and the casein micelles during the acid gelation of skim milk. *Int. Dairy J.* **2011**, *21*, 670–678. [CrossRef]
22. Chai, J.; Jiang, P.; Wang, P.; Jiang, Y.; Li, D.; Bao, W.; Liu, B.; Liu, B.; Zhao, L.; Norde, W.; et al. The intelligent delivery systems for bioactive compounds in foods: Physicochemical and physiological conditions, absorption mechanisms, obstacles and responsive strategies. *Trends Food Sci. Technol.* **2018**, *78*, 144–154. [CrossRef]
23. Destribats, M.; Rouvet, M.; Gehin-Delval, C.; Schmitt, C.; Binks, B.P. Emulsions stabilised by whey protein microgel particles: Towards food-grade Pickering emulsions. *Soft Matter* **2014**, *10*, 6941–6954. [CrossRef] [PubMed]
24. Sarkar, A.; Kanti, F.; Gulotta, A.; Murray, B.S.; Zhang, S. Aqueous lubrication, structure and rheological properties of whey protein microgel particles. *Langmuir* **2017**, *33*, 14699–14708. [CrossRef] [PubMed]
25. Khanji, A.N.; Michaux, F.; Salameh, D.; Rizk, T.; Banon, S.; Jasniewski, J. The study of curcumin interaction with micellar casein and lactic acid bacteria cell envelope. *LWT* **2018**, *91*, 293–302. [CrossRef]
26. Gastaldi, E.; Lagaude, A. Micellar Transition State in Casein Between pH 5.5 and 5.0. *J. Food Sci.* **1996**, *61*, 59–64. [CrossRef]
27. Ozcan-Yilsay, T.; Lee, W.J.; Horne, D.; Lucey, J.A. Effect of trisodium citrate on rheological and physical properties and microstructure of yogurt. *J. Dairy Sci.* **2007**, *90*, 1644–1652. [CrossRef] [PubMed]
28. Kruif, C.G.D.; Anema, S.G.; Zhu, C.; Havea, P.; Coker, C. Water holding capacity and swelling of casein hydrogels. *Food Hydrocoll.* **2015**, *44*, 372–379. [CrossRef]
29. Zouari, A.; Marchesseau, S.; Chevalier-Lucia, D.; Raffard, G.; Ayadi, M.A.; Picart-Palmade, L. Acid gelation of raw and reconstituted spray-dried dromedary milk: A dynamic approach of gel structuring. *Int. Dairy J.* **2018**, *81*, 95–103. [CrossRef]
30. Pang, Z.; Deeth, H.; Sharma, R.; Bansal, N. Effect of addition of gelatin on the rheological and microstructural properties of acid milk protein gels. *Food Hydrocoll.* **2015**, *43*, 340–351. [CrossRef]
31. Lee, C.H.; Rha, C. Microstructure of soybean protein aggregates and its relation to the physical and textural properties of the curd. *J. Food Sci.* **2010**, *43*, 79–84. [CrossRef]
32. Nieuwland, M.; Bouwman, W.G.; Bennink, M.L.; Silletti, E.; de Jongh, H.H. Characterizing length scales that determine the mechanical behavior of gels from crosslinked casein micelles. *Food Biophys.* **2015**, *10*, 416–427. [CrossRef]
33. Serra, M.; Trujillo, A.J.; Jaramillo, P.D.; Guamis, B.; Ferragut, V. Ultra-high pressure homogenization-induced changes in skim milk: Impact on acid coagulation properties. *J. Dairy Res.* **2008**, *75*, 69–75. [CrossRef] [PubMed]
34. Franzoi, M.; Niero, G.; Penasa, M.; Cassandro, M.; De Marchi, M. Technical note: Development and validation of a new method for the quantification of soluble and micellar calcium, magnesium, and potassium in milk. *J. Dairy Sci.* **2018**, *101*, 1883–1888. [CrossRef] [PubMed]
35. Dumpler, J.; Kieferle, I.; Wohlschläger, H.; Kulozik, U. Milk ultrafiltrate analysis by ion chromatography and calcium activity for SMUF preparation for different scientific purposes and prediction of its supersaturation. *Int. Dairy J.* **2017**, *68*, 60–69. [CrossRef]
36. Jenness, R. Preparation and properties of a salt solution which simulated milk ultrafiltrate. *Neth. Milk Dairy J.* **1962**, *16*, 153–164.

37. Zang, J.; Chen, H.; Zhao, G.; Wang, F.; Ren, F. Ferritin cage for encapsulation and delivery of bioactive nutrients: From structure, property to applications. *Crit. Rev. Food Sci. Nutr.* **2017**, *57*, 3673–3683. [CrossRef] [PubMed]
38. Wang, P.; Liu, H.; Wen, P.; Zhang, H.; Guo, H.; Ren, F. The composition, size and hydration of yak casein micelles. *Int. Dairy J.* **2013**, *31*, 107–110. [CrossRef]
39. Wang, P.; Jin, S.; Guo, H.; Zhao, L.; Ren, F. The pressure-induced, lactose-dependent changes in the composition and size of casein micelles. *Food Chem.* **2015**, *173*, 468–474. [CrossRef] [PubMed]
40. Chua, D.; Deeth, H.C.; Oh, H.E.; Bansal, N. Altering the casein to whey protein ratio to enhance structural characteristics and release of major yoghurt volatile aroma compounds of non-fat stirred yoghurts. *Int. Dairy J.* **2017**, *74*, 63–73. [CrossRef]
41. Harbourne, N.; Jacquier, J.C.; O'Riordan, D. Effects of addition of phenolic compounds on the acid gelation of milk. *Int. Dairy J.* **2011**, *21*, 185–191. [CrossRef]
42. Lucey, J.A.; Teo, C.T.; Munro, P.A.; Singh, H. Rheological properties at small (dynamic) and large (yield) deformations of acid gels made from heated milk. *J. Dairy Res.* **1997**, *64*, 591–600. [CrossRef]
43. Mahomud, M.S.; Katsuno, N.; Zhang, L.; Nishizu, T. Physical, rheological, and microstructural properties of whey protein enriched yogurt influenced by heating the milk at different pH values. *J. Food Process. Preserv.* **2017**, *41*, e13236. [CrossRef]
44. Kim, Y.J.; Uyama, H. Biocompatible hydrogel formation of gelatin from cold water fish via enzymatic networking. *Polym. J.* **2007**, *39*, 1040–1046. [CrossRef]
45. Zhao, J.; Chen, Z.; Zou, R.; Zhang, X. The application of agar oligosaccharides in directly acidified milk drinks. *Food Hydrocoll.* **2017**. [CrossRef]

Sample Availability: Samples of the compounds mTGase and micellar casein powder are available/not available from the authors.

molecules

Article

Comparative Reduction of Egg Yolk Cholesterol Using Anionic Chelating Agents

Minerva Bautista Villarreal [1], Claudia T. Gallardo Rivera [1], Eristeo García Márquez [2], José Rodríguez Rodríguez [3], María Adriana Núñez González [1], Abelardo Chávez Montes [4] and Juan G. Báez González [1,*]

1 Departamento de Alimentos, Facultad de Ciencias Biológicas, Universidad Autónoma de Nuevo León, Av. Pedro de Alba s/n, Cd. Universitaria, C.P. 66455 San Nicolás de los Garza, Mexico; minerva.bautistavl@uanl.edu.mx (M.B.V.); claudia.gallardorv@uanl.edu.mx (C.T.G.R.); maria.nunezgn@uanl.edu.mx (M.A.N.G.)
2 Centro de Investigación y Asistencia en Tecnología y Diseño del Estado de Jalisco, A.C. Unidad Noreste, Parque PIIT, vía de la innovación 404, C.P. 66600 Apodaca, Mexico; egarcia@ciatej.mx
3 Instituto Tecnológico de Estudios Superiores de Monterrey (ITESM), Escuela de Ingeniería y Ciencias, Av. Eugenio Garza Sada 2501 Sur, Col. Tecnológico, C.P. 64849 Monterrey, Mexico; jrr@itesm.mx
4 Departamento de Química, Facultad de Ciencias Biológicas, Universidad Autónoma de Nuevo León, Av. Pedro de Alba s/n, Cd. Universitaria, C.P. 66455 San Nicolás de los Garza, Mexico; abelardo.chavezmn@uanl.edu.mx
* Correspondence: juan.baezgn@uanl.edu.mx; Tel.: +011-52-81-83-29-4000 (ext. 3654)

Academic Editors: Lillian Barros and Isabel C.F.R. Ferreira
Received: 1 November 2018; Accepted: 2 December 2018; Published: 5 December 2018

Abstract: Egg yolk is used as an emulsifying agent. Nevertheless, its high concentration of cholesterol is linked to chronic degenerative diseases that cause cardiovascular disease. In this study, three methods for reducing the level of cholesterol in egg yolks were studied. The first method consisted of physical separation of the granules contained in the yolk (Na_G). The second method applied was the use of anionic chelating biopolymers, such as arabic gum solution (AG) and mesquite gum solution (MG), and the third method was extraction with a solvent (S_A). For this purpose, the cholesterol present in egg yolks, the microstructure, particle size, zeta potential, and its emulsifying capacity were determined. The amount of cholesterol removed was 97.24% using 1% mesquite gum ($MG_{1\%}$), and 93.26% using 1% Arabic gum ($AG_{1\%}$). The zeta potential was determined, and the isoelectric point ($\zeta = 0$) of egg yolk was identified as pH 4.6. While, at this pH, the zeta potential of mesquite gum was −14.8 mV, the zeta potential for the arabic gum was −16 mV. The emulsifying capacity of $MG_{1\%}$ was 62.95%, while the emulsifying capacity of $AG_{1\%}$ was 63.57%. The complex obtained can be used in the development of functional foods reduced in cholesterol.

Keywords: egg yolk; cholesterol extraction; granules extraction; anionic chelating biopolymers

1. Introduction

Egg yolk is a good source of lutein, zeaxanthin, proteins, lipids, and vitamins in human nutrition and is made up of practically 50% solids. The major constituents of the solid matter are lipids (65–70% on dry basis) and proteins (30% on dry basis). The proteins present are livetins, lipoproteins [1], and some particles including high-density lipoproteins (HDLs), low-density lipoproteins (LDLs), and phosvitin [2,3].

Egg yolk is an efficient ingredient in many food products, and its functional properties include emulsifying, coagulating, foaming, and gelling properties [4]. Moreover, it contains proteins, vitamins, minerals, essential fatty acids, phospholipids, and other compounds. However, it has high cholesterol content; one simple egg contains between 200 mg and 300 mg of cholesterol/100 g; therefore, it almost meets the dietary intake limit set by the American Heart Association of <300 mg/day [3,5]. Consuming products with a high amount of cholesterol can result in cardiovascular disease. Clinical studies demonstrate that dietary cholesterol may increase serum LDL in certain individuals (hyper-responders). This is generally accompanied by an increase in HDLs [6]. Sichittiano et al. [7] undertook a review of nutraceuticals and functional food ingredients that are beneficial to vascular health. Grape seeds can reduce blood lipid levels, since it includes proanthocyanidins (polyphenols), which seem to play the main role in this process. Proanthocyanidins reduce the levels of triacylglycerol in chylomicrons and in very-low-density lipoproteins (VLDLs) [8]. Other functional ingredients are anthocyanins, which act on LDLs and HDLs. The influence on the lipid profile of anthocyanin supplements obtained from berries was evaluated in dyslipidemic patients. A decrease in LDLs was observed in patients after 12 weeks of treatment [9]. Mesquite gum has some phenolic compounds that are trapped in the gum matrix, and these substances are involved in the defense of the plant; in addition, the polymerization of these compounds produces polyphenols, resulting in brown or yellow gum [10].

Several researchers worked on different ways of decreasing the amount of cholesterol present in egg yolks. Warren et al. [11] used solvents such as hexane. Hexane forms a blend with egg yolk solids, but it requires a process of filtering to remove the solvent and a prolonged drying period. The yield reported was 62.2% cholesterol. Borges et al. [12] used a ratio of 1:12 *w*/*w* (yolk/acetone), and the emulsifying properties were maintained.

Paraskevopoulou et al. [13] extracted cholesterol from egg yolk with ethanol/water 20:80 (*v*/*v*), and 1.5% (*w*/*v*) polysorbate 80; after that, the dispersion was then centrifuged, and the yolk precipitate had 7.1 $\pm$ 0.3 mg of cholesterol. Laca et al. [14] worked on the extraction of egg yolk granules. In this process, egg yolk was mixed with water (1:15 *v*/*v*), the pH was adjusted to 7, and it was kept overnight at 4 °C; the granules had a concentration of 291 mg of cholesterol/100 g of egg yolk, equivalent to a reduction of 77% cholesterol. Another method involved the reduction of cholesterol using β-cyclodextrin (β-CD) due to its affinity for non-polar molecules such as cholesterol [15]. Jeong et al. [16] reported that cholesterol removal was 92.76% when using 25% crosslinked β-CD at 40 °C. Chiu et al. [17] used the immobilization of β-cyclodextrin in chitosan beads (Ch-BCD) by cross-linking with 1,6 hexamethylene diisocyanate (HMDI) reagent for cholesterol absorption from egg yolk, removing 92% of the cholesterol. That cholesterol was removed using 1% *w*/*v* Ch-BCD for 2 h at a proportion 1:30 yolk to water, and a mixture of β-cyclodextrins with chitosan. Garcia et al. [18] used high-methoxyl pectin at a concentration of 3% *w*/*w*, ionic strength 0.39 M, and pH equal to 9.2, and subsequently obtained a reduction of 88.6% cholesterol. Hsieh et al. [19] developed a complex of egg yolk and acacia gum applying the following concentrations: 1%, 3%, 5%, and 10% (*w*/*w*) with cholesterol extraction rates of 70%, 86%, 79%, and 59%, respectively.

Meanwhile, anionic biopolymers, such as gum ghatti, gum tragacanth, gum karaya, xanthan gum, and especially, Arabic gum (AG), are chelating agents, forming complexes with lipoproteins in the yolk. Lipoprotein molecules are positively charged, and anionic polysaccharides are used as chelating agents, controlling pH and temperature [20].

The methods mentioned above for removing cholesterol are complicated because solvents are used, thus prolonging the extraction time. It was shown that the use of biopolymers, including commercial arabic gum, decreases the concentration of cholesterol present in the egg yolk. However, the shortage of arabic gum, due to drought in the regions where it is produced, stimulated the search for other botanical sources that offer greater security of supply and costs [21]. Therefore, the use of mesquite gum as an alternative to arabic gum is proposed in this study. Mesquite gum (MG) has an ability to stabilize colloidal particles (1–100 μm), which it disperses or emulsifies. This property is manifested due to protein arabinogalactans, which allow adsorption in liquid–liquid interfaces [22,23]. The aim of this work was to reduce the amount of cholesterol present in egg yolk by preparing a complex of biopolymer mesquite gum–yolk and arabic gum–yolk, before comparing the results with cholesterol extraction using a solvent, and with physical separation of the granule using sodium chloride.

2. Materials and Methods

2.1. Materials

Eggs were purchased from the local market (commercial brand name "El Dorado"; a box containing 30 white eggs, expiration date 9 April 2018, batch 043501). Salt, sugar, and vinegar were purchased at the supermarket. The separation of the yolk from albumen was undertaken manually. The vitelline membrane was then cut with a scalpel blade, and the content of the yolk was collected in a glass vessel. The yolk was then mixed gently with a glass rod. The solution was maintained at 4 °C in refrigeration (pH 7). The arabic gum and mesquite gum were purchased from Natural Products of Mexico (Yautepec, Morelos, México), and the gum was purified according to the method of Vernon et al. [23]. Sodium chloride, acetone, ethanol, and hexane reagent-grade chemicals were purchased from Desarrollo de Especialidades Químicas (Monterrey, Nuevo León, México); the cholesterol standard was obtained from Sigma-Aldrich Chemical (Toluca, México). Deionized water was used in all the experiments.

2.2. Cholesterol Reduction in Egg Yolk

The egg yolk was treated with physical separation, polysaccharides (arabic gum and mesquite gum), and a solvent for reducing the cholesterol content. The first process was the physical separation of the granules contained in the yolk (Na_G) using aqueous salt solution and separation through centrifugation. Another process was the separation of cholesterol using complexation with biopolymers, arabic gum (AG) and mesquite gum (MG). In the third method, acetone extraction (S_A) was used.

2.2.1. Egg Yolk Granule Extraction (Na_G)

The cholesterol was removed using the method of Laca et al. [14], with a few modifications; 8.5 g of egg yolk and 11.4 g of 0.15 M NaCl solution (1:1.34) was mixed with a vortex for 1 min at 25 °C. The content was centrifuged (Hermle Labnet Z326, Labnet International, Inc., Wehingen, Germany) at 10,000× *g* for 45 min at 25 °C. Finally, the compounds were separated carefully from the aqueous fraction through decantation. The product obtained was lyophilized, and was stored at −20 °C until analysis with GC.

2.2.2. Anionic Polysaccharide/Egg Yolk Complexes

Complex formation was obtained based on the methodology reported by Hsieh et al. [19]. Stock solutions of arabic gum were prepared at 1% ($AG_{1\%}$), 3% ($AG_{3\%}$), and 10% ($AG_{10\%}$), in addition to 1% ($MG_{1\%}$), 3% ($MG_{3\%}$), and 10% ($MG_{10\%}$) mesquite gum. All solutions were maintained in constant agitation all night long. All concentrations are given in ratios of weight/weight (*w*/*w*).

Firstly, 3 g of egg yolk was mixed with 1 g of gum solution (at the concentrations mentioned above) and 4 g of water. The solution was mixed for 1 min in a vortex (Mixer Labnet International, Edison, NJ, USA); it was then centrifuged (Hermle Labnet Z326, Labnet International, Inc., Wehingen, Germany) at 6000× *g* for 15 min at 25 °C. After that, the supernatant was decanted, and then aggregated with 0.5 g of solution (0.9 M NaCl); this was mixed for 1 min in the vortex. Then, 6 g of ethanol was poured and mixed in the vortex for 1 min at 25 °C, and centrifuged at 6000× *g* for 15 min. After that, the solution was decanted; 6 g of ethanol was added to the lipoprotein/anionic biopolymer complexes, and the solution was mixed, before being carefully decanted. The precipitate complex was quantified [19]. Samples were lyophilized and stored at −20 °C until analysis with GC.

2.2.3. Solvent Extraction

Extraction of cholesterol with a solvent (S_A) was performed using the method described by Borges et al. [12] using a ratio (*w*/*w*) of 1:12 (yolk/acetone), and mixing at 100 rpm for 2 min in the stirrer (EURO-ST 60 D S001, IKA, Wilmington, NC, USA). This permitted the separation of the sample after 10 min, and the solvent of the precipitate was carefully decanted. Finally, the precipitate was washed with water. The samples were lyophilized and stored at −20 °C until analysis with GC.

2.3. Quantification of Cholesterol Using Gas Chromatography

Method 26.052 of the Association of Official Agricultural Chemists (AOAC) [24] was used. For the acid hydrolysis, 0.2 g of the sample was mixed with 2 mL of methanol, and 7% H_2SO_4 (v/v). Next, the sample was heated for 90 min at 80 °C; after that, the sample was cooled at 25 °C. Then, 3 mL of hexane was added and mixed for 1 min in the vortex (Mixer Labnet International, Inc., Edison, NJ, USA). The solution was kept for 15 min until the formation of two phases was completed; the process was performed twice. The supernatant recovered was mixed and diluted into a 10-mL flask with hexane. Subsequently, the solution was analyzed with GC (7890B, Agilent Technologies, Santa Clara, CA, USA), coupled to a mass spectrometer (5977A, Agilent Technologies, Santa Clara, CA, USA), and equipped with an HP-5MS capillary column (length: 30 m; inner diameter (ID): 0.25 mm; film thickness: 0.25 µm). The injected sample was 1 µL on split mode. The chromatographic conditions were as follows: column temperature 70 °C, kept for 1 min; increased to 200 °C at 10 °C/min, maintained for 2 min; increased to 300 °C at 10 °C/min, maintained for 7 min. The temperature of the injector was 250 °C, and the temperatures of the ion source and quadruple were 230 °C and 150 °C, respectively. The carrier gas helium flow rate was 1 mL/min. The ionization with electron impact was 70 eV and the scan acquisition mode had a range of 30 to 400 UMA. The calibration curve was done with a cholesterol standard of 20 to 120 ppm.

2.4. Particle Size Measurement

The particle size distribution of the samples ($AG_{1\%}$, $AG_{3\%}$, $AG_{10\%}$, $MG_{1\%}$, $MG_{3\%}$, $MG_{10\%}$, S_A, and Na_G), and the yolk were monitored using a Malvern Mastersizer 3000 (Malvern Instruments, Ltd, Worcestershire, UK) particle size analyzer with a unit of Hydro LV with water as a dispersant. The angular scattering intensity data were analyzed to calculate the size of the particles, creating a scattering pattern using the Mie theory of light scattering. The software calculated the particle size distribution ($D_{(3,2)}$). Optical properties of the sample were defined as a refractive index 1.460 and an absorption of 0.1.

2.5. Zeta Potential (ζ)

The zeta potential was determined using dynamic light scattering equipment Zetasizer Nano ZS90 (Malvern Instruments, Worcestershire, England, UK). The measurements were carried out using a universal dip cell (ZEN 1002, Malvern Instrument, Worcestershire, UK) at 25 °C, using the diluted solutions. The zeta potential is related to the velocity of the solutions in an electric field. The equipment software converts the electrophoretic mobility measurements into zeta potential values using the Smoluchowski model [25]. The zeta potential of egg yolk/polysaccharide solutions at different pH levels was measured with the method of Navidghasemizad et al. [26]. About 0.1 g of sample was diluted to a final volume of 20 mL using distilled water, and the pH was adjusted to values of 2.0, 3.0, 4.0, 5.0, 6.0, 7.0, 8.0, 9.0, and 10.0 using 0.1 M HCl or 0.1 M NaOH solution while stirring the samples. Both complexes, yolk/AG and yolk/MG, were monitored for the zeta potential. The zeta potential was calculated from the average of three measurements of the diluted solutions.

2.6. Emulsifying Capacity

Samples were prepared individually, and 10 g of sample was mixed (Na_G, S_A, $AG_{1\%}$, and $MG_{1\%}$, samples with high yield) with salt (1.59%), sugar (1.06%), and water (11%), separately. Then, the vinegar (3.17%) was added and kept under constant stirring at 300 rpm in the stirrer (IKA Eurostar 60 digital) for 10 min. The test ended when it was not possible to integrate more oil contained in the burette (V_{oil}), and a layer was observed on the surface of the emulsion ($V_{Emulsion}$) [27]. The values were estimated as percentage of emulsified oil (%*EC*) in total emulsion using Equation (1).

$$\%\boldsymbol{EC} = \frac{V_{oil}}{V_{Emulsion}} \times \mathbf{100} \quad (1)$$

2.7. Microstructure Analysis

The topology was analyzed employing scanning electron microscopy (SEM) obtained with the methodology reported by Valverde et al. [28]. Briefly, the samples were fixed overnight in 3% glutaraldehyde in 25 mM phosphate buffer (pH 3.25). After that, the samples were consecutively dehydrated with 20%, 40%, 60%, 80%, and 100% ethanol. Then, the ethanol was consecutively removed with 20%, 40%, 60%, 80%, and 100% acetone, and the samples were analyzed. Finally, they were dried at a pressure of 1×10^{-2} Torr (soft vacuum) in a vacuum desiccator. The dried sample was placed on aluminum SEM stubs and coated with gold/palladium. The microscope used was a JSM-6490LV (JEOL, Tokyo, Japan).

2.8. Color Analysis

The granules were measured for color in the lightness L^*, redness (a^*), and yellowness (b^*) system. Measurements were carried out using a ColorFlex EZ (Hunter Lab, Reston, VA, USA). A fixed amount of sample was poured into the measurement cell, and analyses were conducted in specular exclusion mode.

The color changes are expressed as $\boldsymbol{\Delta E}$ with the color of the egg yolk as a reference sample [29]; hence, $\boldsymbol{\Delta E}$ is the total color change due to the different contributions calculated using Equation (2).

$$\boldsymbol{\Delta E} = \sqrt{(\Delta L^*)^2 + (\Delta a^*)^2 + (\Delta b^*)^2} \quad (2)$$

2.9. Statistical Analysis

All tmeasurements were performed in triplicate, and ANOVA was performed with a confidence level of 95% ($p < 0.05$) using SPSS 20 software (IBM, SPSS Inc, Chicago, IL, USA). To determine the statistically significant difference between values, a one-way variance analysis and a Tukey test were performed.

3. Results and Discussion

3.1. Cholesterol Removal in Egg Yolk

The process of extracting egg yolk granules, and chelating with arabic gum and mesquite gum was used as an alternative to preparing yolk with a high concentration of protein and low cholesterol as a functional ingredient in the food industry, especially due to its functional attributes, such as foaming capacity, high level of phosvitin and high-density lipoprotein, and emulsifying and binding properties [28,30]. Notwithstanding the process, other properties, such as the emulsifying of oil, are reduced. In order of importance, the removal of cholesterol from lower to highest was $Na_G < S_A <$ AG < MG. The cholesterol removed from the egg yolk with Na_G (Table 1) was the least effective method (51.43%). It was previously reported that the Na_G method with ionic solvent removed cholesterol and increased the protein content. The method is relatively fast and inexpensive. The results obtained in the laboratory were compared with those obtained by Strixner and Kulozik [31]. Based on our data, the removal of cholesterol with solvent was the second best method for cholesterol removal (64.15%). However, Martucci and Borges [32] reported a six-stage extraction system of 92% cholesterol removal in a computer simulation study.

Table 1. Cholesterol removal, efficiency, and emulsifying capacity of the different methods.

Method	% Cholesterol Reduction (pH 7)	% Cholesterol Reduction (pH 3)	% Efficiency (pH 7)	% Efficiency (pH 3)	Emulsifying Capacity (% *w*/*w*)
Na_G	51.43 ± 1.86 [e]	–	15.48 ± 1.31 [d]	–	79.52 ± 2.01 [b]
$AG_{1\%}$	93.26 ± 0.55 [ab]	74.74 ± 1.10 [ab]	29.72 ± 0.29 [c]	3.72 ± 0.67 [d]	63.57 ± 1.43 [d]
$AG_{3\%}$	83.85 ± 3.80 [c]	56.34 ± 0.57 [b]	44.17 ± 3.35 [ab]	4.38 ± 0.44 [d]	–
$AG_{10\%}$	89.93±2.31 [b]	86.33 ± 1.45 [a]	37.37 ± 1.88 [b]	16.42 ± 0.84 [bc]	–
$MG_{1\%}$	97.24 ± 1.81 [a]	87.75 ± 1.25 [a]	13.50 ± 0.5 [de]	12.14 ± 0.41 [c]	62.95 ± 0.84 [d]
$MG_{3\%}$	96.68±1.01 [a]	92.55 ± 1.20 [a]	14.85 ± 1.06 [de]	19 ± 2.14 [b]	–
$MG_{10\%}$	96.60 ± 2.04 [a]	80.31 ± 2.30 [a]	11.17 ± 0.72 [e]	46.54 ± 3.38 [a]	–
S_A	64.15 ± 1.29 [d]	–	44.51 ± 0.7 [a]	–	72.33 ± 1.02 [c]
Egg yolk	-	–	–	–	85.06 ± 0.051 [a]

Note: Differing letters within a column are significantly different ($p < 0.05$) ± standard deviation ($n = 3$). Na_G—physical separation of granules; AG—arabic gum; MG—mesquite gum; S_A—solvent extraction.

The removal of cholesterol using acetone in the laboratory was lower than the 81%, as reported by Borges et al. [12]. The removal of cholesterol was lower due to the concentration of water in the fresh yolk, the extraction time, and the acetone ratio. It was reported that the use of non-polar organic solvents prevents protein denaturation, especially in emulsifying activity, which is related to the solubility of the protein. Meanwhile, polar organic solvents reduce the emulsifying activity because they break the hydrophobic interactions between lipids and proteins. Even so, the use of solvents is not completely accepted due to the residues that may remain in the product [33].

The use of arabic gum allowed cholesterol removal between 83.85% and 93.26% (Table 1). The concentrations $AG_{3\%}$ and $AG_{10\%}$ show significant differences ($p < 0.05$). The removal of cholesterol was greater when 1% arabic gum was used. Meanwhile, when 3% and 10% arabic gum was used, the cholesterol removal was almost 83.85% and 89.93%, respectively, despite the fact that pH values were similar to neutral. The structure of arabic gum is a branched heteropolysaccharide with anionic properties. The quantity of arabic gum–yolk depends on the polyanionic properties of the gum, especially of residues of D-glucuronic acid (~2%) and D-glucuronic acid (~21%) [2,34]. The arabic gum has properties of anionic polysaccharides, which may be used as chelating agents, forming insoluble electrostatic complexes (chelating agent/lipoprotein) [18]. Navidghsemizad et al. [35] used a ratio of 50 g of fresh yolk per gram of arabic gum to observe the separation of phases at different pH levels. They concluded that the nature of the polysaccharide and pH had important effects, resulting in the phase separation behavior.

Other anionic polysaccharides were used as chelating agents [19]; however, their low solubility and high viscosity reduce their practical applications. Mesquite gum is a polysaccharide which contains acidic residues of β-D-glucuronic and 4-*O*-methyl-β-D-glucuronic, bound to mono-sugars or oligosaccharide chains [36,37]. The removal of high cholesterol levels when the mesquite gum solution was used was 1% *w*/*w* ($p < 0.05$). From mesquite gum $MG_{1\%}$, $MG_{3\%}$, and $MG_{10\%}$, extractions of cholesterol of 97.24%, 96.68%, and 96.60%, respectively, were obtained. No significant differences were observed in the different concentrations where mesquite gum was used.

In light of our results, it can be stated that sodium chloride or acetone have a lower capacity to remove cholesterol compared to both anionic polysaccharides. The greater efficiency in cholesterol removal was obtained with mesquite gum at 25 °C. Nonetheless, obtaining complexes from mesquite gum–yolk allowed a high removal of bound cholesterol. We think that the complex obtained can be used as functional supplement, necessary for reducing the unwanted effects of cholesterol. Scicchitano et al. [7] mentioned the importance of reducing lipid levels, especially for coronary artery disease.

3.2. Particle Size Measurement

In natural conditions, the yolk is a supramolecular assembly of lipids and proteins, and a highly organized system with approximate size between 0.8 μm and 10 μm [38]. The insoluble structure of yolk has a size range between 0.3 μm and 2 μm [14]. Molecular assembly can be disorganized into individual structures depending on the affinity of the solvent or polysaccharide used to remove cholesterol. The morphology and topology of the particle size determined using the light scattering method can be seen in Figure 1. Figure 1a,b show the granules and the complexes formed with arabic gum and mesquite gum (1%, 3%, and 10%), respectively.

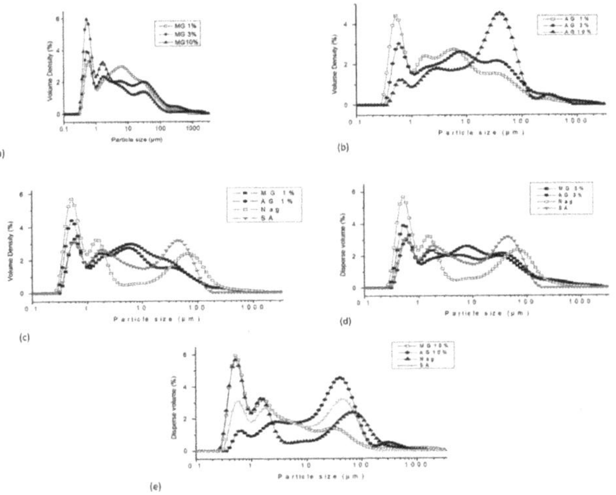

Figure 1. The average particle size: (**a**) –□– $MG_{1\%}$, –●– $MG_{3\%}$, and –▲– $MG_{10\%}$; (**b**) –□– $AG_{1\%}$, –●– $AG_{3\%}$, and –▲– $AG_{10\%}$; (**c**) –■– $MG_{1\%}$, –●– $AG_{1\%}$, –□– Na_G, and –▼– S_A; (**d**) –■– $MG_{3\%}$, –●– $AG_{3\%}$, –□– Na_G, and –▼– S_A and (**e**) –□– $MG_{10\%}$, –●– $AG_{10\%}$,–▲– Na_G, –▼– S_A. MG—mesquite gum; AG—arabic gum; Na_G—physical separation of granules; S_A—solvent extraction.

Regarding the distribution of proteins, the structure of the supramolecular system contains many particles of different sizes, three of which are of special interest: the HDLs, which range from 7 to 20 nm; the micelles formed by LDLs in the egg yolk plasma, which range from 17 to 60 nm; and the LDL sources present in the yolk, which range from 80 to 350 nm [31,38]. Anton [39] and Hsieh et al. [19] mentioned that granules of yolk are composed of 70% HDLs, 16% phosvitin, and 12% LDLs. We believe that the granules obtained from solvents and polysaccharides follow the same distribution pattern. The granules and lipoprotein/anionic polysaccharide complexes had different particle size population profiles; these size profiles may be associated with the process of cholesterol removal, the anionic polysaccharides (arabic gum and mesquite gum), and concentration (Figure 1a,b; 1%, 3%, and 10%), and the ratio. The change in distribution of particle size may be due to the viscosity and the charge density of proteins diverse in egg yolk, between MG–yolk and AG–yolk complexes, including the concentration of reactive groups contained in both biopolymers, which form an electrostatic complex [40]. In macroscopic terms, the distribution of granules was separated into three different sizes, based on the chelate concentration. The concentration of chelating polysaccharides at 10% (arabic gum and mesquite gum) had a range of particle size distribution between 0.3 μm and 600 μm. Similarly, the same profile was obtained with low-molecular-weight chelates. The yolk–chelate ratio of 3% for both polysaccharides produced a range of granule size distribution from 0.3 μm to 300 μm (Figure 1d). Finally, when a high molecular weight at 1% concentration was used, the yolk–chelate reduced the size of granules to between 0.3 μm and 250 μm (Figure 1c). More specifically, the granules and lipoprotein complexes obtained had multimodal distributions, as described below. Figure 1e shows three different population distributions when Na_G, S_A, and AG and MG at 10% were used. The first range was from 0.3 μm to 1 μm, the second was from 1 μm to 4 μm, and the third range of distribution was from 10 μm to 300 μm, when using the Na_G method. The removal of cholesterol using acetone had three different populations of granule size; the range of least distribution was between 0.3 μm and 0.9 μm, and the greatest range showed a variation 0.9 μm to 10 μm. Finally, particle size distribution could be observed in the range of 10 μm to 200 μm. A similar range of distribution was observed when using concentrations of 3% and 1% (both polysaccharides and solvents).

3.3. Zeta Potential (ζ)

The interaction between biopolymers may be segregative, due to steric repulsion or associative interactions such as hydrophobic interactions and hydrogen bonding [41]. Electrostatic interactions are the most common force for the complex formation [42]. The pH affected the charge of biopolymers and proteins of the egg yolk, which influenced the zeta potential (ζ) as a function of pH; $AG_{3\%}$ and $MG_{3\%}$ were studied in the pH range of 2–10.

The zeta potential of egg yolk was positive at pH values of 2 and 4 (+15.5 mV and +8.5 mV, respectively), while it was negative at pH values of 5 and 10 (Figure 2a); this was similar to the report by Navidghasemizad et al. [26], who obtained positive values in the range of pH 3–5, and, at pH 6, it was negative above the zeta potential. The isoelectric point ($\zeta = 0$) of egg yolk was found to be pH 4.6, determined from the zeta potential. While, at this pH, the zeta potential of mesquite gum was −14.8 mV, the zeta potential for the arabic gum was −16 mV. The formation of insoluble complexes appears to occur at pH 3. At this point, the density of the opposite charge between egg yolk and polysaccharides (arabic gum and mesquite gum) has practically the same magnitude. However, the percentage of cholesterol removal was lower than at pH 7. The values are shown in Table 1. The mesquite gum had values of −2.25 mV to −24.61 mV, in the acidic to basic (2 to 10) pH range. We believe that the free and exposed glucuronic acid and protein residues present in mesquite gum reacted in the different media [43] with acidic residues present in the polymer, similar to arabic gum [1]. In Figure 2b, the zeta potential profile of the yolk–3% polysaccharide complex essentially modified the isoelectric point to pH 4 for both biopolymers. For values lower than pH 4, the load profile was positive; however, for values higher than pH 4, the profile was negative.

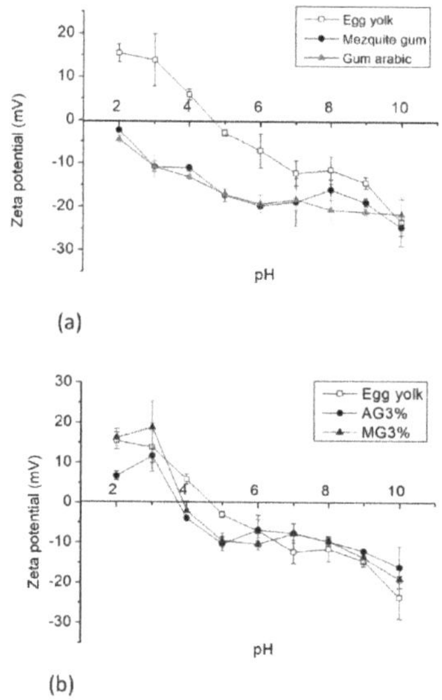

Figure 2. Average zeta potential (ζ) of (**a**) –□– egg yolk, –●– mesquite gum, and –▲– arabic gum, and (**b**) –□– egg yolk, –●– $AG_{3\%}$, and –▲– $MG_{3\%}$.

3.4. Emulsifying Capacity

The emulsifying properties only work in specific cases; these properties cannot be generalized. Therefore, we determined the emulsifying properties of the granules obtained, depending on the different treatments. As shown in Table 1, when comparing the emulsifying capacities of the methods discussed, the egg yolk method supports the highest percentage of oil, followed by Na_G, which is similar to the findings obtained by Laca et al [14]. In $AG_{3\%}$, a lower emulsifying capacity than Na_G was obtained. It was reported that this is because using arabic gum to remove cholesterol allows the loss of 66% of proteins, including the yolk's emulsifying proteins, along with lipids [18].

For the egg yolk–mesquite gum complex, the emulsifying capacity of $MG_{3\%}$ was 62.95%. It is possible that the ethanol used in the washing of biopolymer–yolk complexes dissolves phospholipid cholesterol and diminishes the emulsifying capacity. A monolayer of water prevents the denaturation of the protein [44]; when using S_A, the result obtained was 72.33%. Both products of Na_G were proposed as additives with low cholesterol in products like muffins [4] and salad dressings [14]. The complex of egg yolk–mesquite gum with polyphenols can be used for the development of foods reduced in cholesterol, thereby helping avoid health problems like cardiovascular disease.

3.5. Microstructure Analysis

The microphotograph of Na_G (Figure 3g) shows irregular structures with small aggregates; it is probable these are HDL–phosvitin complexes linked by phosphocalcic bridges between the phosphate groups [45]. The microstructures of treatments with arabic gum ($AG_{1\%}$ $AG_{3\%}$, and $AG_{10\%}$; Figure 3a,b,c) show small aggregates, which are probably due to the interaction of the biopolymer carboxyl groups and the lipoproteins of egg yolk. In the microphotographs of different treatments with mesquite gum ($MG_{1\%}$, $MG_{3\%}$, and $MG_{10\%}$; Figure 3d,e,f), spherical structures within 2 to 5 μm of size were observed; it is probable that they correspond to the interaction between arabinogalactan proteins in mesquite gum and the lipoproteins of egg yolk [20].

The microphotograph of egg yolk shows an irregular structure because it is a complex system with several particles in suspension in a fluid that contains proteins. The microphotograph of egg yolk with acetone (S_A) shows irregular structures with small aggregates.

Figure 3. SEM photographs using different cholesterol extraction methods: (**a**) $AG_{1\%}$, (**b**) $AG_{3\%}$, (**c**) $AG_{10\%}$, (**d**) $MG_{1\%}$, (**e**) $MG_{3\%}$, (**f**) $MG_{10\%}$, (**g**) Na_G, and (**h**) S_A.

3.6. Color Analysis

The values regarding the lightness (L^*), redness (a^*), and yellowness (b^*) values are shown in Table 2. There are no significant differences for lightness L^*, while a^* and b^* values show significant differences ($p < 0.05$). The Na_G shows low (ΔE) total color change, which we can state to be similar to egg yolk; this method removes less pigment than S_A. The redness (a^*) and yellowness (b^*) values for Na_G were 1.7 and 22.73, respectively, greater than those reported for S_A (8.91 and

49.65, respectively). The color of egg yolk is attributed to carotenoids (xanthophylls, including lutein, zeaxanthin, β-cryptoxanthin, and β-carotene) [46]. We suggest that, if a large amount of cholesterol is removed, more carotenoids are also removed.

Table 2. Color measurements of different methods for cholesterol reduction.

Method	L^*	a^*	b^*	ΔE
Egg yolk	56.48 ± 0.01 ^{a}	19.98 ± 0.01 ^{e}	53.57 ± 0.05 ^{d}	–
$AG_{1\%}$	79.76 ± 0.09 ^{d}	4.7 ± 0.03 ^{b}	35.29 ± 0.23 ^{b}	33.32 ± 0.18 ^{b}
$AG_{3\%}$	75.13 ± 3.94 ^{bc}	6.73 ± 1.74 ^{c}	41.29 ± 5.04 ^{c}	26.03 ± 6.13 ^{a}
$AG_{10\%}$	73.41 ± 0.01 ^{b}	1.95 ± 0.00 ^{a}	21.58 ± 0.05 ^{a}	40.44 ± 0.08 ^{c}
$MG_{1\%}$	75.12 ± 0.01 ^{bc}	1.99 ± 0.01 ^{a}	18.63 ± 0.01 ^{a}	43.50 ± 0.05 ^{c}
$MG_{3\%}$	81.92 ± 0.02 ^{d}	1.43 ± 0.01 ^{a}	19.28 ± 0.03 ^{a}	46.56 ± 0.00 ^{c}
$MG_{10\%}$	78.73 ± 0.04 ^{cd}	4.64 ± 0.02 ^{b}	33.21 ± 0.08 ^{b}	33.84 ± 0.03 ^{b}
S_A	82.47 ± 0.01 ^{d}	1.70 ± 0.01 ^{a}	22.73 ± 0.02 ^{a}	42.99 ± 2.20 ^{c}
Na_G	79.55 ± 0.01 ^{d}	8.91 ± 0.01 ^{d}	49.65 ± 0.01 ^{d}	25.89 ± 0.01 ^{a}

Note: Letters varying within a column are significantly different ($p < 0.05$) ± standard deviation ($n = 3$). L^*—lightness; a^*—redness; b^*—yellowness; ΔE—total color change.

4. Conclusions

Separation efficiency of complex lipoproteins (HDLs)–mesquite gum shows a strong dependence on pH. The greatest cholesterol reduction was seen at pH 7.0. The amount of cholesterol removed was 97.24% using 1% mesquite gum ($MG_{1\%}$), and 93.26% using 1% arabic gum ($AG_{1\%}$). This is a consequence of the chemical composition in the chelate (mesquite gum or arabic gum) and yolk. The use of mesquite gum shows structural changes in the form of definite spheres with a low size in comparison with arabic gum, observed using SEM. The high removal of cholesterol contained in the egg yolk using mesquite gum or arabic gum reduced the primary emulsifying capacity of the egg yolk. The use of mesquite gum to remove cholesterol is an alternative method that does not require organic solvents. The use of 3% mesquite gum removed 12.83% more cholesterol than the same concentration of arabic gum. The complex obtained can be used in the development of functional foods reduced in cholesterol.

Author Contributions: M.B.V., J.G.B.G., and E.G.M. designed and led the research, and wrote the paper. M.A.N.G. collaborated in the experimental phase of research. J.R.R. collaborated in the experimental phase of research. A.C.M. collaborated in the experimental phase of research. C.T.G.R. collaborated in the experimental phase of research.

Funding: The authors would like to thank the "Consejo Nacional de Ciencia y Tecnología" (CONACyT) of Mexico for partial financing of this project through grant #277814, as well as through the grant Problemas nacionales-2015-01-1470 and the grant CONACyT CB-157511.

Conflicts of Interest: The authors declare no conflicts of interest.

References

1. Sarika, P.; Pavithran, A.; James, N.R. Cationized gelatin/gum arabic polyelectrolyte complex: Study of electrostatic interactions. *Food Hydrocolloids* **2015**, *49*, 176–182. [CrossRef]
2. Islam, A.; Phillips, G.; Sljivo, A.; Snowden, M.; Williams, P. A review of recent developments on the regulatory, structural and functional aspects of gum arabic. *Food Hydrocolloids* **1997**, *11*, 493–505. [CrossRef]
3. Miranda, J.M.; Anton, X.; Valbuena, C.R.; Saavedra, P.R.; Rodríguez, J.A.; Lamas, A.; Franco, C.M.; Cepeda, A. Egg and egg-derived foods: Effects on human health and use as functional foods. *Nutrients* **2015**, *7*, 706–729. [CrossRef]
4. Marcet, I.; Collado, S.; Paredes, B.; Díaz, M. Rheological and textural properties in a bakery product as a function of the proportions of the egg yolk fractions: Discussion and modelling. *Food Hydrocolloids* **2015**, *54*, 119–129. [CrossRef]

5. Sun, Y.; Yang, H.; Zhong, X.; Zhang, L.; Wang, W. Ultrasonic-assisted enzymatic degradation of cholesterol in egg yolk. *Innovative Food Sci. Emerg. Technol.* **2011**, *12*, 505–508. [CrossRef]
6. Lordan, R.; Tsoupras, A.; Mitra, B.; Zabetakis, I. Dairy fats and cardiovascular disease: Do we really need to be concerned? *Foods* **2018**, *7*, 29. [CrossRef]
7. Scicchitano, P.; Cameli, M.; Maiello, M.; Modesti, P.A.; Muiesan, M.L.; Novo, S.; Palmiero, P.; Saba, P.S.; Pedrinelli, R.; Ciccone, M.M. Nutraceuticals and dyslipidaemia: Beyond the common therapeutics. *J. Funct. Foods* **2014**, *6*, 11–32. [CrossRef]
8. Quesada, H.; Díaz, S.; Pajuelo, D.; Fernández, A.; Garcia, S.; Pujadas, G.; Salvadó, M.J.; Arola, L.; Bladé, C. The lipid-lowering effect of dietary proanthocyanidins in rats involves both chylomicron-rich and VLDL-rich fractions. *Br. J. Nutr.* **2012**, *108*, 208–217. [CrossRef]
9. Qin, Y.; Xia, M.; Ma, J.; Hao, Y.; Liu, J.; Mou, H.; Cao, L.; Ling, W. Anthocyanin supplementation improves serum LDL- and HDL-cholesterol concentrations associated with the inhibition of cholesteryl ester transfer protein in dyslipidemic subjects. *Am. J. Clin. Nutr.* **2009**, *90*, 485–492. [CrossRef]
10. Trejo, J.L. Establecimiento de un cultivo de células en suspensión de *Prosopis laevigata* (Humb. & Bonpl. ex Willd.) M. C. Johnst. para la producción de goma de mezquite. Ph.D. Thesis; Universidad Autónoma Metropolitana, Unidad Iztapalapa: Cd. de México, México, December 2010.
11. Warren, M.; Brown, H.; Davis, D. Solvent extraction of lipid components from egg yolk solids. *J. Am. Oil Chem. Soc.* **1988**, *65*, 1136–1139. [CrossRef]
12. Borges, S.; Martucci, E.; Müller, C. Optimization of the extraction of cholesterol from dehydrated egg yolk using acetone. *LWT-Food Sci. Technol.* **1996**, *29*, 687–690. [CrossRef]
13. Paraskevopoulou, A.; Kiosseoglou, V. Texture profile analysis of heat-formed gels and cakes prepared with low cholesterol egg yolk concentrates. *J. Food Sci.* **1997**, *62*, 208–211. [CrossRef]
14. Laca, A.; Sáenz, M.; Paredes, B.; Díaz, M. Rheological properties, stability and sensory evaluation of low-cholesterol mayonnaises prepared using egg yolk granules as emulsifying agent. *J. Food Eng.* **2010**, *97*, 243–252. [CrossRef]
15. Alonso, L.; Fox, P.; Calvo, M.V.; Fontecha, J. Effect of beta cyclodextrin on the reduction of cholesterol in ewe´s milk manchego cheese. *Molecules* **2018**, *23*, 1789. [CrossRef] [PubMed]
16. Jeong, H.; Sun, H.; Chogsom, C.; Kwak, H. Cholesterol removal from whole egg by crosslinked β-cyclodextrin. *Asian-Australasian J. Anim. Sci.* **2014**, *27*, 537–542. [CrossRef] [PubMed]
17. Chiu, S.H.; Chung, T.W.; Giridhar, R.; Wu, W.T. Immobilization of β-cyclodextrin in chitosan beads for separation of cholesterol from egg yolk. *Food Res. Inter.* **2004**, *37*, 217–223. [CrossRef]
18. García, E.E.; Reis, J.S.; Minim, L.A.; Freitas, J. Cholesterol removal in liquid egg yolk using high methoxyl pectins. *Carbohydr. Polym.* **2007**, *69*, 72–78. [CrossRef]
19. Hsieh, R.J.; Snyder, D.P.; Ford, E.W. Method for Removing Cholesterol and Fat from Egg Yolk by Chelation and Reduced-Cholesterol Egg Product. U.S. Patent 5,302,405, 12 April 1994.
20. López, Y.L.; Goycoolea, F.M.; Valdez, M.A.; Calderón, A.M. Goma de mezquite: Una alternativa de uso industrial. *Interciencia* **2006**, *3*, 183–189.
21. Pérez, J.; Barrios, E.; Róman, A.; Pedroza, R. Interacción goma de mezquite-quitosano en la interfase y su influencia en la estabilidad de emulsiones múltiples W1/O/W2. *Revista Mexicana de Ingeniería Química* **2011**, *10*, 487–499.
22. Moreno, M.B.; Sánchez, M. Mesquite gum as a novel reducing and stabilizing agent for modified tollens synthesis of highly concentrated Ag nanoparticles. *Materials* **2016**, *9*, 817. [CrossRef]
23. Vernon, E.J.; Sherman, P. Rheological properties and applications of mesquite tree (prosopis Juliflora) gum. 1. Rheologycal properties of aqueous mesquite gum solutions. *J. Texture stud.* **1980**, *11*, 339–349. [CrossRef]
24. Liu, K.S. Preparation of fatty acid methyl esters for gas-chromatographic analysis of lipids in biological materials. *J. Am. Oil Chem. Soc.* **1994**, *71*, 1179–1187. [CrossRef]
25. García, E.; Higuera, I.; Espinosa, H. Design of fish oil-in-water nanoemulsion by microfluidization. *Innovative Food Sci. Emerg. Technol.* **2017**, *40*, 87–91. [CrossRef]
26. Navidghasemizad, S.; Temelli, F.; Wu, J. Phase separation behavior of egg yolk suspensions after anionic polysaccharides addition. *Carbohydr. Polym.* **2015**, *117*, 297–303. [CrossRef] [PubMed]
27. McClements, D.J. Characterization of emulsion properties. In *Food Emulsions: Principles, Practice, and Technique*, 3rd ed.; CRC Press: Boca Raton, FL, USA, 2016; pp. 624–625. ISBN 978-1-4987-2669-6.

28. Valverde, D.; Laca, A.; Estrada, L.N.; Paredes, B.; Rendueles, M.; Díaz, M. Egg yolk and egg yolk fractions as key ingredient for the development of a new type of gels. *Int. J. Gastronomy Food Sci.* **2016**, *3*, 30–37. [CrossRef]
29. Santipanichwong, R.; Suphantharika, M. Carotenoids as colorants in reduced-fat mayonnaise containing spent brewer's yeast β-glucan as a fat replacer. *Food Hydrocolloids* **2007**, *21*, 565–574. [CrossRef]
30. Laca, A.; Paredes, B.; Rendueles, M.; Díaz, M. Egg yolk granules: Separation, characteristics and applications in food industry. *LWT-Food Sci. Technol.* **2014**, *59*, 1–5. [CrossRef]
31. Strixner, T.; Kulozik, U. Continuous centrifugal fractionation of egg yolk granules and plasma constituents influenced by process conditions and product characteristics. *J. Food Eng.* **2013**, *117*, 89–98. [CrossRef]
32. Martucci, E.; Borges, S. Extraction of cholesterol from dehydrated egg yolk with acetone: Determination of the practical phase equilibrium and simulation of the extraction process. *J. Food Eng.* **1997**, *32*, 365–373. [CrossRef]
33. Puertas, G.; Vázquez, M. Advances in techniques for reducing cholesterol in egg yolk: A review. *Crit. Rev. Food Sci. Nutr.* **2018**, 1–11. [CrossRef]
34. Gils, P.S.; Ray, D.; Sahoo, P.K. Designing of silver nanoparticles in gum arabic based semi-IPN hydrogel. *Int. J. Biol. Macromol.* **2010**, *46*, 237–244. [CrossRef] [PubMed]
35. Navidghasemizad, S.; Temelli, F.; Wu, J. Physicochemical properties of leftover egg yolk after livetins removal. *LWT-Food Sci. Technol.* **2014**, *55*, 170–175. [CrossRef]
36. Beristain, C.I.; Azuara, E.; Garcia, H.S.; Vernon, E.J. Kinetic model for water/oil absorption of mesquite gum (*Prosopis juliflora*) and gum arabic (*Acacia senegal*). *Int. J. Food Sci. Technol.* **1996**, *31*, 379–386. [CrossRef]
37. Zuidam, N.C.; Nedovic, V.A. *Encapsulation Technologies for Active Food Ingredients and Food Processing*; Springer: New York, NY, USA, 2010; pp. 31–100. ISBN 978-1-4419-1007-3.
38. Strixner, T.; Sterr, J.; Kulozik, U.; Gebhardt, R. Structural study on hen-egg yolk high density lipoprotein (HDL) granules. *Food Biophys.* **2014**, *9*, 314–321. [CrossRef]
39. Anton, M. Recent Advances Concerning the Functional Properties of Egg Yolk Low-Density Lipoproteins. In Proceedings of the EPC Proceedings of 12th European Poultry Conference, Verona, Italy, 10–14 September 2006.
40. Turgeon, S.L.; Schmitt, C.; Sanchez, C. Protein–polysaccharide complexes and coacervates. *Curr. Opin. Colloid Interface Sci.* **2007**, *12*, 166–178. [CrossRef]
41. De Kruif, C.; Tuinier, R. Polysaccharide protein interactions. *Food Hydrocolloids* **2001**, *15*, 555–563. [CrossRef]
42. Samant, S.; Singhal, R.; Kulkarni, P.; Rege, D. Protein-polysaccharide interactions: A new approach in food formulations. *Int. J. Food Sci. Technol.* **1993**, *28*, 547–562. [CrossRef]
43. Orozco, J.; Cruz, F.; Ponce, E.; Vernon, E. Mesquite gum: Fractionation and characterization of the gum exuded from Prosopis laevigata obtained from plant tissue culture and from wild trees. *Carbohydr. Polym.* **2003**, *54*, 327–333. [CrossRef]
44. Beyer, J.D. The Development of a Cholesterol-Reduced Egg Yolk Using Solvent Extraction. Ph.D. Thesis, University of Wisconsin-Madison, Madison, WI, USA, 19 August 1991.
45. Huopalahti, R.; Anton, M.; López-Fandiño, R.; Schade, R. *Bioactive Egg Compounds*; Springer: New York, NY, USA, 2007; pp. 5–6. ISBN 978-3-540-37883-9.
46. Li-Chan, E.C.; Kim, H.O. *Structure and Chemical Composition of Eggs. Egg Bioscience and Biotechnology*; John Wiley and Sons Inc.: Vancuber, BC, Canada, 2007; pp. 1–96. ISBN 9780470181249.

Sample Availability: Not available.

molecules

MDPI

Article

Effect of Beta Cyclodextrin on the Reduction of Cholesterol in Ewe's Milk Manchego Cheese

Leocadio Alonso [1,*], Patrick F. Fox [2], María V. Calvo [3] and Javier Fontecha [3]

1 Instituto de Productos Lácteos de Asturias (CSIC), Paseo Río Linares s/n. 33300 Villaviciosa, Asturias, Spain
2 School of Food and Nutritional Sciences, University College Cork (UCC), T12 Y337 Cork, Ireland; pff@ucc.ie
3 Instituto de Investigación en Ciencias de la Alimentación (CSIC-UAM), 28049 Madrid, Spain; mv.calvo@csic.es (M.V.C.); j.fontecha@csic.es (J.F.)
* Correspondence: lalonso@ipla.csic.es; Tel.: +34-985-893-206; Fax: +34-985-892-233

Academic Editors: Lillian Barros and Isabel C. F. R. Ferreira

Received: 21 June 2018; Accepted: 18 July 2018; Published: 20 July 2018

Abstract: Beta-cyclodextrin (β-CD) is a cyclic oligosaccharide consisting of seven glucose units and is produced from starch using cyclodextrin glycotransferase enzymes to break the polysaccharide chain and forming a cyclic polysaccharide molecule. The use of β-CD in food research for reduction of cholesterol is increasing due to its affinity for non-polar molecules such as cholesterol. The aim of this study was to evaluate the feasibility of using β-CD in cholesterol removal from pasteurized ewe's milk Manchego cheese and evaluate the effect on the main components of the milk, lipids, and flavor characteristics. Approximately 97.6% cholesterol reduction was observed in the cheese that was treated using β-CD. Physicochemical properties (fat, moisture and protein) were not changed by the β-CD treatment, except the soluble nitrogen and non-protein nitrogen that showed slight differences after the treatment. The amount of the different components of the lipid fraction (fatty acids, triglycerides and phospholipids) were similar in cheeses treated and not treated with β-CD. Flavor compound and short chain free fatty acids were not mostly significantly influenced by the effect of the β-CD. β-CD molecules are edible and nontoxic and as a result they can be used safely for cholesterol removal processing in cheese manufacturing. Therefore, the present study suggests that β-CD treatment is an effective process for cholesterol removal from Manchego cheese while preserving its properties.

Keywords: beta cyclodextrin; ewe's milk; cheese; Manchego; lipids; cholesterol

1. Introduction

Although dairy products in general have the image of being healthy foods, this is often not the case for products with a high fat content such as butter, cream and cheeses. The World Health Organization and the American Heart Association have recommended that consumers reduce their consumption of saturated fatty acids and cholesterol to lower the risk of coronary heart disease. This advice, coupled with radical opinions, have created a demand for low-cholesterol products [1]. Nowadays, there is a growing interest in the manufacture of cholesterol-reduced dairy products. Food companies have developed many methods to reduce cholesterol, however, most of these methods are relatively nonselective and remove flavor and nutritional components when cholesterol is removed. Moreover, some methods require high investment and operation costs. Methods for reducing cholesterol in foods have been developed including blending with vegetable oils [2,3], extraction by distillation and crystallization [4,5], adsorption with saponin and digitonin [6,7], assimilation of cholesterol by enzymes from microorganisms [8,9] and removal by supercritical carbon dioxide extraction [10,11]. In the last years, several studies have been published describing the use of β-CD in food applications [12–14].

It has been proved that the β-CD molecule can be used as non-toxic and non-digestible molecule to remove cholesterol effectively from milk and dairy products, egg yolk, and lard [15–20] with much less investment and operation costs. β-CD is a cyclic oligosaccharide consisting of seven glucose units and is produced from starch using cyclodextrin glycotransferase enzymes, to break the polysaccharide chains and form cyclic polysaccharide molecules. The molecule of β-CD is doughnut shaped and its central portion is a circular hydrophobic space similar in diameter to a cholesterol molecule, giving the molecule its affinity for non-polar molecules such as cholesterol [21,22].

Manchego cheese is one of the most representative of the Spanish hard cheeses. It is manufactured in the region of Castilla-La Mancha (Spain) using pure ewe's milk from local herds under conditions regulated by an origin appellation. Manchego cheese is a rich in fat (the fat content in the dry cheese is higher than 50%) [23,24], and possesses a characteristic sharp flavor, which increases with the ripening time. Its texture is smooth but consistent, and a few irregular holes randomly distributed in ivory-colored paste. Although the most of investigations for removing cholesterol in milk using β-CD were performed in cow's milk, no investigations have been reported on the effect of β-CD on reduction of cholesterol in ewe's milk. Therefore, the aims of this study was to evaluate the feasibility of the β-CD in cholesterol removal from pasteurized ewe's milk Manchego cheese and its effect on the main components of milk, focusing especially on the lipidic fractions, and flavor characteristics.

2. Results and Discussion

2.1. Gross Composition

Due to the structural characteristics of β-CD and processing conditions used during cholesterol removal with β-CD, it is possible that some of the milk constituents are also entrapped and removed along with cholesterol. Thus, it is important to investigate the compositional changes occurring during the cholesterol removal process in Manchego cheese.

The chemical composition and cholesterol removal rate of control cheese (CC) without β-CD in milk and the experimental cheese (EC) with 1% of β-CD in milk are presented in Table 1. We used 1% β-CD because in previous studies we studied different concentrations of β-CD in the range (0.1 to 1%) for the elimination of cholesterol in cow's milk fat. We found that in that study the optimal concentration to obtain cholesterol reduction higher than 90% was with a β-CD concentration approx. 0.8% [15]. Fat, moisture and protein content showed similar ratio between the CC and the EC (34.50 ± 1.12% vs. 32.51 ± 1.18%; 36.79 ± 1.65% vs. 38.15 ± 1.93%; 25.68 ± 1.04% vs. 25.10 ± 1.16%) respectively. Fat/dry matter and protein/dry matter (%) were slightly lower in EC with β-CD that the CC as a result of the higher moisture content, as suggested in the study by Seon et al. [20]. The lower fat content of the cholesterol reduced cheese than the control might be attained to the less incorporation with casein via a fat protein network, probably due to modification of the casein matrix by β-CD [25]. Soluble nitrogen (SN) and non-protein nitrogen (NPN) showed differences ($p \leq 0.05$) between CC and EC cheese (4.76 ± 0.23% vs. 5.79 ± 0.32%; 2.41 ± 0.19% vs. 3.95 ± 0.24%), this could be due to the slight increase in the proteolysis in EC cheese that may reflect a higher peptidase activity in the EC by the influence of the β-CD [26]. During the ripening period proteolysis occurs which is an important biochemical event governing the sensory profile. The insoluble caseins are partially converted into polypeptides and amino acids. Treatment of the milk with β-CD from which cheese is manufactured results in modification of caseins matrix and thus altering the SN and NPN and consequently could be accelerate a little the ripening period of the cheese. The cholesterol removal rate of CC related to EC (195.67 ± 6.03 mg/100 g fat vs. 1.37 ± 0.19 mg/100 g fat) reached a reduction of 97.29% (Figure 1). Similar cholesterol removal were also found by Kwak et al. [27] in a study of removal of cholesterol from Cheddar cheese and Kin et al. [28] in blue cheese using β-CD. The remain β-CD showed also differences ($p \leq 0.05$) between CC and EC with value of 0.31%. It confirms that cholesterol removal by β-CD does not affect the proximate chemical composition of Manchego ewe's milk cheese.

Table 1. Gross composition of the control and the experimental Manchego cheese by the effect of the β-CD.

Parameter	CC	EC	REE
Fat (%)	34.50 ± 1.12 [a]	32.51 ± 1.18 [a]	5.77
Moisture (%)	36.79 ± 1.65 [a]	38.15 ± 1.93 [a]	3.70
Protein (%)	25.68 ± 1.04 [a]	25.10 ± 1.16 [a]	8.77
SN (% as protein)	4.76 ± 0.23 [a]	5.79 ± 0.32 [b]	2.26
NPN (% as protein)	2.41 ± 0.19 [a]	3.95 ± 0.24 [b]	6.39
pH	4.87 ± 0.15 [a]	4.85 ± 0.25 [a]	0.41
Cholesterol (mg/100 g fat)	195.67 ± 6.03 [a]	4.72 ± 0.19 [b]	99.30
Cholesterol removal (% fat)	-	97.6 ± 4.56	-
Remain β-CD (%)	-	0.31 ± 0.13	-

CC, control cheese without β-CD in milk; EC, experimental cheese with 1% β-CD in milk; SN, soluble nitrogen (% as protein); NNP, non-protein nitrogen (% as protein); REE (%), relative experimental error; Mean standard deviation (n = 12); [a,b] Different letters in the same row mean significant differences ($p \leq 0.05$).

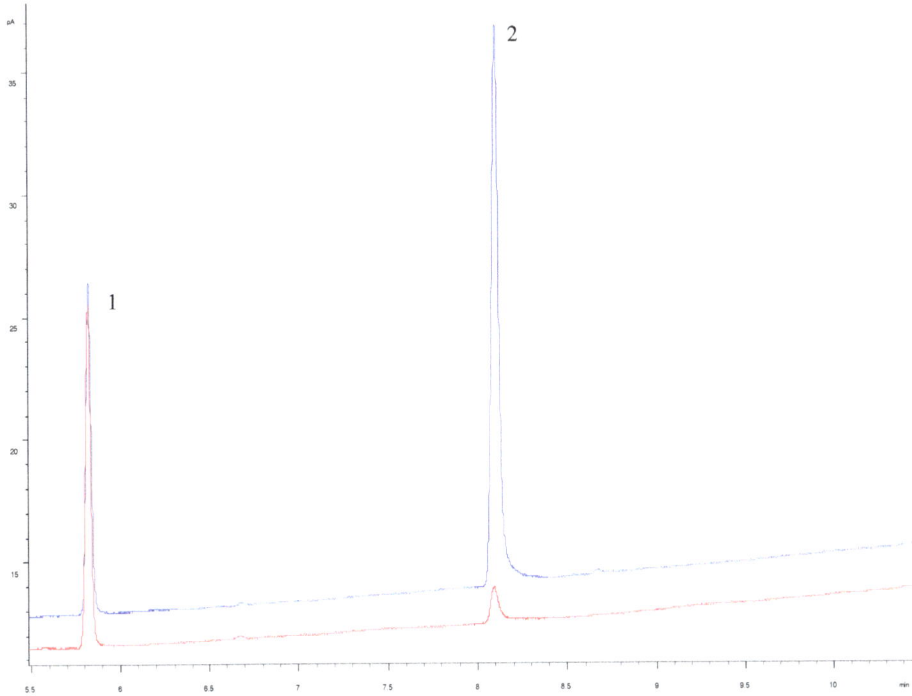

Figure 1. Cholesterol profile by gas chromatography with flame ionization detector in control cheese and experimental cheese with 1% of beta cyclodextrin. Peaks: 1 = 5α-cholestane; 2 = cholesterol. Blue line: control cheese (CC); Red line: experimental cheese (EC).

2.2. *Lipid Characteristics*

Table 2 shows mean values of fatty acids (%) of CC and EC cheeses. Concentrations of individual fatty acids did not exhibit significant differences ($p \leq 0.05$) between fat from the CC and EC cheese with β-CD. There are few reports regarding studies in manufacturing low cholesterol cheeses by β-CD and the effect on the lipidic fraction. Chen et al. [29], using supercritical fluid extraction with carbon dioxide for fractionating milk fat to remove cholesterol, observed that the fractionated milk fat showed considerable differences in fatty acids composition compared with the control cheeses. The amounts for short and medium chain fatty acids reported by these authors were 40% and 10%

less, respectively, in the extracted milk fat compared with the control milk fat. Similar results were found by Gonzalez et al. [10] in a study on solubility of fatty acids in cream from ewe's milk using supercritical fluid carbon dioxide

Table 2. Fatty acids composition (g/100 g fat) from the control and the experimental Manchego cheese by the effect of β-CD.

Fatty Acid	CC	EC	REE
C4:0	2.24 ± 0.19 [a]	2.14 ± 0.26 [a]	4.46
C6:0	1.74 ± 0.06 [a]	1.68 ± 0.05 [a]	3.45
C8:0	1.70 ± 0.05 [a]	1.66 ± 0.08 [a]	2.35
C10:0	5.02 ± 0.15 [a]	4.95 ± 0.13 [a]	1.39
C10:1	0.28 ± 0.03 [a]	0.26 ± 0.07 [a]	7.14
C12:0	3.19 ± 0.11 [a]	3.14 ± 0.18 [a]	1.57
C14.0	9.22 ± 0.84 [a]	9.21 ± 0.51 [a]	0.11
C14:1	0.90 ± 0.03 [a]	0.86 ± 0.06 [a]	4.44
C15:0	0.24 ± 0.02 [a]	0.25 ± 0.05 [a]	4.17
C16:0	27.16 ± 1.52 [a]	27.41 ± 1.18 [a]	0.92
C16:1	0.73 ± 0.12 [a]	0.77 ± 0.17 [a]	5.48
C17:0	0.54 ± 0.07 [a]	0.58 ± 0.07 [a]	7.41
C18:0	13.39 ± 0.55 [a]	13.59 ± 0.52 [a]	1.49
C18:1t	2.62 ± 1.13 [a]	2.65 ± 0.23 [a]	1.15
C18:1c	23.28 ± 0.35 [a]	22.93 ± 1.16 [a]	1.50
C18:2	3.9 ± 0.08 [a]	3.66 ± 0.24 [a]	6.15
C18:3	0.39 ± 0.08 [a]	0.40 ± 0.05 [a]	2.56
C18:2 (c9t11)	0.96 ± 0.06 [a]	0.97 ± 0.06 [a]	1.04

CC, control cheese without β-CD in milk; EC, experimental cheese with 1% β-CD in milk; REE (%), relative experimental error; Mean standard deviation (n = 12); [a] Different letters in the same row mean significant differences ($p \leq 0.05$).

In our study using β-CD for removing cholesterol and the effect on the composition for short-(C4 to C8) (2.24 ± 0.19% vs. 2.14 ± 0.26%; 1.74 ± 0.06% vs. 1.68 ± 0.05%; 1.70 ± 0.05% vs. 1.66 ± 0.08%), medium-(C10 to C12) (5.02 ± 0.15% vs. 4.95 ± 0.13%; 3.19 ± 0.11% vs. 3.14 ± 0.18%), and long chain-(C14 to C18) (9.22 ± 0.84% vs. 9.21 ± 0.51%; 27.16 ± 1.52% vs. 27.41 ± 1.18%; 13.39 ± 0.55% vs. 13.59 ± 0.52%) fatty acids were no significantly different ($p \leq 0.05$) between groups respectively. Similar results were found by Alonso et al. [15], in their study of using β-CD to decrease the level of cholesterol in milk fat.

Table 3 shows the mean values of the individual groups of triglyceride composition of fats of the CC and EC cheeses. The triglycerides of the fat cheese were resolved into 16 groups from C26 to C54. Each group is the sum of the different molecular species of triglycerides that contain the same number of carbon atoms. None of differences between control and experimental cheese with β-CD were observed ($p \leq 0.05$), in the $\sum$ short-(C24–C32) (1.76 ± 0.20% vs. 1.84 ± 0.56%), $\sum$ medium-(C34–C48) (77.45 ± 0.85% vs. 77 ± 0.91%), and $\sum$ long-(C50–C54) (4.53 ± 0.39% vs. 4.43 ± 0.51%). No prior research studies have been reported on the triglycerides in cheeses treated with β-CD for removing cholesterol. Chen et al. [29], Bhaskar et al. [30], and Gonzalez et al. [10], using different techniques, found variations in triglycerides composition between control and experimental milks. The supercritical fluid extraction methods used by these investigators may have caused some variation in triglycerides composition because the triglycerides were removed by solvent extraction, that could have selectively extracted some triglycerides better than other.

Table 3. Triglycerides composition (g/100 g fat) from the control and the experimental Manchego cheese by the effect of the β-CD.

Triglyceride	CC	EC	REE
C24	0.33 ± 0.06 [a]	0.32 ± 0.08 [a]	3.03
C26	0.88 ± 0.09 [a]	0.80 ± 0.05 [a]	9.09
C28	1.64 ± 0.15 [a]	1.54 ± 0.13 [a]	6.09
C30	2.42 ± 0.23 [a]	2.47 ± 0.21 [a]	2.07
C32	3.54 ± 0.40 [a]	3.25 ± 0.39 [b]	8.59
C34	4.89 ± 0.38 [a]	5.04 ± 0.48 [a]	3.07
C36	7.21 ± 0.66 [a]	7.04 ± 0.54 [a]	2.36
C38	10.66 ± 1.11 [a]	10.65 ± 1.30 [a]	0.09
C40	17.35 ± 1.32 [a]	17.89 ± 1.32 [a]	3.11
C42	16.02 ± 1.40 [a]	16.17 ± 1.50 [a]	0.94
C44	8.83 ± 0.77 [a]	8.13 ± 0.66 [a]	8.13
C46	7.14 ± 0.62 [a]	7.04 ± 0.52 [a]	1.40
C48	5.35 ± 0.55 [a]	5.71 ± 0.49 [a]	6.73
C50	4.28 ± 0.35 [a]	4.39 ± 0.51 [a]	2.57
C52	4.57 ± 0.39 [a]	4.31 ± 0.56 [a]	5.69
C54	4.78 ± 0.43 [a]	4.58 ± 0.45 [a]	4.18

CC, control cheese without β-CD in milk; EC, experimental cheese with 1% β-CD in milk; REE (%), relative experimental error. Mean standard deviation (n = 12); [a,b] Different letters in the same row mean significant differences ($p \leq 0.05$).

In relation to the phospholipid fraction, Table 4 shows the composition in phospholipids (%) of CC and EC Manchego cheeses. Analysis of variance did not reveal any significant difference ($p \leq 0.05$) in relative composition of the different phospholipid classes among between groups of cheeses related to the total phospholipids. Phosphatidylethanolamine (42.42 ± 4.05% vs. 38.25 ± 1.40%) was the most predominant phospholipid followed by phosphatidylcoline (27.23 ± 0.74% vs. 1.04 ± 2.21%) and sphyngomyelin (26.70 ± 5.32% vs. 25.20 ± 1.53%). Similar results were obtained by Alonso et al. [31], in a study of the effect of the β-CD on phospholipids of the milk fat in pasteurized milk. These three species of phospholipids represented more than 80% of the total phospholipids in dairy products. One of the reasons why the β-CD did not affect to these components of the milk fat could be based on the fact that β-CD specifically forms an inclusion complex with cholesterol. The central cavity of β-CD is hydrophobic, giving the molecule its affinity for non-polar molecules such as cholesterol. The radius of the cavity is such as to accommodate a cholesterol molecule almost exactly, conferring the highly specific nature of the β-CD ability to form an inclusion complex with cholesterol. They are therefore accessible to β-CD in the aqueous phase forming the insoluble inclusion complex which can be removed by centrifugation [15].

Table 4. Phospholipids composition of the control and the experimental Manchego cheese by the effect of the β-CD.

Phospholipids	CC	EC	REE
Total PLs (mg/100 g fat)	0.12 ± 0.03 [a]	0.11 ± 0.03 [a]	8.83
PE (% of PL)	42.42 ± 4.05 [a]	38.25 ± 1.40 [a]	9.83
PI (% of PL)	1.93 ± 1.31 [a]	2.46 ± 0.62 [a]	0.27
PS (% of PL)	1.75 ± 0.53 [a]	3.21 ± 1.94 [a]	1.20
PC (% of PL)	27.23 ± 0.74 [a]	31.04 ± 2.21 [a]	0.14
SM (% of PL)	26.70 ± 5.32 [a]	25.20 ± 1.53 [a]	0.06

CC, control cheese without β-CD in milk; EC, experimental cheese with 1% β-CD in milk; PLs, Phospholipids; PE, phosphatidylethanolamine; PI, phosphatidylinositol; PS, phosphatidylserin; PC, phosphatidylcoline; SM, sphyngomyelin; REE (%), relative experimental error; Mean standard deviation (n = 12); [a] Different letters in the same row mean significant differences ($p \leq 0.05$).

2.3. Flavor Characteristics

Flavor compounds isolated from CC and EC cheeses with three months of ripening are shown in Table 5. A total of 13 flavor compounds were isolated in both cheeses and some differences were observed between samples. In all cheeses, 13 flavor compound were detected, including five ketones, three aldehydes and five alcohols. Analysis of the variance did not reveal any significant difference in the total amount of ketones (2505.61 ± 36.40 ppm vs. 2314.95 ± 26.07 ppm) aldehydes (1139.63 ± 18.68 ppm vs. 1377.45 ± 24.94 ppm) and alcohols (4235.77 ± 17.13 ppm vs. 4808.87 ± 23.79 ppm) between CC and EC cheeses. 3-methylbutanal (1121.42 ± 48.32 vs. 1358.96 ± 70.32) and ethanol (4107.60 ± 62.30 ppm vs. 4685.30 ± 95.79 ppm) were the only compounds significantly different ($p \leq 0.05$) found in CC and EC cheeses, ethanol production was the highest among flavor compounds measured, similar to those found by Kwak et al. [27] in Cheddar cheese treated with β-CD. In the study by Jeon et al. [32] of the removal of cholesterol of cream cheese by β-CD, no differences were found in the overall flavour compounds in the treated cheese compared to the regular cream cheese.

Ketones with odd carbon number have typical odor characteristics and low perception thresholds. These compounds are formed by β-oxidation and decarboxylation of fatty acids. It is known that aldehydes are not the major compounds in cheeses, as they are rapidly converted to alcohols or their corresponding acids. Branched chain aldehydes like 3-methylbutanal are formed by the catabolism of branched chain amino acids by an aminotransferase [33], and this compound was the only statistically different ($p \leq 0.05$) in EC comparing with the CC cheese together with the ethanol. 3-Methylbutanal is an intermediate in the catabolism of leucine. Lactic acid bacteria present in the cheese together with some yeast are involved in the formation of 3-methylbutanal and alcohols (ethanol) during ripening of the cheese [33], and in our study there is a high proteolysis in the EC with a high content in non-protein nitrogen (include aminoacids as leucine), and this is the main reason why the content of 3-methylbutanal is higher in EC than in the CC. Ethanol was also higher in EC than in CC, due that this compound is also an intermediate in the catabolism of aminoacids and in the fermentation of the residual lactose by the yeast and lactic acid bacteria [32].

Table 5. Volatile compounds (ppm) of the control and the experimental Manchego cheese by the effect of the β-CD.

Compounds	CC	EC	REE
Ketones			
2-Propanone	420.38 ± 32.39 ^{a}	381.05 ± 26.89 ^{a}	9.03
2-Butanone	27.65 ± 4.51 ^{a}	25.16 ± 4.21 ^{a}	9.01
2,3-Butanedione	1271.54 ± 48.45 ^{a}	1145.81 ± 56.38 ^{a}	9.89
2-Heptanone	562.30 ± 29.49 ^{a}	512.18 ± 22.78 ^{a}	8.91
3-Hydroxy-2-butanone	186.12 ± 18.66 ^{a}	248.70 ± 20.09 ^{a}	3.34
Aldehydes			
3-Methylbutanal	1121.42 ± 48.32 ^{a}	1358. 96 ± 70.32 ^{b}	21.18
Hexanal	14.16 ± 6.50 ^{a}	13.54 ± 4.09 ^{a}	4.38
Nonanal	4.55 ± 1.21 ^{a}	4.95 ± 1.19 ^{a}	8.79
Alcohols			
2-Propanol	13.50 ± 3.56 ^{a}	12.64 ± 3.70 ^{a}	6.31
Ethanol	4107.60 ± 62.30 ^{a}	4685.30 ± 95.79 ^{b}	14.07
2-Methyl-1-propanol	49.18 ± 7.11 ^{a}	45.66 ± 7.80 ^{a}	7.16
2-Butanol	29.31 ± 6.85 ^{a}	26.69 ± 5.56 ^{a}	8.94
2-Heptanol	36.18 ± 6.04 ^{a}	39.57 ± 6.12 ^{a}	9.37

CC, control cheese without β-CD in milk; EC, experimental cheese with 1% β-CD in milk; REE (%), relative experimental error; Mean standard deviation (n = 12); ^{a,b} Different letters in the same row mean significant differences ($p \leq 0.05$).

The amounts of short chain free fatty acids (SCFFAs), acetic, propionic, butyric and caproic acids in the control and cholesterol reduced cheeses are shown in Table 6. There was no significant difference ($P \leq 0.05$) in total and individual amounts of FFAs (149.14 ± 5.86 ppm vs. 154.70 ± 6.12 ppm) at the end of the three month ripened, between the CC and EC cheeses. These results indicate that there was no differences in the amounts of short chain FFAs between the control and the cholesterol reduced cheese made by β-CD. Similar results in the amount of short chain SCFFAs in the control and cholesterol reduced process and cheddar cheese made by β-CD were found by [27,34]. The release of butyric and caproic acid at the three months ripening contribute to the backbone characteristics of Manchego cheese [35,36].

Table 6. Short chain free fatty acids (SCFFA) (ppm) of the control and the experimental Manchego cheese by the effect of the β-CD.

SCFFA	CC	EC	REE
Acetic	92.91 ± 7.19 ^{a}	95.06 ± 6.19 ^{a}	2.31
Propionic	35.28 ± 5.65 ^{a}	38.36 ± 4.96 ^{a}	8.73
Butyric	17.10 ± 3.96 ^{a}	17.32 ± 3.60 ^{a}	1.29
Caproic	13.85 ± 2.52 ^{a}	13.96 ± 3.12 ^{a}	0.79
Total	159.14 ± 5.86 ^{a}	164.70 ± 6.12 ^{a}	3.49

CC, control cheese without β-CD in milk; EC, experimental cheese with 1% β-CD in milk; REE (%), relative experimental error; Mean standard deviation (n = 12); ^{a} Different letters in the same row mean significant differences ($p \leq 0.05$).

The sensory attributes of CC and EC cheese for a maximum of 5 score are shown in Table 7. No significant differences ($p \leq 0.05$) were observed in flavor (3.32 ± 0.44 vs. 3.07 ± 0.89), arome (3.59 ± 0.49 vs. 3.28 ± 0.83), color (3.69 ± 0.68 vs. 3.49 ± 0.73) and acceptability (3.45 ± 0.60 vs. 3.22 ± 0.76) between CC and EC cheese. These attributes are correlated with the production of SCFFAs acids and methyl ketones during ripening (3 months) in the CC and EC cheese, which were not affected the treatment with β-CD. Texture was significantly different ($p \leq 0.05$) in the EC with respect to the CC (3.70 ± 0.57 vs. 3.29 ± 0.72). This could be due than in the experimental cheese resulted in a higher proteolysis due to a greater peptidase activity in the cholesterol reduced cheese, that is higher in the EC by the treatment with β-CD and an slight high moisture in the cheese treated, increased by β-CD, which resulted in a slow drainage, as suggested Metzge et al. [37]. The overall preference was maintained over the ripening period of three months and no differences were found between CC and EC for flavor, aroma, color and acceptability. This study indicates that even though some differences were observed, most of the sensory characteristics and overall preferences were comparable to those of the control and three months cheese ripened treated with β-CD. Therefore, we may suggest the possibility of cholesterol reduced Manchego cheese manufactured by β-CD.

Table 7. Sensory analysis of the control and the experimental Manchego cheese by the effect of the β-CD. Flavor, arome, color, texture and acceptability were evaluated on a five point scale (1 = poor to 5 = excellent).

Attribute	CC	EC	REE
Flavor	3.32 ± 0.44 ^{a}	3.07 ± 0.89 ^{a}	7.53
Aroma	3.59 ± 0.49 ^{a}	3.28 ± 0.83 ^{a}	8.63
Color	3.69 ± 0.68 ^{a}	3.49 ± 0.73 ^{a}	5.42
Texture	3.70 ± 0.57 ^{a}	3.29 ± 0.72 ^{b}	11.12
Acceptability	3.45 ± 0.60 ^{a}	3.22 ± 0.76 ^{a}	6.65

CC, control cheese without β-CD in milk; EC, experimental cheese with 1% β-CD in milk; REE (%), relative experimental error; Mean standard deviation (n = 12); ^{a,b} Different letters in the same row mean significant differences ($p \leq 0.05$).

3. Materials and Methods

3.1. Chemicals

α-Cyclodextrin (α-CD), β-cyclodextrin (β-CD) and all reagents grade were supplied by Sigma (St. Louis MO, USA). Deionized water was prepared by a water purification system (Millipore Co., Burlington, MA, USA).

3.2. Manchego Manufacture

Ewe's milk was previously treated with 1% β-CD by the method described by Alonso et al. [15]. One hundred L volumes of whole pasteurized milk (74 °C for 15 s) milk containing 1.0% *wt/vol* of β-CD were placed in a cold room at 4 °C and mixed by a stirrer (430 rcf) during 30 min. After mixing, the treated milk was left standing overnight at 4 °C (to allow time for binding the cholesterol) and precipitate the cholesterol-β-CD complex at the bottom of the tank. The upper layer without the complex was separated for making the cheeses. Manchego cheese was made by the procedure described by Fernández-García et al. [35]. Cheeses were ripened at a temperature of 12–14 °C with relative humidity of 85–90% during 3 months. The cheese-making experiment was carried out in triplicate for control and cheeses treated with 1% of β-CD.

3.3. Gross Composition

Fat, moisture and protein contents and nitrogen fractions were determined using the method by Alonso et al. [38].

3.4. Beta Cyclodextrin Analysis

β-CD was analysed by the method proposed by Alonso et al. [39]. Ten g of cheese was mixed with 5 mg of α-CD dissolved in one mL of water (internal standard for quantitative analysis). After shaking for 2 min at 40 °C it was centrifuged at room temperature for 40,000 rpm for 30 min, the upper layer was separated and filtered through a 0.45 μm membrane (Millipore Co.). A 30 μL aliquot of the supernatant spiked with the internal standard were transferred to the autosampler. A 10 μL aliquot of the supernatant was injected onto column for HPLC analysis.

The apparatus used for HPLC analysis was a Waters Alliance 2695 separation module coupled to a 410 refractive index (RI) detector, data acquisition and analysis were performed using the Empower 2 chromatography data software (Waters, Milford, MA, USA). Separation was carried out on YMC ODS-AQ column (Teknochroma, Miami, FL, USA). The mobile phase composition was a mixture of methanol and water (7:93) in isocratic condition at a flow rate 1 mL/min. The standard solutions were prepared in water to establish elution time and the quantification of β-CD was conducted by comparing sample peak area of β-CD with α-CD as the internal standard.

3.5. Lipid Extraction

Lipids were extracted from samples following a procedure described by an International Standard Method for milk and milk products [40]. Briefly, it consisted of an addition of an ammonia-ethanol solution to a test portion followed by lipid extraction using diethyl ether and hexane. Then, the upper layer was removed, and the solvent completely evaporated. The lipid extracts obtained were placed into amber glass vials, flushed with a stream of nitrogen and stored at −20 °C until analyzed.

3.6. Determination of Cholesterol

The technique chosen for cholesterol determination was as described by Alonso et al. [41] using direct injection of milk fat by capillary gas chromatography (GC). Approximately 30 mg anhydrous milk fat and 0.1 mL 5-α-cholestane as internal standard (3.5 mg/mL in hexane) was dissolved in 1 mL of hexane; 0.5 μL of the resulting solution was injected for GC analysis. For GC analysis for free cholesterol by this direct method we used an Agilent Technology 6890 chromatograph (Palo Alto, CA, USA) equipped with flame ionization detector. Analyses were performed using a HP-5 fused silica capillary column (30 m × 0.32 mm i.d. 0.25 μm thickness). Experimental chromatographic conditions were: He carrier gas at 17 psi head pressure; initial column temperature 280 °C, held for 1 min, increased to 355 °C at 3 °C/min. Injector temperature 350 °C and detector temperature was 360 °C. Peak identification was done by comparison of relative retention times with retention times of standards. Quantification of cholesterol was conducted by comparing sample peak area with of the 5 α-cholestane internal standard. The percentage of cholesterol reduction in milk fat was calculated by the formula [(100 − amount of cholesterol in milk fat) × 100]/amount of cholesterol in untreated milk).

3.7. Fatty Acids and Triglycerides Analysis

Fatty acids methyl esters (FAMES) were prepared by alkaline catalyzed methanolysis of the extracted lipids using 2 N KOH in methanol. The FAMES were analyzed on an Agilent Technology 6890 chromatograph (Palo Alto, CA, USA) with FID detector. Fatty acids were separated using CP-Sil 88 fused-silica capillary column (50 m × 0.25 mm i.d. × 0.2 μm film thickness, Chrompack, CA, USA) using the method described by Alonso et al. [42]. GC analysis of triglycerides by direct injection was performed on an Agilent gas chromatograph 6890 (Palo Alto, CA, USA) equipped with flame ionization detector. Analyses were performed using a WCOT fused silica capillary column (25 m × 0.25 mm × 0.1 μm film thickness) coated with OV 17 TRI (J.W. Scientific, Polson, CA, USA) using the method described by Alonso [43].

3.8. Phospholipids Analysis

Extractions of cheese fat were carried out with an Accelerated Solid Extraction ASE-200 extractor (Dionex Corp., Sunnyvale, CA, USA) using 2 g of freeze-dried cheese sample that was well mixed with 2 g of sea sand and loaded into a stainless steel extraction cell covered with filters on both sides. For the maximum cheese fat yield, the extraction included the use of dichloromethane-methanol solution (2:1, *vol*/*vol*) as solvent mixture and 10.3 MPa of pressure as fixed conditions described by Castro-Gómez et al. [44].

Separation of lipid classes was accomplished in an HPLC system (model 1260; Agilent Technologies Inc., Santa Clara, CA, USA) coupled with an evaporative light scattering detector (SEDEX 85 model; Sedere SAS, Alfortville CEDEX, France) using prefiltered compressed air as the nebulizing gas at a pressure of 350 kPa at 60 °C; the gain was set at 3. Two columns in series (250 × 4.5 mm Zorbax Rx-SIL column with 5-μm particle diameter; Agilent Technologies Inc.) and a precolumn with the same packing were used [44].

3.9. Analysis of Volatile Compounds

Analysis of volatile fraction was performed by headspace gas chromatographic mass spectrometric (GC-MS) method described by Alonso et al. [45]. To 10 g of previously homogenized cheese, 80 μL of aqueous solution of propionic acid ethyl methyl ester (1.14 mg/mL) as internal standard and anhydrous sodium sulphate (10 g) to retain water 176 were added. Individual standard dilutions in aqueous solution were prepared and were stored hermetically in sealed vials at 20 °C until their use. Prior to be analyzed in a static headspace apparatus (Model HSS 19395; Hewlett Packard), the samples were maintained at 80 °C for 60 min until the sample and gaseous phase reached the thermodynamic equilibrium. Apparatus was programmed as follows: 5 s pressurization, 18 equilibrium and filling and 2 min for injection. Helium was employed as carrier gas at a flow 18 rate of 17.5 mL/min. A Hewlett Packard GC Model 5890 coupled to selective MS Model 5972 was employed for volatile compounds analysis. Samples were injected in the split mode (split 18 rate of 7:1) on a capillary silica column with polyethylene glycol (HP Innovas, 60 m, 0.25 mm 18 ID, 0.25 μm film thickness, Hewlett Packard). Helium was used as carrier gas, at a flow rate of 18 36.5 cm/s. The column temperature program was: 33 °C for 5 min, increase at 1 °C/min up to 38 °C and then at 7 °C/min up to 210 °C, and held for

10 min. Injection was carried out at 200 °C and the interface line of MS at 280 °C. Electronic ionisation energy and photomultiplier voltage 18 were 70 eV and 1647 V, respectively.

3.10. Short Chain Free Fatty Acids

For the analysis of SCFFAs, cheese sample (1 g) was homogenized in 20 mL of distilled water, centrifuge at 10,000 rpm for 10 min and filtered by 0.40 μm filter. A Hewlett-Packard model 5890 A equipped with a flame ionization detector on a capillary silica column (HP FFAP, 30 m, 0.25 mm ID, 0.25 μm film thickness, Agilent J & W) was used for analysis. Quantitative analysis were done using 2-ethylbutanoic acid as internal standard.

3.11. Sensory Analysis

Samples of Manchego cheese were cut in slices of approximately 8 × 8 cm of a thickness of approx. 1 cm and placed on white plates. Samples were tempered at ambient temperature (20 ± 2 °C) and then presented to the panelists. Twenty two trained sensory panelists from the members of the research Institute which, trained in sensory analysis of cheese, evaluated randomly coded cheeses. The testing conditions of the room for the sensory analysis were in conformity with the ISO requirements [46]. Flavor, aroma, color, texture and acceptability were evaluated on a five point scale (1 = poor to 5 = excellent).

3.12. Statistical Analysis

Experimental data were treated by analysis of variance (ANOVA) using the statistical software SAS (version 8.02, SAS Institute Inc., Cary, NC, USA). Differences among treatments were determined by statistical analysis using a Student t-test where $p \leq 0.05$ was considered statiscally significant.

4. Conclusions

Approximately 97.6% cholesterol reduction was observed in the cheese that was treated using β-CD. Physicochemical properties (fat, moisture and protein) were not changed by the β-CD treatment, except the NS and NNP that showed slight differences attributed to the treatment. The amount of the different components of the lipid fraction (fatty acids, triglycerides and phospholipids) were similar in both, treated and untreated cheese with β-CD. Flavor compounds and short chain free fatty acids were mostly not significantly influenced by the β-CD. Although, the β-CD molecules are edible and nontoxic and a results they can be used safely for cholesterol removal processing. Therefore, the present study suggested that the treatment with the β-CD was an effective process for cholesterol removal from Manchego cheese, while preserving its nutritional properties. Further studies to evaluate the effect of the intake of the control and low cholesterol Manchego cheeses on the concentration of serum cholesterol would be of interest.

Author Contributions: L.A. conceived, designed the experimental and performed the experiments. P.F.F. revised the manuscript. M.V.C. designed the experimental and performed the experiments. J.F. designed the experimental and performed the experiments.

Funding: Authors thank to the Ministry of Economy and Competitiveness from Spain (grant number AGL-2014-56464; AGL-2017-84878).

Acknowledgments: The authors thanks to the Monte Toledo dairy cheese factory (Toledo, Castilla-La Mancha. Spain).

Conflicts of Interest: The authors declare that there are no conflicts of interest.

References

1. Hansel, B.; Nicolle, C.; Lalanne, F.F.; Brucket, E. Effect of low-fat, fermented milk enriched with plant sterols on serum lipid profile and oxidative stress in moderate hypercholesterolemia. *Am. J. Clin. Nutr.* **2007**, *86*, 790–796. [CrossRef] [PubMed]

2. Hariharan, K.; Kurien, S.; Rao, S.V. Effect of supplementation of milk fat with peanut oil on blood lipids and lipoproteins in infants. *Int. J. Food Sci. Nutr.* **1995**, *46*, 309–317. [CrossRef] [PubMed]
3. Krause, A.J.; Lopetcharat, K.; Drake, M.A. Identification of the characteristics that drive consumer liking of butter. *J. Dairy Sci.* **2007**, *90*, 2091–2102. [CrossRef] [PubMed]
4. Arul, J.; Boudreau, A.; Makhlouf, J.; Tardif, R.; Bellavia, T. Fractionation of anhydrous milk fat by short path destillation. *J. Am. Oil Chem. Soc.* **1988**, *65*, 1642–1646. [CrossRef]
5. Arul, J.; Boudreau, A.; Makhlouf, J.; Tardif, R.; Bellavia, T. Distribution of cholesterol in milk fat fractions. *J. Dairy Res.* **1988**, *55*, 361–371. [CrossRef] [PubMed]
6. Micich, T.J. Behaviours of polymers supported digitonin with cholesterol in the absence and presence of butter oil. *J. Agric. Food Chem.* **1990**, *38*, 1839–1843. [CrossRef]
7. Micich, T.J.; Foglia, T.A.; Holsinger, V.H. In vitro studies on saponin-vitamin complexation. *J. Agric. Food Chem.* **1992**, *40*, 1321–1325. [CrossRef]
8. Gilliland, S.E.; Nelson, C.R.; Maxwell, C. Assimilation of cholesterol by *Lactobacillus acidophilus*. *Appl. Environ. Microbiol.* **1985**, *49*, 377–389. [PubMed]
9. Walkek, K.W.; Gilliland, S.E. Relationships among bile tolerance, bile salt deconjugation, and assimilation of cholesterol by *Lactobacills acidophilus*. *J. Dairy Sci.* **1993**, *76*, 956–961. [CrossRef]
10. Gonzalez-Hierro, M.T.; Ruiz-Sala, P.; Alonso, L.; Santamaria, G. Extraction of ewe's milk cream with supercritical carbon dioxide. *Z. Lebensm. Unters. Forsch.* **1995**, *200*, 297–300. [CrossRef] [PubMed]
11. Sangbin, L.; Mi-Kyung, J.; Hae-Soo, K. Cholesterol removal from milk fat by supercritical carbon dioxide extraction in coupled with adsorption. *Korean J. Food Sci. Technol.* **1998**, *30*, 574–580.
12. Kwak, H.S.; Ahn, J. Optimization cholesterol removal in cream using β-cyclodextrin and response surface methodology. *J. Food Sci.* **1999**, *64*, 629–632.
13. Lee, D.K.; Ahn, J.; Kwak, H.S. Cholesterol removal from homogenized milk with β-cyclodextrin. *J. Dairy Sci.* **1999**, *82*, 2327–2330. [CrossRef]
14. Byung-Sung, P.; Aera, J. Dietary β-cyclodextrin reduces the cholesterol levels in meats and backfat of fibishing pigs. *J. Sci. Food Agric.* **2008**, *88*, 813–818.
15. Alonso, L.; Cuesta, P.; Fontecha, J.; Juárez, M.; Gilliland, S.E. Use of beta cyclodextrin to decrease the level of cholesterol in milk fat. *J. Dairy Sci.* **2009**, *92*, 1941–1946. [CrossRef] [PubMed]
16. Park, B.S.; Kang, H.K.; Jang, A. Influence of feeding β-cyclodextrin hens on the egg production and cholesterol content of egg yolk. *Asian Australas. J. Anim. Sci.* **2005**, *18*, 835–840. [CrossRef]
17. Yen, G.C.; Tsui, L.T. Cholesterol removal from a lard water mixture with β-cyclodextrin. *J. Food Sci.* **1995**, *60*, 561–564. [CrossRef]
18. Alonso, L.; Fontecha, J.; Cuesta, P.; Juárez, M.; Gilliland, S.E. Industrial application of beta-cyclodextrin for manufacturing low colesterol butter. *Milchwiss.-Milk Sci. Int.* **2010**, *65*, 36–37.
19. Lee, S.J.; Hwang, J.H.; Lee, S.; Ahn, J.; Kwak, H.S. Property changes and cholesterol lowering effects in evening primose oil-enriched and cholesterol-reduced yoghurt. *Int. J. Dairy Technol.* **2007**, *60*, 22–30. [CrossRef]
20. Seon, K.H.; Ahn, J.; Kwak, H.S. The accelerated ripening of cholesterol-reduced Cheddar cheese by cross-linked β-cyclodextrin. *J. Dairy Sci.* **2009**, *92*, 49–57. [CrossRef] [PubMed]
21. Reineccius, T.A.; Reineccius, G.A.; Peppard, T.L. Potential for β-cyclodextrin as partial fat replacer in low-fat foods. *J. Food Sci.* **2004**, *69*, 334–341. [CrossRef]
22. Szente, L.; Szejtli, J. Cyclodextrin as food ingredients. *Trends Foods Sci. Technol.* **2004**, *15*, 137–142. [CrossRef]
23. Haenlein, G.F.W. The nutritional value of sheep milk. *Int. J. Anim. Sci.* **2001**, *16*, 253–268.
24. Park, Y.W.; Juárez, M.; Ramos, M.; Haenlein, G.F.W. Physico-chemical characteristeics of goat and sheep milk. *Small Rumin. Res.* **2007**, *68*, 88–113. [CrossRef]
25. Ibrahim, A.H.; Soryal, K.A. Effect of using cross linked β-cyclodextrin on physicochemical properties of Domiati goat's cheese. *J. Food Dairy Sci.* **2014**, *5*, 581–600.
26. Smit, G.; Verheul, A.; van Kranenburg, R.; Ayad, E.; Siezen, R.; Engels, W. Cheese flavor development by enzymatic conversions of peptides and amino acids. *Food Res. Int.* **2000**, *33*, 153–160. [CrossRef]
27. Kwak, H.S.; Jung, C.S.; Seok, J.S.; Ahn, J. Cholesterol removal and flavor development in cheddar cheese. *Asian Australas. J. Anim. Sci.* **2003**, *16*, 409–416. [CrossRef]
28. Kim, H.Y.; Bae, H.Y.; Kwak, H.S. Development of cholesterol-reduced Blue cheese made by crosslinked β-cyclodextrin. *Milchwiss* **2008**, *63*, 53–56.

29. Chen, H.; Schwartz, S.; Spanos, G.A. Fractionation of butter oil by supercrital carbon dioxide. *J. Dairy Sci.* **1992**, *75*, 2659–2669. [CrossRef]
30. Bhashar, A.R.; Rizvi, S.S.H.; Sherbon, J.W. Anhydrous milk fat fractionation with continous countercurrent supercritical carbon dioxide. *J. Food Sci.* **1993**, *56*, 748–752. [CrossRef]
31. Alonso, L.; Fontecha, J.; Cuesta, P. Effect of beta-cyclodextrin on phospholipids and cholesterol of the milk fat globule membrane. *J. Adv. Dairy Res.* **2015**, *3*, 3–7.
32. Jeon, S.S.; Joo, S.; Ganesan, P.; Kwak, H.S. Comparative study of flavor, texture, and sensory in cream cheese and cholesterol removed cream cheese. *Food Sci. Biotechnol.* **2012**, *21*, 159–165. [CrossRef]
33. Smit, G.; Smit, B.A.; Engel, J.M. Flavour formation by lactic acid bacteria and biochemical flavour profiling of cheese products. *FEMS Microbiol. Rev.* **2005**, *29*, 591–610. [CrossRef] [PubMed]
34. Kim, S.Y.; Park, S.Y.; Ahn, J.; Kwak, H.S. Properties of cholesterol reduced block type process cheese made by crosslinked β-CD-cyclodextrin. *Korean J. Food Anim. Resour.* **2008**, *28*, 463–469. [CrossRef]
35. Fernández García, E.; López Fandiño, R.; Alonso, L. Effect of food grade enzyme preparation from *Aspergillus oryzae* on free fatty acid realease in Manchego type cheese from ovine and bovine milk. *Eur. Food Res. Tecnol.* **1994**, *199*, 262–264.
36. Fernández-García, E.; López-Fandiño, R.; Alonso, L.; Ramos, M. The use of lipolytic and proteolytic enzymes in the manufacture of Manchego cheese type cheese from ovine and bovine milk. *J. Dairy Sci.* **1994**, *77*, 2139–2149. [CrossRef]
37. Metzge, N.S.; Mistry, V.V. A new approch using homogenization of cream in the manufacture of reduced Cheddar cheese. Manufacture, composition, and yield. *J. Dairy Sci.* **1994**, *77*, 3506–3515. [CrossRef]
38. Alonso, L.; Juárez, M.; Ramos, M.; Martín Alvarez, P.J. Overall composition, nitrogen fraction and fat characteristics of Cabrales cheese during ripening. *Eur. Food Res. Technol.* **1987**, *185*, 481–486. [CrossRef]
39. Alonso, L.; Cuesta, P.; Gilliland, S.G. Development of a HPLC method for determining residual beta-cyclodextrin in milk, cream and butter as a novel prebiotic in dairy foods. *Chromatographia* **2008**, *69*, 1089–1092.
40. ISO-IDF. *Milk and Milk Products. Extraction Methods for Lipids and Liposoluble Compounds*; ISO 14152; International Dairy Federation Standard: Brussels, Belgium, 2001; p. 172.
41. Alonso, L.; Lozada, L.; Fontecha, J.; Juarez, M. Determination of cholesterol in milk fat by gas chromatography with direct injection and sample saponification. *Chromatographia* **1995**, *41*, 23–25. [CrossRef]
42. Alonso, L.; Fontecha, J.; Lozada, L.; Fraga, M.J.; Juárez, M. Fatty acid composition of caprine milk: Major, branched-chain and *trans* fatty acids. *J. Dairy Sci.* **1999**, *82*, 878–884. [CrossRef]
43. Alonso, L. Capillary gas chromatography of some triglycerides in cheese using programmed temperature injection. *Chromatographia* **1993**, *35*, 9–12. [CrossRef]
44. Castro-Gómez, M.; Rodriguez Alcalá, L.M.; Calvo, M.V.; Romero, J.; Mendiola, J.A.; Ibáñez, E.; Fontecha, J. Total milk fat extraction and quantification of polar and neutral lipids of cow, goat, and ewe milk using a pressurized liquid system and chromatographic techniques. *J. Dairy Sci.* **2014**, *97*, 6719–6728. [CrossRef] [PubMed]
45. Alonso, L.; Fontecha, J.; Juárez, M. Development of a headspace gas chromatographic mass spectrometric method for determining methyl-ketones and secondary alcohols in blue cheese. *J. Chromatogr. Sci.* **1999**, *37*, 108–112. [CrossRef]
46. International Organization for Standardization. *Sensory Analysis. General Guidance for the Design of Test Rooms*; ISO 8589; International Organization for Standardization: Brussels, Belgium, 2007.

Sample Availability: Samples of the compounds are not available from the authors.

molecules

MDPI

Article

Effect of Enzyme Modified Soymilk on Rennet Induced Gelation of Skim Milk

Kaixin Li [1,2], Jianjun Yang [1,2], Qigen Tong [1,2], Wei Zhang [1,2] and Fang Wang [1,2,*]

1 Beijing Laboratory for Food Quality and Safety, Food Science and Engineering College, Beijing University of Agriculture, Beijing 102206, China; joycelee_1108@163.com (K.L.); 18811626398@163.com (J.Y.); tongqigen@163.com (Q.T.); zhangwei300109@sina.com (W.Z.)

2 Beijing Key Laboratory of Agricultural Product Detection and Control for Spoilage Organisms and Pesticides, Food Science and Engineering College, Beijing University of Agriculture, Beijing 102206, China

* Correspondence: fangfang1031edna@126.com; Tel.: +86-010-8079-9170

Academic Editors: Lillian Barros and Isabel C.F.R. Ferreira

Received: 29 October 2018; Accepted: 23 November 2018; Published: 26 November 2018

Abstract: In this study, soymilk was hydrolyzed to different degrees with flavourzyme, and then soymilk and enzyme modified soymilk at various levels were added to skim milk respectively, to generate a mixed gel using rennet. Rheological properties, scanning electron microscopy imaging, and physical and chemical indexes were examined to reveal the effect of enzyme modified soymilk on rennet induced gelation of skim milk. Results showed that soymilk inhibited the aggregation of skim milk, led to a decrease in storage modulus (G′), significantly increased moisture content and curd yield, and the resulting network was coarse. Enzyme modified soymilk with a molecular weight below 20 kDa led to a more uniform curd distribution, which counteracted the reduction of G′ and allowed for the formation of a stronger gel. Both the moisture content and the curd yield increased with the addition of soymilk and enzyme modified soymilk, and overall the effect of adding a high degree of hydrolysis of enzyme modified soymilk was superior. Compared to untreated soymilk, the addition of a certain amount of enzyme modified soymilk resulted in a new protein structure, which would improve the texture of blend cheese.

Keywords: soymilk; enzyme modified soymilk; skim milk; rennet induced gelation; cheese; rheological properties

1. Introduction

In recent years, with rising food price levels, increased dietary awareness and health concerns, people are becoming more interested in soy products and novel value-added food products. Consuming foods that contain casein and soy protein has many beneficial health effects, and gels made from these two proteins can achieve dual health effects of two products [1].

The gelation of milk often occurs in two ways. In the first mechanism, during acidification, colloidal calcium phosphate is dissolved, and caseins formed a self-supporting network near their isoelectric point [2]. In the other mechanism, rennet specifically hydrolyzes k-casein at the Phe105–Met106 bond, decreasing steric and electrostatic stabilization and resulting in casein aggregation [3,4]. Soymilk gels are typically prepared by heating soymilk, and gelation can also be induced by adding magnesium chloride or glucono-δ-lactone [5].

Many studies have focused on mixed skim milk and soymilk gels. Previous studies have shown that under acidification conditions, the viscoelastic properties and microstructure of generated protein gels of skim milk and soymilk mixtures are dependent on the concentration of skim milk powder and soy protein concentrate [6]. In addition, adding rennet to skim milk and soymilk systems, rennet will only hydrolyze the cow milk components [7]. It has been suggested that combined acid and rennet induced gelation of cow milk and soymilk mixtures would lead to the simultaneous aggregation of proteins [8].

Soy protein hydrolysate has been used in specialized adult nutrition formulations [9]. Many studies have shown when soy protein isolate was enzymed, its functional properties can be improved [10–14]. Rinaldoni et al. [15] added soy protein to skimmed cow milk to develop a spreadable cheese-like product. Their results showed that soy protein concentrate improved cheese yield, but the elastic properties were significantly reduced. Gao et al. [16] studied the enzymatic hydrolysis process of soymilk, and mixed enzyme modified soymilk with cow milk to make mozzarella cheese. It was found that the stretchability, elasticity and hardness of mozzarella cheese had been greatly improved. For most cheeses, rennet induced gelation is the key process affecting cheese yield and quality [17–19]. Therefore, the objective of this study was to explore the effect of different amounts and hydrolysis degrees of enzyme modified soymilk on renneting of skim milk, to provide a theoretical basis for the production of a new blend cheese.

2. Results

2.1. Protein Structural Characteristics

The effect of enzymatic hydrolysis on protein profiles of skim milk, soymilk and enzyme modified soymilk is shown in Figure 1, and lanes 1–5 are the result of the marker, skim milk, soymilk, and high and low degree of hydrolysis of enzyme modified soymilk, respectively. Compared to soymilk, the distribution of bands of enzyme modified soymilk changed dramatically, indicating the decomposition of soy protein. After enzymatic hydrolysis of soy protein, the bands were mostly concentrated below 20 kDa, and its β-conglycinin, and acidic and basic subunits of glycinin almost disappeared entirely. Compared to the low degree of hydrolysis of enzyme modified soymilk, the molecular weight of soy protein further reduced after the high degree of hydrolysis.

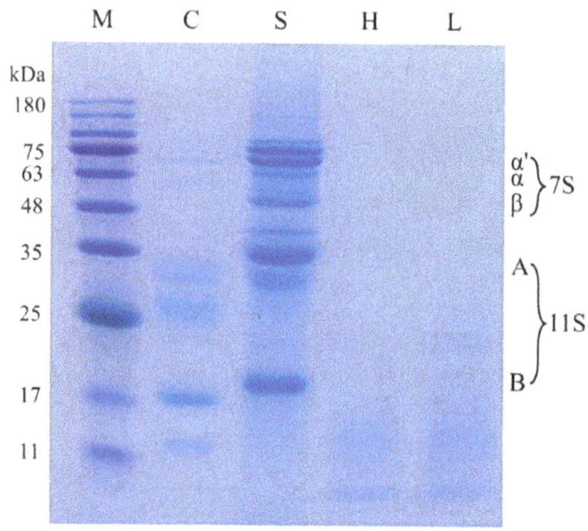

Figure 1. Sodium dodecyl sulfate polyacrylamide gel electropheresis for the control, soymilk and enzyme modified soymilk. M, marker; C, skim milk; S, soymilk; H, high degree of hydrolysis of soymilk; L, low degree of hydrolysis of soymilk. 7S and 11S indicate the relative proteins in soymilk.

2.2. Rheological Properties

Rheological properties were used to monitor the effect of soymilk and enzyme modified soymilk on the coagulation process of skim milk. After adding rennet, the storage modulus of the control increased rapidly, and then changed slowly with gelation time (Figure 2a–c). The storage modulus of experimental samples (except S2 and S3) showed the same trend. However, the storage modulus of experimental samples was lower than that of the control sample at the same renneting time ($p < 0.05$). The trend of the changes of loss modulus was consistent with that of storage modulus (data not shown). The loss tangent (the ration of loss modulus to storage modulus) decreased with renneting (Figure 2d–f). Compared to the control, gelation time of experimental samples lagged significantly ($p < 0.05$) (Table 1), and the delay effect in gelation time was more significant ($p < 0.05$) with the increase in soymilk and enzyme modified soymilk. The order of gelation time of experimental samples at the same concentration was L group $\geq$ H group > S group. At the end of gel formation, compared with control sample, the final strength of the curd determined by the storage modulus value at 60 min (G'_{60min}) in experimental samples decreased significantly ($p < 0.05$) (Table 1), and the decrease was more significant with the increase in soymilk and enzyme modified soymilk ($p < 0.05$). The order of G'_{60min} of the experimental samples at the same concentration was L group > H group > S group.

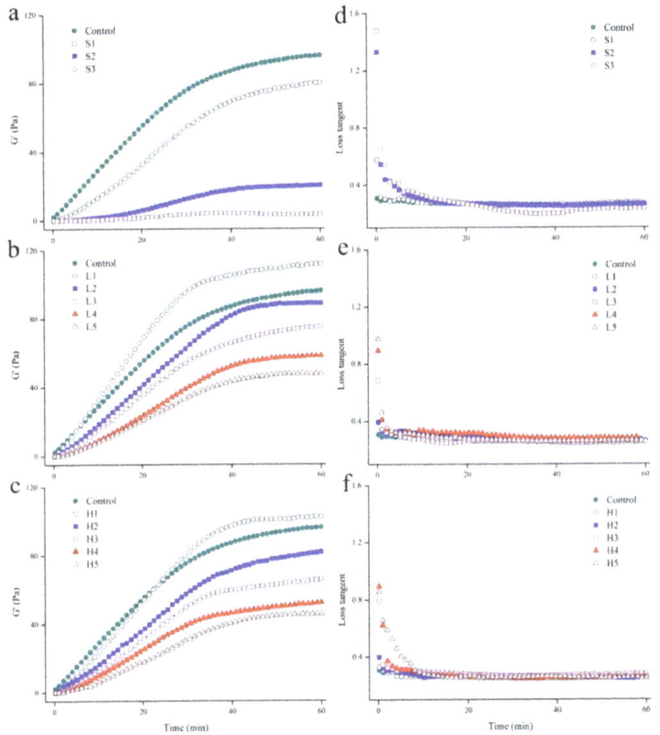

Figure 2. The changes in storage modulus (G') (**a**–**c**) and loss tangent (**d**–**f**) during rennet induced gelation. Curves are representative runs.

Table 1. The rheological parameters of mixtures during renneting.

Sample Number	G'_{60min} [1] (Pa)	Gelation Time [2] (min)
Control	96.26 ± 1.56 [k]	0.36 ± 0.13 [a]
S1	80.33 ± 1.86 [i]	1.07 ± 0.16 [c]
S2	20.49 ± 1.40 [b]	8.14 ± 0.15 [i]
S3	3.53 ± 0.95 [a]	16.28 ± 0.11 [j]
L1	111.85 ± 1.68 [m]	0.78 ± 0.08 [b]
L2	88.86 ± 1.77 [j]	1.02 ± 0.20 [bc]
L3	75.32 ± 1.33 [h]	1.53 ± 0.13 [d]
L4	58.14 ± 1.94 [f]	2.28 ± 0.14 [e]
L5	47.88 ± 1.40 [d]	3.04 ± 0.17 [g]
H1	102.47 ± 1.46 [l]	0.83 ± 0.09 [bc]
H2	81.74 ± 1.24 [i]	1.07 ± 0.13 [c]
H3	65.84 ± 0.88 [g]	1.72 ± 0.13 [d]
H4	52.31 ± 1.78 [e]	2.68 ± 0.11 [f]
H5	45.14 ± 1.18 [c]	3.55 ± 0.18 [h]

Means with different letters within the same column are significantly different ($p < 0.05$). [1] G'_{60min} means storage modulus value of the gel at the end of renneting at 60 min. [2] Gelation time means the time point when storage modulus value of the gel was ≥1 Pa.

2.3. Physical and Chemical Indicators

The addition of soymilk and enzyme modified soymilk significantly increased the moisture content and curd yield (Table 2). With the increase in soymilk, curd yield showed a significant increase ($p < 0.05$), but the moisture content did not change significantly ($p > 0.05$). With the increase in enzyme modified soymilk, the moisture content and curd yield increased significantly ($p < 0.05$). At 5% and 10% addition, the moisture content and curd yield of the S group were significantly higher than those of L group and H group ($p < 0.05$). However, no significant differences were found in moisture content and curd yield among the three groups at 15% addition ($p > 0.05$), and moisture content and curd yield showed no significant differences between L group and H group at 20% addition ($p > 0.05$). Under 25% addition, the curd yield was higher in L group than in H group, and moisture content between the two samples showed no significant difference ($p > 0.05$).

Table 2. The moisture content and curd yield of rennet induced curds.

Sample Number	Moisture Content (%)	Curd Yield (%)
Control	75.25 ± 3.17 [a]	13.52 ± 0.32 [a]
S1	78.78 ± 0.78 [cd]	15.20 ± 0.20 [de]
S2	78.70 ± 0.51 [cd]	15.71 ± 0.12 [f]
S3	78.01 ± 0.65 [bcd]	15.40 ± 0.15 [ef]
L1	75.16 ± 1.37 [a]	14.34 ± 0.17 [bc]
L2	75.90 ± 1.74 [ab]	14.99 ± 0.34 [d]
L3	76.42 ± 0.84 [abc]	15.61 ± 0.18 [f]
L4	77.24 ± 1.40 [abcd]	16.50 ± 0.18 [g]
L5	78.74 ± 0.94 [cd]	17.06 ± 0.28 [h]
H1	76.89 ± 0.19 [abc]	14.05 ± 0.12 [b]
H2	75.76 ± 0.31 [ab]	14.58 ± 0.11 [c]
H3	77.52 ± 0.49 [abcd]	15.51 ± 0.21 [ef]
H4	77.68 ± 1.49 [abcd]	16.41 ± 0.22 [g]
H5	79.87 ± 2.29 [d]	16.34 ± 0.25 [g]

Means with different letters within the same column are significantly different ($p < 0.05$).

2.4. Microstructure

The above results showed that rennet induced gelation was affected by soymilk and enzyme modified soymilk. Therefore, the ratio of skim milk and soymilk/enzyme modified soymilk selected here was 85:15, to ensure their comparability. The caseins in the control were relatively smooth (Figure 3a). The soy protein in the sample with soymilk attached to the surface of caseins, formed rough and small clusters, and therefore resulted in a coarse structure (Figure 3b). Compared with the control and S3, the addition of enzyme modified soymilk caused a more uniform protein network (Figure 3c,d). Compared to the control, S3 and L3, the casein aggregates in H3 were smaller, and the small voids became denser. The soy protein in S3 and enzyme modified soy protein in L3 mainly existed in large caseins, but the distribution of enzyme modified soy protein in H3 was more uniform.

Figure 3. Scanning electron microscopy micrographs of rennet induced curds. **a**, control; **b**, S3; **c**, L3; **d**, H3 (magnification: ×15,000) (The arrow denotes soy protein and its hydrolysate).

3. Discussion

In our study, the pH values during renneting among mixtures showed no significant differences (data no shown), and the gel formation was the action of rennet. Under the conditions of our study, the addition of rennet to soymilk did not result in gelation, as monitored using a rheometer with no change in storage modulus (data not shown), which was agreed with previous studies [9,20]. The increase in the storage modulus and the decrease in loss tangent indicated that there was an interaction among caseins, and the increased contact area of caseins therefrom led to the formation of a gel network [21]. Previous studies have shown that rheological properties were mainly affected by casein concentration [22]. In experimental groups, the decrease in skim milk, which was partly replaced with soymilk or enzyme modified soymilk, resulted in the decrease in k-casein and therefore the number of action sites for rennet. Moreover, soymilk and enzyme modified soymilk set a barrier to the accumulation of caseins and further affected their aggregation (Figure 4). Thus, we speculated that these two effects would lead to a lag in gelation time and a decrease in final strength, which was also reported in previous literature [23,24]. Meanwhile, the delay in gelation time was more significant and final strength further decreased with the increase of soymilk and enzyme modified soymilk. Compared to the control with the final strength of 96.26 Pa, when soymilk content increased to 15%, the final strength was reduced to 3.55 Pa (Table 2), indicating that aggregation of caseins was strongly hindered [8]. However, this effect can be overcome to a certain extent by adding enzyme modified soymilk, showing that the final strength of L3 and H3 was 75.32 Pa and 65.84 Pa, respectively. Compared with the control and S1, the decrease of k-casein in L1 and H1 had little effect on casein aggregation. The main contributor may be the reduction of protein size in enzyme modified soymilk as monitored by sodium dodecyl sulfate polyacrylamide gel electropheresis (SDS-PAGE), which was similar to the report of Luo et al. [25]. They studied the effect of native fat globule size on gel formation, and found storage modulus of the curd with small fat globules was higher than that of the curd with large fat globules. The high level of enzymatic hydrolysis increased the number of peptides, which may shield enzymatic sites and inhibit the aggregation of caseins (Figure 4), and therefore final strength in H group was significantly lower than that of L group.

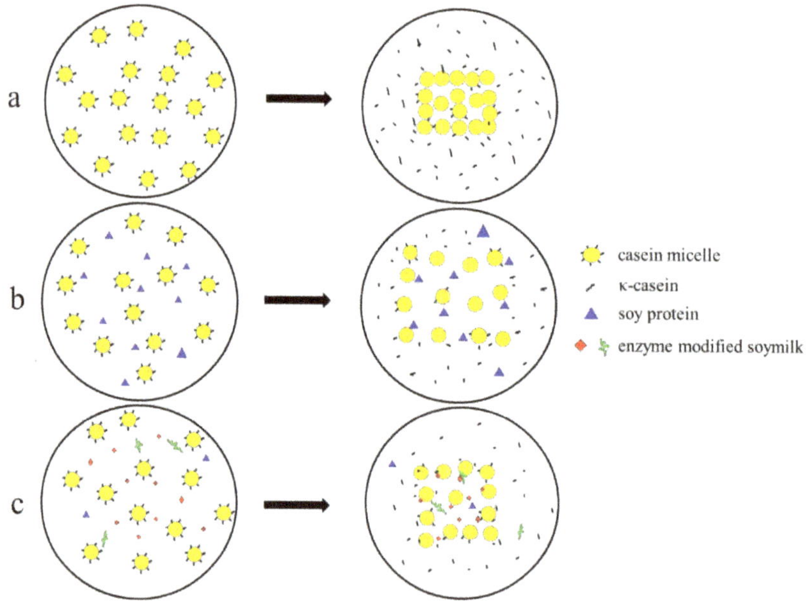

Figure 4. Schematic diagram of rennet induced gelation of different mixtures. **a**, skim milk; **b**, skim milk and soymilk; **c**, skim milk and enzyme modified soymilk.

Previous studies have shown that the addition of soy protein to milk can increase the moisture content of the resulting cheese [26], which was consistent with our result. Soy protein had water-holding properties [9], and therefore we speculated that soy protein trapped in the gel structure can retain more moisture through hydrogen bonds and other forces. Furthermore, our rheological results have shown that soymilk significantly inhibited rennet induced aggregation of caseins, affected dehydration of curds and increased moisture content and curd yield, which was in agreement with other studies [27–29]. Utsumi [30] indicated that the water-holding property of soy protein after enzymatic hydrolysis can be improved. However, the moisture content in L group and H group was lower than that in S group in our study, which may be ascribed to the decreased inhibition of gelation and the loss of small substances after enzymatic hydrolysis (Figure 4). As the amount of soymilk increased, the hydrophilicity of soy protein and the inhibition effect on casein aggregation were more significant, and therefore moisture content and curd yield further increased. There was no significant difference in moisture content between L group and H group under the same addition, but the curd yield of H group was lower than that of L group, and the losing of smaller peptide segments or amino acids with whey after a high degree of hydrolysis may be the major contributor.

The control curd showed a compact protein matrix, which was composed of thick chains and large clusters of caseins (Figure 3a), and it was consistent with previous report [31]. The soy protein adhered or was bound to the surface of caseins, formed a clustered structure and therefore inhibited the aggregation of caseins, causing the increased moisture content and curd yield. The research of Ingrassia et al. showed a similar phenomenon [32]. Both glycinin and β-conglycinin in soymilk can be effectively degraded by flavourzyme [33], which may decrease the inhibition of gelation and therefore contribute to a more compact and orderly gel structure. The increase in the number of small substances after a high degree of hydrolysis further inhibited the aggregation of caseins, and thus decreased the size of casein aggregates.

4. Materials and Methods

4.1. Skim Milk Preparation

Fresh milk was supplied by a local farm (Fuchun Farm, Beijing Sanyuan Food Co., Ltd., Beijing, China) and sodium azide (0.02%) was added immediately. The milk was centrifuged at 4000× g for 20 min at 4 °C (LYNX 2000, Thermo Scientific, Waltham, MA, USA) and then filtered three times through filters (Fisher Scientific, Whitby, ON, Canada), as far as possible to remove fat. The resulting skim milk was pasteurized at 63 °C for 30 min, cooled to room temperature, and stored in a refrigerator until use. The fat content of skim milk was 0.04 ± 0.01% by the AOAC method [34], and protein content was 3.21 ± 0.05% measured using the Kjeldahl method [19].

4.2. Enzyme Modified Soymilk Preparation

Soybeans (Helen soybean, Heilongjiang Black Soil Town Modern Agriculture Development Co., Ltd., Haerbin, Heilongjiang, China) containing 36% protein and 16% fat were obtained from a local market. Soymilk was prepared as previously reported [35], with slight modifications. Soybeans were soaked overnight in deionized water for hydration and mixed with a certain amount of deionized water to obtain the desired protein content. Subsequently, samples were passed through a household soymilk maker (RM-125, Ruimei, Wuxi, Jiangsu, China). After soymilk maker cycle was completed, soymilk was passed through a filter (Fisher Scientific, Whitby, ON, Canada) and passed through cheesecloth to remove okara. Then soymilk was boiled at 100 °C for 10 min and rapidly cooled to room temperature. The protein content of the obtained soymilk was 3.32 ± 0.07% measured using the Kjeldahl method [19].

Flavourzyme (0.28%; HP202474, Novozymes, Tianjin, China) was added to soymilk for 2 h and 4 h at 45 °C, respectively. The subsequent degree of hydrolysis of soymilk was 5.92% and 9.88% according to the Ninhydrin method [36]. L and H were used to represent the low (L) and high (H) degree of hydrolysis. The enzyme modified soymilk was then incubated at 85 °C for 15 min to inactivate the enzyme. When samples were cooled to room temperature, sodium azide (0.02%) was added. The obtained samples were stored at 4 °C until use.

4.3. Electrophoresis

The effect of enzyme treatments on protein profiles of soymilk and enzyme modified soymilk was determined using SDS-PAGE with separating and stacking gels containing 15 and 4% acrylamide, respectively. A molecular weight marker ranging from 11 to 180 kDa (PR1910, Solarbio, Beijing, China) was used as a standard. Skim milk, soymilk and enzyme modified soymilk were dissolved with sample buffer (10 mM DTT, pH 6.8, 1 mM EDTA, 1% SDS, 10% glycerol, and 0.01% bromphenol blue) and then boiled for 5 min. After centrifugation, the electrophoresis was carried out following the method of Lamsal et al. [37].

4.4. Mixture Preparation

The mixtures were prepared according to Table 3, and skim milk was selected as the control. For convenient reading, samples with different ratios of skim milk and soymilk/enzyme modified soymilk are indicated by **S1–S3**, **L1–L5** and **H1–H5**. The mass ratio selected were to ensure that rennet induced gelation would occur. After the mixtures of skim milk and soymilk/enzyme modified soymilk had been prepared, the samples were stirred at room temperature for 30 min, and the pH was adjusted to 6.7 with 0.1 M NaOH. All the samples were stored in a refrigerator until use.

Table 3. The mixtures with different mass ratios of skim milk and soymilk/enzyme modified soymilk.

Mass Ratio	Soymilk	Low Degree of Hydrolysis of Soymilk	High Degree of Hydrolysis of Soymilk
95:5	S1	L1	H1
90:10	S2	L2	H2
85:15	S3	L3	H3
80:20	-	L4	H4
75:25	-	L5	H5

4.5. Curd Making

All samples (weighed and recorded as W_1) were first warmed at 32 °C for 30 min and then incubated with rennet (0.01%). Rennet (CHY-MAX powder Extra) was got from Chr. Hansen (Beijing, China) and the coagulation strength was ~2235 IMCU/g. After 1 h of incubation, the resulting gels were cut manually into small pieces (1 × 1 × 1 cm), and then set for 5 min to promote syneresis. Curds were subsequently centrifuged at 4000× *g* for 15 min at room temperature. The upper whey was removed and the curd was collected carefully and weighed (recorded as W_2). The curd yield was calculated using following Equation:

$$\text{Curd yield} = W_2/W_1 \times 100\%$$

The moisture content of the curds was analyzed using an oven method [34]. All measurements were obtained in triplicate.

4.6. Rheological Properties

The coagulation process of the mixtures was monitored as described previously [38]. All samples were first maintained at 32 °C for 30 min. The gelation time was defined as the time point when storage modulus value of the gel was ≥1 Pa [6]. All measurements were determined in triplicate.

4.7. Microstructure Determination

The samples for microstructure determination were prepared as described earlier [39]. Then samples were critical point dried in a Technics Critical Point Dryer (CPD030, Leica Mikrosysteme GmbH, Wien, Austria), fixed on the sample table (MC1000, Hitachi, Ibarakiken, Japan) for spraying operation and then examined using a scanning electron microscope (SU8010, Hitachi, Ibarakiken, Japan) operated at 25 kV. Representative micrographs were selected for presentation.

4.8. Statistical Analysis

Statistical analysis of the data was performed using SPSS 18.0 (SPSS Inc., Armonk, NY, USA). The differences were compared at a significance level of $p < 0.05$.

5. Conclusions

Soymilk and enzyme modified soymilk have a significant effect on renneting of skim milk. Compared to skim milk, a certain proportion of soymilk and enzyme modified soymilk would decrease storage modulus of the mixed gel, and significantly increase moisture content and curd yield. In addition, with the increase in soymilk and enzyme modified soymilk, the inhibition of gel formation was stronger, and moisture content and curd yield further increased. Compared to soymilk, enzyme modified soymilk with more small molecular substances decreased the impediment of rennet induced gelation, improved storage modulus of the gel, and the resulting curd had a more uniform structure.

Author Contributions: Data curation, K.L.; Funding acquisition, F.W.; Investigation, K.L., J.Y. and W.Z.; Project administration, Q.T. and F.W.; Writing—original draft, K.L.; Writing—review and editing, F.W.

Funding: This project was financially supported by the Beijing Municipal Commission of Education Research Program (KM201710020013).

Acknowledgments: We would like to thank Donghua Yu for assisting with the scanning electron micrographs at the Institute of Atomic Energy Utilization, Chinese Academy of Agricultural Sciences.

Conflicts of Interest: The authors declare no conflict of interest.

References

1. Schaafsma, G. The protein digestibility-corrected amino acid score (pdcaas)—A concept for describing protein quality in foods and food ingredients: A critical review. *J. Aoac Int.* **2005**, *88*, 988–994. [PubMed]
2. Lucey, J.A.; Singh, H. Formation and physical properties of acid milk gels: A review. *Food Res. Int.* **1997**, *30*, 529–542. [CrossRef]
3. Dalgleish, D.G. Proteolysis and aggregation of casein micelles treated with immobilized or soluble chymosin. *J. Dairy Res.* **1979**, *46*, 653–661. [CrossRef]
4. Alexander, M.; Dalgleish, D.G. Application of transmission diffusing wave spectroscopy to the study of gelation of milk by acidification and rennet. *Colloid. Surface B* **2004**, *38*, 83–90. [CrossRef] [PubMed]
5. Poysa, V.; Woodrow, L. Stability of soybean seed composition and its effect on soymilk and tofu yield and quality. *Food Res. Int.* **2002**, *35*, 337–345. [CrossRef]
6. Roesch, R.; Juneja, M.; Monagle, C.; Corredig, M. Aggregation of soy/milk mixes during acidification. *Food Res. Int.* **2004**, *37*, 209–215. [CrossRef]
7. Grygorczyk, A.; Alexander, M.; Corredig, M. Combined acid- and rennet-induced gelation of a mixed soya milk–cow's milk system. *Int. J. Food Sci. Technol.* **2013**, *48*, 2306–2314. [CrossRef]
8. Lin, C.; Hill, A.; Corredig, M. Gelation of mixtures of soymilk and reconstituted skim milk subjected to combined acid and rennet. *J. Texture Stud.* **2012**, *43*, 468–476. [CrossRef]
9. Mahnoud, M.I. Physicochemical and functional properties of protein hydrolysates in nutritional products. *Food Technol.* **1994**, *48*, 89–94.
10. Ortiz, S.E.M.; Wagner, J.R. Hydrolysates of native and modified soy protein isolates: Structural characteristics, solubility and foaming properties. *Food Res. Int.* **2001**, *35*, 511–518. [CrossRef]
11. Achouri, A.; Zhang, W. Effect of succinylation on the physicochemical properties of soy protein hydrolysate. *Food Res. Int.* **2001**, *34*, 507–514. [CrossRef]
12. Tsumura, K.; Saito, T.; Tsuge, K.; Ashida, N.; Kugimiya, W.; Inouye, K. Functional properties of soy protein hydrolysates obtained by selective proteolysis. *LWT-Food Sci. Technol.* **2005**, *38*, 255–261. [CrossRef]
13. Were, L.; Hettiarachchy, N.S.; Kalapathy, U. Modified soy proteins with improved foaming and water hydration properties. *J. Food Sci.* **1997**, *62*, 821–824. [CrossRef]
14. Suro´wka, K.; Zmudzin´ski, D.; Suro´wka, J. Enzymic modification of extruded soy protein concentrates as a method of obtaining new functional food components. *Trends Food Sci. Tech.* **2004**, *15*, 153–160. [CrossRef]
15. Rinaldoni, A.N.; Palatnik, D.R.; Zaritzky, N.; Campderrós, M.E. Soft cheese-like product development enriched with soy protein concentrates. *LWT-Food Sci. Technol.* **2014**, *55*, 139–147. [CrossRef]
16. Gao, Y.; Fang, Z.; Cao, X.; Lin, J.; Xie, J.; Xu, X. Optimization of process of analog Mozzarella cheese containing limit hydrolyzed soy milk. *Food Sci Technol.* **2010**, *35*, 94–100.
17. Ikonen, T.; Ruottinen, O.; Syväoja, E.L.; Saarinen, K.; Pahkala, E.; Ojala, M. Effect of milk coagulation properties of herd bulk milks on yield and composition of Emmental cheese. *Agric. Food Sci.* **1999**, *8*, 411–422. [CrossRef]
18. Pandey, P.K.; Ramaswamy, H.S.; Stgelais, D. Water-holding capacity and gel strength of rennet curd as affected by high-pressure treatment of milk. *Food Res. Int.* **2000**, *33*, 655–663. [CrossRef]

19. Wang, F.; Tong, Q.; Luo, J.; Xu, Y.; Ren, F. Effect of carrageenan on physicochemical and functional properties of low-fat colby cheese. *J. Food Sci.* **2016**, *81*, E1949–E1955. [CrossRef] [PubMed]
20. Stanley, D.W.; Deman, J.M. Structural and mechanical properties of textured proteins. *J. Texture Stud.* **1978**, *9*, 59–76. [CrossRef]
21. Lucey, J.A. Formation and physical properties of milk protein gels. *J. Dairy Sci.* **2002**, *85*, 281–294. [CrossRef]
22. Ferrer, M.A.; Hill, A.R.; Corredig, M. Rheological properties of rennet gels containing milk protein concentrates. *J. Dairy Sci.* **2008**, *91*, 959–969. [CrossRef] [PubMed]
23. Haratifar, S.; Corredig, M. Interactions between tea catechins and casein micelles and their impact on renneting functionality. *Food Chem.* **2014**, *143*, 27–32. [CrossRef] [PubMed]
24. Logan, A.; Leis, A.; Li, D.; Øiseth, S.K.; Puvanenthiran, A.; Augustin, M.A. Rennet gelation properties of milk: Influence of natural variation in milk fat globule size and casein micelle size. *Int. Dairy J.* **2014**, *46*, 71–77. [CrossRef]
25. Luo, J.; Wang, Y.; Guo, H.; Ren, F. Effects of size and stability of native fat globules on the formation of milk gel induced by rennet. *J. Food Sci.* **2017**, *82*, 670–678. [CrossRef] [PubMed]
26. Mahdy, A.; Xia, W.; Zhang, G. Effect of soy protein supplementation on the quality of ripening Cheddar-type cheese. *J. Dairy Sci.* **2004**, *57*, 209–214.
27. Robinson, R.K.; Tamime, A.Y. The role of protein in yoghurt. In *Developments in Food Proteins*; Hudson, B.J.F., Ed.; Elsevier Applied Science Publishers LTD: London, UK, 1986; pp. 1–35.
28. Martindiana, A.B.; Janer, C.; Pelaez, C.; Requena, T. Development of a fermented goat's milk containing probiotic bacteria. *Int. Dairy J.* **2003**, *13*, 827–833. [CrossRef]
29. Herrero, A.M.; Requena, T. The effect of supplementing goats milk with whey protein concentrate on textural properties of set-type yoghurt. *Int. J. Food Sci Tech.* **2006**, *41*, 87–92. [CrossRef]
30. Utsumi, S.; Matsumura, Y.; Mori, T. Structure-function relationships of soy proteins. In *Food Proteins and Their Applications*; Damodaran, S., Paraf, A., Eds.; Marcel Dekker: New York, NY, USA, 1997; pp. 257–291.
31. Domagała, J.; Najgebauer-Lejko, D.D.; Wieteska-Śliwa, I.; Sady, M.; Wszołek, M.; Bonczar, G.; Filipczak-Fiutak, M. The influence of milk protein cross-linking by transglutaminase on the rennet coagulation time and the gel properties. *J. Sci. Food Agric.* **2015**, *96*, 3500–3507. [CrossRef] [PubMed]
32. Ingrassia, R.; Costa, J.P.; Hidalgo, M.E.; Canales, M.M.; Castellini, H.; Riquelme, B.; Risso, P. Application of a digital image procedure to evaluate microstructure of caseinate and soy protein acid gels. *LWT-Food Sci. Technol.* **2013**, *53*, 120–127. [CrossRef]
33. Hrckova, M.; Rusnakova, M.; Zemanovic, J. Enzymatic hydrolysis of defatted soy flour by three different proteases and their effect on the functional properties of resulting protein hydrolysates. *Czech. J. Food Sci.* **2002**, *20*, 7–14. [CrossRef]
34. Wang, F.; Zhang, X.; Luo, J.; Guo, H.; Zeng, S.; Ren, F. Effect of proteolysis and calcium equilibrium on functional properties of natural cheddar cheese during ripening and the resultant processed cheese. *J. Food Sci.* **2011**, *76*, E248–E253. [CrossRef] [PubMed]
35. Mullin, W.J.; Fregeau-Reid, J.A.; Butler, M.; Poysa, V.; Woodrow, L.; Jessop, D.B.; Raymond, D. An interlaboratory test of a procedure to assess soybean quality for soymilk and tofu production. *Food Res. Int.* **2001**, *34*, 669–677. [CrossRef]
36. Awolumate, E.O. Accumulation and quality of storage protein in developing cowpea, mung bean and soya bean seeds. *J. Sci. Food Agric.* **2010**, *34*, 1351–1357. [CrossRef]
37. Lamsal, B.P.; Jung, S.; Johnson, L.A. Rheological properties of soy protein hydrolysates obtained from limited enzymatic hydrolysis. *LWT-Food Sci. Technol.* **2007**, *40*, 1215–1223. [CrossRef]
38. Wang, F.; Liu, X.; Hu, Y.; Luo, J.; Lv, X.; Guo, H.; Ren, F. Effect of carrageenan on the formation of rennet-induced casein micelle gels. *Food Hydrocolloid.* **2014**, *36*, 212–219. [CrossRef]
39. Lobato-Calleros, C.; Robles-Martinez, J.C.; Caballero-Perez, J.F.; Vernon-Carter, E.J.; Aguirre-Mandujano, E. Fat replacers in low-fat mexican manchego cheese. *J. Texture Stud.* **2001**, *32*, 1–14. [CrossRef]

Sample Availability: Samples of skim milk and soymilk/enzyme modified soymilk are available from the authors.

molecules

MDPI

Article

Bark Extract of the Amazonian Tree *Endopleura uchi* (Humiriaceae) Extends Lifespan and Enhances Stress Resistance in *Caenorhabditis elegans*

Herbenya Peixoto [1], Mariana Roxo [1], Emerson Silva [2], Karla Valente [2], Markus Braun [1], Xiaojuan Wang [1] and Michael Wink [1,*]

1 Institute of Pharmacy and Molecular Biotechnology, Heidelberg University, INF 364, D-69120 Heidelberg, Germany; hspeixoto1@gmail.com (H.P.); marianaroxocorreia@gmail.com (M.R.); markus.braun@gmx.ch (M.B.); wxjsz@hotmail.com (X.W.)

2 Faculty of Pharmaceutical Science, Federal University of Amazonas (UFAM), 6200 General Rodrigo, Manaus 69077-000, Brazil; eslima75@gmail.com (E.S.); karlaacussena@gmail.com (K.V.)

* Correspondence: wink@uni-heidelberg.de; Tel.: +49-62-2154-4880

Academic Editors: Lillian Barros and Isabel C.F.R. Ferreira

Received: 6 February 2019; Accepted: 1 March 2019; Published: 6 March 2019

Abstract: *Endopleura uchi* (Huber) Cuatrec (Humiriaceae), known as uxi or uxi-amarelo in Brazil, is an endemic tree of the Amazon forest. In traditional medicine, its stem bark is used to treat a variety of health disorders, including cancer, diabetes, arthritis, uterine inflammation, and gynecological infections. According to HPLC analysis, the main constituent of the bark extract is the polyphenol bergenin. In the current study, we demonstrate by in vitro and in vivo experiments the antioxidant potential of a water extract from the stem bark of *E. uchi*. When tested in the model organism *Caenorhabditis elegans*, the extract enhanced stress resistance via the DAF-16/FOXO pathway. Additionally, the extract promoted an increase in the lifespan of the worms independent from caloric restriction. It also attenuated the age-related muscle function decline and formation of polyQ40 plaques, as a model for Huntington's disease. Thus, these data support anti-aging and anti-oxidant properties of *E. uchi*, which has not yet been described. More studies are needed to assess the real benefits of *E. uchi* bark for human health and its toxicological profile.

Keywords: *Caenorhabditis elegans*; antioxidants; bergenin; stress resistance; lifespan; Huntington; uxi; *Endopleura uchi*

1. Introduction

Endopleura uchi (Huber) Cuatrec, popularly known in Brazil as uxi or uxi-amarelo, is an endemic tree found throughout the entire Brazilian part of the Amazon basin [1]. The species belongs to the family Humiriaceae and is the only member of its genus. This valuable tree is locally used for its wood, bark, fruit, and seeds [2].

Traditional medicinal applications of the stem bark of *E. uchi* include the treatment and prevention of cancer, diabetes, high cholesterol, arthritis, diarrhea, and genitourinary disorders, especially uterine inflammations and infections [3]. A recent ethnobotanical survey has reported a high demand for uxi bark in regional markets due to its popular therapeutic claims [4–7]. However, few studies have investigated the bioactivities of *E. uchi*.

Silva and Teixeira [8] reported the in vitro antioxidant and antibacterial activity of the bark, as well as inhibition of cholinesterase (AChE, BuChE) and α-glucosidase. The authors associated the inhibition of α-glucosidase with the traditional use of the bark to treat diabetes. Additionally, no cytotoxic effect was observed when tested in human colorectal adenocarcinoma cells (Caco-2). When tested in HeLa cells, a polysaccharide fraction of *E. uchi* barks significantly reduced proliferation and cell viability [9]. Sá et al. [10] demonstrated that the subchronic administration of *E. uchi* bark extract has no toxic effects on male and female Wistar rats. Politi et al. [11] also assessed the safety profile of *E. uchi* bark and reported the absence of oral acute toxicity.

Previous phytochemical investigations of *E. uchi* bark have revealed the presence of tannins, terpenoids (saponins and steroids), and coumarins [12–14]. The isocumeric secondary metabolite bergenin has been reported by several researchers as the major compound in *E. uchi* bark [8,15–17].

In the current study, we investigated a water extract from the stem bark of *E. uchi* regarding its potential antioxidant and anti-aging properties using the nematode *Caenorhabditis elegans* as an experimental model, which is widely used in this context.

2. Material and Methods

2.1. Plant Material and Extract

Endopleura uchi extract (EU) was obtained from stem bark purchased from a local trader in Manaus-AM (Brazil). The bark material was weighed, milled, and exhaustively extracted with distilled water (5 × 1 L) at room temperature during an overall extraction period of 5 days. Using a rotary evaporator, the water extract was concentrated at low pressure at 40 °C, frozen at −80 °C, and finally lyophilized to obtain a fine dried powder. The plant material used in this study is deposited in the sample collection of IPMB (Institut für Pharmazie und Molekulare Biotechnologie, Heidelberg, Germany) under the accession number IPMB P8636.

2.2. Antioxidant Activity

In a 96-well microplate, 100 µL of sample were added to 100 µL of 200 µM DPPH. After 30 min, the absorbance was measured in a microplate reader (Tecan Trading AG, Männedorf, Switzerland) at 517 nm [18]. All measurements were performed in triplicate. The EC_{50} is presented in µg/mL.

2.3. Total Phenolic Content

In a 96-well microplate, 20 µL of sample were added to 100 µL of Folin-Ciocalteu reagent; after 5 min, 80 µL of sodium carbonate (7.5% solution) were added to the wells. The reaction ran for 2 h protected from the light and at room temperature; the absorbance was measured at 750 nm. The assay was carried out in triplicate and repeated three times. The phenolic content is expressed as gallic acid equivalents (GAE/g of sample).

2.4. Chemical Characterization and Quantification of Bergenin

Bergenin content of the uchi extract was determined by high performance liquid chromatography (HPLC) in a Shimadzu Proeminence Chromatograph with a UV-Vis detector SPD-10A. The method used was adapted from Tacon and Nunomura [17,19]. The chromatography was run in gradient mode with methanol: formic acid 0.1% as the mobile phase A, and aqueous formic acid 0.1% as the mobile phase B. The column C-18 SphereClone 5 µ ODS (150 × 4.60 mm and particle size 5 µm) and the detector was set to wavelength of 272 nm. The flow rate of the mobile phase was 0.8 mL/min. The calibration curve was constructed using bergenin (Sigma, St Louis, MO, USA), ranging from 0.04 to 1.5 mg/mL, obtaining a linear correlation coefficient of 0.9995.

2.5. C. elegans Strains and Maintenance

The worms were cultivated on NGM plates inoculated with living *E. coli* OP50 as food source and incubated at 20 °C, except when mentioned. For the current work we used the strains N2 (wt), CF1038 (daf-16(mu86)), GR1307 (daf-16(mgDf50)), CF1553 (muIs84 [(pAD76) sod-3p::GFP + rol-6]), AM141 (rmIs133[P(unc-54)Q40::YFP]), TJ375 (gpIs1[hsp-16-2::GFP]), and BA17 [fem-1(hc17) IV)].

Age synchronous cultures were obtained by treating the adult hermaphrodites with a lysis solution (5 M NaOH and 5% NaOCl) for 5 min and separating the eggs from the debris by density gradient centrifugation using 60% sucrose solution [20]. The collected eggs were allowed to hatch in M9 buffer [21].

2.6. Survival Assay under Oxidative Stress

For this assay, age synchronized L1 larvae (N2, CF1038, and GR1307 strains) were grown in *S*-medium. The larvae were sorted into groups of 75 individuals and treated with the extract for 48 h. Subsequently, 80 μM of the pro-oxidant juglone (5-hydroxy-1,4-naphthalenedione) were added to each group and 24 h later the number of live and dead worms were scored. We considered a worm to be dead when it did not respond to a gentle touch with a platinum wire [20]. The assay was carried out in triplicate and is presented as mean ± SEM compared by one-way ANOVA followed by Bonferroni (*post-hoc*).

2.7. Intracellular ROS Accumulation

For this assay, we used age synchronized L1 larvae (N2 strain) grown in *S*-medium. The larvae were sorted into groups and treated with the extract for 48 h. Subsequently, the ROS-sensitive fluorescent dye H2DCFDA (50 μM) was added to each group. The staining took 1 h, protected from the light, at 20 °C; afterwards the worms were mounted onto a glass slide and paralyzed with a drop of 10 mM sodium azide. Using a BIOREVO BZ-9000 fluorescence microscope (Keyence Deutschland GmbH, Neu-Isenburg, Germany), live images of at least 30 worms per group were captured with an excitation filter set to 480/20 nm and emission filter set to 510/38 nm [20]. The relative fluorescence was determined densitometrically using the software Image J (National Institute of Health, Bethesda, Bethesda, MD, USA). The results are presented as mean fluorescence intensity (mean ± SEM) and compared by one-way ANOVA followed by Bonferroni (*post-hoc*). The assay was repeated three times.

2.8. Quantification of Gene Expression Using GFP Reporter

For this assay, we used age-synchronized worms grown in S-medium. To evaluate *sod-3*:GFP expression, L1 larvae (CF1553 strain) were treated with the extract for 48 h and analyzed under a fluorescence microscope, as described above. To evaluate *hsp-16.2*::GFP expression, L4 larvae (TJ375 strain) were treated with the extract for 48 h and subsequently exposed to 20 μM juglone; 24 h later, the worms were analyzed by fluorescence microcopy, as described. The assays were repeated three times and the results are presented as fluorescence intensity (mean ± SEM) compared by one-way ANOVA followed by Bonferroni (*post-hoc*).

2.9. Longevity Assay

For this assay, we used age-synchronized worms at day 1 of adulthood (BA17 strain) grown in S-medium. The adults were treated with the extract, except the control group. Throughout the entire period of observation, the worms were incubated at 25 °C and transferred every second day to fresh medium supplemented with extract following their treatment groups. Dead worms were scored during the transfer and removed from the assay. Worms exhibiting extruded gonads or internally hatched progeny were scored as censored and removed from the assay. We considered a worm to be dead when it did not respond to a gentle touch with a platinum wire [22]. The assay was repeated three times and the results are presented as percentage of survival. The statistical significance was determined by Log-rank (Mantel-Cox) tests followed by Gehan-Breslow-Wilcoxon Test.

2.10. Quantification of PolyQ40::GFP Aggregate Formation

For this assay, we used age synchronized L1 larvae (AM141 strain) grown in *S*-medium. This strain serves as a model for Huntington disease. The larvae were treated with the extract for 48 h and subsequently submitted to fluorescence microscopy, as described above [22]. The results are presented as number of PolyQ40::GFP aggregates (mean ± SEM) and compared by one-way ANOVA followed by Bonferroni (*post-hoc*).

2.11. Pharyngeal Pumping Rate

For this assay, we used age synchronized worms (N2 strain) grown on NGM agar plates. The adult worms were daily transferred to fresh plates, supplemented following their treatment groups, throughout the entire reproductive period. At day 5 and day 10 of adulthood, the worms were observed under a stereomicroscope for 1 min to score the pumping activity of the pharynx, which can serve as a measure for muscle impairment during aging [23]. The results are presented as pumps/min (mean ± SEM) and compared by two-way ANOVA followed by Bonferroni (*post-hoc*).

2.12. Body Length

If *C. elegans* is kept under dietary restriction (DR), its body length will decrease. To assess the body length, age synchronized L4 larvae (N2 strain) were treated with the extract for 24 h; subsequently, the worms were mounted onto a glass slide and submitted to bright field microscopy. Live images were taken from at least 30 worms per group; the length was measured from head to tail using the software ImageJ (version 1.48, National Institute of Health, Bethesda, MD, USA). The results are presented as body length in µm (mean ± SEM) and compared by one-way ANOVA followed by Bonferroni (*post-hoc*).

2.13. Antimicrobial Activity

Susceptibility of *Escherichia coli* strain OP50 to the extracts was assessed by means of well diffusion test according to CLSI (2014) with slight modifications [24]. Briefly, bacteria were grown on Müller-Hinton agar (MHA) and the cell suspension was adjusted to 0.5 McFarland standard. Wells with 6 mm in diameter were punched out and loaded with 70 µL of 10 mg/mL sample dissolved in sterile water. Ampicillin and ciprofloxacin (256 µg/mL) were used as positive controls. Diameters of the zones of inhibition (ZI) were assessed 24 h after incubation at 35 °C. The assay was repeated three times. The bacteria were purchased from the Caenorhabditis Genetics Center (CGC, University of Minnesota, Minneapolis, MN, USA).

3. Results

3.1. Antioxidant Activity in Vitro and Chemical Characterization of the Bark Extract

The bark extract obtained from *E. uchi* showed an antioxidant capacity in vitro as powerful as standard dietary antioxidants, such as vitamin C and EGCG, when tested in DPPH assay (Table 1). Correspondingly, a high content of phenolics was observed using Folin-Ciocalteu method (850 GAE/g extract). Through HPLC, the isocoumeric bergenin was found to be the major compound in the extract (4.5 g/100 g of dry extract; Figure 1).

Table 1. Antioxidant activity of *Endopleura uchi* (EU) assessed by DPPH assay.

SAMPLE	EC50 (µG/ML)
EU	8.0
EGCG	1.2
VITAMIN C	2.1

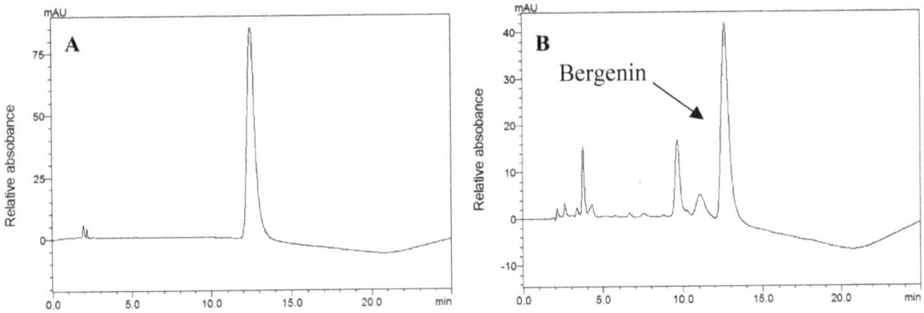

Figure 1. HPLC profile of the bergenin standard (**A**) and the aqueous bark *Endopleura uchi* extract (**B**) analyzed at 272 nm.

3.2. Effect of the Extract on Intracellular ROS Accumulation

Endogenous intracellular ROS production was investigated in wild type (N2 (wt)) worms under stress-free conditions. The result obtained indicated a significant decrease in ROS accumulation among worms treated with the EU as compared with the untreated control group. The decrease was up to 80% when the worms were treated with 200 μg/mL EU (adjusted p-value < 0.0001; Figure 2).

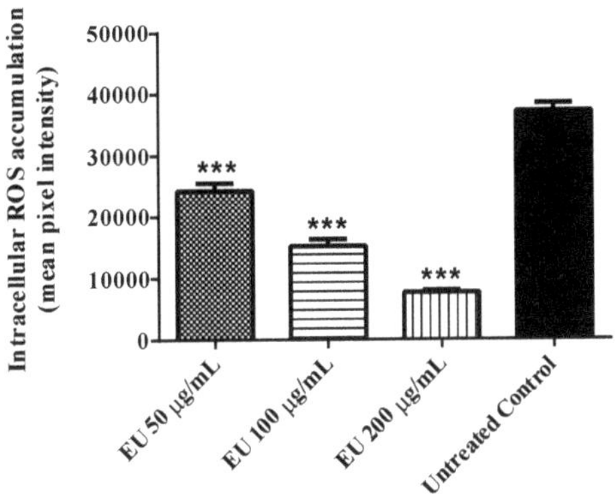

Figure 2. Quantification of intracellular ROS in N2 worms using DCFDA after treatment with *Endopleura uchi* extract (EU). Worms treated with EU showed lower levels of ROS compared to the control group. Data are presented as mean pixel intensity ± SEM (n = 40, replicated 3 times). *** $p < 0.001$, compared to the untreated control by one-way ANOVA followed by Bonferroni (*post-hoc*).

3.3. Protection against Oxidative Stress

Protection of the worms against oxidative stress by the extract was assessed by comparing the survival rate of wild type worms (N2 (wt)) after juglone-induced oxidative stress. The results indicated a significant higher survival rate among EU treated worms. At the highest tested concentration (200 μg/mL EU), 77% of the worms remained alive after juglone exposure compared to 27% scored in the extract-free group (adjusted p-value = 0.0017; Figure 3a).

In order to find out if DAF16-FOXO pathway plays a role in the stress resistance observed after EU treatment, we performed the same protocol using mutant strains, in which DAF16 was inactivated (strains: CF1038 (*daf-16*(mu86)I) and GR1307 (*daf-16*(mgDf50)). As illustrated in Figure 3b,c, these mutants did not benefit from the antioxidant properties of EU as did the wild type worms (N2 (wt)).

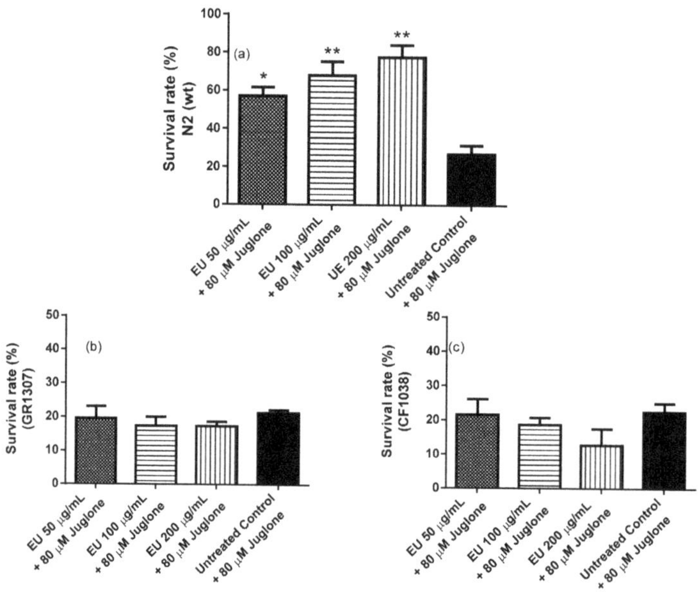

Figure 3. Survival of *C. elegans* after juglone-induced oxidative stress. Survival rate of N2 worms was significantly enhanced in the groups treated with the bark extract (EU) (**a**). However, the survival rate of DAF-16 mutants (GR1307 [*daf-16*(mgDf50) I] and CF1038 [*daf-16*(mu86) I]) did not differ between the groups (**b**) and (**c**), respectively. Each bar represents the mean ± SEM from three independent assays. Note: * $p < 0.05$ and ** $p < 0.01$ compared to the untreated control by one-way ANOVA followed by Bonferroni's method (*post-hoc*).

3.4. Effect of the Extract on the Expression of Stress Response Genes (hsp-16.2::GFP and sod-3::GFP)

The expression of *sod-3* was investigated using mutant worms (strain CF1553), in which *sod-3* has been fused with a GFP reporter. From analyses of the emitted fluorescence, we observed a significant increase in *sod-3*::GFP expression by 44% among EU treated worms compared with the untreated control group (adjusted *p*-value < 0.0001; Figure 4a).

The expression of *hsp-16.2* was assessed using the mutant strain TJ375, in which *hsp-16.2* is fused with GFP. After mild oxidative stress, induced by adding a low concentration of juglone to the medium, we observed a significant fluorescence in the nematodes. The intensity was reduced among EU treated worms as compared with untreated worms. The decrease was up 40% at a concentration of 200 μg/mL EU (adjusted *p*-value < 0.0001; Figure 4b).

3.5. Effect of the Extract on Longevity

A long-term assay was performed to test whether EU can influence longevity in *C. elegans*. The results obtained indicated extension of lifespan by 33% among BA17 worms treated with EU as compared to untreated control group (*p*-value $p < 0.0001$; Figure 5a). However, when the assay was performed with the *daf*-16 null mutants (CF1038 strain) no significant difference in the mean lifespan was observed between treated and untreated worms (Figure 5b), indicating that the transcription factor DAF16 plays a role in this context.

3.6. Effect of the Extract on the Pharyngeal Pumping Rate

During aging, muscle activity is impaired. As a marker for muscle activity, the pharyngeal pumping activity can be monitored in *C. elegans*. The pumping activity of the pharynx was scored in wild type (N2 (wt)) worms at day 5 and 10 of adulthood and revealed a significant difference among the groups. Worms cultured in medium supplemented with EU exhibited an improved pumping function of the pharynx. At day 10, the pumping rate among EU treated worms was 128% higher than that scored among untreated worms (p-value < 0.01; Figure 6). The data indicates that the muscle function is better preserved in worms under EU treatment as they age. Such a result also indicates that EU treated worms did not starve during their lifetime, so caloric restriction effect can be ruled out.

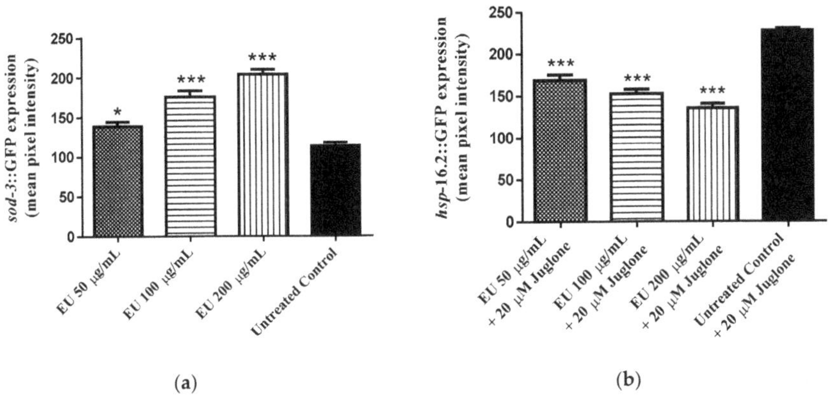

Figure 4. Expression of stress response genes. CF1553 worms [(pAD76)*sod-3*p::GFP + rol-6] treated with *Endopleura uchi* (EU) showed significant higher levels of SOD-3::GFP compared to the untreated control group (**a**), and mutant worms TJ375 [*hsp-16.2*::GFP(gplsI)] exposed to 20 µM juglone presented significant lower levels of HSP-16.2::GFP when compared with the untreated control worms similarly exposed to 20 µM juglone (**b**). Data are presented as mean pixel intensity (mean ± SEM) from three independent experiments. Note: * $p < 0.05$ and *** $p < 0.001$ related to the control, analyzed by one-way ANOVA followed by Bonferroni (*post-hoc*).

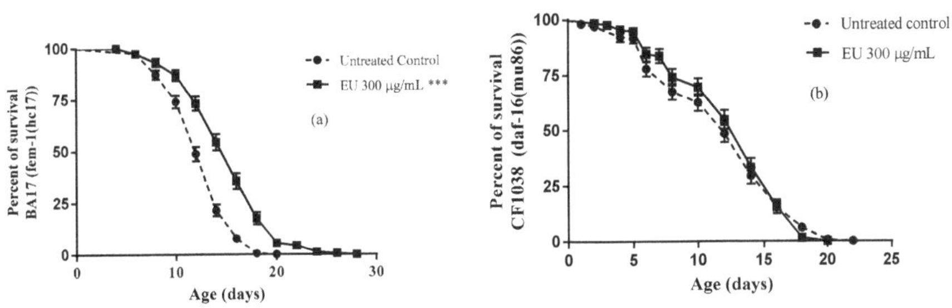

Figure 5. Longevity of *C. elegans* after treatment with *Endopleura uchi* extract (EU). BA17 worms treated with EU 300 µg/mL presented significantly longer lifespan compared to untreated control group (**a**). However, lifespan of *daf-16* null mutants (CF1038) were no significantly different between EU treated and untreated worms (**b**). The results are presented as percentage of surviving worms and the statistical significance determined by Log-rank (Mantel-Cox) tests followed by Gehan-Breslow-Wilcoxon Test. Note: *** $p < 0.001$.

3.7. Body Length

Body length is an important measurement to evaluate the possible deleterious effect of DR in *C. elegans*. In the current study, the body length of adult wild type (N2 (wt)) worms was compared between those who had been under EU treatment and the untreated ones. The data obtained indicated no differences between the groups. The treatment group had a length of 1.339 ± 0.018 mm as compared to controls with of 1.261 ± 0.018 mm. The result indicates that the worms did not undergo caloric restriction or any toxic effect able to impair body development while treated with EU extract.

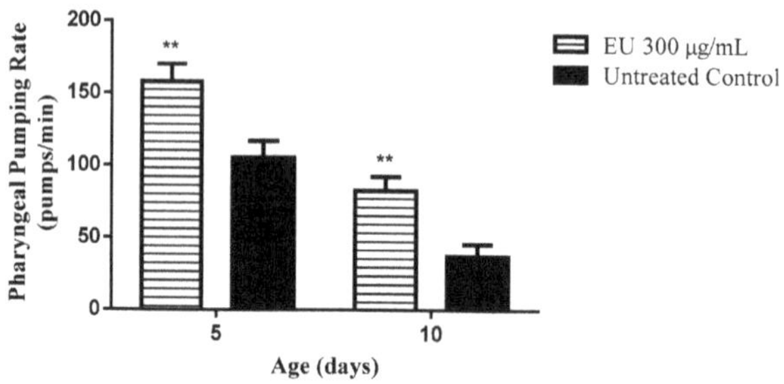

Figure 6. Pharyngeal pumping rate in *C. elegans* after treatment with the bark extract (EU). The treatment of wild type worms with EU 300 μg/mL significantly attenuated the age-associated decline in the muscle function of pharynx. Data are presented as mean ± SEM. Note: ** $p < 0.01$ related to the control by a two-way ANOVA.

3.8. Antimicrobial Activity

The bark extract was tested against *E. coli* OP50, bacterial strains used to feed *C. elegans*, and no bactericidal effect from EU was observed (Table 2). This finding further indicates that the worms did not undergo caloric restriction during the treatment due to a reduction of food source.

Table 2. Antimicrobial activity of the *Endopleura uchi* (EU) bark extract assessed by well diffusion test.

	Ampicillin (256 μg/mL)	Ciprofloxacin (256 μg/mL)	EU (10 mg/mL)
E. coli OP50	27.3 ± 1.2 *	42.8 ± 0.3 *	NI

* Zone of inhibition (mm); NI: no inhibition.

3.9. Effect of the Bark Extract on the Formation of polyQ40 Aggregates

In Huntington's disease, the disease-causing Huntington gene is mutated and carries many glutamate repeats. The formation of polyQ40 aggregates was assessed in AM141 mutants, which produce polyQ fused with GFP. The results obtained indicated a significant lower number of polyQ40::GFP aggregates among EU treated worms. At the highest tested concentration, 300 μg/mL EU, the number of fluorescent aggregates scored was reduced by 60% when compared with the untreated control group (adjusted p-value < 0.0001; Figure 7).

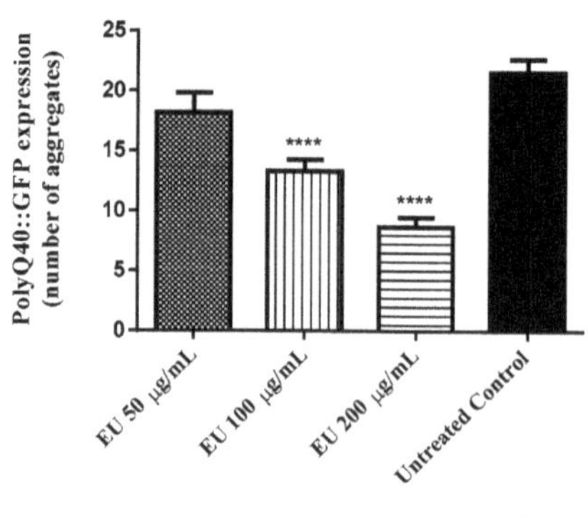

Figure 7. PolyQ40::GFP aggregate formation in mutant worms (AM141) after treatment with *Endopleura uchi* extract (EU). Worms treated with EU exhibited significant lower number of polyQ40::GFP aggregates compared to the control group. Data are presented as mean ± SEM. Note: *** $p < 0.001$ related to the control by a one-way ANOVA followed by Bonferroni (*post-hoc*).

4. Discussion

The Folin-Ciocalteu assay indicated a high phenolic content in EU water extract. HPLC UV/VIS analyses of our extract identified the phenolic bergenin as a major component, in line with the literature [17,19]. Bergenin has been reported as an antioxidant, anti-HIV, gastroprotective, neuroprotective, hepatoprotective, and immunomodulatory agent [25–29], effects that could explain the traditional uses of the barks of uxi by the locals in Amazonia.

The high antioxidant activity in vitro was also observed in vivo using *C. elegans* as a model organism. Notably, wild type worms treated with EU exhibited a higher survival rate after induced oxidative stress as compared to untreated worms submitted to identical conditions. These data demonstrate the capacity of the extract to counteract oxidative damage promoted by exogenous sources such as the pro-oxidant juglone [30]. In agreement, in EU treated worms we found lower accumulation of endogenous cellular ROS and lower pattern of expression for *hsp*-16.2, the gene that codes for HSP-16.2, a small heat shock protein whose expression is induced in response to harsh cellular conditions such as heat stress and oxidative damage [31,32].

Polyphenolic rich extracts are proposed to enhance cellular stress resistance through modulation of stress response genes in addition to free radical scavenging activities [33]. In the current study, *sod-3*, the gene coding for the mitochondrial antioxidant enzyme superoxide dismutase 3, showed higher expression among worms treated with EU. The upregulation of *sod*-3 suggests the participation of the transcription factor DAF-16, the *C. elegans* orthologue for the mammalian FOXO transcription factor, whose target genes are mainly involved in stress resistance, metabolism, and longevity [34,35]. When testing EU in DAF-16 null mutants (CF1038 and GR1307 strains), we noticed that the protecting effect of the extract, previously demonstrated in wild type worms submitted to the survival assay, was absent. The data, therefore, confirms the requirement of DAF-16 transcription factor to promote the antioxidant effect of EU.

Plant extracts with a high content of polyphenolic compounds, such as those obtained from *Camellia sinensis*, *Calycophyllum spruceanum*, and *Paullinia cupana*, have been shown to extend lifespan in *C. elegans* due to their capacity to modulate molecular mechanisms that drive cellular stress resistance and metabolism, in line with the free radical theory of aging [22,23,36]. Considering the pronounced in vivo antioxidant activity of EU elicited by its capacity to modulate stress response genes in DAF-16 pathway, we also investigated whether this extract could affect longevity in *C. elegans* and have found a positive result. The mean lifespan of the EU treated worms was increased by 33%, an effect that was absent in DAF-16 null mutants, indicating a molecular basis underlying it.

However, lifespan extension is not essentially followed by an extension of healthspan, the period of life free from diseases. The fundamental mechanisms underlying both are distinct and complex. Authors highlight that the sole extension of lifespan might not be desirable if it represents just an extended period of frailty, where individuals are vulnerable to aging-related diseases [37,38].

In literature, the role of oxidative stress on the onset of aging-related diseases is well documented, thus several antioxidant compounds are claimed to be capable of attenuating or preventing the impact of aging [39]. In agreement, we obtained evidence for an anti-aging effect of EU in *C. elegans* by studying its muscle function, which works as marker of aging [40,41]. Analyzing the results, we observed a higher contraction rate among worms treated with the extract at all scored timepoints. This data indicates that the treatment can attenuate the age-related muscle function decline, which is considered one important aspect of the healthspan to be maintained to achieve the so-called successful aging [42].

Caloric restriction is a well-known pro-longevity stimulus [43]. To investigate whether the worms could have faced caloric restriction during the period when they were under treatment, we tested the extract against *E. coli* OP50 and did not find a bactericidal effect. Moreover, the pharyngeal contractile capacity of the treated worms was higher as compared to untreated worms, and when we measured the body length we found no differences between treated and untreated worms. The data indicate that the worms did not starve at any timepoint of their lifespan. Thus, we assume that EU can effectively extend lifespan in *C. elegans* by molecular mechanisms other than caloric restriction.

Neurodegenerative diseases are more prevalent in elderly and dramatically impair life quality. In this context, we decided to treat mutant worms expressing polyQ40, an expanded series of glutamine residues involved in the pathophysiology of Huntington's disease [44,45]. Our data indicate that EU treatment is able to attenuate the formation of polyQ40 aggregates, another result supporting the anti-aging properties of EU which needs to be studied in more detail.

In conclusion, the polyphenol-rich water extract from the stem bark of *Endopleura uchi* exhibited substantial antioxidant activity in vitro and in vivo. The extract was able to enhance the stress resistance in *C. elegans* through the modulation of the DAF-16/FOXO pathway. Additionally, the extract exhibited anti-aging properties being able to extend lifespan and to attenuate markers of aging, such as age-related muscle function decline and the formation of polyQ40 aggregates. Considering the traditional application of the bark of uxi and its extensive use by the local population, more studies are needed to elucidate the biological activities as well as its toxicological profile in detail.

Author Contributions: H.P. and M.R. were responsible for designing and performing the methods, analyzing data and writing the manuscript. M.B. was responsible for testing the antimicrobial properties of the extract. E.S. and K.V. were responsible for identifying and quantifying bergenin in the extract. X.W. participated in the correction and formatting of the manuscript. M.W. was the supervisor of this work, corrected data analyses and the manuscript.

Funding: This work was supported by the Brazilian National Counsel of Technological and Scientific Development (CNPq) through the program Science Without Borders (CsF). The authors received financial support from the Deutsche Forschungsgemeinschaft and Ruprecht-Karls-Universität Heidelberg within the funding program Open Access Publishing.

Conflicts of Interest: The authors declare no conflict of interest.

References

1. *Food and Fruit-Bearing Forest Species, 3: Examples from Latin America/F2911*; Food and Agriculture Organization: Rome, Italy, 1986; p. 44.
2. Shanley, P. *Fruit Trees and Useful Plants in Amazonian Life*; Food and Agriculture Organization of the United Nations (FAO): Rome, Italy, 2011.
3. Correa, M.; Penna, L.d.A. *Dicionario das plantas uteis do Brasil e das exoticas cultivadas*: Volume 5. MR. Rio de Janeiro: Instituto Brasiliero de Desenvolvimento Florestal 687p.-illus. *Por Icones. Geog* **1974**, *4*.
4. Rodrigues, E. Plants and animals utilized as medicines in the Jaú National Park (JNP), Brazilian Amazon. *Phytother. Res.* **2006**, *20*, 378–391. [CrossRef] [PubMed]
5. Scudeller, V.V.; Veiga, J.B.d.; Araújo-Jorge, L.d. *Etnoconhecimento de plantas de uso medicinal nas comunidades São João do Tupé e Central (Reserva de Desenvolvimento Sustentável do Tupé)*; UEA: Manaus, Brazil, 2009; pp. 185–199.
6. Pinto, A.Z.d.L.; De Assis, A.F.S.; Pereira, A.G.; Pasa, M.C. Levantamento etnobotânico de plantas medicinais comercializadas no mercado do porto em Cuiabá, Mato Grosso, Brasil. *FLOVET-Boletim do Grupo de Pesquisa da Flora, Vegetação e Etnobotânica* **2013**, *1*.
7. Caetano, R.S.; de Souza, A.C.R.; Feitoza, L.F. O Uso de plantas medicinais utilizadas por frequentadores dos ambulatórios Santa Marcelina, Porto Velho-RO. *Saúde e Pesquisa* **2014**, *7*.
8. Silva, L.R.; Teixeira, R. Phenolic profile and biological potential of *Endopleura uchi* extracts. *Asian Pac. J. Trop. Med.* **2015**, *8*, 889–897. [CrossRef] [PubMed]
9. Bento, J.F. Caracterização de polissacarídeos e metabólicos secundários da casca de *Endopleura uchi* (Huber) cuatrec: Avaliação dos efeitos do decocto e de frações polissacarídicas em células hela e macrófagos. Ph.D. Thesis, Federal University of Paraná, Curitiba, PR, Brazil, November 2013.
10. Sá, B.M.; Lima, C.S.; Silva, U.D.A.; Carvalho, H.O.; Fernandes, C.P.; Resque, R.L.; de Oliveria, T.T.; Carvalho, J.C.T. Subchronic toxicity evaluation of the hydroethanolic extract from *Endopleura uchi* (Huber) Cuatrec in Wistar rats. *Afr. J. Pharm. Pharmacol.* **2015**, *9*, 223–229.
11. Politi, F.A.S.; Moreira, R.R.D.; Salgado, H.R.N.; Pietro, R. Preliminary tests on acute oral toxicity and intestinal motility with extract of pulverized bark of *Endopleura uchi* (Huber) Cuatrec. (Humiriaceae) in mice. *Rev. Pan-Amaz. Saude* **2010**, *1*, 187–189. [CrossRef]

12. Luna, J.; Silva, T.; Bento, E.; Santana, A. Isolamento e identificação estrutural dos constituintes químicos de *Endopleura uchi* (Humiriaceae). *Reunião Anual da Sociedade Brasileira de Química, Minas Gerais, Brasil* **2000**, 2, 123.
13. Politi, F.A.S. Estudos farmacognósticos e avaliação de atividades biológicas de extratos obtidos das cascas pulverizadas de *Endopleura uchi* (Huber) Cuatrec. (Humiriaceae). Master's Thesis, Universidade Estadual Paulista, Araraquara, SP, Brazil, February 2009.
14. Abreu, V.G.d.C.; Corrêa, G.M.; Lagos, I.A.d.S.; Silva, R.R.; Alcântara, A.F.d.C. Pentacyclic triterpenes and steroids from the stem bark of uchi (*Sacoglottis uchi*, Humiriaceae). *Acta Amazonica* **2013**, *43*, 525–528. [CrossRef]
15. De Abreu, H.A.; Lago, I.A.d.S.; Souza, G.P.; Piló-Veloso, D.; Duarte, H.A.; Alcântara, A.F.d.C. Antioxidant activity of (+)-bergenin—a phytoconstituent isolated from the bark of *Sacoglottis uchi* Huber (Humireaceae). *Org. Biomol. Chem.* **2008**, *6*, 2713–2718. [CrossRef] [PubMed]
16. Borges, J.C.; Ripardo Filho, H.d.S.; Guilhon, G.; Carvalho, J.C.; Santos, L.S.; Sousa, P.J. Antinociceptive activity of acetylbergenin in mice. *Lat. Am. J. Pharm.* **2011**, *30*, 1303–1308.
17. Tacon, L.A.; Freitas, L.A. Box-Behnken design to study the bergenin content and antioxidant activity of *Endopleura uchi* bark extracts obtained by dynamic maceration. *Revista Brasileira de Farmacognosia* **2013**, *23*, 65–71. [CrossRef]
18. Blois, M.S. Antioxidant determinations by the use of a stable free radical. *Nature* **1958**. [CrossRef]
19. Nunomura, R.; Oliveira, V.G.; Da Silva, S.L.; Nunomura, S.M. Characterization of bergenin in *Endopleura uchi* bark and its anti-inflammatory activity. *J. Braz. Chem. Soc.* **2009**, *20*, 1060–1064. [CrossRef]
20. Peixoto, H.; Roxo, M.; Krstin, S.; Röhrig, T.; Richling, E.; Wink, M. An anthocyanin-rich extract of acai (*Euterpe precatoria* Mart.) increases stress resistance and retards aging-related markers in *Caenorhabditis elegans*. *J. Agric. Food Chem.* **2016**, *64*, 1283–1290. [CrossRef] [PubMed]
21. Stiernagle, T. Maintenance of *C. elegans*. *WormBook* **2005**. [CrossRef] [PubMed]
22. Peixoto, H.; Roxo, M.; Röhrig, T.; Richling, E.; Wang, X.; Wink, M. Anti-Aging and antioxidant potential of *Paullinia cupana* var. sorbilis: Findings in *Caenorhabditis elegans* indicate a new utilization for roasted seeds of Guarana. *Medicines* **2017**, *4*, 61. [CrossRef] [PubMed]
23. Peixoto, H.; Roxo, M.; Koolen, H.; da Silva, F.; Silva, E.; Braun, M.S.; Wang, X.; Wink, M.J.M. *Calycophyllum spruceanum* (Benth.), the amazonian "tree of youth" prolongs longevity and enhances stress resistance in *Caenorhabditis elegans*. *Molecules* **2018**, *23*, 534. [CrossRef] [PubMed]
24. Ashour, M.L.; El-Readi, M.Z.; Hamoud, R.; Eid, S.Y.; El Ahmady, S.H.; Nibret, E.; Herrmann, F.; Youns, M.; Tahrani, A.; Kaufmann, D. Anti-infective and cytotoxic properties of *Bupleurum marginatum*. *Chin. Med.* **2014**, *9*, 1–10. [CrossRef] [PubMed]
25. Piacente, S.; Pizza, C.; De Tommasi, N.; Mahmood, N. Constituents of *Ardisia japonica* and their in vitro anti-HIV activity. *J. Nat. Prod.* **1996**, *59*, 565–569. [CrossRef] [PubMed]
26. Goel, R.; Maiti, R.; Manickam, M.; Ray, A. Antiulcer activity of naturally occurring pyrano-coumarin and isocoumarins and their effect on prostanoid synthesis using human colonic mucosa. *Ind. J. Exp. Biol.* **1997**, *35*, 1080–1083.
27. Takahashi, H.; Kosaka, M.; Watanabe, Y.; Nakade, K.; Fukuyama, Y. Synthesis and neuroprotective activity of bergenin derivatives with antioxidant activity. *Bioorg. Med. Chem.* **2003**, *11*, 1781–1788. [CrossRef]
28. Lim, H.-K.; Kim, H.-S.; Choi, H.-S.; Oh, S.; Choi, J. Hepatoprotective effects of bergenin, a major constituent of *Mallotus japonicus*, on carbon tetrachloride-intoxicated rats. *J. Ethnopharmacol.* **2000**, *72*, 469–474. [CrossRef]
29. Nazir, N.; Koul, S.; Qurishi, M.A.; Taneja, S.C.; Ahmad, S.F.; Bani, S.; Qazi, G.N. Immunomodulatory effect of bergenin and norbergenin against adjuvant-induced arthritis—A flow cytometric study. *J. Ethnopharmacol.* **2007**, *112*, 401–405. [CrossRef] [PubMed]
30. Aithal, B.K.; Kumar, M.S.; Rao, B.N.; Udupa, N.; Rao, B.S. Juglone, a naphthoquinone from walnut, exerts cytotoxic and genotoxic effects against cultured melanoma tumor cells. *Cell Biol. Int.* **2009**, *33*, 1039–1049. [CrossRef] [PubMed]
31. Strayer, A.; Wu, Z.; Christen, Y.; Link, C.D.; Luo, Y. Expression of the small heat-shock protein HSP 16-2 in *Caenorhabditis elegans* is suppressed by *Ginkgo biloba* extract EGb 761. *FASEB J.* **2003**, *17*, 2305–2307. [CrossRef] [PubMed]
32. Swindell, W.R. Heat shock proteins in long-lived worms and mice with insulin/insulin-like signaling mutations. *Aging* **2009**, *1*, 573. [CrossRef] [PubMed]

33. Scalbert, A.; Johnson, I.T.; Saltmarsh, M. Polyphenols: Antioxidants and beyond. *Am. J. Clin. Nutr.* **2005**, *81*, 215–217. [CrossRef] [PubMed]
34. Braeckman, B.P.; Vanfleteren, J.R. Genetic control of longevity in *C. elegans*. *Exp. Gerontol.* **2007**, *42*, 90–98. [CrossRef] [PubMed]
35. Mukhopadhyay, A.; Oh, S.W.; Tissenbaum, H.A. Worming pathways to and from DAF-16/FOXO. *Exp. Gerontol.* **2006**, *41*, 928–934. [CrossRef] [PubMed]
36. Abbas, S.; Wink, M. Epigallocatechin gallate from green tea (*Camellia sinensis*) increases lifespan and stress resistance in *Caenorhabditis elegans*. *Planta Med.* **2009**, *75*, 216–221. [CrossRef] [PubMed]
37. Hansen, M.; Kennedy, B.K. Does longer lifespan mean longer healthspan? *Trends Cell Biol.* **2016**, *26*, 565–568. [CrossRef] [PubMed]
38. Bansal, A.; Zhu, L.J.; Yen, K.; Tissenbaum, H.A. Uncoupling lifespan and healthspan in *Caenorhabditis elegans* longevity mutants. *Proc. Natl. Acad. Sci. USA* **2015**, *112*, 277–286. [CrossRef] [PubMed]
39. Cui, H.; Kong, Y.; Zhang, H. Oxidative stress, mitochondrial dysfunction, and aging. *J. Signal Transduct.* **2011**, *2012*. [CrossRef] [PubMed]
40. Peixoto, H.; Roxo, M.; Krstin, S.; Wang, X.; Wink, M. Anthocyanin-rich extract of Acai (*Euterpe precatoria* Mart.) mediates neuroprotective activities in *Caenorhabditis elegans*. *J. Funct. Foods* **2016**, *26*, 385–393. [CrossRef]
41. Collins, J.J.; Huang, C.; Hughes, S.; Kornfeld, K. The measurement and analysis of age-related changes in *Caenorhabditis elegans*. *WormBook* **2005**. [CrossRef]
42. Anton, S.D.; Woods, A.J.; Ashizawa, T.; Barb, D.; Buford, T.W.; Carter, C.S.; Clark, D.J.; Cohen, R.A.; Corbett, D.B.; Cruz-Almeida, Y.; et al. Successful aging: Advancing the science of physical independence in older adults. *Ageing Res. Rev.* **2015**, *24*, 304–327. [CrossRef] [PubMed]
43. Fontana, L.; Partridge, L.; Longo, V.D. Extending healthy life span—from yeast to humans. *Science* **2010**, *328*, 321–326. [CrossRef] [PubMed]
44. Finkbeiner, S. Huntington's disease. *Cold Spring Harb Perspect. Biol.* **2011**, *3*, 7476. [CrossRef] [PubMed]
45. Cohen, E.; Dillin, A. The insulin paradox: Aging, proteotoxicity and neurodegeneration. *Nat. Rev. Neurosci.* **2008**, *9*, 759–767. [CrossRef] [PubMed]

Sample Availability: Samples of the compounds are available from the authors.

MDPI
St. Alban-Anlage 66
4052 Basel
Switzerland
Tel. +41 61 683 77 34
Fax +41 61 302 89 18
www.mdpi.com

Molecules Editorial Office
E-mail: molecules@mdpi.com
www.mdpi.com/journal/molecules

www.ingramcontent.com/pod-product-compliance
Lightning Source LLC
Chambersburg PA
CBHW061818210926
51850CB00006B/741
* 9 7 8 3 0 3 6 5 4 1 0 6 8 *